Communications
in Computer and Information Science

2714

Series Editors

Gang Li , *School of Information Technology, Deakin University, Burwood, VIC, Australia*

Joaquim Filipe, *Polytechnic Institute of Setúbal, Setúbal, Portugal*

Zhiwei Xu, *Chinese Academy of Sciences, Beijing, China*

Rationale

The CCIS series is devoted to the publication of proceedings of computer science conferences. Its aim is to efficiently disseminate original research results in informatics in printed and electronic form. While the focus is on publication of peer-reviewed full papers presenting mature work, inclusion of reviewed short papers reporting on work in progress is welcome, too. Besides globally relevant meetings with internationally representative program committees guaranteeing a strict peer-reviewing and paper selection process, conferences run by societies or of high regional or national relevance are also considered for publication.

Topics

The topical scope of CCIS spans the entire spectrum of informatics ranging from foundational topics in the theory of computing to information and communications science and technology and a broad variety of interdisciplinary application fields.

Information for Volume Editors and Authors

Publication in CCIS is free of charge. No royalties are paid, however, we offer registered conference participants temporary free access to the online version of the conference proceedings on SpringerLink (http://link.springer.com) by means of an http referrer from the conference website and/or a number of complimentary printed copies, as specified in the official acceptance email of the event.

CCIS proceedings can be published in time for distribution at conferences or as post-proceedings, and delivered in the form of printed books and/or electronically as USBs and/or e-content licenses for accessing proceedings at SpringerLink. Furthermore, CCIS proceedings are included in the CCIS electronic book series hosted in the SpringerLink digital library at http://link.springer.com/bookseries/7899. Conferences publishing in CCIS are allowed to use our online conference service (Meteor) for managing the whole proceedings lifecycle (from submission and reviewing to preparing for publication) free of charge.

Publication process

The language of publication is exclusively English. Authors publishing in CCIS have to sign the Springer CCIS copyright transfer form, however, they are free to use their material published in CCIS for substantially changed, more elaborate subsequent publications elsewhere. For the preparation of the camera-ready papers/files, authors have to strictly adhere to the Springer CCIS Authors' Instructions and are strongly encouraged to use the CCIS LaTeX style files or templates.

Abstracting/Indexing

CCIS is abstracted/indexed in DBLP, Google Scholar, EI-Compendex, Mathematical Reviews, SCImago, Scopus. CCIS volumes are also submitted for the inclusion in ISI Proceedings.

How to start

To start the evaluation of your proposal for inclusion in the CCIS series, please send an e-mail to ccis@springer.com

Nguyen Thai-Nghe · Thanh-Nghi Do ·
Salem Benferhat

Editors

Intelligent Systems and Data Science

Third International Conference, ISDS 2025
Can Tho City, Vietnam, October 18–19, 2025
Proceedings, Part II

 Springer

Editors
Nguyen Thai-Nghe
Campus II
Can Tho University
Can Tho, Vietnam

Thanh-Nghi Do
Campus II
Can Tho University
Can Tho, Vietnam

Salem Benferhat
Faculté des Sciences Jean Perrin
University of Artois
Lens, France

ISSN 1865-0929   ISSN 1865-0937 (electronic)
Communications in Computer and Information Science
ISBN 978-981-95-3357-2   ISBN 978-981-95-3358-9 (eBook)
https://doi.org/10.1007/978-981-95-3358-9

This Springer imprint is published by the registered company Springer Nature Singapore Pte Ltd.
The registered company address is: 152 Beach Road, #21-01/04 Gateway East, Singapore 189721, Singapore

If disposing of this product, please recycle the paper.

Preface

In today's rapidly evolving digital age, we face the challenge of managing vast and diverse information from numerous sources and domains. This heterogeneous data presents significant difficulties in collection, integration, storage, machine learning, fusion, and query processing. Addressing these challenges requires the development of innovative, interpretable, and explainable data science and artificial intelligence technologies to harness and utilize this data effectively.

Following the success of the first International Conference on Intelligent Systems and Data Science (ISDS) in 2023 at Can Tho University and the second ISDS in 2024 at Nha Trang University, these two proceedings volumes present the papers from the third ISDS (ISDS 2025), held on October 18–19, 2025, at Can Tho University. ISDS 2025 served as a dynamic forum for researchers to discuss challenges, share findings, identify emerging issues, and foster collaborations in the fields of Intelligent Systems and Data Science.

We received 181 submissions from 8 countries to the main conference and two special sessions. To handle the review process, we invited 67 expert reviewers. Each paper was reviewed by at least three reviewers. We followed a single-blind process in which the identities of the reviewers were not known to the authors. This year, we used the EasyChair conference management service to manage the submission and selection of papers. After a rigorous review process, followed by discussions among the program chairs, 67 papers were accepted as long papers and 2 as short papers, resulting in acceptance rates of 37.02% and 1.10% for long and short papers, respectively.

The conference program included six sessions: Intelligent Systems; AI in Health Care Analytics; AI in E-Commerce, Education, and Society; AI in Agriculture, Aquaculture, and Environments; Big Data, IoT, and Cloud Computing; and Natural Language Processing.

We wish to thank the other members of the organizing committee, the reviewers, and the authors for the immense amount of hard work that went into making ISDS 2025 a success. The achievement of the conference was also contributed to by the kind devotion of many sponsors and volunteers.

We hope you enjoyed the conference!

October 2025

Nguyen Thai-Nghe
Thanh-Nghi Do
Salem Benferhat

Organization

Honorary Chairs

Thanh-Thuy Nguyen	University of Engineering and Technology, VNU Hanoi, Vietnam
Trung-Tinh Tran	Can Tho University, Vietnam

General Chairs

Huu-Hoa Nguyen	Can Tho University, Vietnam
Ngoc-Hai Tran	Can Tho University, Vietnam

Program Chairs

Nguyen Thai-Nghe	Can Tho University, Vietnam
Thanh-Nghi Do	Can Tho University, Vietnam
Salem Benferhat	University of Artois, France

Publication Chairs

Minh-Khiem Nguyen	Can Tho University, Vietnam
Thanh-Dien Tran	Can Tho University, Vietnam

Technical Program Committee

Salem Benferhat	Cril, CNRS UMR8188, Université d'Artois, France
Vo Quoc Bao Bui	Can Tho University, Vietnam
Phan Anh Cang	Vinh Long University of Technology Education, Vietnam
Jeong-Hyun Chang	Kyonggi University, South Korea
Duc-Hoang Chu	National Technology Innovation Fund, Vietnam
Bob Dao	Monash University, Australia
Thanh-Nghi Do	Can Tho University, Vietnam

Thanh-Thai Do	Ho Chi Minh City University of Technology, VNU-HCM, Vietnam
Thanh-Nghi Doan	An Giang University, Vietnam
Van-Hieu Duong	Tien Giang University, Vietnam
Trung-Nghia Duong	German Research Center for Artificial Intelligence (DFKI), Germany
Tan Nghia Duong	Hanoi University of Science and Technology, Vietnam
Vatcharaporn Esichaikul	Asian Institute of Technology, Thailand
Dewan Farid	United International University, Bangladesh
Masayuki Fukuzawa	Kyoto Institute of Technology, Japan
Myongggi Hong	Kwandong University, South Korea
Trong-Minh Hoang	Posts and Telecommunications Institute of Technology, Vietnam
Tomas Horvath	Eötvös Loránd University, Hungary
Cong-Phap Huynh	Vietnam-Korea University of ICT, Vietnam
Quang Nghi Huynh	Can Tho University, Vietnam
Xuan Hiep Huynh	Can Tho University, Vietnam
Phuoc Hai Huynh	An Giang University, Vietnam
Trung-Hieu Huynh	Industrial University of Ho Chi Minh City, Vietnam
Kwanghoon Kim	Kyonggi University, South Korea
Hoai-Bao Lam	Can Tho University, Vietnam
Thi-Bao-Thu Le	Ho Chi Minh City University of Technology, Vietnam
Thi-Phuong Le	EBI School of Industrial Biology, France
Thanh Van Le	Ho Chi Minh City University of Technology, Vietnam
Hong-Trang Le	Ho Chi Minh City University of Technology, VNU-HCM, Vietnam
Danh Luong	Can Tho University, Vietnam
Thanh Ma	Can Tho University, Vietnam
Tan-Ha Mai	Ho Chi Minh City University of Technology, Vietnam
Xuan-Trang Mai	Phenikaa University, Vietnam
Hung Ba Ngo	Can Tho University, Vietnam
Duc-Luu Ngo	Bac Lieu University, Vietnam
Thai-Nghe Nguyen	Can Tho University, Vietnam
Huu-Phat Nguyen	Hanoi University of Science and Technology, Vietnam
Thi-Ai-Thao Nguyen	Ho Chi Minh City University of Technology, Vietnam
Huu-Hoa Nguyen	Can Tho University, Vietnam

Khiem Nguyen	Can Tho University, Vietnam
Van-Hoa Nguyen	An Giang University, Vietnam
Hai Thanh Nguyen	Can Tho University, Vietnam
Bao-An Nguyen	Tra Vinh University, Vietnam
Dinh Hung Nguyen	Nha Trang University, Vietnam
Chanh Nghiem Nguyen	Can Tho University, Vietnam
Vinh Nguyen	Can Tho University, Vietnam
Thanh-Khoa Nguyen	Can Tho University, Vietnam
Huu Van Long Nguyen	Can Tho University, Vietnam
Vu-Lam Nguyen	Kien Giang Community College, Vietnam
Hanh Nguyen	Can Tho University, Vietnam
Chi-Ngon Nguyen	Can Tho University, Vietnam
Khac Cuong Nguyen	Nha Trang University, Vietnam
Manh-Cuong Nguyen	Nha Trang University, Vietnam
Sinh Van Nguyen	International University - Vietnam National University HCMC, Vietnam
Quang-Hung Nguyen	Posts and Telecommunications Institute of Technology, Vietnam
Quang-Vu Nguyen	Vietnam-Korea University of ICT, Vietnam
Sichoon Noh	Namseoul University, South Korea
Atsushi Nunome	Kyoto Institute of Technology, Japan
JunHo Park	Kyonggi University, South Korea
SooJin Park	Kyonggi University, South Korea
Dinh-Lam Pham	Kyonggi University, South Korea
Phi Pham	Can Tho University, Vietnam
Truong Hong Ngan Pham	Can Tho University, Vietnam
Hoang-Anh Pham	Ho Chi Minh City University of Technology, Vietnam
Nguyen-Khang Pham	Can Tho University, Vietnam
Van-Nam Pham	Nha Trang University, Vietnam
Thi-Ngoc-Diem Pham	Can Tho University, Vietnam
Thi-Thu-Thuy Pham	Nha Trang University, Vietnam
Thuong-Cang Phan	Can Tho University, Vietnam
Phuong-Lan Phan	Can Tho University, Vietnam
Trong Nhan Phan	HCMC University of Technology, Vietnam
Trung-Nghia Phung	Thai Nguyen University of ICT, Vietnam
Minh-Tuan Thai	Can Tho University, Vietnam
Minh-Quang Tran	Ho Chi Minh City University of Technology, Vietnam
Thanh-Dien Tran	Can Tho University, Vietnam
Nguyen Minh Thu Tran	Can Tho University, Vietnam
Hoang Viet Tran	Can Tho University, Vietnam

Nguyen Minh Thai Tran	Lincoln University, New Zealand
An C. Tran	Can Tho University, Vietnam
Thien Tran	HCMC University of Foreign Languages-Information Technology, Vietnam
Duc-Tan Tran	Phenikaa University, Vietnam
Quoc Dinh Truong	Can Tho University, Vietnam
Quang Vinh Truong	Ho Chi Minh City University of Technology, Vietnam
Quoc-Bao Truong	Can Tho University, Vietnam
Chi Truong	Ho Chi Minh City University of Technology, Vietnam
Phuoc-Hung Vo	Tra Vinh University, Vietnam
Van-Tai Vo	Can Tho University, Vietnam
Thi-Ngoc-Chau Vo	Ho Chi Minh City University of Technology, Vietnam

Secretaries

Quoc Bao Le Huynh	Can Tho University, Vietnam
Cam Nhung Mai Thi	Can Tho University, Vietnam
Hai Thanh Nguyen	Can Tho University, Vietnam

Finance Chairs

Phuong Lan Phan	Can Tho University, Vietnam
Lam Mai Chi Dinh	Can Tho University, Vietnam

Contents – Part II

AI in Agriculture, Aquaculture, and Environments

LAVA: Leveraging Self-attention to Learn Video Features for Wastewater
Pipe Anomaly Detection 3
    Ti-Hon Nguyen, Minh-Thu Tran-Nguyen, Carole Delenne,
    and Salem Benferhat

Tomato Leaf Classification Using GAN-Generated Images 19
    Huynh Ngoc Tuyet, Nguyen Tri Phuc, and Nguyen Thai-Nghe

AquaNetCT: Detecting and Representing Urban Drainage Networks
from Video Analysis 31
    Thanh Ma, Salem Benferhat, Minh-Thu Tran-Nguyen,
    and Thanh-Nghi Do

A Portable Multi-spectral Sensor-Based System with Machine Learning
Models for Non-destructive Sweetness Assessment of Cherry Tomatoes 46
    Hoang Thien-Phu Tran, Le Dinh-Kha Vo, Quoc-Hung Pham,
    Thanh-Nhan Nguyen, Huy-Hoang Vo, Tuan-Kiet Tran,
    and Nhut-Thanh Tran

AI in E-Commerce, Education, and Society

Student Engagement Estimation in Classroom Videos Using Video Swin
Transformer 61
    Nha Tran, Hung Nguyen, Dat Ly, Hung Q. Nguyen, Anh Tran,
    and Hien D. Nguyen

ROPEVISION: A Computer Vision-Based System for Jump Rope
Technique Recognition and Counting 77
    Nguyen Hoang Pham, Thanh Nguyen Thi Ngoc, and Thao Nguyen Pham

CAHIM: Constructing Auto-generated HIstorical Mindmaps Based
on Clustering and Ontology Structuring 90
    Thanh Ma, Hieu Nguyen, Phu-An Thai, Xuan Nguyen, Ky Nguyen,
    Thy Le, Le-Diem Bui, Nguyen-Khang Pham, and Won Ho

Applying the Temporal Fusion Transformer Model in Stock Price
Forecasting: The Cases of VIC, VRE, and VHM 105
    Dinh Quoc Thai, Bao-An Nguyen, and Tai Vo-Van

A Dual-Stage AI Model for Personalised Major Advising 120
Ba Duy Nguyen, Diem Trinh Bui Thi, and Quoc Dinh Truong

AI in Health Care Analytics

SAM Meets U^2-Net: Minimal-Supervision Skin Lesion Segmentation 137
Nguyen Ngoc Dung and Doan Van Thang

Retrospective Contribution Analysis of Intrinsic 3D Breast Features
for Esthetic Outcome Evaluation of Breast Reconstruction Surgery 146
*Nam Phong Duong, Takumi Sonoi, Yoshihiro Sowa,
and Masayuki Fukuzawa*

Gestational Diabetes Prediction Using Classification Methods 157
*Al Maruf Hassan, The-Phi Pham, Md. Maruf Hassan,
Abdul Kadar Muhammad Masum, and Dewan Md. Farid*

GeKAN: Enhancing High-Dimensional Gene Expression Classification
with Feature-Selected Kolmogorov-Arnold Networks 176
Bich-Chung Phan, Thanh Ma, and Huu-Hoa Nguyen

Fuzzy-EDTrans: A Fuzzy Ensemble Learning Approach for Brain Tumor
Classification 191
Anh-Cang Phan, Chan-Khoa Ho, and Khac-Tuong Nguyen

A Machine Learning-Based Tool for Autism Screening Using
Psychological Medical Records 204
*Hao Nguyen Thi Bich, Thanh Nguyen Van Quoc, Thuan Nguyen Dinh,
and Nhut Nguyen Minh*

A Deep Learning Framework with Squeeze-and-Excitation Attention
for Guava Disease Classification 217
Tuong Le, Duc-Tan Nguyen, Hoang-Duy Nguyen, and Thang Cap

Big Data, IoT, and Cloud Computing

Towards End-to-End Traffic Congestion Estimation Using Learned ROI
and Vehicle Object Dynamics 231
Khang Le, Anh Duc Tran, Thi Minh Tam Tran, and Quang Tran Minh

Simultaneous Low-Light Image Enhancement and Deblurring
for Surveillance Systems Using Intelligent Cameras 246
*Minh-Thien Duong, Van-Ho Tran, Vi-Do Tran, Hoai-An Trinh,
Thien M. Tran, and My-Ha Le*

Design of a Solar-Powered AIoT System for Safety Corridor Violation
Detection in High-Voltage Power Transmission Networks 258
   Huu-Phuoc Nguyen, Tan-Phat Tran, and Ngoc-Luat Pham

A Real-Time Adaptive Traffic Signal Control Framework Using
Graph Neural Networks and Multi-Agent Reinforcement Learning
with Crowdsourced Event Integration 270
   Do Thanh Thai and Quang Tran Minh

Intelligent Systems

Towards Local Learning Algorithm on Apache Spark for Large Datasets 289
   Quoc-Bao Bui-Vo and Thanh-Nghi Do

PM2.5 Prediction Model Using CNN with Depthwise Separable
Convolution and Bi-LSTM 301
   Thi-Phuong-Trang Nguyen, Duc-Cuong Nguyen, and Tuong Le

Improving Data Quality Post-anonymization in K-Anonymity Models
Using a Hybrid Artificial Rabbit and Arctic Puffin Optimization 312
   *Le Hoang Minh Nhat, Bach Ngoc Vy, Phung Thi Ly,
   and Dinh Nguyen Trong Nghia*

FMN-OE: Evaluation And Elimination Of Hyperbox Overlap 328
   *Dinh-Minh Vu, Thanh-Son Nguyen, Trinh-Hoang Vu, Thi-Duong Vu,
   Duc-Luu Nguyen, Quang-Huy Hoang, and Ba-Dung Le*

Enhancing Cryptocurrency Forecasting Using Temporal Features On
High-Frequency Data 343
   *Hoang-Sang Le, Dinh-Chuong Vu, Hoang-Hai Pham,
   and Duy-Hoang Tran*

Edge-Optimized CNN Accelerator: A Flexible and Lightweight
Architecture 357
   *Tuan-Kiet Tran, Minh-Tuyen Huynh, Duc-Hung Pham,
   Cong-Kha Pham, and Huu-Thuan Huynh*

DWACE: Enhancing Graph Clustering via Autoencoder and Contrastive
Learning Refinement of DeepWalk Embeddings 372
   *Nguyen Xuan Thao Mai, Cong Phap Huynh, Supeeti Kunchan,
   and Dai Tho Dang*

Defective and Non-defective Printed Lot Information Classification Based
on Optical Character Recognition   384
   Viet-Phuong Le, Thi-Kim-Ngoc Tran, Tan-Viet-Khoa Nguyen,
   and Van-Toi Nguyen

A Method for Enhancing Images Quality Based on Machine Learning   398
   Sinh Van Nguyen, Vinh Xuan Nguyen, and Hanh Le Thi Ngoc

Natural Language Processing

Vietnamese Certificate Analysis: Addressing Hybrid Document
Challenges Through Multimodal Layout-Text Fusion   419
   Nam Duong Ho, Duy Tran Ngoc Bao, Quan Thi Khac, and Dang Le Binh

Reflective Thought and Code for Solving Numerical Problems with Large
Language Models   433
   Long Tri Thai Son, Ngo Viet Anh, Nguyen Thi Thuy Linh,
   and Nguyen Viet Ha

RCM-MSA: Robust Contextual Modeling for Multi-modal Sentiment
Analysis Using Transformer and Attention Mechanism   448
   Hoai-Phuong Nguyen-Cao, Phuoc-Hung Vo, and Nhut-Lam Nguyen

Integrating Information Retrieval and LLMs: A Document Retrieval
Chatbot in Education Settings   461
   Duy Dang Khoa Nguyen, Vi Kiet Mach, Thang Le Dinh,
   and Cuong Pham-Nguyen

Graph–Beam Retriever: A Lightweight Framework for Evidence Selection
in Vietnamese Fact Verification   476
   Tieu Phung Mai Suong, Nguyen Tran Thanh Nha, Bay Vo,
   Dinh Minh Hoa, and Thien Khai Tran

Benchmarking Conventional, Deep, and Large Language Models
on Vietnamese Clickbait Detection   489
   Nguyen Phuoc Dai, Huynh Ly Tan Khoa, Luu Van Nhat Hao,
   Y. Nguyen Minh, Bay Vo, and Thien Khai Tran
   Author Index

Author Index   497

Contents – Part I

AI in Agriculture, Aquaculture, and Environments

Evaluating Spatial Coverage in UAV-Based Rice Growth Stage
Classification with Deep Learning Models 3
    Van-Hoa Nguyen, Duy-Phuong Nguyen-Ly, and Thanh-Nghi Do

Improving Rice Disease Segmentation via Iterative Pseudo-label Correction . . . 16
    Nhat-Tinh-Anh Nguyen, Van-Hoa Nguyen, and Thanh-Phong Le

Explainable AI for Agriculture: Teacher-Student Learning in Rice Pest
Classification 31
    *Luyl-Da Quach, Duc-Trong Le, Chi-Ngon Nguyen,
    and Nguyen Thai-Nghe*

Optimizing EfficientNetV2 for Insect Classification 46
    Quoc-Anh Nguyen, Thanh-Nghi Doan, and Huu-Hoa Nguyen

AI in E-Commerce, Education, and Society

Semantic Similarity for User-Based Collaborative Recommendation
Systems 63
    Tran-Diem-Hanh Nguyen, Hoang-Bao Nguyen, and Quoc-Nghia Phan

Reforming Attention in Transformer-Based Graph Neural Networks
for Fashion Recommendation 77
    Tin T. Tran, Manh Mai Van, Nguyen Khac Anh Tai, and Le Tuan Thanh

Decision Support System for Selecting Majors in IT Training Programs 92
    Tam Nghe Thi Thanh, Sinh Van Nguyen, and Son Thanh Le

An Explainable Biased Multi-relational Matrix Factorization Model Using
XAI for Recommendation Systems 105
    Duy-Quang Tran, Nguyen Minh Khiem, and Nguyen Thai-Nghe

An Efficient Lightweight Multimodal Framework for Autism Behavior
Recognition on the AV-ASD Dataset 117
    Van-Minh Luong and Truc Nguyen

AI in Health Care Analytics

Painless and Accurate AI-Driven Method for Personal Glucose Tracking
Based on Multi-spectral Sensing 133
Thanh-Nhan Nguyen, Quoc-Hung Pham, Duy-Khanh Nguyen,
Hien-My Nguyen, Thanh-Linh Le Ho, Quoc-An Le, and Nhut-Thanh Tran

MS-CXR: Improving Multi-label Chest X-Ray Classification
via a Multi-architecture Soft Voting 147
Quoc-Khang Tran and Nguyen-Khang Pham

ESQ-YOLO: An Efficient Method for Blood Cell Detection Based
on Improved YOLOv8 162
Tien Le Trac, Nghi Pham Thi Phuong, Tri Nguyen Phan Minh,
and Vinh Le Van

Breast Cancer Classification Using Ensemble Deep Learning
with RIDOPT-Net 175
Ho-Dat Tran, Thuong-Cang Phan, Van-Kha Vo, and Anh-Cang Phan

A Semi-Automated MRI-Image Processing System for Exploratory
Analysis of Radiomics Features Toward Clinical Prediction of Meningioma
Growth 190
Haruki Minamoto, Yuta Oi, Ichita Taniyama, Koji Sakai,
and Masayuki Fukuzawa

A Meta-Classifier Built on Self-supervised Models for Improving Chest
X-Ray Image Classification 201
Tri-Thuc Vo and Thanh-Nghi Do

Big Data, IoT, and Cloud Computing

Secure Offloading for Terahertz-Band UAV-Aided MEC Backscatter
Communication Networks 215
Anh-Tuan Tran, Quang Nhat Tran, Xuan An Bui, Van Chien Nguyen,
and Dac-Binh Ha

Enhancing Model Privacy and Security for Blockchain-Based Federated
Learning 228
Minh-Tuan Thai, Thu Do Ly Anh, and Huu-Hoa Nguyen

An Improved Low-Power Wearable Sensor for Dairy Cow Behavior
Classification Using Deep Learning Algorithms 243
Nhat Minh Tran, Thang Viet Tran, Ngo Minh Tri Nguyen,
and Chi-Ngon Nguyen

A Time-Frequency Signal Analysis System for Drastic Improvement
of Sensitivity and Detection Duration of Biosensors 257
Kanato Adachi, Takumi Kinoshita, Minoru Noda,
and Masayuki Fukuzawa

Intelligent Systems

Vibration-Based Diagnosis and Monitoring of Bearing Degradation 271
Ngoc-Tu Nguyen

Random Forest with Z-score in Supervised Machine Learning 285
Abdullah Rahman, Sumaiya Rahman, Huu-Hoa Nguyen,
Md. Maruf Hassan, Abdul Kadar Muhammad Masum,
and Dewan Md. Farid

Optimizing Image Colorization: Innovations with CVAE 297
Ngoc-Giau Pham, Van-Hieu Duong, Thanh-Hai Tong-Le,
Khac-Hoang Nguyen, and Phu-Thanh Ha-Quach

Ontology-Enhanced Semantic Image Search with Deep Learning
and CLIP Embedding 312
Pham Thi Thu Thuy and Kim Hwa Soo

Deep Ant Colony Optimization for Electric Pickup and Delivery Problem 326
Thanh-Nhan Truong, Thuy-Anh Ma, and Duy-Hoang Tran

Automatic 3D Space Reconstruction System from Video Using 3D
Gaussian Splatting 341
Anh-Khoi Ngo-Ho, Anh-Khoa Ngo-Ho, and Trung-Linh Nguyen

Automated Trace Clustering: Selecting Feature Encodings and Clustering
Algorithms Based on Process Model Quality 355
Pham Thi Kim Hue, Nguyen Thi Thuy Linh, Long Tran Quoc,
and Quang-Thuy Ha

AI-Agent-Powered Video Object Contextualization & Retrieval System:
Architecture and System 370
Dinh-Lam Pham, Jeong-hyeon Chang, Eun-bi Jo, Hyun Yoo,
Kyunghee Sun, Kyong-Sook Kim, and Kwanghoon Pio Kim

AGSEU-Net: An Optimal Deep Learning Framework for Noise Filtering 383
Nguyen Kim Quoc, Tran Chau Thanh Thien, and Ha Minh Tan

Natural Language Processing

Enhancing Vietnamese Hate Speech Detection with Multi-agent Debate 401
    Thanh Le, Luu Duc Ngo, Tan Quang Nguyen, Tuong Le,
    and Thien Khai Tran

Enhanced Chart Detection in Document Images with Segmentation
Backbone 414
    Dang Hien Long Tran, Ho Duc An Nguyen, Van Thong Huynh,
    Tuan-Anh Tran, Xuan Toan Mai, and Hong Tai Tran

ChartLite: Simplified End-to-End Extraction of Chart Data for Enhanced
Visual Understanding 429
    Phuc Thanh Danh Nguyen, Nhu Tinh Anh Nguyen, Hong Tai Tran,
    Van Thong Huynh, Tuan-Anh Tran, and Xuan Toan Mai

Applying Large Language Model in Survey Extraction 444
    Quang Hung Nguyen, Phan Nhat Minh Le, Ngoc Long Nguyen,
    and Thi Vinh Thanh Le

AI-Powered Investment Advisor: Enhancing Financial Decisions
with NLP and Predictive Analytics 457
    Loc Tan Dinh, Huy Nguyen Thanh, Huong Nguyen Thi Thanh,
    and Chi Tran Thi Kim

A Novel Pipeline for Automatic UML Sequence Diagram Synthesis
and Multimodal Scoring 473
    Van-Viet Nguyen, Huu-Khanh Nguyen, Kim-Son Nguyen,
    Hue Luong Thi Minh, The-Vinh Nguyen, and Duc-Quang Vu
    Author Index

Author Index 487

AI in Agriculture, Aquaculture, and Environments

LAVA: Leveraging Self-attention to Learn Video Features for Wastewater Pipe Anomaly Detection

Ti-Hon Nguyen[1], Minh-Thu Tran-Nguyen[1(✉)], Carole Delenne[2,3], and Salem Benferhat[4]

[1] College of Information Technology, Can Tho University, Can Tho 92000, Vietnam
{nthon,tnmthu}@ctu.edu.vn
[2] IUSTI, AMU, CNRS, Marseille, France
carole.delenne@univ-amu.fr
[3] Previously at HSM, Univ Montpellier, CNRS, IRD, Montpellier, France
[4] CRIL, CNRS UMR 8188, Universite d'Artois, Lens, France
benferhat@cril.fr

Abstract. Urban network data often originate from multiple sources and are represented in diverse formats, which can make their processing and analysis complex. In this work, we focus specifically on video data, particularly inspection television (ITV) videos of wastewater pipes. These videos play a crucial role in the management and maintenance of urban networks. On one hand, they help identify anomalies that may affect the pipes, such as obstructions or degradations. On the other hand, they provide essential information about the structural properties of the pipes and networks, including their diameter and the direction of wastewater flow. In this paper, we propose a classification algorithm for ITV videos, with a particular focus on detecting diameter changes, internal cracks, chemical attacks, and turbid-colored water within the pipes. This task is essential for predictive maintenance and hydraulic modeling of wastewater networks. We build on Video Vision Transformer (ViViT) and TimeSformer-based methodologies for video classification, which allow for the effective capture of both spatial and temporal relationships between the different frames in the video data. We specifically describe various mechanisms for generating training datasets from a subset of manually annotated images. The experimental study shows promising results on real-world ITV videos of wastewater pipe networks.

Keywords: Wastewater pipe anomalies · Video classification · Transformer · Video feature

1 Introduction

The exploitation of geospatial data is a major challenge in many applications today. The development of artificial intelligence technologies, particularly

© The Author(s), under exclusive license to Springer Nature Singapore Pte Ltd. 2026
N. Thai-Nghe et al. (Eds.): ISDS 2025, CCIS 2714, pp. 3–18, 2026.
https://doi.org/10.1007/978-981-95-3358-9_1

machine learning, has provided new opportunities for managing such geospatial data. This development is further reinforced by open science policies, which make a large number of datasets on geographic or georeferenced information publicly available. This work fully aligns with this perspective, with a particular focus on the automatic exploitation of video data related to wastewater network management. The automatic processing of these video data has many potential applications. The most obvious is the detection of anomalies, such as pipe degradation or blockages. Another, less obvious but equally important, application is the enrichment of geographic information system (GIS) data, particularly for attributes or properties related to the pipes in these networks. Indeed, GIS contains vast amounts of data on various components of wastewater networks, which are essential both for mapping these networks and for supporting hydraulic modeling. However, one significant challenge for optimal hydraulic modeling is incomplete or missing data. This work aims to exploit video inspection data of pipes to detect and provide more precise information.

In this study, we focus on four types of changes in wastewater pipes: changes in pipe sizes (BAA), internal cracks (BAB), chemical attacks (BAF), and turbid-colored water (BDD). A properly maintained sewage pipe system is crucial for effective urban water management. As the amount of data—particularly from sensors—continues to grow, professionals in the wastewater sector face increasing challenges in analyzing this large volume of information and detecting network anomalies [12]. To effectively leverage video data for these applications, advanced machine learning models are required, among them, the Transformer architecture has emerged as a powerful tool capable of capturing complex sequential patterns. The Transformer architecture, originally introduced by Vaswani et al. [14], is a sequence-to-sequence model initially developed for machine translation tasks [11]. Since its introduction, the Transformer has demonstrated remarkable success across a wide range of natural language processing (NLP) applications, particularly in the development of large pre-trained language models such as BERT [5]. Building on this success, Dosovitskiy et al. [6] proposed the Vision Transformer (ViT), a pure Transformer architecture applied to sequences of image patches, achieving competitive performance on image classification benchmarks. Extending this approach to the video domain, Arnab et al. [1] introduced ViViT (Video Vision Transformer), a Transformer-based architecture designed specifically for video classification, which directly models spatio-temporal dependencies to extract high-level video representations. To tackle the challenges of anomaly detection in video inspections, we employ video classification models based on ViViT [1] and TimeSformer [2]. These models leverage the self-attention mechanisms to automatically learn rich and discriminative video features, making them particularly suited for complex video analysis tasks. The rest of this paper is organized as follows. Section 2 reviews previous work on anomaly detection in wastewater inspection videos. Section 3 briefly presents the algorithm used to construct an experimental video dataset from wastewater network data for training video classification models, as well as the model for detecting anomalies in wastewater pipes. Section 4 presents the experimental results on real inspection video data, followed by conclusions and directions for future work in Sect. 5.

2 Related Work

In 2022, Li et al. [8] conducted a comprehensive review of methodologies, datasets, and challenges related to wastewater network pipe inspection since 2000. They proposed a systematic taxonomy of sewer inspection algorithms, categorizing them into three main approaches: classification, detection, and segmentation. They highlighted the critical need for publicly available datasets and open-source code to facilitate reproducible research. A total of 32 datasets from various countries were summarized, with the United States contributing the largest share—12 datasets. These datasets exhibit significant variability in size, from 32 to 2,202,582 images, and encompass a wide range of resolutions—from 64×64 to $4000 \times 46,000$ pixels. Their study also identified key challenges, primarily arising from poor dataset quality, as reported in the existing literature. While many researchers have achieved promising results, most studies focus on a limited subset of damage types, leaving the effectiveness of these models on comprehensive datasets largely unaddressed. To overcome limitations due to insufficient data quality, Jung et al. (2024) [7] proposed a sensor system capable of capturing both images and point clouds in a standardized way, facilitating the creation of datasets suitable for developing deep learning (DL) models. Although point clouds are not yet widely used in current inspection practices, they offer considerable potential for detecting geometric anomalies not easily identifiable in conventional images. Their approach involves a multi-sensor robotic system integrated with a deep learning framework. They also synthesized several datasets related to anomaly detection in sewer pipelines. Besides image datasets, video data is also becoming increasingly available, with new datasets derived from video recordings collected in China since 2022. In this study, we also aim to leverage video-based models to enhance the accuracy and interpretability of anomaly detection and prediction in sewer inspection tasks. Li et al. [9] proposed a novel instance segmentation model, Pipe-SOLO, designed to detect six common types of sewer defects: cracks, faulty joints, open joints, protruding laterals, broken pipes, and surface damage. The model builds upon the original SOLOv2 architecture by integrating a more efficient backbone module (Res2Net-Mish-BN-101) and adopting an enhanced Bi-directional Feature Pyramid Network (EBiFPN) as the neck to improve feature fusion. To address the challenge of blurred CCTV images from sewer inspections, a GAN-based dehazing model was applied during preprocessing. The dataset used in the study was manually collected and annotated from CCTV sewer footage recorded in various locations across Seoul, South Korea, and provided by the Seoul Digital Foundation. Experimental results demonstrated that Pipe-SOLO outperformed existing segmentation methods, achieving a 7.3% improvement in mean Average Precision (mAP) compared to state-of-the-art approaches. Zhang et al. (2023) [16] proposed an enhanced sewer anomaly detection model based on YOLOv4 for identifying four common defect types: cracks, deposits, roots, and stagger. A key improvement of their method is the use of a spatial pyramid pooling (SPP) module to expand the receptive field and enhance the model's ability to fuse contextual features across different receptive fields. The effectiveness of combining the DIoU loss function with the SPP module was also validated. They utilized 2,700 relabeled images from the publicly available Sewer-ML dataset, formatted in a multi-label structure. Experimental results showed that, compared to the original YOLOv4 model, the improved version achieved a 4.6% increase in mean Average Precision (mAP), reaching 92.3%, with a recall of 89.0%.

Chen et al. [4] proposed a modified deep learning architecture, RegNet+, for the automated detection and classification of sewer defects using CCTV inspection footage. The model builds upon the original RegNet by incorporating dropout layers and replacing the standard ReLU activation with LeakyReLU in the squeeze-and-excitation (SE) block to enhance generalization and reduce over-fitting. From 7,733 CCTV sewer videos, 20 distinct defect categories were manually annotated, and 12,000 images were selected through data augmentation, with 600 samples per class. Experimental results demonstrated that RegNet+ consistently outperformed several widely used deep learning models, including VGG16, VGG19, GoogleNet, ResNet50, and the original RegNet, in both single-class and multi-class classification scenarios.

Biswas et al. [3] processed data from 221 German municipalities, including 10,205 video files, and extracted 156,028 images of wastewater pipe anomalies. These were classified using standard German inspection codes, corresponding to the European pipe inspection standard DIN EN 13508-2 [7], consistent with the database used here. They tested six deep learning architectures: HRNet, ResNet-152, EfficientNet, MobileNetV3-Large, InceptionV4, and DenseNet-264. The models achieved balanced accuracy above 85% on 19 categories, including BAF (chemical attacks) and BDD (turbid and colored waters), with remaining categories around 80%. BAF was the most prevalent, with 8,629 instances and 91.5% balanced accuracy. On the Sewer-ML dataset, for BAF and BDD (OB and VA), true positive recall dropped to 2.49% and 2.2%, respectively.

Liu et al. [10] introduced the VideoPipe Challenge to advance video anomaly analysis for urban pipe inspection, crucial for maintaining real-world sewer infrastructure. They launched a challenge website (https://videopipe.github.io) and released two benchmark datasets—QV-Pipe and CCTV-Pipe—on February 28, 2022, supporting Video Anomaly Classification and Temporal Anomaly Localization. The video data, authorized by Shenzhen Bwell Robotics Co., Ltd., is shared under the Creative Commons Attribution-NonCommercial-ShareAlike 4.0 License. Challenge participants demonstrated strong performance using deep learning architectures and advanced loss functions. Video-based backbones consistently outperformed image-based models. Transformer-based architectures excel at modeling long-range dependencies across frames, and the super-image approach, rearranging frames into a single composite image, effectively captures temporal relationships. Focal loss also helps address class imbalance in long-tailed datasets.

3   Methodology

In this work, we start with a dataset composed of wastewater pipe inspection videos and corresponding anomaly images annotated by domain experts. Based on this dataset, we propose an approach to construct experimental video datasets tailored for training video classification models such as TimeSformer [2] and ViViT [1], with the objective of detecting anomalies in sewer inspection videos. Algorithm 1 focuses on identifying defective anomalies in inspection videos by employing a pretrained video classification model. To support this task, Algorithm 2 is introduced to generate video datasets by selecting n consecutive frames in close proximity to a manually annotated target frame. The constructed datasets serve as inputs for training and evaluating the anomaly detection performance of Algorithm 1.

Algorithm 1. Detection of anomalies in inspection video

1: Given: inspected video ins_video, pre-trained video classification model $classifier$, the set of anomaly code $anomalies\_code$, length of sub video by second sub_length.

2: Initialize: an empty list of tuple $anomalies\_found$ ▷ This empty list is used to store detected anomalies; each element will be a sublist containing: the index of the sub-video, the video segment itself, and the predicted class

3: $set\_of\_sub\_video \leftarrow split\_video\_function(ins\_video, sub\_length)$ ▷ Usually sub_length is 5 seconds

4: **for** $i \in range(len(set\_of\_sub\_video))$ **do**

5:     $class \leftarrow classifier.predict(set\_of\_sub\_video[i])$

6:     **if** $class \in anomalies\_code$ **then**

        $anomalies\_found.append([i, set\_of\_sub\_video[i], class])$

7:     **end if**   ▷ We ignore sub-videos that are classified as having no anomalies.

8: **end for**

9: **return** $anomalies\_found$

The goal of Algorithm 1 is to identify defective segments in a sewer inspection video using a pre-trained video classification model. The inputs are: a video denoted as ins_video, a trained classification model $classifier$, a set of known anomaly codes $anomalies\_code$, and a defined segment duration sub_length used to divide the video into smaller sub-videos. In the initialization step, an empty list $anomalies\_found$ is created to store sub-video segments identified as containing anomalies, along with their index and associated predicted class (line 2 of the algorithm). The algorithm proceeds by splitting the input video into a collection of smaller sub-videos, each of length sub_length seconds, using the function $split\_video\_function$ (line 3). This produces the set $set\_of\_sub\_video$ containing all the sub-segments generated from the original video. The main processing loop then iterates over each sub-video (line 4). For each segment at index i, the classifier predicts a class label $class$ (line 5). If the predicted class matches one of the predefined anomaly codes in $anomalies\_code$, the sub-video is flagged as problematic. The algorithm records the index i, the corresponding sub-video, and the predicted class by appending them as a sub-list to *anomalies_found* (line 6). Finally, the algorithm returns the complete list of detected anomalies, enabling targeted analysis of specific segments in the inspection video (line 9).

3.1   Video Classification Models

TimeSformer. TimeSformer [2] takes as input a video clip $X \in \mathbb{R}^{H \times W \times 3 \times F}$, where H and W denote the height and width of each frame, 3 is the number of RGB channels, and F is the number of frames in the clip. Each frame is decomposed into N non-overlapping patches (similar to ViT [6]), each of size $P \times P$, such that the N patches cover the entire frame:

$$N = \frac{H \cdot W}{P^2}$$

TimeSformer then flattens each patch into a vector

$$x_{(p,t)} \in \mathbb{R}^{3P^2}$$

where $p = 1, \ldots, N$ indexes spatial locations within a frame, and $t = 1, \ldots, F$ indexes the temporal frames. Each vector contains the RGB values of the corresponding patch. Then TimeSformer linearly map each patch $x_{(p,t)}$ into an embedding vector $\mathbf{z}_{p,t}^{(0)} \in \mathbb{R}^D$ as Eq. (1):

$$\mathbf{z}_{(p,t)}^{(0)} = E\mathbf{x}_{(p,t)} + \mathbf{e}_{(p,t)}^{pos} \tag{1}$$

where $\mathbf{e}_{(p,t)}^{pos} \in \mathbb{R}^D$ is the linear embedding added to encode the spatiotemporal position of each patch. The resulting sequence of embedding vector $\mathbf{z}_{p,t}^{(0)}$ for $p = 1, \ldots, N$, and $t = 1, \ldots, F$ presents the input to the Transformer and plays a role similar to the sequences of embedded words that are fed to text Transformer in NLP. Like BERT Transformer [5], TimeSformer adds in the first position of the sequence a special learnable vector $\mathbf{z}_{(0,0)}^{(0)} \in \mathbb{R}^D$ representing the embedding of the classification token.

Divided Space-Time Attention in TimeSformer is both more efficient and accurate than other types of attention. In this mechanism, within each encoder block, temporal attention is first computed by comparing each patch (p, t) with all patches at the same spatial location p across other frames. The resulting temporal attention is then used to compute spatial attention by comparing each patch with all other patches within the same frame at time t.

The Video Vision Transformer (ViViT) [1] introduces a straightforward method for converting a video $V \in \mathbb{R}^{T \times H \times W \times C}$ into a sequence of tokens $\hat{z} \in \mathbb{R}^{n_t \times n_h \times n_w \times d}$. This process, known as **Tubelet embedding**, involves extracting non-overlapping spatio-temporal "tubelets" from the input video and projecting them linearly into $\mathbb{R}^d$. The resulting sequence is then reshaped into $\mathbb{R}^{N \times d}$ and enriched with positional embeddings to produce $\boldsymbol{z}$, which serves as the input to the Transformer. As illustrated in Fig. 1, this approach generalizes the standard 2D patch-based embedding in ViT to three dimensions, analogous to applying 3D convolutional operations for spatio-temporal representation learning. Formally, for a tubelet of size $t \times h \times w$, the number of tokens along each axis is defined as $n_t = \frac{T}{t}$, $n_h = \frac{H}{h}$, and $n_w = \frac{W}{w}$, where T, H, W and t, h, w are the temporal, height and width dimensions of the video and tubelet, respectively. Thus, reducing the tubelet size leads to a higher number of tokens and increased computational complexity. Unlike models such as TimeSformer, which extract temporal information independently across frames through "patch sampling", ViViT integrates both spatial and temporal information during tokenization, enabling a more unified spatio-temporal feature representation.

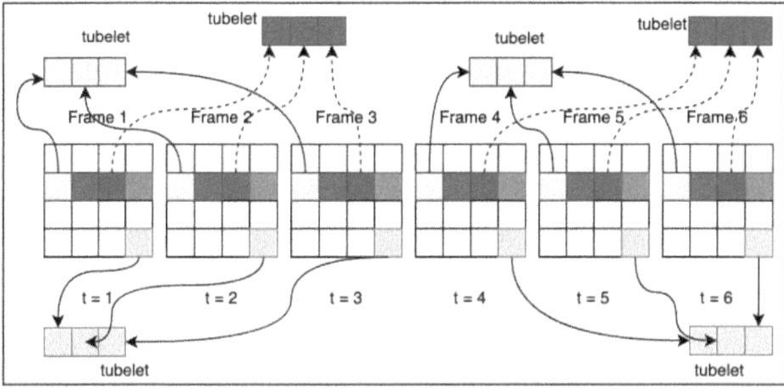

Fig. 1. Tubelet embedding. Tubelet is the combination of k patches ($k = 3$) that have the same position p of k continuous frame.

3.2 Building Experiment Datasets

The data used in this paper were obtained from "Montpellier Méditerranée Métropole" (3M), a public organization responsible for water management throughout the Montpellier metropolitan area, located in the south of France. Different portions of the network were studied, and for each portion, 897 videos of wastewater pipe inspections were recorded. Based on these videos, PDF and image reports of anomalies were created, containing information such as the details of wastewater pipe anomalies: positions of anomalies, associated codes and names, as well as maps showing the locations of manholes, pipes, and manhole identifiers. In addition to these reports, relationships between images and videos, and the problem codes of images, are stored in TXT and XML metadata files.

The following section outlines the process of constructing the experimental dataset, focusing first on the BAA anomaly (reduction of pipe height) as an example, with the same methodology applied to other anomalies. A total of 3,583 labeled images were manually extracted from 897 inspection videos, including 268 BAA-labeled images and 3,315 images labeled with other defect codes such as BAB (internal crack) and BAF (chemical attack). Among these videos, 173 contain at least one BAA image, while the remaining 724 do not. Some videos include multiple BAA-labeled frames, indicating the presence of the anomaly at multiple locations along the pipeline. The models [1,2] use short-labeled videos in training and testing sets. Therefore, we have to build the experiment dataset with short videos. Specifically, sub-videos centered around the manually annotated BAA frames were extracted and labeled as BAA samples, forming the positive examples in the classification dataset.

However, the metadata does not let us know how long the problem occurred in the video. We can split the entire video into a sub-video with equal window time, such as one or two seconds. However, if the sub-video contains too low-number BAA frames problem, it will be hard for the model to learn the feature to discriminate the BAA problem. We used Algorithm 2, named "Vnframe", for getting a sub-video with all frames close to the manually labeled BAA image. This algorithm tries to find n consecutive frames in the full video and meets the condition that these frames are close to the manually labeled BAA image (extracted in that full video). Then, n of these frames will be selected as a sub-video with the BAA label (positive class). We have three primary sources to take the sub-video for the none-BAA video (negative class).

- The first source is to get the sub-video in the same full video, which has manually extracted BAA images. Getting a none-BAA sub-video in the same full video with a BAA sub-video helps the model learn the difference between the BAA and none-BAA features in the same pipeline. To select a sub-video in this source, we can use Algorithm 2 to find n consecutive frames in the full video that are not close to the manually labeled BAA image, constructing these n frames as a sub-video and label it as none-BAA video.
- The second source to get a none-BAA sub-video is in the videos that do not contain any manually extracted BAA images, which may cause a lot of none-BAA sub-video; therefore, we could use random numbers to identify if we should get that considered sub-video. Additionally, we need to control the balance number of BAA and non-BAA in the dataset to help train the video classification model more easily. Getting a none-BAA sub-video in the full video that contains no BAA images can help the classification model to learn rich general features between all the pipelines we have in the data.
- The third source is to get positive sub-videos for all labeled images of other classes (except BAA images) and use them as non-BAA videos.

The sources for creating the none-BAA video are much richer than the source of the BAA video. Therefore, we proposed to create two experiment datasets; the first dataset, BAA-1, gets the none-BAA video from the second source, and the second dataset, BAA-2, gets the none-BAA video from the third source. Both datasets also include none-BAA videos from the first source, which helps the model learn useful features for clarifying the BAA problem instead of learning to classify the pipeline with other pipelines. Both BAA-1 and BAA-2 included three sub-sets: the training, testing and validation sets. We used the training to train the models, the validation set for evaluating models during the training process, and the testing set for final numerical test results.

We also used the manually labeled images as an additional testing set to increase confidence in the test results. We converted these images into videos, each with only one frame. The BAA images will become BAA videos, and the images of other classes will become none-BAA videos. We named this dataset T1 and used T1 as a testing dataset. To ensure that no BAA images are in the none-BAA class of these experiment datasets, after construction, we use a small brute-force algorithm to eliminate the sub-video that has at least one close frame with one of the baa images in the original dataset. In the same way, we also do not keep images of other classes close to one of the BAA images because those images may contain a BAA problem.

3.3   Vnframe Algorithms

Transfer Learning [13, 15, 17] is the method where the model is initially trained using one dataset, typically a large one, and subsequently employed to derive features for a related task involving the same data type (text or image data type). In this work, we used the pre-trained image model ViT [6] to transfer the image and frame of the video into the feature vector.

The Similarity of a video frame and an image is identified based on the cosine of their feature vector. Given two vectors $\boldsymbol{a} = [a_1, a_2, ..., a_n]$ and $\boldsymbol{b} = [b_1, b_2, ..., b_n]$, with n is the dimension of $\boldsymbol{a}$ and $\boldsymbol{b}$, θ the angle formed by $\boldsymbol{a}$ and $\boldsymbol{b}$, the cosine similarity of $\boldsymbol{a}$ and $\boldsymbol{b}$ is determined by the equation (2):

$$\text{cosine\_similarity}(\boldsymbol{a}, \boldsymbol{b}) = \cos(\theta) = \frac{\boldsymbol{a} \cdot \boldsymbol{b}}{\| \boldsymbol{a} \| \| \boldsymbol{b} \|} = \frac{\Sigma_{i=1}^{n} a_i b_i}{\sqrt{\Sigma_{i=1}^{n} a_i^2} \sqrt{\Sigma_{i=1}^{n} b_i^2}} \quad (2)$$

Algorithm 2 is named **Vnframe**, which refers to the algorithm for creating a video experimental dataset by using n consecutive frames closest to a target frame. Vnframe get as **input** a set of labeled images A and corresponding video V, the min and max number of frames in new video n and m, the min and max cosine similarity min_cos and max_cos, the max number allowed of frames that do not fit conditions $max\_penalty$, the pre-trained vision transformer model vit_model. The **output** is the set of videos with the positive label (same with set A) or negative (none of set A) depending on the value we set for min_cos and max_cos. In the Algorithm 2, for each image, the corresponding video is retrieved and segmented into individual frames. The image is encoded into a 768-dimensional feature vector using the ViT model. Subsequently, the algorithm iterates through all frames of the video, extracting feature vectors and computing the cosine similarity between each frame and the reference image. Frames satisfying the similarity threshold are appended to the $frame\_index\_list$, and the $penalty$ is reset. Conversely, if a frame does not meet the threshold, the *penalty* is incremented. When the *penalty* exceeds *max_penalty*, the algorithm checks whether the number of valid frames collected so far is greater than or equal to n. If the condition is met, a sub-video is constructed from the collected frame indices. Regardless of whether the sub-video is saved, the list and penalty counter are reset to allow evaluation of the next potential segment. Additionally, if the maximum number of frames m is reached, or if the end of the video is encountered and the collected frames meet the minimum requirement (n), the current frame list is similarly used to construct and save a $sub-video$. This dynamic frame selection strategy ensures that each sub-video exhibits high semantic alignment with the reference image, while tolerating limited frame inconsistencies. The integration of a penalty mechanism enables robust handling of $frame-level$ noise and allows for adaptive segmentation without requiring rigid temporal boundaries. The Algorithm 2 uses the variables *max_penalty* and *penalty* because the content of the videos in the dataset is affected by many factors such as water height, robot speed, etc. That means we accept a few frames in the sequence of frames that mismatch the condition of being the frame of the sample. The *frame_segment* and *construct_video* functions are implemented using the OpenCV library.

Figure 2 presents an example illustrating the application of Algorithm 2 for constructing a video dataset corresponding to the BAA anomaly class. In this approach, each sub-video labeled as BAA is generated by selecting n consecutive frames that are temporally and visually closest to a manually annotated BAA image. Figure 2 further depicts the process of extracting and labeling such sub-videos. When a BAA anomaly is identified within a video, the algorithm extracts n continuous frames centered around the anomaly's timestamp. These frames are then aggregated to form a sub-video, which is subsequently labeled as belonging to the BAA class.

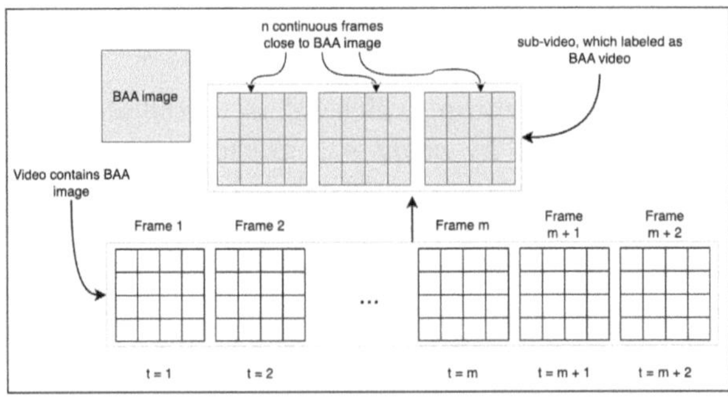

Fig. 2. Extracting and labeling sub-video as the sample of the BAA class.

Algorithm 2. Constructing *sample video* from *image* and *corresponding video*

1: Given: a set of labeled images A and corresponding video V, the min and max numbers of frames in new video n and m, the min and max cosine similarity min_cos and max_cos, the max allowed number of frames that do not fit conditions $max\_penalty$, the pre-trained vision transformer model vit_model.

2: Initialize:

       an empty list of frame index $frame\_index\_list$,

       $penalty = 0$

3: **for** $image \in A$ **do**

4:    $v \leftarrow$ corresponding video of $image$ in V

5:    $video\_frame\_set \leftarrow frame\_segment(v)$

6:    $vt\_image \leftarrow vit\_model.get\_feature(image)$

7:    **for** $i \in [0,\ len(video\_frame\_set))$ **do**

8:       $vt\_frame\_i \leftarrow vit\_model.get\_feature(video\_frame\_set[i])$

9:       $cosine\_value \leftarrow cosine(vt\_image, vt\_frame\_i)$

10:      **if** $cosine\_value \geq min\_cos$ and $cosine\_value \leq max\_cos$ **then**

          $frame\_index\_list.append(i)$

          $penalty = 0$

11:      **else**

          $penalty\ + = 1$

12:      **end if**

13:      **if** $penalty > max\_penalty$ **then**

14:        **if** $len(frame\_index\_list) \geq n$ **then**

          $construct\_video(frame\_index\_list[0], frame\_index\_list[$
$len(frame\_index\_list) - 1], video\_frame\_set)$

          $frame\_index\_list.empty()$

          $penalty = 0$

15:        **else**

          $frame\_index\_list.empty()$

          $penalty = 0$

16:        **end if**

17:      **end if**

18:      **if** $len(frame\_index\_list) == m$

        or $(i == len(video\_frame\_set)$

        and $len(frame\_index\_list) \geq n)$ **then**

        $construct\_video(frame\_index\_list[0], frame\_index\_list[len($
$frame\_index\_list) - 1], video\_frame\_set)$

        $frame\_index\_list.empty()$

        $penalty = 0$

19:      **end if**

20:    **end for**

21: **end for**

22: **return** 0

---

4 Experimentation

4.1 Parameters and Dataset

We fine-tuned TimeSformer and ViViT on our experimental datasets. ViViT requires substantial RAM; therefore, we used 4 NVIDIA A100-SXM4 GPUs with 40 GB of memory each. The software environment consists of Python 3.12.4, managed via Anaconda along with the necessary libraries. The parameters for TimeSformer and ViViT in our experiments are as follows: the *per-device train batch size* is 2 on 4 GPUs, yielding a *total batch size* of 8; the *max training steps* is calculated as (*number of videos/total batch size*) $\times 10$; and the *learning rate* is 5×10^{-5}. In creating the BAA video (and other anomalies), the parameters of the Vnframe algorithm are as follows: *max_cos* is 1.0, *min_cos* is 0.9, *max_penalty* is 3. The value of *cosine similarity* will never be higher than 1. However, this parameter is needed for the algorithm to work when creating non-BAA videos. In creating the none-BAA video from *the first source*, the parameters of the Vnframe algorithm are as follows: *max_cos* is 0.8, *min_cos* is 0.5, *max_penalty* is 3. We set value for *max_cos* because we want none-BAA videos to be different from the labeled BAA image in the same full video, while *min_cos* helps ensure that the none-BAA class doesn't have excessively noisy frames, such as pipe connection frames. Table 1 shows the number of videos in the experiment datasets. It includes training, testing and validation sets of the BAA-1, BAA-2 and T1 datasets (See Sect. 3.2 for the dataset construction methods). Besides BAA, we also evaluate three other problems, the BAF, BAB and BDD. The T1 for testing the classifier of other problems are also different from each other, therefore, we name them with additional problem codes, T1-BAA, T1-BAF, T1-BAB and T1-BDD.

Table 1. Number of videos in the experiment dataset.

Dataset	Class	Train	Test	Val	Total	Dataset	Class	Train	Test	Val	Total
BAA-1	BAA	1475	185	184	1844	BAB-1	BAB	343	43	43	429
	None	3256	407	407	4070		None	926	116	116	1158
BAA-2	BAA	1475	185	184	1844	BAB-2	BAB	343	43	43	429
	None	10995	1375	1374	13744		None	11149	1394	1394	13937
T1-BAA	BAA	–	268	–	268	T1-BAB	BAB	–	64	–	64
	None	-	3103	–	3103		None	–	3449	–	3449
BAF-1	BAF	688	86	86	860	BDD-1	BDD	842	106	105	1053
	None	1824	229	228	2281		None	2121	266	265	2652
BAF-2	BAF	688	86	86	860	BDD-2	BDD	842	106	105	1053
	None	10639	1330	1330	13299		None	11115	1390	1389	13894
T1-BAF	BAF	–	80	–	80	T1-BDD	BDD	–	309	–	309
	None	-	3365	–	3365		None	–	3082	–	3082

4.2 Results

Table 2 presents the numerical test results in terms of accuracy, precision, recall and F_1 score for TimeSformer and ViViT models across various datasets. These datasets were constructed using the Vnframe algorithm (videos composed of n frames) and by converting anomalous images into videos (single-frame videos). The results in Table 2 show that both TimeSformer and ViViT achieve very high accuracy, mostly above 93% and even exceeding 99% when trained and tested on the same dataset. Both models demonstrate strong performance on the video datasets (VnFrames), with ViViT slightly outperforming TimeSformer in most cases, and all reported F1 scores are above 86%, ranging from the highest

Table 2. Numerical test result (%).

Dataset	Model	Accuracy	Precision	Recal	F_1
Trained/tested on BAA-1 (standard test)	TimeSformer	97.30	92.54	99.47	95.88
	ViViT	96.45	91.09	98.40	94.60
Trained on BAA-1/ Tested on T1-BAA	TimeSformer	95.25	76.26	60.73	67.61
	ViViT	93.44	56.55	84.73	67.83
Trained/tested on BAA-2 (standard test)	TimeSformer	97.88	91.30	90.81	91.06
	ViViT	98.27	93.41	91.89	92.64
Trained on BAA-2/ Tested on T1-BAA	TimeSformer	94.90	94.02	40.00	56.12
	ViViT	96.41	93.26	60.36	73.29
Trained/tested on BAF-1 (standard test)	TimeSformer	95.24	93.62	90.72	92.15
	ViViT	98.10	94.17	100.00	97.00
Trained on BAF-1/ Tested on T1-BAF	TimeSformer	97.68	55.00	61.11	57.89
	ViViT	97.56	52.24	77.78	62.50
Trained/tested on BAF-2 (standard test)	TimeSformer	98.87	98.53	81.71	89.33
	ViViT	99.01	98.57	84.15	90.79
Trained on BAF-2/ Tested on T1-BAF	TimeSformer	98.03	86.67	28.89	43.33
	ViViT	98.35	92.31	40.00	55.81
Trained/tested on BAB-1 (standard test)	TimeSformer	98.74	98.11	98.11	98.11
	ViViT	98.11	100.00	94.34	97.09
Trained on BAB-1/ Tested on T1-BAB	TimeSformer	97.92	40.78	77.78	53.50
	ViViT	93.51	17.05	83.33	28.30
Trained/tested on BAB-2 (standard test)	TimeSformer	99.86	100.00	96.23	98.08
	ViViT	99.79	96.30	98.11	97.20
Trained on BAB-2/ Tested on T1-BAB	TimeSformer	99.12	96.00	44.44	60.76
	ViViT	99.43	90.48	70.37	79.17
Trained/tested on BDD-1 (standard test)	TimeSformer	98.66	97.96	96.97	97.46
	ViViT	96.77	96.77	90.91	93.75
Trained on BDD-1/ Tested on T1-BDD	TimeSformer	94.60	94.38	46.46	62.27
	ViViT	93.78	95.24	36.92	53.22
Trained/tested on BDD-2 (standard test)	TimeSformer	98.26	96.51	78.30	86.46
	ViViT	99.67	99.03	96.23	97.61
Trained on BDD-2/ Tested on T1-BDD	TimeSformer	93.51	96.46	33.54	49.77
	ViViT	93.81	100.00	35.38	52.27

of 98.11% (TimeSformer on BAF-1) to the lowest of 86.46% (TimeSformer on BDD-2). Both models demonstrate high performance when trained and tested on the video datasets (VnFrames), with ViViT slightly outperforming TimeSformer in most cases. All reported F_1 scores exceed 86%, with the highest being TimeSformer on the BAF-1 dataset($F_1 = 98.11\%$) and the lowest being TimeSformer on the BDD-2 dataset ($F_1 = 86.46\%$). Notably, the BAF and BAB codes consistently demonstrate strong performance across all types of datasets in our experiments.

The effectiveness of using a single frame versus a short video segment composed of n frames for detecting anomalies such as BAA, BAB, BAF, and BDD on the $T1$ testing dataset was compared. The evaluation demonstrates that using video segments (n frames) yields significantly better performance. As shown in Fig. 3 and Table 2, both accuracy and F_1 score are notably lower when the input consists of only a single frame. These results confirm that incorporating temporal context enhances the model's ability to identify defects in sewer inspection videos.

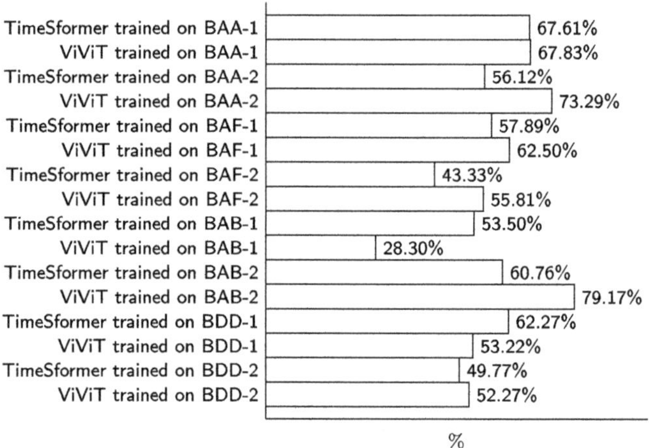

Fig. 3. F_1 of tested on T1

5   Conclusion and Future Works

In this paper, we build upon Video Vision Transformer (ViViT) and TimeSformer-based methodologies for video classification, which enable the effective capture of both spatial and temporal relationships across frames in video data. Our work specifically targets the BAA, BAF, BAB and BDD anomaly categories. Furthermore, we propose the Vnframe algorithm for constructing experimental video datasets from expert-labeled images, leveraging a pre-trained large vision model to facilitate this process. Numerical experiments conducted on the Prades-le-Lez dataset demonstrate that our approach achieves highly promising detection performance. In future work, we aim to extend our methodology to address additional types of defects within the pipeline infrastructure. We will also investigate further model optimization strategies and explore multilabel classification frameworks, as certain inspection videos may contain multiple co-occurring anomalies. Ultimately, our long-term objective includes developing models capable of supporting automatic video annotation for large-scale sewer inspection tasks

Acknowledgments. This research has received support from the European Union's Horizon research and innovation programme under the MSCA-SE (Marie Skłodowska-Curie Actions Staff Exchange) grant agreement 101086252; Call: HORIZON-MSCA-2021-SE-01; Project title: STARWARS (STormwAteR and WastewAteR networkS heterogeneous data AI-driven management). We thank "Montpellier Méditerranée Métropole" for providing essential data. This research also received support from the French national project ANR (Agence Nationale de la Recherche, ANR-21-CE23-0004) CROQUIS (Collecte, représentation, complétion, fusion et interrogation de données hétérogènes et incertaines de réseaux d'eaux urbains).

References

1. Arnab, A., Dehghani, M., Heigold, G., Sun, C., Lučić, M., Schmid, C.: Vivit: a video vision transformer. In: Proceedings of the IEEE/CVF International Conference on Computer Vision, pp. 6836–6846 (2021)
2. Bertasius, G., Wang, H., Torresani, L.: Is space-time attention all you need for video understanding? In: ICML, vol. 2, p. 4 (2021)
3. Biswas, R., Mutz, M., Pimplikar, P., Ahmed, N., Werth, D.: Sewer-ai: sustainable automated analysis of real-world sewer videos using dnns. In: International Conference on Pattern Recognition Applications and Methods (2023). https://api.semanticscholar.org/CorpusID:257356921
4. Chen, Y., et al.: Deep learning based underground sewer defect classification using a modified regnet. Comput. Mater. Continua **75**, 5455–5473 (2023). https://doi.org/10.32604/cmc.2023.033787
5. Devlin, J.: Bert: pre-training of deep bidirectional transformers for language understanding. arXiv preprint arXiv:1810.04805 (2018)
6. Dosovitskiy, A.: An image is worth 16x16 words: transformers for image recognition at scale. arXiv preprint arXiv:2010.11929 (2020)
7. Jung, J.T., Reiterer, A.: Improving sewer damage inspection: development of a deep learning integration concept for a multi-sensor system. Sensors **24**(23) (2024). https://doi.org/10.3390/s24237786. https://www.mdpi.com/1424-8220/24/23/7786
8. Li, Y., Wang, H., Dang, L.M., Song, H.K., Moon, H.: Vision-based defect inspection and condition assessment for sewer pipes: a comprehensive survey. Sensors **22**(7) (2022). https://doi.org/10.3390/s22072722. https://www.mdpi.com/1424-8220/22/7/2722
9. Li, Y., Wang, H., Dang, L., Jalil Piran, M., Moon, H.: A robust instance segmentation framework for underground sewer defect detection. Measurement **190**, 110727 (2022) . https://doi.org/10.1016/j.measurement.2022.110727. https://www.sciencedirect.com/science/article/pii/S0263224122000318

10. Liu, Y., et al.: Videopipe 2022 challenge: real-world video understanding for urban pipe inspection. In: 2022 26th International Conference on Pattern Recognition (ICPR), pp. 4967–4973 (2022). https://doi.org/10.1109/ICPR56361.2022.9956055
11. Sutskever, I.: Sequence to sequence learning with neural networks. arXiv preprint arXiv:1409.3215 (2014)
12. Therrien, J.D., Nicolaï, N., Vanrolleghem, P.A.: A critical review of the data pipeline: how wastewater system operation flows from data to intelligence. Water Sci. Technol. **82**(12), 2613–2634 (2020). https://doi.org/10.2166/wst.2020.393
13. Torrey, L., Shavlik, J.: Transfer learning. In: Handbook of Research on Machine Learning Applications and Trends: Algorithms, Methods, and Techniques, pp. 242–264. IGI global (2010)
14. Vaswani, A.: Attention is all you need. Adv. Neural Inf. Process. Syst. (2017)
15. Weiss, K., Khoshgoftaar, T.M., Wang, D.D.: A survey of transfer learning. J. Big Data **3**(1), 1–40 (2016). https://doi.org/10.1186/s40537-016-0043-6
16. Zhang, J., Liu, X., Zhang, X., Xi, Z., Wang, S.: Automatic detection method of sewer pipe defects using deep learning techniques. Appl. Sci. **13**, 4589 (2023). https://doi.org/10.3390/app13074589
17. Zhuang, F., et al.: A comprehensive survey on transfer learning. Proc. IEEE **109**(1), 43–76 (2020)

Tomato Leaf Classification Using GAN-Generated Images

Huynh Ngoc Tuyet, Nguyen Tri Phuc, and Nguyen Thai-Nghe[✉]

Can Tho University, Can Tho, Vietnam
{hntuyet,ntphuc}@ctu.edu.vn, ntnghe@cit.ctu.edu.vn

Abstract. Agriculture plays a vital role in economic growth, and the rapid advancement of Artificial Intelligence (AI) has enabled increasingly automated and efficient agricultural practices. In this study, we focus on the classification of tomato leaf diseases using image data from 10 classes. In addition, we employed a Conditional Generative Adversarial Network (cGAN) to generate synthetic images for data augmentation, achieving an optimal average FID score of 11.21, which indicates high-quality generated images. We then proposed a hybrid classification framework that extracts deep features from a custom-designed network, DCNN-GAPNet, and uses them as input to traditional machine learning models such as SVM and kNN. Experimental results demonstrate that this hybrid approach outperforms deep CNNs alone. Notably, the DCNN-GAPNet + SVM model trained on the cGAN-augmented dataset achieved the highest classification performance, significantly surpassing models trained on the original dataset.

Keywords: Conditional GAN · DCNN-GAPNet · Machine Learning · Tomato Leaf Disease · Hybrid Deep Learning

1 Introduction

Recognizing signs of disease on crop leaves is an important step in taking care of plants in modern farming. Today, detecting leaf diseases by visual inspection still has many limitations, especially because most farmers do not have deep knowledge of plant pathology. This can lead to inaccurate guesses that may affect the yield and health of crops. In this context, combining Machine Learning (ML) and Deep Learning (DL) for detecting leaf diseases from images has become a promising research direction. However, most existing datasets suffer from issues such as a limited number of images and an unbalanced distribution of samples across disease classes.

One of the good methods for increasing image data today is using Generative Adversarial Networks (GANs) - a powerful tool for creating high-quality fake images. In this study, we use a GAN model to generate more tomato leaf images for 10 classes (9 diseases and 1 healthy class). Then, we compare the performance of models that combine machine learning and deep learning to classify disease

N. Thai-Nghe et al. (Eds.): ISDS 2025, CCIS 2714, pp. 19–30, 2026.
https://doi.org/10.1007/978-981-95-3358-9_2

leaves. These models are trained on the original dataset and on the dataset that includes GAN-generated images. Our results show that combining deep learning (to extract features from images) with machine learning (to classify based on those features) gives better performance than using each model alone. These results prove that using both deep learning and machine learning, together with GAN-based image generation, can improve the classification of leaf diseases in smart farming systems. These results prove that using both deep learning and machine learning, together with GAN-based image generation, can improve the classification of leaf diseases in smart farming systems.

The rest of this paper is organized as follows: Sect. 2 presents related works, Sect. 3 describes the proposed method, Sect. 4 shows the experimental results, and Sect. 5 discusses the contributions and future directions of this research.

2   Related Work

In agriculture, plant diseases and insect pests are major threats to crop production. Early detection and diagnosis of these problems are very important. Today, the rapid development of deep learning methods has provided powerful tools with high accuracy for detecting plant diseases. Many studies have been carried out on this topic and have achieved promising results. However, the accuracy of deep learning models strongly depends on the amount and quality of labeled data used for training. In most cases, datasets face problems such as class imbalance or a shortage of training samples.

To address this problem in the case of tomato leaf diseases, Abbas et al. [1] proposed a deep learning method for detecting tomato diseases using a Conditional Generative Adversarial Network to create synthetic images of tomato leaves. The data before and after augmentation with cGAN were trained using a transfer learning model based on DenseNet121. As a result, the proposed method achieved classification accuracies of 99.51%, 98.65%, and 97.11% for tomato leaf image classification into 5 classes, 7 classes, and 10 classes, respectively.

In another approach, Deshpande and Patidar [2] proposed a Deep Convolutional Neural Network (DCNN) to enhance feature representation, combined with GAN-based data augmentation to handle class imbalance in a tomato leaf dataset (consisting of 10 classes). The proposed method achieved an accuracy of 99.74%, precision of 99%, recall of 99%, and F1-score of 99%, showing a significant improvement compared to traditional classification methods on plant disease datasets.

Based on recent trends showing that combining deep learning feature extraction with machine learning models brings positive results, Sujatha et al. [3] proposed a hybrid method using both DL and ML to detect diseases on agricultural leaves. The goal was to improve accuracy and scalability compared to traditional manual methods. The datasets used in their study included Banana Leaf, Custard Apple Leaf and Fruit, Fig Leaf, and Potato Leaf. Experimental results showed that for the Banana Leaf dataset, the combination of Inception v3 and SVM gave the best performance with an accuracy of 91.9% and an AUC of 99.6%. For the Custard Apple Leaf and Fruit dataset, the combination of VGG19 and kNN achieved the highest accuracy of 99.1% and an AUC of 91.1%. In addition, for the Fig Leaf dataset, the hybrid model also performed well with an accuracy of 86.5% and an AUC of 93.3%. Finally, on the Potato Leaf dataset, Inception v3 combined with SVM showed the best results with an accuracy of 62.6% and an AUC of 89%.

In addition to international studies, there have also been research efforts in Vietnam applying machine learning and deep learning to plant disease detection. Truong Thi Phuong Thanh and Nguyen Thai Nghe [4] used transfer learning methods to identify several diseases on rice leaves and achieved promising results on real-world datasets. Similarly, Bui Van Hau et al. [5] developed a plant disease recognition system for smart agriculture, which allows automatic disease detection from leaf images using integrated artificial intelligence techniques. These efforts reflect the growing domestic interest in applying Artificial Intelligence (AI) to agriculture, contributing to more effective monitoring of crop health for farmers.

3   Method

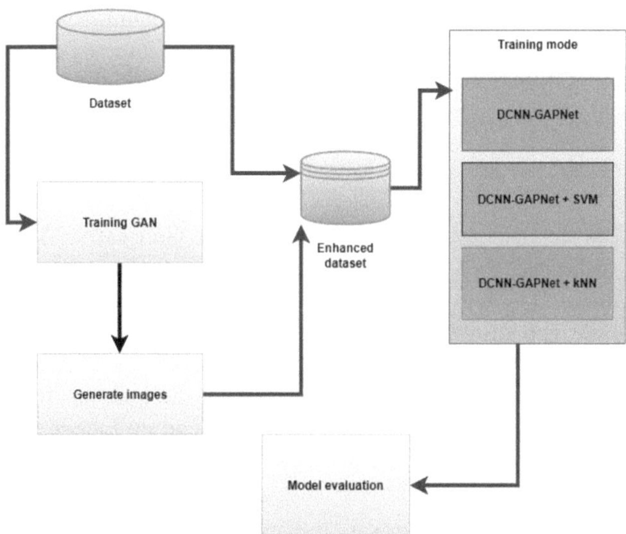

Fig. 1. The proposed method combining GAN-based data generation and hybrid deep learning model

Figure 1 illustrates the overall workflow of the proposed method. The original dataset is first used to train a cGAN model, which generates synthetic leaf images to augment the data. Both the original dataset and the enhanced dataset (original + synthetic images) are then used to train three types of models: DCNN-GAPNet, DCNN-GAPNet combined with Support Vector Machine (SVM) with, and DCNN-GAPNet combined with k-Nearest Neighbors (kNN). All models are trained and evaluated on both datasets to compare their classification performance and to assess the impact of GAN-based data augmentation.

For the classification task, we employed an SVM with the RBF kernel, as it effectively captures non-linear relationships in the data. The regularization parameter was set to $C = 1$, providing a balance between minimizing classification errors and preventing overfitting. The kernel coefficient was configured as $\gamma = $ "scale", following the default recommendation to adapt to the characteristics of the dataset. Furthermore, the probability output option was enabled to allow for a more comprehensive evaluation of the model's predictions.

In addition to SVM, we also experimented with the KNN classifier, where we selected $k = 3$ neighbors. This value is small enough to ensure that the model remains sensitive to local data patterns, yet sufficiently large to mitigate the influence of noise from individual samples. The Euclidean distance metric was kept as the default choice to maintain simplicity and ensure consistency across comparisons.

3.1   Generative Adversarial Nets (GAN)

Generative Adversarial Networks (GANs) were first introduced by Goodfellow et al. in 2014 [7] as a novel framework for generative modeling through adversarial training.

Equation 1. Objective function $V(D, G)$ of the original GAN.

$$\min_G \max_D V(D, G) = \mathbb{E}_{x \sim p_{\text{data}}(x)}[\log D(x)] + \mathbb{E}_{z \sim p_z(z)}[\log(1 - D(G(z)))] \quad (1)$$

Equation 1 is the original objective function of a Generative Adversarial Network (GAN), modeled as a two-player minimax game between a generator G and discriminator D. The generator learns to produce realistic data samples from noise, while the discriminator attempts to distinguish between real and synthetic samples. The objective function encourages D to correctly classify real and fake inputs, and G to generate samples that fool D. he training process reaches equilibrium when G produces data that D can no longer reliably classify as real or fake

3.2   Conditional Generative Adversarial Network

Conditional Generative Adversarial Network (cGAN) is a variant of the GAN introduced in [6]. In cGAN, both the Generator and the Discriminator are conditioned on additional information, typically a class label or specific input features, with the goal of controlling the type of image generated based on specific input.

Equation 2. Objective function $V(D, G)$ of the original cGAN.

$$\min_G \max_D V(D, G) = \mathbb{E}_{x \sim p_{\text{data}}(x)}[\log D(x|y)] + \mathbb{E}_{z \sim p_z(z)}[\log(1 - D(G(z|y)))] \quad (2)$$

The objective function of Conditional GAN (Eq. 2) extends the original GAN by incorporating auxiliary information y, such as class labels. Both the generator G and the discriminator D are conditioned on y. This allows the generator to produce samples specific to a given class and helps the discriminator distinguish real versus fake samples with respect to that class. The training process follows a two-player minimax game where D aims to correctly classify real and generated samples conditioned on y, and G learns to generate realistic data that can fool D.

In this study, cGAN is applied to generate additional synthetic images for all classes of the tomato leaf dataset based on the original training data.

3.3 Deep Convolutional Neural Network with Global Average Pooling Network DCNN-GAPNet

```
Model: "DCNN-GAPNet"
```

Layer (type)	Output Shape	Param #
input_img (InputLayer)	(None, 256, 256, 3)	0
conv1 (Conv2D)	(None, 256, 256, 96)	2,688
relu1 (ReLU)	(None, 256, 256, 96)	0
pool1 (MaxPooling2D)	(None, 128, 128, 96)	0
conv2 (Conv2D)	(None, 128, 128, 256)	221,440
relu2 (ReLU)	(None, 128, 128, 256)	0
pool2 (MaxPooling2D)	(None, 64, 64, 256)	0
conv3 (Conv2D)	(None, 64, 64, 384)	885,120
relu3 (ReLU)	(None, 64, 64, 384)	0
pool3 (MaxPooling2D)	(None, 32, 32, 384)	0
gap (GlobalAveragePooling2D)	(None, 384)	0
fc1 (Dense)	(None, 128)	49,280
output (Dense)	(None, 10)	1,290

```
Total params: 1,159,818 (4.42 MB)
Trainable params: 1,159,818 (4.42 MB)
Non-trainable params: 0 (0.00 B)
```

Fig. 2. Proposed DCNN-GAPNet architecture

The architecture consists of three convolutional blocks with a similar structure: each includes a Conv2D layer, a ReLU activation function, and a MaxPooling2D layer, with the number of filters increasing in each subsequent block. After these three blocks, the model employs a GlobalAveragePooling2D layer to reduce the feature dimensions instead of using Flatten commonly found in traditional CNN architectures. Finally, two fully connected layers are used to classify tomato leaf images. The detailed structure of the proposed architecture is illustrated in Fig. 2.

3.4 Hybrid Model for Classification

To improve classification accuracy, we propose a second approach that involves extracting deep features from the trained DCNN-GAPNet model. Specifically, the final fully connected output layer is removed, and the output of the fc1 layer (with 128 dimensions) is used as feature vectors. These vectors are subsequently used to train traditional machine learning classifiers such as SVM and kNN. By leveraging the rich hierarchical features learned by DCNN-GAPNet, these classifiers can benefit from more discriminative representations than raw image inputs. This hybrid approach combines the representational power of deep learning with the robustness and interpretability of classical machine learning methods.

3.5 Fréchet Inception Distance (FID)

In this study, we use the Fréchet Inception Distance (FID) to evaluate the similarity between the feature distributions of real and generated images. A pretrained InceptionV3 model is employed to extract features from the images. For real images, we collect up to 100 samples for each class from the original dataset. Images are taken in the order they appear in the dataset, and the process stops once the required number for all classes is reached. This set of real images is kept fixed as a reference throughout the training process. For generated images, at each epoch, the model produces a certain number of new images for each class.

To ensure the highest quality of generated samples, we adopt a per-class model selection strategy based on FID scores. First, for each trained model $M_i \in \mathcal{M}$, we compute the FID score $\text{FID}(M_i, c_j)$ between the generated images and the fixed set of real images for every class $c_j \in \mathcal{C}$. This results in a matrix of FID scores across all models and classes. Next, for each class c_j, we compare the FID values obtained from all candidate models and select the model that achieves the lowest score. Formally, the optimal generator for class c_j is defined as:

$$M^*_{c_j} = \arg\min_{M_i \in \mathcal{M}} \text{FID}(M_i, c_j), \qquad \text{FID}^*_{c_j} = \min_{M_i \in \mathcal{M}} \text{FID}(M_i, c_j).$$

By following this procedure, each class is associated with the generator that produces the most realistic images according to the FID metric. This class-specific selection ensures that the final augmented dataset leverages the best-performing model for each category, leading to higher-quality and more balanced synthetic data overall.

4 Experimental Results

In this section, we conduct experiments on a dataset obtained from[1]. The experimental scenarios include: generating synthetic images using a cGAN, and performing classification on both the original dataset and the augmented dataset.

---

[1] https://www.kaggle.com/datasets/kaustubhb999/tomatoleaf/data.

The classification methods evaluated include: (1) a custom deep learning model named DCNN-GAPNet; (2) a hybrid model combining DCNN-GAPNet and SVM, and (3) a hybrid model combining DCNN-GAPNet and kNN.

4.1 Dataset

Table 1. Statistics of original dataset per class

Class	Original Images
healthy	1,000
Bacterial_spot	1,000
Early_blight	992
Late_blight	1,000
Leaf_Mold	990
Septoria_leaf_spot	1,000
Spider_mites_Two-spotted_spider_mite	1,000
Target_Spot	1,000
Tomato_mosaic_virus	618
Yellow_Leaf_Curl_Virus	1,000
Total	**9,600**

Table 1 summarizes the distribution of original dataset across different tomato leaf disease classes. Most categories contain around 1,000 samples, except for Tomato_mosaic_virus, which is notably underrepresented with only 618 images. In total, the dataset consists of 9,600 original images.

The table presents the statistics of image quantities for each tomato leaf disease class, divided into four groups:

Table 2. Statistics of train, GAN-generated, test and validation images per class

Class	Train	GAN	Validation	Test
Healthy	2,744	300	804	306
Bacterial_spot	2,543	300	732	283
Early_blight	2,209	300	643	246
Late_blight	2,801	300	792	312
Leaf_Mold	2,478	300	739	276
Septoria_leaf_spot	2,593	300	746	289
Spider_mites_Two-spotted_spider_mite	1,572	300	435	175
Target_Spot	1,644	300	457	183
Tomato_mosaic_virus	1,937	300	584	216
Yellow_Leaf_Curl_Virus	1,835	300	498	204
Total	**22,356**	**3,000**	**6,430**	**2,490**

- Train Images: the number of original images collected from the PlantVillage dataset, after applying rotation and flipping augmentation.
- GAN-Generated Images: the number of synthetic images generated using the cGAN model to augment the data for each class.
- Validation Images: the number of images set aside for validation purposes and not used during training.
- Test Images: used for the final evaluation of the model and are kept completely separate from the training and validation sets to ensure the objectivity of the evaluation results.

Table 2 shows the distribution of the dataset after applying basic augmentation techniques to the training set, specifically horizontal flipping and rotations of $+15°$ and $-15°$. As a result, the training set was expanded to 22,356 images, while the validation and test sets remained unchanged to ensure a fair and unbiased evaluation.

4.2   Environment Setup

The models in this study were trained on a workstation equipped with an NVIDIA GeForce RTX 4090 GPU with 24 GB of memory, using CUDA version 12.2 and driver version 535.247.01.

4.3   Synthetic Image Generation with cGAN

(a) Original tomato leaf images

(b) GAN-generated tomato leaf images

Fig. 3. Examples of original and GAN-generated tomato leaf images across all classes

Figure 3 illustrates the original tomato leaf images for each disease class (top) and the synthetic images generated by the cGAN model (bottom). The cGAN was trained for 50 epochs and achieved the best FID score (Fréchet Inception Distance) of 11.69, indicating that the generated images closely resemble the real data in terms of visual quality.

4.4   Classification Results

Table 3. Comparison of performance between DCNN-GAPNet and DCNN-GAPNet with cGAN

Metric	DCNN-GAPNet	DCNN-GAPNet + cGAN
Train_accuracy	0.9384	0.9646
Train_loss	0.1785	0.1024
Val_accuracy	0.9261	0.9549
Val_loss	0.2344	0.1395
F1-score	0.9275	0.9561
Precision	0.9317	0.9580
Recall	0.9288	0.9569
AUC	0.9603	0.9760

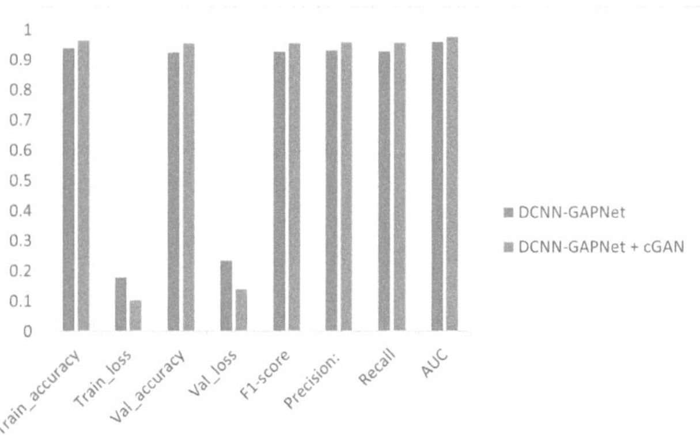

Fig. 4. Comparison of performance between DCNN-GAPNet and DCNN-GAPNet with cGAN

Performance of DCNN-GAPNet vs DCNN-GAPNet with cGAN. Figure 4 illustrates the performance comparison between the two CNN-GAPNet models on the original dataset and the dataset augmented with cGAN. It can be observed that CNN-GAPNet trained on the cGAN-augmented data shows improved performance compared to the model trained on the original dataset. Table 3 presents the detailed performance metrics of both models.

Table 4. Comparison of performance metrics between SVM and kNN models with input as feature vectors extracted from DCNN-GAPNet, with and without GAN-based augmentation

Metric	SVM	SVM + GAN	kNN	kNN + GAN
Accuracy	0.9731	**0.9866**	0.9798	0.9810
F1 Score	0.9748	**0.9876**	0.9809	0.9822
AUC Score	0.9994	**0.9998**	0.9954	0.9970
MCC Score	0.9700	**0.9850**	0.9775	0.9788
Loss	0.0880	**0.0486**	0.3505	0.2328

Comparison of Performance Metrics Between Hybrid Models with and Without GAN. Table 4 presents the classification performance on the test set using two machine learning models, SVM and kNN, with input features extracted from the DCNN-GAPNet model, both before and after data augmentation using cGAN. The results show that both DCNN-GAPNet + kNN and DCNN-GAPNet + SVM models achieved improved performance when trained on augmented data. Notably, the SVM model demonstrated superior results after augmentation, with an accuracy of 0.9866, F1-score of 0.9876, AUC score of 0.9998, and MCC score of 0.9850.

Figure 5 presents the confusion matrices of the DCNN-GAPNet models combined with SVM and kNN, evaluated on both the original and GAN-augmented datasets. The results demonstrate a noticeable improvement in classification performance, particularly for the DCNN-GAPNet + SVM model on the augmented dataset. This is evident in the higher number of correct predictions for under-represented classes such as class 2, class 4, and class 8. In contrast, the models trained on the original dataset still exhibit some classification errors, especially for classes with visually similar characteristics. While the DCNN-GAPNet + kNN model performs well on both datasets, it does not achieve the same level of optimal performance as the SVM. This suggests that kNN is less sensitive to complex data structures, even when augmented using GAN.

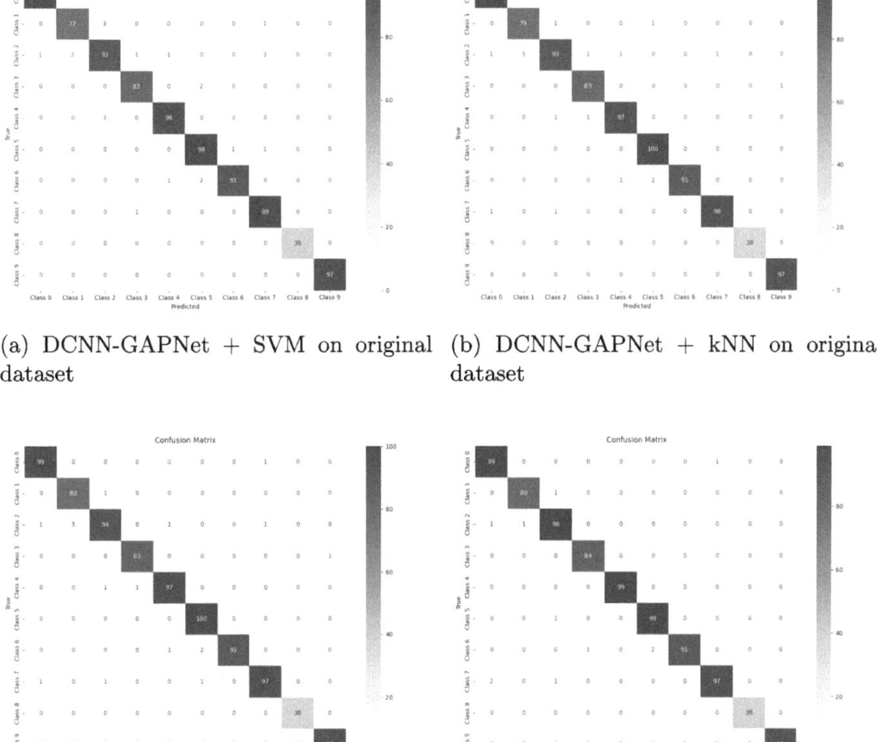

(a) DCNN-GAPNet + SVM on original dataset

(b) DCNN-GAPNet + kNN on original dataset

(c) DCNN-GAPNet + SVM on GAN-augmented dataset

(d) DCNN-GAPNet + kNN on GAN-augmented dataset

Fig. 5. Confusion matrices of models

5 Conclusion

In this study, we proposed a method for classifying diseases on tomato leaves using a hybrid approach that combines a custom-designed deep learning model, CNN-GAPNet, with traditional machine learning models such as SVM and kNN. To address the issues of data imbalance and scarcity, we applied a conditional GAN to generate synthetic images as a data augmentation strategy.

Experimental results demonstrate that cGAN-based augmentation significantly improves model performance and classification accuracy on the test set, as reflected in evaluation metrics such as Accuracy, F1-score, AUC, and MCC. Notably, the hybrid model DCNN-GAPNet + SVM achieved outstanding results with an F1-score of 0.9876 and an AUC of 0.9998 on the augmented dataset. The proposed approach validates the effectiveness of combining deep feature extraction with traditional machine learning classifiers, alongside leveraging cGAN for data augmentation, as a promising solution to enhance plant leaf disease classification performance. In the future, we will continue to investigate the application of more advanced GAN architectures to generate higher-quality augmented data. Additionally, we aim to evaluate the proposed method on other agricultural datasets to further generalize its effectiveness.

References

1. Abbas, A., Jain, S., Gour, M., Vankudothu, S.: Tomato plant disease detection using transfer learning with C-GAN synthetic images. Comput. Electron. Agric. **187**, 106279 (2021). https://doi.org/10.1016/j.compag.2021.106279
2. Deshpande, R., Patidar, H.: Tomato plant leaf disease detection using generative adversarial network and deep convolutional neural network. Imaging Sci. J. **70**(1), 1–9 (2022). https://doi.org/10.1080/13682199.2022.2161696
3. Sujatha, R., Krishnan, S., Chatterjee, J.M., et al.: Advancing plant leaf disease detection integrating machine learning and deep learning. Sci. Rep. **15**, 11552 (2025). https://doi.org/10.1038/s41598-024-72197-2
4. Truong, T.P.T., Nguyen, T.N.: Nhan dang benh tren la lua bang phuong phap hoc chuyen giao. Can Tho Univ. J. Sci. **58**(4), 1–7 (2022). https://doi.org/10.22144/ctu.jvn.2022.157 https://doi.org/10.22144/ctu.jvn.2022.157
5. Bui, V.H., Nguyen, T.T., Pham, A.T., Hoang, T.M.: He thong nhan dang benh cay trong hieu qua ung dung trong nong nghiep thong minh. J. Sci. Technol. HaUI **60**(6) (2024). https://doi.org/10.57001/huih5804.2024.206
6. Mirza, M., Osindero, S.: Conditional generative adversarial nets. arXiv preprint arXiv:1411.1784 (2014)
7. Goodfellow, I., et al.: Generative adversarial nets. In: Advances in Neural Information Processing Systems (NeurIPS), vol. 27 (2014). https://doi.org/10.48550/arXiv.1406.2661

AquaNetCT: Detecting and Representing Urban Drainage Networks from Video Analysis

Thanh Ma[1][(✉)], Salem Benferhat[2][(✉)], Minh-Thu Tran-Nguyen[1][(✉)],
and Thanh-Nghi Do[1][(✉)]

[1] Can Tho University, Can Tho, Vietnam
{mtthanh,tnmthu,dtnghi}@ctu.edu.vn
[2] CRIL, CNRS-Artois University, Lens, France
benferhat@cril.fr

Abstract. Effective management of urban drainage networks is vital for resilience and infrastructure reliability. This paper introduces AquaNetCT, an automated framework for detecting, analyzing, and visualizing manholes in Can Tho city, Vietnam. The system employs computer vision for manhole detection, extracting geospatial and textual metadata based on OCR. Deep learning models (i.e., CNN-DenseNet121) classifies manhole types with over 95% F1-Score accuracy, enabling prioritized maintenance. The dataset extracted from images/frames is structured and stored into formats (i.e., JSON, CSV, ...), and depicted using a graph-based algorithm designed to model the drainage network. This framework streamlines inspection workflows and enhances the operational efficiency and sustainability of smart city drainage management.

Keywords: Image Classification · Manhole Graph · Object Detection · GIS · Video Analysis

1 Introduction

Flooding is a recurring and significant challenge[1] in the Mekong Delta[2], a region heavily influenced by its vast network of rivers and seasonal monsoon rains. Known as the rice bowl of Vietnam (VN), the Delta faces annual floods that, while vital for replenishing soil fertility, also pose severe risks to urban areas. Among these, Can Tho [11,27], the economic and cultural heart of the region, is particularly vulnerable. Rapid urbanization, coupled with inadequate drainage infrastructure, has exacerbated waterlogging and intensified the impact of seasonal floods on the city's daily life and economic activities [11].

To mitigate these challenges, an efficient drainage system [26,29] is indispensable. At the core of this system lies a network of strategically placed manholes,

---

[1] https://www.mrcmekong.org/flood-and-drought/.
[2] https://nhess.copernicus.org/articles/24/3627/2024/.

which provide essential access for maintenance, cleaning, and preventing blockages. These manholes not only ensure the functionality of the drainage network but also serve as a critical defense against urban flooding. Strengthening and modernizing Can Tho's drainage infrastructure is vital to enhancing the city's resilience to flooding, safeguarding its development, and securing the livelihoods of its growing population.

Despite the critical role of the drainage system in managing urban flooding, its management in Can Tho still faces numerous challenges [8]. One major issue is the overlapping functions of various drainage types, including household wastewater systems, road drainage, and industrial or institutional networks [22,24]. The lack of clear delineation and integration among these systems complicates maintenance and efficient operation, creating bottlenecks in stormwater management [6,18]. Furthermore, repeated modifications and updates to the underground drainage infrastructure over the years have led to inconsistencies in system records. Multiple versions of subterranean networks [4,21] remain undocumented or partially integrated, making it difficult to visualize and effectively manage the entire system. As a result, efforts to address flooding often focus on visible surface-level drainage [19], while the underlying complexities of the underground network remain neglected. This disjointed management approach poses significant obstacles to achieving a comprehensive and sustainable solution for Can Tho's drainage and flood resilience.

The diverse shapes and designs of manholes [31] scattered across Can Tho's roads and urban areas provide valuable visual cues that can help reconstruct a comprehensive drawing of the city's drainage system [7,14]. By analyzing the placement, size, and types of these manholes, it is possible to infer the underlying structure of the drainage network, offering insights into its connectivity and functionality [12]. While this method may not yield absolute precision, it provides an effective and intuitive means of addressing the problem using *"visible surface knowledge"*. Building on this concept, an AI-driven system called AquaNetCT[3] has been proposed to leverage these surface-level-view pieces of knowledge[4]. By integrating images of manholes with their geographic coordinates (longitude and latitude), the system aims to create a graph-based representation of the drainage network [10,16]. This network model visualizes the relationships and connections between different nodes (manholes) and edges (pipes), enabling a clearer understanding of the system's architecture. The integration of advanced computer vision techniques, such as Object Detection [5,25] and Image Classification [1,30], combined with spatial mapping, empowers AquaNetCT to extract and analyze critical information from visual dataset (i.e., video/images).

The primary contributions of this research are multifaceted, addressing key challenges in urban drainage system management through following solutions: (1) *Dataset Collection*: Images and videos of manholes are gathered via a mobile app with integrated media capture and geolocation; *Information Extraction*:

---

[3] Aqua: water or urban drainage; Net: network/graph; CT: Can Tho city, Vietnam.

[4] We utilize the information of manhole extracted from input videos instead of the knowledge of the underground water pipeline.

Using OCR and object detection, key details such as location, street, and structural attributes are automatically extracted for analysis; *Type Classification*: Manholes are categorized by function, type, and design through analysis of physical features and surface markings; *Graph Construction*: Based on extracted data and coordinates, a graph model of the drainage system is created to represent connections and support efficient management.

2   Background

In this section, we outline the methodology for dataset collection and establish the foundational steps for constructing the manhole graph/network, detailing the process from data extraction to the necessary components required to build the manhole graph.

2.1   Collecting the Dataset

The dataset is collected in the form of images (denoted as I) and videos (denoted as V) using the *GPS Map Camera: Geotag Photos* application by *Susamp Infortech* (see Fig. 1). This application enabled us to capture images with geotagging information, including longitude (Lon), latitude (Lat), address, city, country, time, and date. The geotagging feature ensures that each image contains location-based data with reasonable accuracy, providing sufficient spatial context for analysis. The collected data provides valuable insights for further analysis and study, especially in relation to the geographical distribution and space/time aspects of the content. We will provide the amount of collection in the experimental result section.

Fig. 1. Pieces of information extracted from an image of the dataset collected.

2.2   Extracting Geo-Information

Geographic information extraction (GIE) [9] using OCR [20] involves identifying and digitizing textual data related to location coordinates (longitude and latitude) from images. The process begins by isolating regions of interest containing textual information, followed by OCR techniques [28] to convert the visual text into machine-readable formats. This extracted data, including longitude, latitude, address and others , is then structured into standardized formats (e.g., JSON, Csv) for seamless integration into mapping and visualization[5]. By automating geographic information extraction, this method ensures precise localization and facilitates the management of urban drainage systems from images.

Definition 1 (GIE Model). *Let t be a text region in an input image x (i.e., $t \in x$). A GIE model is a function $h^{\phi}(t)$ that maps t to coordinates $G = \{(lon_i, lat_i, addr_i, pic_i, s_i)\}$, where lon_i and lat_i are longitude and latitude, pic_i is the url of x and s_i is the confidence score. The model, parameterized by ϕ, uses OCR for precise geo-spatial data extraction.*

Example 1. Let $t_1 \in x$ be a text region in image I. We have $G = h^{M_1}(t_1)$, where $M_1 =$ *"paddleocr"*, $G = \{g_1, g_2, \ldots\}$. Therein, for example in Fig. 1, we obtain $g_j = (105.764181, 10.011567,$ "32 Tran Hoang Na, Hung Loi, Ninh Kieu, Can Tho, Viet Nam", *"./directory/img1.png"*, $92.6)$ is a coordinate and address extracted from I, i.e., 105.764181 is a longitude; 10.011567 is a latitude; *"32 Hoang Na, Hung Loi, Ninh Kieu, Can Tho, Viet Nam"* is the address; 92.6 is the accuracy.

2.3 Detecting the Manholes

Object detection (OD for short) [2,33] plays a crucial role in identifying and localizing manholes within urban drainage systems. In this study, we investigate the performance of various YOLO (You Only Look Once) [5,13] versions for real-time and accurate manhole detection. The models are single-stage object detectors that divide an input image into a grid, predicting bounding boxes, class labels, and confidence scores simultaneously. By leveraging the strengths of OD's different iterations, including improvements in accuracy, speed, and feature extraction capabilities, we aim to evaluate their effectiveness in detecting manholes under diverse environmental conditions. The results will provide insights into the optimal OD version [25] for practical implementation in urban infrastructure management.

Definition 2 (Manhole Detection). *Let x be an input image. A manhole detection model is a function $g^{\theta}(x)$ that maps x to a set of detections $D = \{(b_i, c_i, s_i)\}_{i=1}^{n}$, where b_i denotes the bounding box, c_i is the class label, and s_i is the confidence score. The model, parameterized by θ, performs detection in a single forward pass, optimizing for real-time and accurate localization of manholes.*

Example 2. Let I_1 be an input image and $\theta =$ *"Yolov8"*. We have $D = g^{Yolov8}(I_1)$, where $D = \{d_1, d_2, \ldots\}$. Therein, $d_j = (b_j, c_j, s_j)$ is a bounding box covering one manhole detected.

---

[5] presented in the next session.

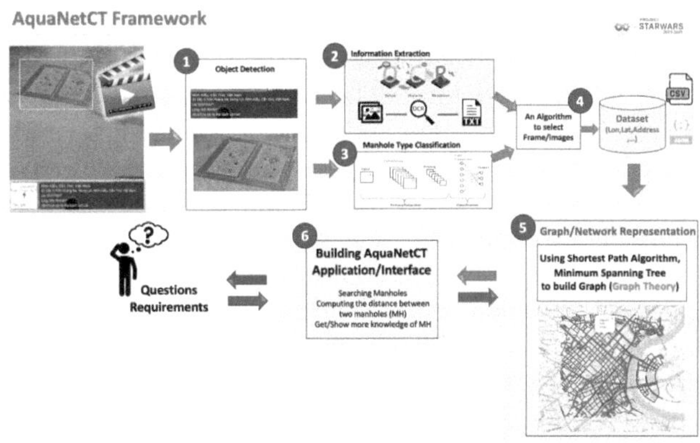

Fig. 2. AquaNetCT Framework for managing drainage networks in Can Tho.

2.4   Identification of Manhole Types

Manhole classification models based on Convolutional Neural Networks (CNNs) [17] provide an advanced approach for automatically categorizing various types of manholes from images/videos in Can Tho city. CNNs excel in learning hierarchical spatial features, allowing for highly accurate recognition and differentiation between distinct manhole types. In this study, we investigate several state-of-the-art architectures, including MobileNet [23], DenseNet [32], ResNet [15], and Xception [3], among others, to identify the most effective model for manhole classification.

Definition 3 (Manhole Type Classification). *Let x be an input image. An manhole classification model using a CNN is a function f^μ that maps an input image x, represented as a multi-dimensional pixel array, to a class $y = f^\mu(x)$ is either a categorical label or a probability vector over the class set Y. The image x is assigned to the class Y_j corresponding to the highest probability in y, where μ denotes the CNN architecture.*

Example 3. Let us have $Y = \{Y_1, Y_2, \ldots\}$ and $\mu \in \{$MobileNet, DenseNet, $\ldots\}$. In this case, we assume that an input image I_1 and $\mu = MobileNet$. Then, we obtain $Y_k = f^{MobileNet}(I_1)$, where $Y_k \in Y$.

3   AquaNetCT Framework

We introduce the AquaNetCT framework, designed to streamline the identification, analysis, and visualization of manholes within urban drainage networks. Its operation is organized into two stages (see Fig. 2): *Extracting Information from Videos and Building a Dataset:* This stage begins with Object Detection (Step 1), using computer vision to identify manholes and capture essential metadata like GPS coordinates, timestamps, and addresses, forming the foundation for mapping and analysis. Next, Information Extraction (Step 2) applies OCR to digitize textual details (e.g., longitude, latitude, names), ensuring critical information is preserved for classification and analysis. Manhole Type Classification (Step 3) utilizes CNNs to categorize manholes by function (e.g., drainage, utilities). This one is necessary to tailor policies and maintenance actions for specific types of manholes. Finally, the data is organized into a structured Dataset (Step 4) in formats like JSON or CSV, optimized for advanced analysis. Exporting the dataset ensures version control and historical tracking, creating a solid base for graph representation and query-answering; *Representing the Dataset as a Graph (or Network):* At this stage, the dataset is transformed into a graph representation to capture spatial and functional links between manholes. Graph-based algorithms, including the Shortest Path and Minimum Spanning Tree, are applied to optimize inspection routes and reduce redundancy, ensuring an efficient yet accurate network model This representation integrates seamlessly with GIS systems, enhancing visualization and enabling data-driven decision-making for infrastructure management.

By converting video data into structured datasets, mapping them into graphs, and enabling intelligent queries, the framework turns raw inputs into actionable insights. Integrating deep learning, graph theory, and language models makes it both precise and versatile for infrastructure management. The next section introduces our algorithm for representing the manhole network via graph theory.

4 Proposed Algorithms

First of all, we define formally the manhole graph $\mathcal{G}$. This graph concentrates on storing the pieces of information of manholes. It includes type of manhole, longitude and latitude, the address of the manhole, and the URL pointed to the captured manhole image. It will be defined as follows:

Definition 4 (Manhole Graph). *A Manhole Graph is a weighted, undirected graph $\mathcal{G} = (V, E)$, where: V is the set of nodes, with each node $v \in V$ represented as a tuple: $v = (type, lon, lat, street, image)$, where: type: The type of the manhole, lon, lat : The geographic coordinates (longitude and latitude), street: The name of the street (ward, province, city) where the manhole is located, image: A visual representation of the manhole (e.g., a photo or frames of video); E is the set of edges, where each edge $e \in E$ is represented as: $e = (v_i, v_j, w)$, with $v_i, v_j \in V$ being two connected nodes and w representing the shortest path (nearest) distance between v_i and v_j, calculated based on their geographic coordinates.*

To construct our graph, we propose two algorithms presenting in two following sub-sections.

4.1 Selecting the Knowledge from Video's Short Segment

The algorithm is designed to efficiently select the most representative frames from an input video $\mathcal{V}$ by identifying those containing manholes. The underlying idea is straightforward and effective. In particular, prioritize frames where the detected manhole occupies the largest area. This approach ensures that the selected frame's coordinates are closest to the actual location of the manhole in the video. To this end, we propose Algorithm 1.

Algorithm 1: Algorithm for selecting plausible knowledge.

Input : $\mathcal{V}$: An input video; θ : The object detection model; $\alpha = 85$: Threshold of confidence.
Output: $\mathcal{F}$: List of images containing manholes with plausible knowledge.

```
1  begin
2      F ← ∅, maxArea ← 0, currentFrame ← ∅
3      for f ∈ V do
4          D ← gθ(f)  // Detect manholes using OD model (e.g., Yolo)
5          if D ≠ ∅ then
6              area ← max({ComputeArea(b) | (b,c,s) ∈ D ∧ s > α})
7              if area > maxArea then
8                  maxArea ← area
9                  currentFrame ← f
10         else if maxArea > 0 then
11             F ← F ∪ {currentFrame}
12             maxArea ← 0
13             currentFrame ← ∅
14     if maxArea > 0 then
15         F ← F ∪ {currentFrame}  // Ensure the last valid frame is added.
16     return F
```

The complexity of Algorithm 1 is primarily influenced by the number of frames n in the input video and the computational cost d of the object detection model $g^{\theta}(f)$. For each frame, the algorithm applies object detection and computes areas for detected objects, leading to an overall complexity of $O(n \cdot d)$. This demonstrates that the algorithm's scalability is heavily dependent on the efficiency of the object detection model, making it suitable for optimization to enhance performance in large-scale applications. After obtaining list of "valid" images, we construct our manhole graph in the next sub-section.

4.2   Constructing the Manhole Graph

In this session, we provide representing the manhole network based on the combination of graph theory algorithms. Drawing from expert insights and practical observations, the following principles/conditions are proposed to guide the modeling and management of drainage networks from a surface-level perspective: (1) The network must be acyclic, as drainage systems are designed to prevent interconnections or circular flows that could disrupt water movement. (2) In the absence of detailed underground mapping data, it is reasonable to assume that the two nearest manholes are directly connected by a drainage path, providing a practical approximation for network construction. (3) Differentiating between manhole types is crucial for efficient maintenance and operation, as each type is typically associated with a specific drainage system, ensuring proper functionality and ease of management. These principles aim to balance theoretical rigor with practical feasibility in the design and analysis of drainage networks. Algorithm 2 is implemented complying the above principles.

This algorithm constructs a graph representation of a manhole network using image-based data. Starting with a folder of images ($\mathcal{F}$), extracted by Algorithm

---

Algorithm 2: Algorithm for representing the manhole network.

Input   : $\mathcal{F}$: A folder of input images (From Algorithm 1); ϕ: The OCR model; θ: The object detection
model; $\xi = 80$: Threshold of confidence; μ: The CNN architecture; α, β: Lower and upper bounds
for valid distances; $type$: Type of file.
Output: $\mathcal{G}$: The manhole network underlying the graph form.

```
 1  begin
 2  |   V ← ∅, E ← ∅ // Initialize set of nodes, edges
 3  |   for I ∈ F do
 4  |   |   t ← ExtractRegion(I) // Extract regions of interest in the image
 5  |   |   G ← h^φ(t) // Perform OCR to extract geographic information
 6  |   |   Y ← {g^μ(b) | b ∈ D} // Classify manhole types
 7  |   |   for (lon, lat, addr, pic, s) ∈ G do
 8  |   |   |   if s > ξ then
 9  |   |   |   |   V ← V ∪ {Node(y, lon, lat, addr, pic) | y ∈ Y}
10  |   |   |   |   S ← S ∪ {Name(I) : (y, lon, lat, addr, pic) | y ∈ Y}
11  |   MakeFile(S, type)
12  |   Max ← 0 // Initialize maximum distance
13  |   for v1 ∈ V do
14  |   |   for v2 ∈ V do
15  |   |   |   w ← Dijkstra(v1, v2) // Compute distance based on (lon, lat)
16  |   |   |   if (α ≤ w ≤ β) ∧ (w > Max) then
17  |   |   |   |   Max ← w
18  |   |   |   |   P ← (v1, v2, w) // Potential edge
19  |   |   E ← E ∪ P // Add potential edge to E
20  |   G ← MakeGraph(V, E) // Construct the graph
21  |   G ← MST(G) // Compute the Minimum Spanning Tree (MST)
22  |   return G // Return the final network
```

1, it isolates regions of interest and applies an OCR model (ϕ) to extract geographic metadata, such as longitude, latitude, addresses and other. A node $v_i \in V$ contains these pieces of information. Regarding the $t \longleftarrow ExtractRegion(I)$, for each image $I \in \mathcal{F}$, regions of interest t are defined as subsets of the image: $t = \{r \mid r \subseteq I$ and $\phi(r) > \tau\}$, where r represents a spatial region, $\phi(r)$ is a scoring function based on the OD model (i.e. YOLO), and τ is a predefined threshold to filter meaningful regions. Simultaneously, an object detection model (θ) identifies manholes, while a classification model (μ) assigns them to specific types. Only detections with confidence scores exceeding a threshold ($\xi = 80$ for default) are added as nodes (V) in the network, each enriched with metadata for downstream analysis. Moreover, to define edges (E), the algorithm calculates pairwise distances between nodes using Dijkstra's algorithm[6] and establishes connections only if the distances (w) lie within the range $[\alpha, \beta]$. Distances that are too small may indicate overlapping or redundant nodes, while excessively large distances violate the practical constraints of Condition/Principle 2. Thus, the selection of $[\alpha, \beta]$ ensures that the connectivity between manholes is both realistic and consistent with practical requirements. The resulting graph $\mathcal{G}$ is then refined by computing its MST[7] (MST), which optimally eliminates redundant cycles while preserving essential connectivity. By computing shortest paths with Dijkstra over a road-network graph, rather than straight-line distances, we ensure every proposed pipe segment follows existing streets and public easements, avoiding the legal and logistical impossibility of digging through private property. Subsequently extracting a Minimum Spanning Tree prunes redundant links while preserving connectivity, thereby minimizing total trench length and construction cost. In Vietnam, where sewer lines almost invariably trace road

---

[6] https://www.w3schools.com/dsa/dsa_algo_graphs_dijkstra.php.

[7] https://cp-algorithms.com/graph/mst_kruskal.html.

corridors to leverage existing drainage channels and obviate disruptive excavation, this Dijkstra + MST framework will mirrors on-ground reality and delivers an optimally economical and implementable network design.

The algorithm's complexity is primarily driven by two factors: collecting information from the input images and constructing the graph. Iterating through $\mathcal{F}$ (input images) has a linear complexity of $O(|\mathcal{F}|)$. However, the double loop over nodes V for pairwise distance calculation using Dijkstra's algorithm incurs a complexity of $O(|V|^2 \cdot T_D)$, where T_D is the complexity of Dijkstra's algorithm. The graph construction and the computation of the MST contribute an additional $O(|E| \log |V|)$. Overall, the algorithm's complexity is dominated by $O(|V|^2)$, making it computationally expensive for large networks. To reduce computational expense for large networks, the algorithm can limit the search space by managing drainage systems at the ward level instead of the district or city level. By dividing the network into smaller clusters for each ward, computations such as graph construction and optimization are performed locally, significantly reducing complexity. These local networks can later be integrated for broader analysis if needed, ensuring both efficiency and scalability.

5   Experiment and Results

This section analyzes the dataset, compares classification and detection models, and presents the Python-based user interface with its deployment. It also discusses the system's strengths and limitations, with the dataset and source code publicly available on GitHub[8].

5.1   Dataset and Configurations

We train a CNN model for classifying types of manhole into four types of manholes. The dataset is showed in Table 1. The dataset presented in this study encompasses two primary categories: classification and object detection, each tailored to distinct tasks within the domain. The classification subset comprises 3,853 images, distributed among four specific classes: DouRec, Rec, RoundSqr, and Sqr, with respective sample sizes of 870, 1,020, 920, and 1,043 images. This distribution ensures a balanced representation of classes, enabling robust training and evaluation of classification models. In contrast, the object detection subset contains 2,124 images, accompanied by 5,187 annotations. These annotations were meticulously labeled using RoboFlow, a powerful platform that facilitates precise and efficient labeling for complex datasets. The object detection annotations aim to localize and identify multiple objects within each image, reflecting real-world scenarios and enhancing the system's applicability. Together, these datasets provide a comprehensive foundation for developing and evaluating machine learning models, with a strong emphasis on accuracy and versatility in both classification and detection tasks. For the classification model, we investigate the manhole dataset with various architechtures of CNN, i.e., DenseNet121, Lenet, InceptionV2, Xception, MobileNet.

---

[8] https://github.com/Vannguyen0312/Manholes-App/.

5.2 Detecting the Manhole and Type Classification

The detection of manhole involves utilizing computer vision models to identify the manhole in the images (Y8, Y9, Y11) based on various configurations. The table summarizes the results across different configurations, evaluating performance using metrics such as Recall, mAP@50, and Testing Time. Each configuration represents a combination of batch size (BS) and input resolution, influencing the model's ability to accurately detect manholes (Table 2).

Table 1. Dataset of Manhole Type Classification and Manhole Detection

No	Name	Nbr of Images	Annotation
1	**Classification**	3853	–
	DouRec	870	
	Rec	1020	
	RoundSqr	920	
	Sqr	1043	
2	**Object Detection**	2124	5187

Table 2. Results of Object Detection for Manhole Types (Epochs = 30)

ID	Config	Recall			mAP@50			Testing Time (s)		
		Y8	Y9	Y11	Y8	Y9	Y11	Y8	Y9	Y11
1	BS32, 448×448	0.836	0.835	0.808	0.866	0.867	0.850	0.327	0.442	0.388
2	BS64, 448×448	0.841	0.828	0.797	0.828	0.864	0.847	**0.325**	0.394	0.365
3	BS32, 480×480	0.842	**0.847**	0.807	0.857	**0.876**	0.855	0.362	0.445	0.425
4	BS64, 480×480	0.844	0.824	0.814	0.870	0.863	0.850	0.348	0.428	0.402

Table 3. Performance Comparison of Different Models for Manhole Types. Early-stopping = 0.5

Id	CNN Models	Acc (%)	Prec (%)	Rec (%)	F1 (%)	AVG Time (s)
1	Lenet (BS = 32, 32×32)	94.67	94.73	94.40	94.43	0.0023
2	Densenet121 (BS = 64, 64×64)	97.46	97.46	97.21	97.25	0.0182
3	Xception (BS = 64, 48×48)	94.53	94.53	94.10	94.21	0.1352
4	Resnet (BS = 64, 64×464)	92.34	92.45	92.61	92.56	0.1753
5	Mobilenet (BS = 32, 32×32)	97.01	96.34	96.78	96.69	0.0141

The results show that input resolution strongly affects detection performance. Higher resolutions, such as "BS32, 480×480", yield the best recall (0.847) and

Table 4. Summary of Results

Nbr of Videos	Nbr of Manholes (MH)	Manhole Detection	MH Type Classification	Computation Time (s)
5	82	82/82	82/82	2.43
10	177	177/177	176/177	5.67
15	291	291/291	288/288	8.56
20	402	399/402	394/399	11.28
25	468	463/468	455/463	14.12
30	578	573/578	565/573	17.52
37	663	654/663	644/654	24.34

mAP@50 (0.876) with Y9, confirming their effectiveness in manhole detection. Lower resolutions (e.g., "BS32, 448×448") achieve competitive accuracy with faster inference (0.327 s), making them suitable when speed is prioritized. Batch size also influences performance: smaller batches (BS32) enhance accuracy, while larger ones (BS64) slightly reduce recall and mAP@50 but improve efficiency. Based on this trade-off, we select Y9 with BS32 and 480×480 resolution, achieving 84.47% recall.

To classify manhole types for knowledge representation, we experimented with several CNN architectures, with results summarized in Table 3. Densenet121 achieved the best overall performance (97.46% accuracy, 97.25% F1), balancing precision and recall, and was therefore selected for the AquaNetCT system. Mobilenet offered slightly lower accuracy (97.01%, F1 96.69%) but faster inference (0.0141 s), while Lenet traded lower accuracy (94.67%) for the fastest runtime (0.0023 s), making it suitable for real-time use. Xception (94.53%, 0.1352 s) and Resnet (92.34%, 0.1753 s) lagged behind in both accuracy and efficiency. From the results, we select Densenet121 to implement AquaNetCT system.

5.3   Results on Videos

In this session, we provide a comprehensive analysis of the performance of a system in detecting and classifying manholes across varying numbers of videos, alongside the average time required for processing. Overall, the data highlights a strong correlation between the number of videos processed and the system's accuracy and efficiency. As the number of videos increases, the detection and classification performance remain consistent, albeit with minor fluctuations (Table 4).

Initially, with a smaller dataset of 05 videos, the system operates at peak performance, achieving flawless accuracy as the number of identified, detected, and classified manholes remains consistent at 82. The processing time is minimal at just 2.43 s, reflecting the system's high efficiency when dealing with limited computational demands. However, as the dataset size grows, small but noticeable discrepancies emerge. For instance, with 20 videos, the system identifies 402 manholes, detects 399, and classifies 394. By the time the dataset expands to 37 videos, these figures diverge further to 663, 654, and 644, respectively. The increasing gap between detection and classification indicates that the latter task is more computationally intensive and prone to errors, potentially due to overlapping features between manhole types or reduced data quality in larger datasets. The steady increase in average processing time further underscores the impact of workload on system efficiency. While processing 5 videos takes only 2.43 s, the time required for 37 videos rises sharply to 24.12 s. This proportional increase suggests that the system struggles to maintain computational efficiency as the dataset size escalates, likely due to the linear or exponential growth in data complexity and the resources required for accurate analysis.

The underlying reasons for these trends can be attributed to both algorithmic and practical limitations. The system's ability to maintain near-perfect detection accuracy even with larger datasets demonstrates its robustness in identifying manholes. However, the decline in classification accuracy suggests that distinguishing between manhole types involves more sophisticated computations, which may be adversely affected by noise, overlapping characteristics, or insufficient feature differentiation. Similarly, the increasing processing time reflects the growing computational burden, indicating a need for system optimization to ensure scalability without compromising speed or accuracy.

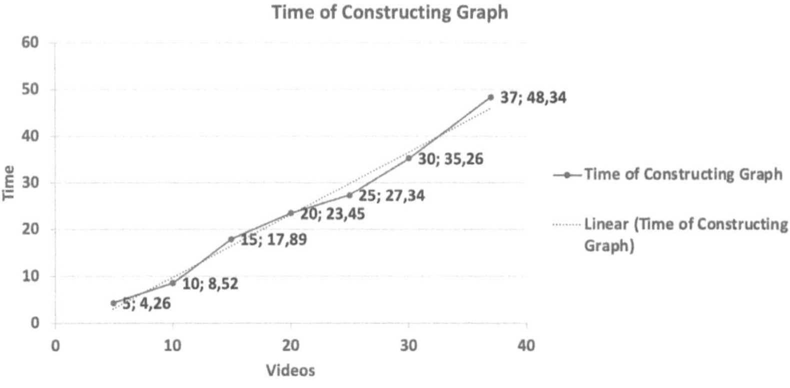

Fig. 3. Time of building Manhole Graph from videos.

Figure 3 demonstrates a clear upward trend in the time required to construct graphs as the number of videos increases. The time rises almost proportionally, starting at 4.26 s for 5 videos and climbing steadily to 48.34 s for 37 videos. This progression indicates that constructing graphs becomes increasingly computationally intensive with larger datasets. The sharp rise in time, particularly between 30 and 37 videos, suggests potential inefficiencies in the graph construction algorithm as it scales. Optimization strategies, such as improved algorithms or parallel processing, may be necessary to handle larger datasets more efficiently. Now, we provide an interface of our proposal in the next section.

5.4   Interface of AquaNetCT

The AquaNetCT system interface is designed using the Streamlit platform[9] with Python as the core programming language, offering a user-friendly and interactive experience. The mapping component of the interface leverages OpenCage[10], providing geospatial visualization capabilities. Libraries utilized in the system include networkx, osmnx, paddleocr, ultralytics, among others, ensuring robust functionality and performance (Fig. 4).

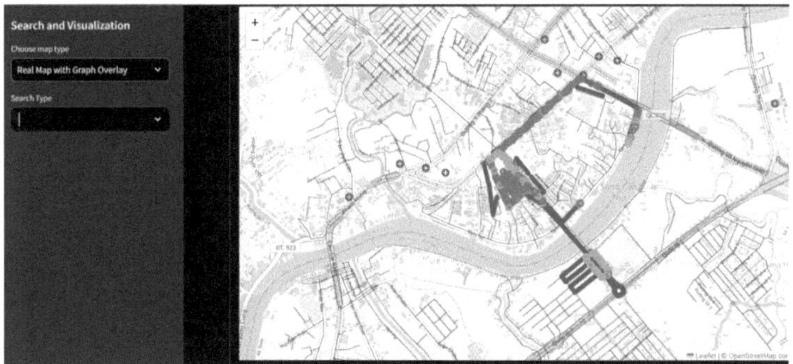

Fig. 4. Interface of Our System.

The system consists of two main components: graph representation and interface. The graph representation section follows the workflow defined in the Framework, covering steps 1 through 5, while our application operates as step 6, facilitating user interactions and queries. Users can search for specific data using graph types, street names, or manhole IDs. Additionally, the system computes and displays the distance between two manholes using the Haversine formula, ensuring accurate geospatial calculations. This combination of features and tools creates a powerful platform for efficient data analysis and visualization in urban infrastructure management.

6   Conclusion and Future Work

This study presents a structured approach to manhole analysis by integrating dataset collection, information extraction, classification, and graph-based modeling. By leveraging OCR and object detection, key details such as street names, geographic coordinates, and structural attributes are extracted, enabling automated processing. Moreover, manhole classification enhances understanding of their function, while graph construction provides a clear visualization of the drainage network, facilitating better management and optimization. For future

---

[9] https://streamlit.io/.
[10] https://opencagedata.com/.

work, we aim to develop query-based analysis for intuitive data retrieval, enhance graph-based spatial analysis, and integrate GIS functionalities to precisely map manhole locations. Additionally, incorporating detailed manhole attributes such as diameter and material will improve predictive modeling and infrastructure assessment, advancing data-driven urban drainage management.

Acknowledgements. This study is received support from the European Union's Horizon research and innovation program under the MSCA-SE (Marie Skłodowska-Curie Actions Staff Exchange) grant agreement 101086252; Call: HORIZON-MSCA-2021-SE-01; Project title: STARWARS.

References

1. Alzubaidi, L., et al.: Review of deep learning: concepts, cnn architectures, challenges, applications, future directions. J. Big Data **8**, 1–74 (2021)
2. Amit, Y., Felzenszwalb, P., Girshick, R.: Object detection. In: Computer Vision: A Reference Guide, pp. 875–883. Springer, Heidelberg (2021)
3. Chollet, F.: Xception: deep learning with depthwise separable convolutions. In: CVPR'17, pp. 1251–1258 (2017)
4. Dang, T., Khattak, S., Mascarich, F., Alexis, K.: Explore locally, plan globally: a path planning framework for autonomous robotic exploration in subterranean environments. In: ICAR'19, pp. 9–16. IEEE (2019)
5. Diwan, T., Anirudh, G., Tembhurne, J.V.: Object detection using yolo: challenges, architectural successors, datasets and applications. Multimedia Tools Appl. **82**(6), 9243–9275 (2023)
6. Duin, B., et al.: Toward more resilient urban stormwater management systems–bridging the gap from theory to implementation. Front. Water **3**, 671059 (2021)
7. García, L., Barreiro-Gomez, J., Escobar, E., Téllez, D., Quijano, N., Ocampo-Martínez, C.: Modeling and real-time control of urban drainage systems: a review. Adv. Water Res. **85**, 120–132 (2015)
8. Hoa, L.T.V., Shigeko, H., Nhan, N.H., Cong, T.T.: Infrastructure effects on floods in the Mekong river delta in Vietnam. Hydrological Process. Int. J. **22**(9), 1359–1372 (2008)
9. Hu, X., Hu, Y., Resch, B., Kersten, J.: Geographic information extraction from texts (geoext). In: European Conference on Information Retrieval, pp. 398–404. Springer, Heidelberg (2023). https://doi.org/10.1007/978-3-031-28241-6_44
10. Huang, P.C., Lee, K.T.: An alternative for predicting real-time water levels of urban drainage systems. J. Environ. Manag. **347**, 119099 (2023)
11. Huong, H.T.L., Pathirana, A.: Urbanization and climate change impacts on future urban flooding in Can Tho city, Vietnam. Hydrol. Earth Syst. Sci. **17**(1), 379–394 (2013)
12. Jarvis, R.S.: Drainage network analysis. Prog. Phys. Geogr. **1**(2), 271–295 (1977)
13. Jiang, P., Ergu, D., Liu, F., Cai, Y., Ma, B.: A review of yolo algorithm developments. Procedia Comput. Sci. **199**, 1066–1073 (2022)
14. Kleidorfer, M., Mikovits, C., Jasper-Tönnies, A., Huttenlau, M., Einfalt, T., Rauch, W.: Impact of a changing environment on drainage system performance. Procedia Eng. **70**, 943–950 (2014)

15. Koonce, B., Koonce, B.: Resnet 50. Convolutional neural networks with swift for tensorflow: image recognition and dataset categorization, pp. 63–72 (2021)
16. Kwon, S.H., Kim, J.H.: Machine learning and urban drainage systems: state-of-the-art review. Water **13**(24), 3545 (2021)
17. Li, Z., Liu, F., Yang, W., Peng, S., Zhou, J.: A survey of convolutional neural networks: analysis, applications, and prospects. IEEE Trans. Neural Netw. Learn. Syst. **33**(12), 6999–7019 (2021)
18. Mohammadiun, S., et al.: Effects of bottleneck blockage on the resilience of an urban stormwater drainage system. Hydrol. Sci. J. **65**(2), 281–295 (2020)
19. Nasello, C., Tucciarelli, T.: Dual multilevel urban drainage model. J. Hydraul. Eng. **131**(9), 748–754 (2005)
20. Nguyen, T.T.H., Jatowt, A., Coustaty, M., Doucet, A.: Survey of post-ocr processing approaches. ACM Comput. Surv. (CSUR) **54**(6), 1–37 (2021)
21. Otsu, K., et al.: Supervised autonomy for communication-degraded subterranean exploration by a robot team. In: 2020 IEEE Aerospace Conference, pp. 1–9. IEEE (2020)
22. Piro, P., Palermo, S.A., Tropea, M., Saleh, M.M., De Rango, F.: IoT and artificial intelligence integration for a stormwater monitoring and management system. In: International Conference on Numerical Computations: Theory and Algorithms, pp. 290–297. Springer, Heidelberg (2023). https://doi.org/10.1007/978-3-031-81244-6_28
23. Qin, Z., Zhang, Z., Chen, X., Wang, C., Peng, Y.: Fd-mobilenet: improved mobilenet with a fast downsampling strategy. In: IEEE ICIP'18, pp. 1363–1367. IEEE (2018)
24. Ramovha, N., Chadyiwa, M., Ntuli, F., Sithole, T.: The potential of stormwater management strategies and artificial intelligence modeling tools to improve water quality: a review. Water Res. Manag. 1–34 (2024)
25. Sapkota, R., Meng, Z., Churuvija, M., Du, X., Ma, Z., Karkee, M.: Comprehensive performance evaluation of yolo11, yolov10, yolov9 and yolov8 on detecting and counting fruitlet in complex orchard environments. arXiv preprint arXiv:2407.12040 (2024)
26. Tri, V.K.: Hydrology and hydraulic infrastructure systems in the Mekong delta, Vietnam. In: The Mekong Delta System: Interdisciplinary Analyses of a River Delta, pp. 49–81. Springer, Heidelberg (2012)
27. Long, N., Cheng, Y., Le, T.D.N.: Flood-resilient urban design based on the indigenous landscape in the city of Can Tho, Vietnam. Urban Ecosyst. **23**(3), 675–687 (2020)
28. Wankhede, P.A., Mohod, S.W.: A different image content-based retrievals using ocr techniques. In: ICECA'17, vol. 2, pp. 155–161. IEEE (2017)
29. Yazdanfar, Z., Sharma, A.: Urban drainage system planning and design-challenges with climate change and urbanization: a review. Water Sci. Technol. **72**(2), 165–179 (2015)
30. Zhang, D., Han, D., Zhong, Q., Chen, Q.: Classification and description of the drainage state of manholes in urban drainage systems. Water Sci. Technol. **89**(1), 146–159 (2024)
31. Zhou, B., et al.: Smartphone-based road manhole cover detection and classification. Autom. Constr. **140**, 104344 (2022)
32. Zhu, Y., Newsam, S.: Densenet for dense flow. In: the IEEE ICIP'17, pp. 790–794. IEEE (2017)
33. Zou, Z., Chen, K., Shi, Z., Guo, Y., Ye, J.: Object detection in 20 years: a survey. Proc. IEEE **111**(3), 257–276 (2023)

A Portable Multi-spectral Sensor-Based System with Machine Learning Models for Non-destructive Sweetness Assessment of Cherry Tomatoes

Hoang Thien-Phu Tran[1,2], Le Dinh-Kha Vo[1,2], Quoc-Hung Pham[1,2(✉)], Thanh-Nhan Nguyen[1,2], Huy-Hoang Vo[1,2], Tuan-Kiet Tran[1,2], and Nhut-Thanh Tran[3]

[1] Faculty of Computer Engineering, University of Information Technology, Ho Chi Minh City, Vietnam
`hungpq@uit.edu.vn`
[2] Vietnam National University, Ho Chi Minh City, Vietnam
[3] Faculty of Automation Engineering, Can Tho University, Can Tho, Vietnam

Abstract. Sweetness is one of the critical quality indicators in cherry tomatoes. Traditionally, measuring sweetness requires a destructive method: the fruit must be cut open, the juice extracted, and then analyzed using a testing device. However, destructive analysis and high-resolution spectroscopy have notable drawbacks, such as high investment costs and non-portable equipment, making them unsuitable for on-site measurements or industrial applications. This study employed a combination of spectroscopy with wavelengths in the visible and near-infrared (VIS-NIR) range and an AI-driven method to predict the sweetness of whole cherry tomatoes. Before modeling, the spectral data were preprocessed using the SNV method. Multiple Linear Regression, Support Vector Regression, and Gaussian Process Regression were applied to model the complex relationship between spectral features and the sweetness values obtained through destructive testing. The GPR model achieved the highest performance for the raw spectral dataset, with a 0.86 coefficient of determination (R2) and a 0.33 brix Root Mean Square Error (RMSE). In contrast, SVR combined with SNV preprocessing yielded the highest accuracy, with an R2 of 0.88 and an RMSE of 0.30 brix. The proposed system demonstrates strong potential for non-destructive sweetness assessment in cherry tomatoes, particularly in on-site settings and industrial environments where rapid, cost-effective, and portable analysis is essential.

Keywords: Quantitative prediction · Sweetness · Cherry Tomatoes · Multispectral sensor

1   Introduction

The tomato is among the most popular and nutritious fruits globally, demonstrating significant economic benefits and high consumption. The first signs that affect consumer choice are based on freshness and color; however, flavor is the actual value of the tomato, particularly in terms of sweetness. The Soluble Solids Content (SSC) is measured in $°Brix$, which is also used to evaluate the quality of tomatoes. This unit represents the ratio of sugar content, including sucrose, glucose, and fructose, by weight in a solution. Glucose and fructose mainly contribute to the SSC, with about *65%* of the total soluble solids [1]. There is a direct linear relationship between $°Brix$ and sugar content, which is related to tomatoes' sweetness and overall flavor. The refractometry technique is the traditional method for measuring the $°Brix$ of a sample by evaluating the solution's refractive index [2]. This approach has been widely used in the laboratory and portable digital devices because of its high accuracy [3]. However, this approach requires extracting the liquid from the sample, which is unsuitable for continuous or nondestructive analysis cases. Therefore, an approach without destructive sampling is needed for advanced alternative technologies. VIS-NIR spectroscopy, from *410* to *2500* nm wavelength, is highly informative for non-destructive quantitative analysis [4]. The VIS-NIR spectrum results from light absorption by chromophores and weak, high-frequency absorption bands of molecular vibrations, such as C-H, O-H, and N-H bonds [5]. Sugars contain both C-H and O-H bonds, and their bonds interact with light in the VIS-NIR range, resulting in a rich, informative, and patterned spectrum of the sample. The intensity of these absorption bands increases as the sugar concentration increases. The VIS-NIR has more advantages over traditional analytical techniques, such as refractometers. Previous studies have limitations that rely on expensive, high-resolution, laboratory-based spectrometers [6]. Ruizendaal et al. used hyperspectral imaging to map quality attributes, which have also been explored. However, the system is typically even more costly and complex, making it infeasible for widespread use in packinghouses or in the field. These challenges highlight a gap in an accurate but also low-cost, portable, robust, and user-friendly solution.

The VIS-NIR spectral data is complex due to the wide absorption bands and overlapping, so advanced computational methods are applied to extract meaningful quantitative information [7]. VIS-NIR spectroscopy technology has been extensively used to determine the soluble solids content, such as total soluble solids (TSS) or Brix. The application of VIS-NIR yields a spectrum of performance results. For instance, Annelisa et al. utilized a waveband cover from *840* nm to *1050* nm to predict $°Brix$ in market tomatoes [8]; their results R^2 of *0.67* and an RMSE of *0.32*. Similarly, Evia et al. utilized a wider range of wavelengths from *400* nm to *1000* nm to implement a non-destructive system for TSS in tomatoes, with results showing an R^2 of *0.64* and a lower Root Mean Square Error (RMSE) of *0.17* [9]. A high predictive performance for TSS, reporting R^2 of *0.93* and an RMSE of *0.41*, was achieved by Jens et al. within the *760-1080* nm range [10]. These studies confirm the viability of applying VIS-NIR spectroscopy for SSC prediction. The model accuracy significantly depends on the specific instrumentation, waveband range, and spectral dataset characteristics. An essential consideration when applying spectral technology in agriculture is the trade-off between the performance of high-resolution spectrometers and the low cost of multispectral sensors. High-resolution spectrometers can capture detailed and continuous data across numerous narrow wavelength bands, providing rich information and enabling the development of highly accurate analytical models. However, their high cost, operational complexity, and intensive data analysis limit their use in field applications. On the other hand, multispectral sensors offer a lower-cost solution for practical implementation due to their use of a limited number of discrete and broader bands [11]. The main drawback of a multispectral sensor is its potential to miss subtle features for accurate quantification. Consequently, a typical pipeline uses high-cost spectroscopy to define key wavelengths, designing targeted and cost-effective solutions for high performance in real-world applications [12]. Machine learning algorithms are used to build a model that links the spectral data to the corresponding $°Brix$ values measured by the destructive method. The characteristics of VIS-NIR spectra are complex due to the broad absorption features that merge, making it challenging to convert complex spectral data into quantitative values [13]. Therefore, the motivation of this study is to implement a low-cost device based on an 18-channel VIS-NIR multispectral sensor and an embedded Raspberry Pi computer. Several pre-processing methods are used to reduce light scatter effects and scale VIS-NIR spectral data to have unit variance, such as SNV and Standard Scaler. Several algorithms, such as MLR, SVR, and GPR, define an association between the VIS-NIR spectral data and the destructively measured $°Brix$ values.

2   Materials and Methods

2.1   Proposed System

Hardware Implementation. Figure 1 illustrates the block diagram of the proposed system for real-time Brix prediction in cherry tomatoes. The proposed system was designed to be compact and portable for spectral collecting. It includes several components: a *12 W* Halogen light source, a SparkFun AS7265x 18-channel multispectral sensor, a Raspberry Pi 4B, two *3.7 V* lithium-ion cells, a relay, and an OLED screen. The halogen light source and the multispectral sensor are positioned on opposite sides and have space for a cherry tomato sample, as shown in Fig. 1. For consistent measurements, the halogen lamp emits light through the cherry tomato sample. In contrast, the multispectral sensor captures the transmittance light part of the sample. The multispectral sensor covers a range of wavebands from *410* nm to *940* nm. The embedded computer controls all operations and executes the predictive models. This operation is precisely managed by a relay module connected to the embedded computer's GPIO pins. These components are housed within a mica-based enclosure to avoid ambient light, ensuring the spectral data is accurate for use in the field. Finally, the OLED screen was used to display the predicted Brix value.

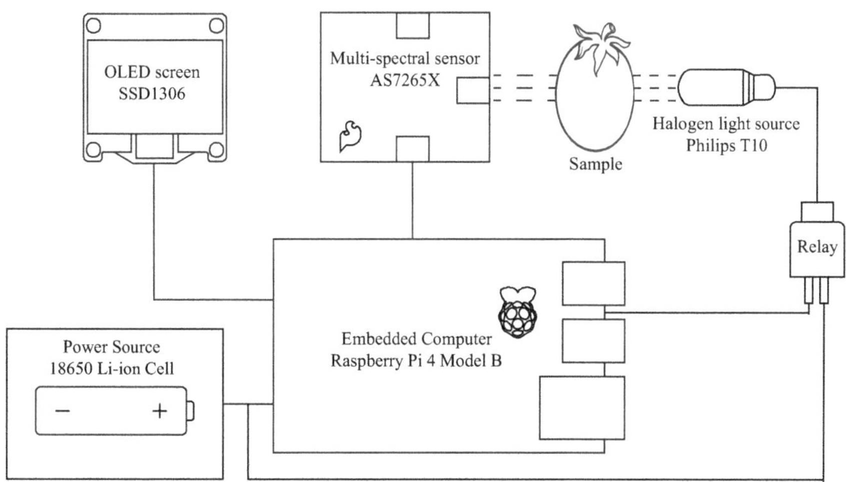

Fig. 1. A block diagram illustrating the hardware configuration.

Software Designing. Most of the system's software was developed in Python, leveraging its extensive libraries for scientific computing and seamless hardware interfacing. As illustrated in Fig. 2, the overall workflow commences with an initial model-building stage. During this phase, 18 VIS-NIR input features acquired from each tomato sample were integrated with its corresponding reference Brix value, meticulously measured using a digital refractometer. The dataset was partitioned into a training (*80%*) subset and a test (*20%*) subset. To improve the performance and robustness of the machine learning models, various preprocessing methods, including SNV transformation and scaling, were applied for data correction and standardization. Subsequently, several machine learning algorithms were trained and rigorously evaluated to identify the best-performing model based on predefined metrics. System seamlessly transitions into a real-time prediction program upon selecting the optimal model. When a new tomato sample is introduced in this operational mode, the software automatically acquires its spectral data, applies the identical SNV and scaling transformations established during the training phase, and feeds the processed data into the selected model. This process allows instant prediction of the tomato's Brix value, which is displayed on an OLED screen.

2.2 Data Acquisition

A dataset of *1,170* cherry tomatoes was established to investigate the correlation between VIS-NIR spectra and sweetness. The samples were sourced from Phong Thuy Agro Production and Trading Co., Ltd. in Da Lat, Vietnam, and all experiments were performed at a controlled ambient temperature of *25 °C*.

For each sample, spectral data were first acquired non-destructively. A halogen lamp illuminated the tomato, and an AS7265x multispectral sensor captured

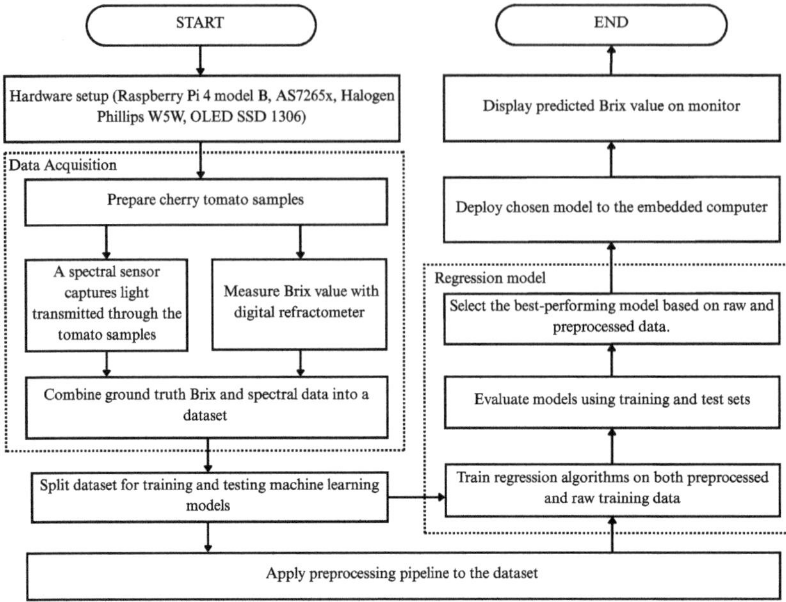

Fig. 2. Processing Flow for Brix Value Prediction

the attenuated light intensity across 18 discrete wavelengths: *410, 435, 460, 485, 510, 535, 560, 585, 610, 645, 680, 705, 730, 760, 810, 860, 900*, and *940* nm. Subsequently, the ground truth Brix value for each tomato was determined through destructive analysis. The tomato was crushed and filtered to extract its juice, which was then analyzed with a digital refractometer (Brix Trans Instruments, RBX0032). This process created a final dataset where each spectral measurement was mapped to a corresponding reference Brix value. The measured values spanned a range from *1.50* brix to *8.20* brix, providing a basis for training regression-based machine learning algorithms.

2.3 Pre-processing Techniques and Machine Learning Models

Several data preparation methods and machine learning models were employed and evaluated to build a robust predictive model. SNV is widely used as a scatter correction technique for spectral data [14]. For each spectrum, SNV corrects for baseline shifts by subtracting that spectrum's mean and reducing multiplicative effects by dividing by its standard deviation. After SNV, Standard Scaler was applied to each spectral band across all samples to remove the mean and scale to unit variance. It ensures that each spectral channel contributes equally to the model training process, preventing features with larger variances from dominating the model [15]. Machine learning regression models are used to find the correlation between the VIS-NIR wavelength and the Brix value. MLR, SVR, and GPR were used in this study to compare effectiveness and compatibility and choose the most suitable models to apply on embedded computers.

2.4 Performance Evaluation

R^2 and RMSE are used to benchmark the regression models. R^2 quantifies the proportion of the variance in the dependent variable explained by the independent variables. Its formula is shown in Eq. 1

$$R^2 = 1 - \frac{\sum_{i=1}^{n}(X_i - Y_i)^2}{\sum_{i=1}^{n}(\overline{Y} - Y_i)^2} \tag{1}$$

$$\text{RMSE} = \sqrt{\frac{1}{n}\sum_{i=1}^{n}(X_i - Y_i)^2} \tag{2}$$

where n represents the number of samples in the dataset, X_i is the predicted value of the ith sample, and Y_i is the actual corresponding value. $\overline{Y}$ denotes the mean of all observed values. In addition to evaluating how well the predicted value matches the corresponding value, RMSE measures the error of the predicted value with actual values on average. Its formula is shown in Eq. 2.

3 Results and Discussions

3.1 Descriptive Statistics of Reference Brix Values

The foundation for a robust and reliable predictive model is the quality and distribution of the calibration data. Table 1 summarizes the target variable used in this study.

Table 1. Spectral Characteristics of Cherry Tomatoes

Items	Training dataset	Testing dataset
Sample quantity	936	234
Minimum values	1.5 brix	3.1 brix
Maximum values	8.2 brix	7.9 brix
Mean Value	5.24 brix	5.18 brix
Standard Deviation	0.92 brix	0.88 brix

The dataset was meticulously split into a training set including 936 samples and a test set with 234 samples. A crucial observation supporting the integrity of this split is the close match in both the mean Brix values (*5.24* brix for the training set vs. *5.18* brix for the test set) and their respective standard deviations (*0.92* brix vs. *0.88* brix). This strong resemblance between the training and test distributions is vital, as it indicates a well-executed and unbiased data division, suggesting that any model developed using the training data should exhibit comparable performance when assessed on the unseen data. Furthermore, the training subset was established with a more expansive target variable range (*1.5* brix to *8.2* brix) compared to the test set (*3.1* brix to *7.9* brix). This strategic configuration ensures that the model is trained across the full breadth of expected Brix variation within cherry tomatoes, and its subsequent validation involves interpolation within these established boundaries, thereby mitigating the less reliable process of extrapolation.

3.2   Spectral Characteristics and Pre-Processing

Figure 3a depicts the original transmitted data of samples ranging from *410* to *940* nm. The spectra exhibit significant variations in baseline height and intensity. These variations are partly affected by non-chemical factors, such as fruit size, density, and positional deviations of samples.

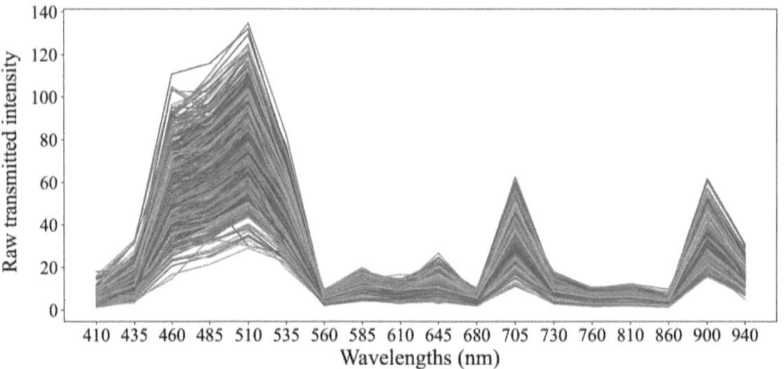

(a) Visualization of raw spectral data

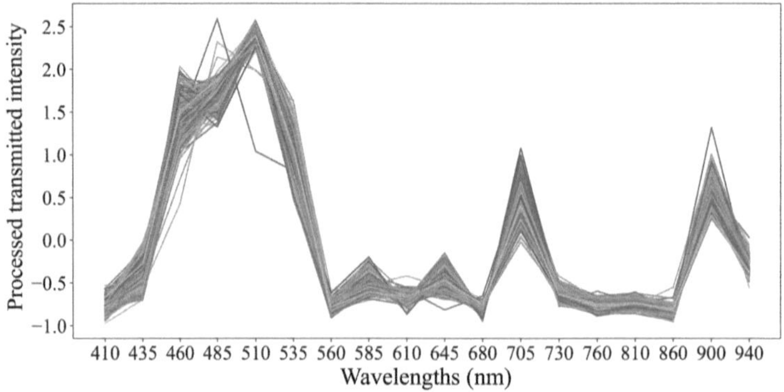

(b) Visualization of pre-processed spectral data with SNV method

Fig. 3. Comparison of spectral data before and after pre-processing.

A prominent, broad absorption is visible at *460* and *510* nm wavelengths. At the same time, smaller peaks are also observed near *705* nm and *900* nm wavelengths. However, pronounced baseline fluctuation and overlapping signals hinder the direct extraction of reliable quantitative information from raw data.SNV was applied to the raw spectral data to correct for multiplicative light scattering effects. As illustrated in Fig. 3b, this pre-processing step successfully minimized baseline variations and enhanced the comparability of spectral profiles across samples. The resulting spectra exhibits more clearly delineated absorption peaks while preserving their relative intensities. This result is an essential prerequisite for developing a robust calibration model, as it ensures that the model correlates spectral features with chemical composition rather than physical artifacts.

3.3 Evaluating Model Performance

The performance of three models, MLR, SVR, and GPR, was then evaluated on raw and SNV pre-processed datasets, with both standardized using a standard scaler. The results, summarized in Table 2, lead to two critical findings. First, the association between the VIS-NIR spectra and the target variable is highly nonlinear. The linear MLR model performed poorly on raw and pre-processed data, yielding R^2 scores on the test set of only *0.18* and *0.20*, respectively.

Table 2. Training results of machine learning models

	Training set		Test set	
	R^2 Score	RMSE	R^2 Score	**RMSE**
Raw Data				
MLR	0.26	0.79 brix	0.18	0.80 brix
SVR	0.94	0.22 brix	0.85	0.34 brix
GPR	**0.98**	**0.11 brix**	**0.86**	**0.33 brix**
Raw data pre-processed with SNV				
MLR	0.28	0.78 brix	0.20	0.78 brix
SVR	**0.98**	**0.16 brix**	**0.88**	**0.30 brix**
GPR	0.98	0.12 brix	0.85	0.34 brix

In contrast, the nonlinear SVR and GPR models demonstrated substantially higher predictive power. Second, SNV pre-processing proved crucial for training reliable models. While GPR was the top-performing model on the raw data, the SVR model achieved the best overall performance after SNV was applied, with its test set R^2 increasing from *0.85* to *0.88* and the RMSE decreasing from *0.34* to *0.30*. By removing the noise associated with light scattering, SNV refines the feature space, enabling the SVR algorithm to identify optimal decision boundaries more effectively.

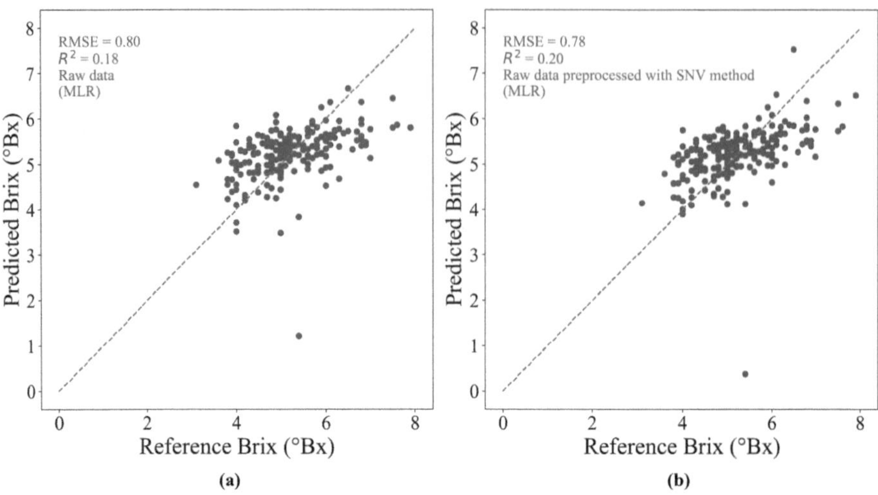

Fig. 4. Scatter plot of MLR model trained with: (a) Raw dataset, (b) Pre-processed dataset with SNV

This quantitative superiority is visually validated in the scatter plots of Fig. 4, 5, and 6, which plot references Brix values against model predictions.

The plots for the MLR models (Fig. 4a, b) show a widely dispersed cloud of points, confirming their poor predictive ability. In contrast, the plots for SVR and GPR (Fig. 5a, b, and Fig. 6a, b) exhibit tight clustering around the line of perfect agreement. The tightest cluster belongs to the SVR model trained on SN pre-processed data (Fig. 5b). Compared to the highly effective GPR model with SNV (Fig. 6b), the SVR model shows slightly less variance, confirming its superior accuracy. This graphical and quantitative evidence demonstrates that the SNV pre-processing and an SVR model are this study's most effective strategies for predicting Brix content.

To contextualize these findings, the proposed low-cost system's performance is highly competitive compared to other recent studies in the field. Our best model (SVR with SNV) achieved an R^2 of *0.88* and an RMSE of *0.30* brix on the test set. This result surpasses the accuracy reported in several studies that likely used more expensive laboratory equipment. For example, our R^2 of *0.88* is higher than that achieved by Zunita et al. [9] ($R^2 = $ *0.64*), de Brito et al. [8] ($R^2 = $ *0.67*), and Ino et al. [16] ($R^2 = $ *0.81*). Furthermore, our prediction error (RMSE $= $ *0.30*) is lower than that of Wold et al. [17] (RMSE $= $ *0.41*) and comparable to de Brito et al. (RMSE $= $ *0.32*). While some studies report higher R^2 values, such as Wold et al. ($R^2 = $ *0.93*) and Li et al. [18] ($R^2 = $*0.92*), our model demonstrates a more balanced performance. For instance, the model by Wold et al. has a higher RMSE, suggesting less precision despite the high correlation. The exceptionally high RMSE reported by Li et al. (*6.57*) suggests a different

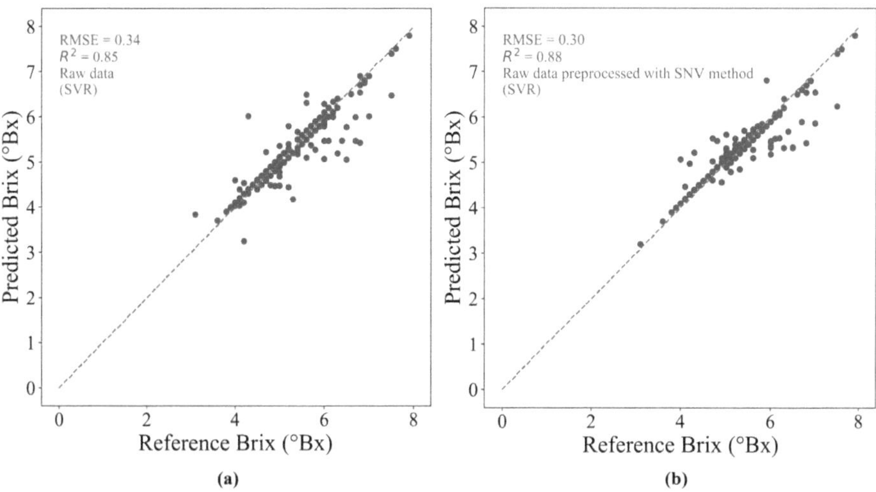

Fig. 5. Scatter plot of SVR model trained with: (a) Raw dataset, (b) Pre-processed dataset with SNV

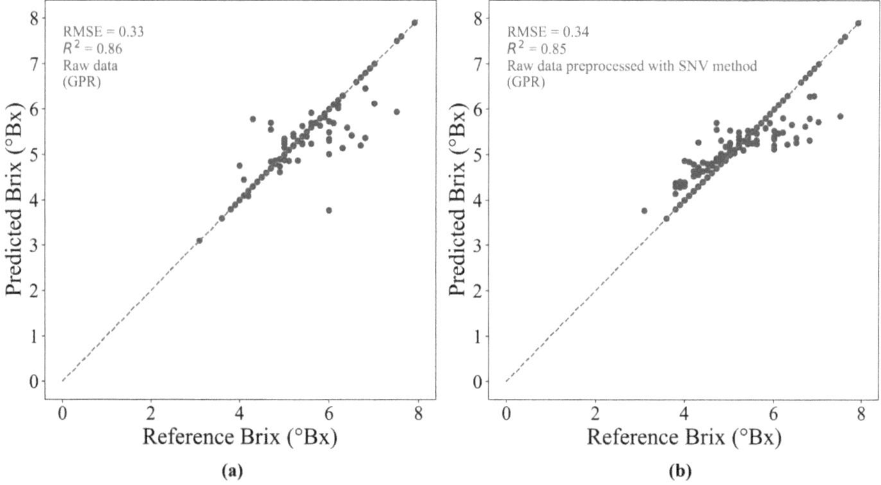

Fig. 6. Scatter plot of GPR model trained with: (a) Raw dataset, (b) Pre-processed dataset with SNV

scale or methodology, making a direct comparison difficult. Notably, the study by Zunita et al. achieved a very low RMSE of *0.17*, coupled with a low R^2 of *0.64*, indicating that their model may not be robust across a broader range of data. Crucially, this study demonstrates that by combining a low-cost multispectral sensor with appropriate nonlinear modeling (SVR) and pre-processing (SNV), it is possible to achieve a level of accuracy and precision that is not only suitable for practical application but also rivals or exceeds the performance of more complex and costly laboratory-grade systems.

4   Conclusion

This study successfully presents a device that utilizes a multispectral sensor in the VIS-NIR range for non-destructive Brix measurement of cherry tomatoes. The SNV method and SVR model combination provide an accurate prediction, achieving an R^2 of *0.88* and an RMSE of *0.30*. The research confirmed the necessity of appropriate spectral pre-processing methods to remove physical noise and the use of regression algorithms to explain the complex, nonlinear relationship between spectral data. The broader impact of this work is significant for the horticultural industry, which proposes a viable technological solution that enables it to move beyond subjective or destructive quality assessment. The following study will consider a broader range of cherry tomato cultivars in different seasons and geographical regions. Although the SVR model is the most suitable for this study, exploring other machine learning models, such as Neural Networks (NNs), could improve the system's trustworthiness. Finally, expanding the dataset with a broader range of samples would help to create an even more generalized and robust prediction model for industrial use.

Acknowledgment. This research is funded by University of Information Technology-Vietnam National University Ho Chi Minh City under grant number D1-2025-15.

References

1. Amr, A., Raie, W., et al.: Tomato components and quality parameters. a review. Jordan J. Agric. Sci. **18**(3), 199–220 (2022)
2. Todaro, M., Gannuscio, R., Mancuso, I., Ducato, B., Scatassa, M.: The use of brix refractometer as a simple and economic device to estimate the protein content of sheep milk. Int. Dairy J. **154**, 105940 (2024)
3. Uzun-Viruzab, M., Voicilas, D.R., Iacobescu, A., Rădulescu, L., Megyesi, C.I.: Characterization of some jams obtained from exotic fruits. J. Agroaliment. Process. Technol. **30**, 399–404 (2024)
4. Cevoli, C., Iaccheri, E., Fabbri, A., Ragni, L.: Data fusion of FT-NIR spectroscopy and vis/NIR hyperspectral imaging to predict quality parameters of yellow flesh "jintao" kiwifruit. Biosys. Eng. **237**, 157–169 (2024)
5. Ozaki, Y., Huck, C., Tsuchikawa, S., Engelsen, S.B.: Near-Infrared Spectroscopy: Theory, Spectral Analysis, Instrumentation, and Applications. Springer, Cham (2020)
6. Ruizendaal, J., Polder, G., Kootstra, G.: Automated and non-destructive estimation of soluble solid content of tomatoes on the plant under variable light conditions. Biosys. Eng. **242**, 80–90 (2024)

7. Pham, Q.H., Nguyen, T.N., Ba, A.Q.H., Ngo, H.H., Vo, H.H., Tran, N.T.: An embedded system for non-invasive glucose monitoring. In: 2023 International Conference on Multimedia Analysis and Pattern Recognition (MAPR), pp. 1–6. IEEE (2023)
8. de Brito, A.A., et al.: Determination of soluble solid content in market tomatoes using near-infrared spectroscopy. Food Control **126**, 108068 (2021)
9. Pratiwi, E.Z.D., Pahlawan, M.F., Rahmi, D.N., Amanah, H.Z., Masithoh, R.E.: Non-destructive evaluation of soluble solid content in fruits with various skin thicknesses using visible-shortwave near-infrared spectroscopy. Open Agric. **8**(1), 20220183 (2023)
10. Chicco, D., Warrens, M.J., Jurman, G.: The coefficient of determination r-squared is more informative than SMAPE, MAE, MAPE, MSE and RMSE in regression analysis evaluation. Peerj Comput. Sci. **7**, e623 (2021)
11. Tran, N.T., Fukuzawa, M.: A portable spectrometric system for quantitative prediction of the soluble solids content of apples with a pre-calibrated multispectral sensor chipset. Sensors **20**(20), 5883 (2020)
12. Tran, N.T., Phan, Q.T., Nguyen, C.N., Fukuzawa, M.: Machine learning-based classification of apple sweetness with multispectral sensor. In: 2021 21st ACIS International Winter Conference on Software Engineering, Artificial Intelligence, Networking and Parallel/Distributed Computing (SNPD-Winter), pp. 23–27. IEEE (2021)
13. Vitorino, R., Barros, A.S., Guedes, S., Caixeta, D.C., Sabino-Silva, R.: Diagnostic and monitoring applications using near infrared (NIR) spectroscopy in cancer and other diseases. Photodiagn. Photodyn. Ther. **42**, 103633 (2023)
14. Hemmateenejad, B., Mobaraki, N., Baumann, K.: Robust multiplicative scatter correction using quantile regression. J. Chemom. **38**(11), e3589 (2024)
15. Boteju, G., Tang, L., Brown, M.S.: Support vector machine for predicting student dropout under different normalization methods. In: 2024 IEEE International Conference on Big Data (BigData), pp. 8633–8636. IEEE (2024)
16. Ino, M., Ono, E., Shimizu, Y., Omasa, K.: Verification of commercial near-infrared spectroscopy measurement and fresh weight diversity modeling in brix% for small tomato fruits with various cultivars and growth conditions. Sensors **23**(12), 5460 (2023)
17. Wold, J.P., Sanden, K.W., Skaret, J., Carlehøg, M., Tjåland, M., Hansen, A.: Non-contact interactance NIR spectroscopy for estimating TSS and sensory sweetness in conveyor-belt transported cherry tomatoes (lycopersicon esculentum'piccolo'). Spectrochim. Acta Part A Mol. Biomol. Spectrosc. **335**, 125962 (2025)
18. Li, X., Tsuta, M., Hayakawa, F., Nakano, Y., Kazami, Y., Ikehata, A.: Estimating the sensory qualities of tomatoes using visible and near-infrared spectroscopy and interpretation based on gas chromatography-mass spectrometry metabolomics. Food Chem. **343**, 128470 (2021)

AI in E-Commerce, Education, and Society

Student Engagement Estimation in Classroom Videos Using Video Swin Transformer

Nha Tran[1,2,3], Hung Nguyen[1], Dat Ly[1(✉)], Hung Q. Nguyen[1], Anh Tran[1], and Hien D. Nguyen[2,3]

[1] Faculty of Information Technology, Ho Chi Minh City University of Education, Ho Chi Minh City, Vietnam
`4901103017@student.hcmue.edu.vn`
[2] Faculty of Computer Science, University of Information Technology, Ho Chi Minh City, Vietnam
[3] Vietnam National University, Ho Chi Minh City, Vietnam

Abstract. Monitoring student facial expressions for interest is crucial for modern learning, offering insights into their cognitive and emotional states so instructors can adapt teaching and provide support. Traditional methods are slow and subjective. The existing deep learning methods often relies on internet-based datasets and simplistic emotion-to-engagement mappings. They fail to capture temporal dynamics, and limits their real-world classroom applicability. This study presents a framework integrating a Video Swin Transformer to extract spatio-temporal facial features and classify seven basic emotions. Subsequently, a frequency-weighted mapping function is applied to convert the resulting emotion sequences into four discrete levels of student engagement. Besides, this study also introduces an open dataset, *HCMUE-SEED*, comprising video recordings of 35 students participating in authentic in-class learning activities. The proposed emotion recognition model achieves 83.41% accuracy, and the end-to-end engagement estimation reaches 78.83% accuracy. It is effective and potential to inform real-time, objective classroom analytics for instructors.

Keywords: Student engagement · Facial expression recognition · Spatial - Temporal model · Engagement mapping function

1 Introduction

In modern education, teaching objectives extend beyond knowledge transmission to the holistic development of learners' competencies and require active, proactive engagement in all activities [1,2]. Accordingly, evaluation practices must shift from periodic summative tests to continuous tracking of student progress, as mandated by the Ministry of Education and Training Circulars 22/2021/TT-BGDĐT and 27/2020/TT-BGDĐT [3,4]. Consequently, instructors must infer

N. Thai-Nghe et al. (Eds.): ISDS 2025, CCIS 2714, pp. 61–76, 2026.
https://doi.org/10.1007/978-981-95-3358-9_5

engagement from nonverbal cues especially facial expressions to adapt teaching strategies and sustain motivation. Automated analysis of these expressions enables timely, scalable feedback, overcoming the subjectivity and disruption associated with in class questioning or self-report surveys [5,6].

With the advent of deep learning, numerous models have been developed to detect learner emotions [7,8], yet they typically operate on static images from e-learning benchmarks, employ simplistic emotion-to-engagement mappings, and lack exposure to real-world classroom variability in lighting, camera angles, and student density [9,10]. Although custom in classroom datasets annotated with engagement levels directly from facial expressions have been introduced to clarify the emotion–engagement link, these resources are often private and annotated subjectively, limiting reproducibility and robustness [11]. Despite recent deep learning approaches have advanced emotion recognition, they remain reliant on static images, employ overly simplistic engagement mappings, and lack real-world classroom video data. To address these gaps, this study presents a dual-stage framework that leverages a Video Swin Transformer for spatio–temporal feature extraction and a novel frequency-weighted mapping function, trained and evaluated on a newly collected in-classroom video corpus to infer four discrete engagement levels.

The main contributions of this study can be summarized as follows:

- A dual-stage framework that uses a Video Swin Transformer and a frequency-weighted mapping to infer four discrete engagement levels from seven emotion classes.
- The HCMUE Student Emotion & Engagement Dataset (HCMUE-SEED), an in-classroom video corpus annotated with seven basic emotions under diverse lighting, occlusion, and viewing conditions.
- Validation against learners' self-reports confirms the method's robustness and its capacity to deliver real-time, objective feedback for adaptive teaching.

2   Related Work

In recent years, deep learning–based emotion recognition has been widely explored in educational contexts [12–14]. For instance, Tang et al. [12] proposed the MLG-PN model based on the Vision Transformer (ViT) to learn the relationships between local and global facial features, thereby calculating the Engagement Level Estimation (ELE) value to measure student engagement. In another study, Ashwin et al. [13] proposed a hybrid CNN architecture to analyze student engagement by classifying emotions into three states: Engaged, Boredom, and Neutral. Despite achieving considerable success, these approaches operate on individual frames and thus cannot leverage temporal dynamics such as changes in expression over time or subtle facial movements, which are critical factors for accurately distinguishing between similar emotional states.

To overcome the aforementioned limitation, recent studies have developed spatio-temporal models to assess student engagement more directly from video sequences [15–17]. For examples, Sandeep Mandia et al. [15] integrate a Graph Convolutional Network with BiLSTM and an attention module to model dynamic facial cues and infer engagement levels. Meanwhile, Selim et al. [16] proposed a model combining EfficientNetB7 and BiLSTM to assess student engagement in online classrooms. Despite significant improvements in results, indirect assessment methods that map emotions to engagement remain quite simplistic, while direct classification approaches are hampered by their dependence on the quality and quantity of labeled data, which is a scarce and expensive resource to collect and is often imbalanced.

Additionally, a major challenge facing current research is the scarcity of public data, especially datasets collected in real-world classroom settings for assessing learner engagement. The majority of existing benchmark datasets [18–20] for emotion assessment are largely collected in laboratory environments, which do not fully reflect the complexity and diversity of interactions in actual learning settings. As for datasets on learner engagement levels, they primarily focus on online learning contexts, where each student is typically recorded individually via webcam [21–23]. This leads to a significant data gap for in-person classroom environments, which are considered more challenging for data collection and analysis. Table 1 summarizes the size, data types, and key limitations of these representative emotion and engagement datasets.

Table 1. Overview of Public Emotion and Engagement Datasets

Dataset	Size	Data Type	Limitations
AffectNet [20]	~450,000	Single Image (in-the-wild)	Web-sourced labels are noisy; no sequential frames; no classroom context
DAiSEE [21]	9,068	Video (online)	Online-only, single-user webcams; limited group interaction
EngageNet [22]	474	Video (online classroom)	Very small; imbalanced classes; lacks in-classroom diversity
CMOSE [23]	12,193	Video (online lecture)	Controlled settings; lacks occlusion and multi-view classroom scenarios

In summary, despite advances from static-image classifiers to spatio–temporal architectures, current methods still rely on coarse emotion-to-engagement mappings and lack large in-classroom video datasets, limiting their real-world applicability. This work addresses these gaps by integrating temporal context with a frequency-aware mapping function and introducing a new classroom video corpus.

3   Dataset Construction

3.1   Data Collection

Fig. 1. Illustrations of expressions in the HCMUE-SEED dataset.

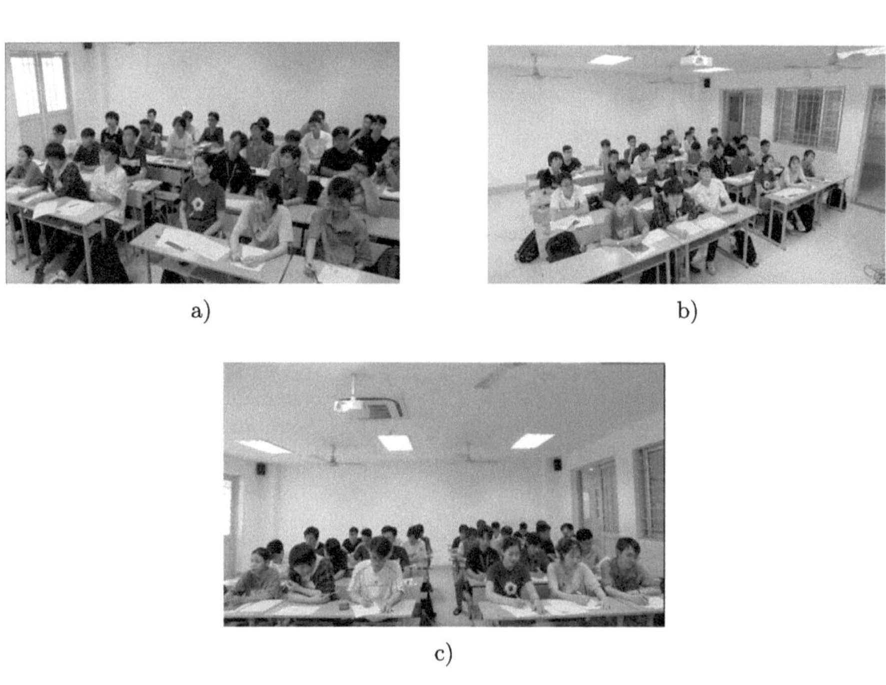

Fig. 2. Illustrating camera angles in the dataset: (a) left, (b) right, (c) center.

As presented in the related work section, the datasets in this class still have many limitations. To address this problem, we built the HCMUE-SEED dataset. This dataset includes seven common emotions that occur in the classroom: surprise, excited, angry, happy, sad, calm, and bore, as illustrated in Fig. 1. The selection of these emotion labels is based on the study [10]. The dataset was collected with the participation of 35 volunteers aged 20 to 23 (detailed information is described in Table 2). The data collection process was conducted over several days in classrooms at the Ho Chi Minh City University of Education. Prior to participation, all volunteers were clearly informed about the research objectives and signed a consent form. We organized multiple recording sessions on different days and deliberately varied seating arrangements and camera viewpoints, thereby increasing the diversity of lighting, occlusion, and behavioral contexts. This design choice ensured that the dataset captured a broad range of natural expressions and classroom scenarios, enhancing its representativeness for model training and evaluation. Each recording session lasted 45 min, comprising three learning activities, with camera angles illustrated in Fig. 2. The recording was performed using cameras with Full HD 1080p resolution and a frame rate of 30 FPS. After each activity, volunteers were given a 5-min break to self-report their emotional state during the activities.

Table 2. Statistics on student age, gender, and academic year

Age	Quantity	Gender	Academic Year
21	30	26 males, 04 females	3
22	4	02 males, 02 females	4
23	1	01 males	4

3.2 Data Pre-processing

The collected videos were segmented into short clips, each with a duration of 2 s. Each clip was then processed using the RetinaFace [24] model to detect and determine the bounding box of every face appearing in each frame. The positional information of these faces was passed to the DeepSort-Realtime [25] tracking algorithm to assign a unique identifier (ID) to each face and track it continuously across subsequent frames. Once a face was stably tracked, the system extracted the region containing that face from the current frame. The cropped facial images were then saved into separate folders, with each folder corresponding to a unique ID. Subsequently, a manual labeling process was performed. For each ID, a set of 15 frames representing the typical emotion was selected. In this step, substandard images-such as those that were blurry, too small, showed only a partial face, or were occluded-were discarded. The number of valid samples after labeling is summarized and presented in Table 3. The dataset was divided into three distinct subsets: 72% for training, 20% for validation, and 8% for testing. In addition, the splitting process ensured that there was no overlap between the samples in the three subsets, preserving the independence of the training, validation, and testing data.

Table 3. Number of samples for each angle in the dataset

	Left angle	Center angle	Sight angle	Total
Surprise	70	100	56	226
Excited	101	130	122	353
Angry	108	116	109	333
Happy	119	150	135	404
Sad	108	111	99	318
Calm	111	148	125	384
Bored	64	69	51	184
Total	681	824	697	2202

4 Facial Expression Recognition Model and Engagement Analysis

4.1 Overview

We propose a framework for analyzing student engagement via facial expression recognition, as illustrated in Fig. 3. First, classroom videos are ingested and decoded into individual frames. Next, each frame is processed by RetinaFace for face detection and DeepSORT-Realtime for tracking, ensuring that every learner is assigned a unique ID. Each detected bounding box is then cropped and aggregated into temporal face sequences per ID. These sequences are fed into an emotion recognition model to classify each into one of seven basic emotions. Finally, the resulting time-stamped emotion labels for each individual are recorded and aggregated into engagement levels using our proposed mapping function.

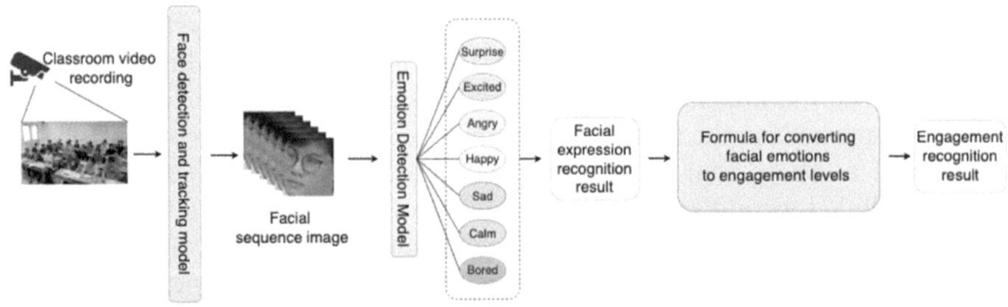

Fig. 3. Proposed method overview.

4.2 Facial Emotion Recognition with RGB Sequences

In this study, we employ the Video Swin Transformer architecture [26] for the task of emotion recognition from RGB image sequences, leveraging its ability to effectively learn spatio–temporal relationships and capture subtle temporal variations in facial expressions. Specifically, we choose the Video Swin Transformer because its hierarchical feature representation and shifted window self-attention mechanism allow the model to focus both on local facial patches enabling the detection of micro expressions and fine grained muscle movements and on global context across long frame sequences. Figure 4 illustrates the overall architecture of the Video Swin Transformer.

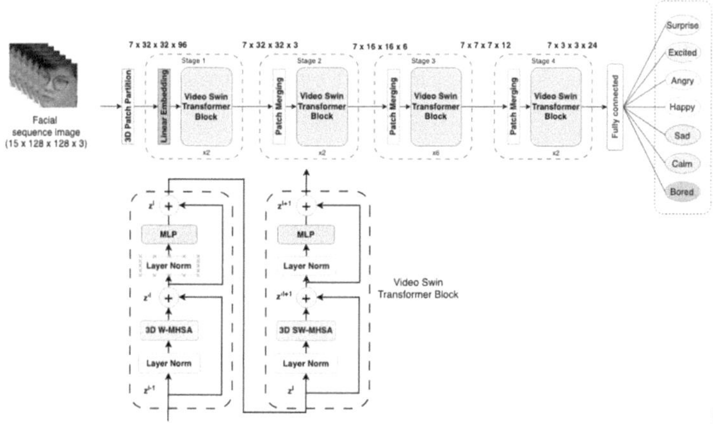

Fig. 4. Architecture of the Video Swin Transformer.

Specifically, for each sequence X consisting of 15 consecutive, extracted, and normalized face frames, we define

$$X = \{\, x_t \in \mathbb{R}^{128 \times 128 \times 3} \mid t = 1, \ldots, 15 \},\tag{1}$$

where each x_t is a face-cropped image resized to 128×128 and normalized for model input.

Each sequence X is partitioned into 3D patches and projected via a linear embedding layer:

$$Z = \mathrm{LinearEmbed}\big(\mathrm{PatchPartition}(X)\big) \in \mathbb{R}^{T' \times D},\tag{2}$$

where T' denotes the number of patches and D the embedding dimension.

The embedded feature Z is then processed through four stages of the Video Swin Transformer. In each stage, the model alternates between 3D Window Multi-Head Self-Attention and 3D Shifted Window Multi-Head Self-Attention mechanisms to jointly capture local and global spatio–temporal dependencies. The output of the final stage is the tensor H:

$$H = \mathrm{Swin3D}(Z) \in \mathbb{R}^{T'' \times D'},\tag{3}$$

where T'' and D' denote the number of patches and embedding dimension of the tensor after the attention stages.

Tensor $H \in \mathbb{R}^{T'' \times D'}$ is first flattened into a vector of length $T''D'$, then passed through a fully connected layer with weights $W_{\text{fc}} \in \mathbb{R}^{d \times (T''D')}$ and bias $b_{\text{fc}} \in \mathbb{R}^d$, followed by a ReLU activation to yield the feature vector:

$$F_{\text{RGB}} = \text{ReLU}\big(W_{\text{fc}}\, \text{vec}(H) + b_{\text{fc}}\big) \in \mathbb{R}^d, \tag{4}$$

where $\text{vec}(H)$ denotes the flattening operation and d is the dimensionality of the feature space.

Next, F_{RGB} is fed into an output fully connected layer with weight matrix $W_{\text{out}} \in \mathbb{R}^{7 \times d}$ and bias $b_{\text{out}} \in \mathbb{R}^7$, followed by a Softmax to generate the probability vector

$$z = W_{\text{out}}\, F_{\text{RGB}} + b_{\text{out}} \in \mathbb{R}^7, \quad \hat{p} = \text{Softmax}(z) \in \mathbb{R}^7. \tag{5}$$

The component $\hat{p}_k$ corresponds to the probability of the k-th emotion class. The predicted class is determined by

$$\hat{k} = \arg \max_{1 \leq k \leq 7} \hat{p}_k. \tag{6}$$

4.3 Engagement Analysis

To map the seven emotion labels to four engagement levels, we extend the formulation of Altuwairqi *et al.* [10] by incorporating the temporal frequency of each emotion, thereby capturing both the diversity and intensity of affective states.

Given E is the set of seven facial emotion labels.

$$E = \{\, e_k \mid k \in \{\text{Surprise, Excited, Angry, Happy, Sad, Calm, Bored}\}\}. \tag{7}$$

Similarly, let EL denote the set of four engagement levels:

$$EL = \{\, EL_j \mid j \in \{\text{Strong, High, Medium, Low}\}\}. \tag{8}$$

First, we initialize the base impact weight $\text{Im}_{e_k \in EL(j)}$ for each emotion e_k in engagement level EL_j as:

$$\text{Im}_{e_k \in EL(j)} = \frac{1}{N_{EL(j)}}, \tag{9}$$

where $N_{EL(j)}$ is the total number of emotion labels in group $EL(j)$. The computed impact weights $\text{Im}_{e_k \in EL(j)}$ for the seven emotions examined in this study are presented in Table 4.

Table 4. Statistics on the number of emotions and the Im_{e_k} impact value for each emotion across 4 levels of engagement [10]

Engagement Level	$N_{EL(j)}$	Emotion	Impact Value
Strong Engagement	1	Surprise	1.000
High Engagement	6	Excited	0.166
		Angry	0.166
Medium Engagement	8	Happy	0.125
		Sad	0.125
Low Engagement	5	Bored	0.200
		Calm	0.200

Next, to incorporate the temporal frequency of each emotion within minute i for learner id, we multiply the base impact weight $\text{Im}_{e_k \in EL(j)}$ by the occurrence count $M_{oe_k \in \text{minute}i_{id}}$—the number of times emotion e_k is detected during minute i:

$$\text{Im}_{oe_k \in \text{minute}i_{id}} = \text{Im}_{e_k \in EL(j)} \times M_{oe_k \in \text{minute}i_{id}}. \tag{10}$$

Unlike the original formulation that simply sums the base impact weights $\text{Im}_{e_k \in EL(j)}$, incorporating the frequency term $M_{oe_k \in \text{minute}i_{id}}$ enables the model to capture emotional intensity—emotions that recur more often within the same time window contribute more strongly.

Finally, the engagement level $EL_{id,i}$ for learner id at minute i is determined as the level j that maximizes the total weighted impact:

$$EL_{id,i} = \arg\max_{j} \left(\sum_{e_k \in EL(j)} \text{Im}_{oe_k \in \text{minute}i_{id}} \times \frac{\sum_{e_k \in EL(j)} M_{oe_k \in \text{minute}i_{id}}}{\sum_{e_k} M_{oe_k \in \text{minute}i_{id}}} \right). \tag{11}$$

The sequence $\{EL_{id,i}\}_{i=1}^{T}$ is visualized as individual time-series line plots, depicting each learner's engagement trajectory over time and thereby enabling both trend analysis and the early detection of passive or disengaged behavior in the classroom.

5   Experiments and Results

5.1   Experiments Setup

The experimental environment was set up on Google Colab, equipped with an Intel® Xeon® CPU @2.20 GHz, 56 GB of RAM, and a 16 GB Tesla A100 GPU. The entire workflow was developed using Python 3.10.13, combined with Bash scripts to automate the process from data preprocessing to model training, thereby optimizing the speed and efficiency of experimental replication. The training parameters included the Adam optimizer with beta coefficients of (0.9, 0.99), an initial learning rate of 0.001 (with a minimum of 1e-7), training for 100 epochs, a dropout rate of 0.2, and the categorical cross-entropy loss function.

5.2   Results of Facial Expression Recognition

We conducted experiments to compare basline architectures for processing RGB image sequences on HCMUE-SEED dataset. The evaluated models include ResNet50-BiLSTM, ResEmoNet (State-of-the-Art on FER-2013) [27], a Hybrid EfficientNetB7+Bi-LSTM (State-of-the-Art on DAiSEE) [16], and the Video Swin Transformer (pre-trained on ImageNet+Kinetics-400). The results are presented in Table 5 which summarizes the performance metrics, and Fig. 5 which displays the confusion matrices for the five models on our dataset.

It can be observed in Table 5 that the EfficientNetB7+Bi-LSTM and ResNet50-BiLSTM baselines achieve moderate performance 66.81% and 69.43% accuracy (68.80% and 69.90% F1), respectively but both models exhibit limited capacity to capture long-range temporal dependencies and lack an explicit attention mechanism, which constrains their ability to model dynamic facial

Table 5. Experimental results of emotion recognition models

Model	Accuracy	F1-Score	Precision	Recall
Hybrid EfficientNetB7 + Bi-LSTM	66.81%	68.80%	71.54%	67.89%
ResNet50-BiLSTM	69.43%	69.90%	70.36%	71.06%
ResEmoNet	81.22%	82.77%	84.80%	82.65%
Video Swin Transformer (pre-train)	83.41%	82.34%	82.94%	82.26%

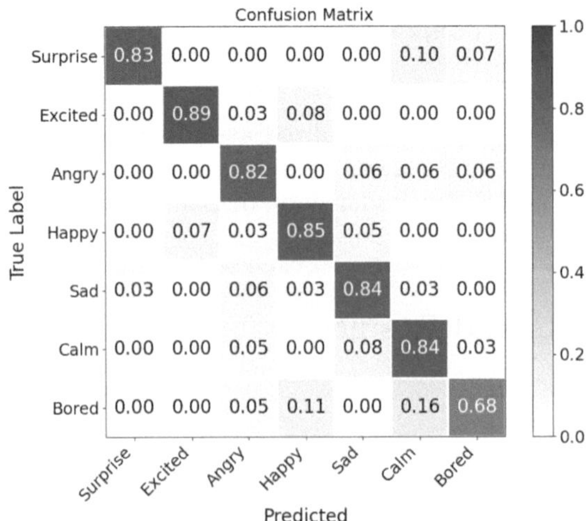

Fig. 5. Confusion matrix of Video Swin Transformer.

expressions. In contrast, the ResEmoNet advances performance substantially to 81.22% accuracy (82.77% F1) by leveraging a task-optimized backbone and FER2013 pretraining. Finally, the Video Swin Transformer attains the best results—83.41% accuracy and an 82.34% F1-score demonstrating that its hierarchical shifted-window self-attention and large-scale ImageNet+Kinetics pretraining robustly capture fine grained spatio–temporal patterns in classroom videos.

Figure 5 presents the confusion matrix of the Video Swin Transformer on the HCMUE-SEED dataset, illustrating class-wise performance. The model achieves high recognition rates for Excited, Happy, Sad, and Calm (all above 84%), with slightly lower accuracy on Surprise (83%) and Bored (68%). Misclassifications predominantly occur between semantically similar emotions. For example, Surprise is occasionally classified as Calm, Excited as Happy, and Calm is confused with Sad indicating that, although the spatio–temporal self-attention mechanism effectively captures dynamic facial cues, distinguishing between emotions with comparable arousal or valence under varied classroom conditions remains the principal challenge.

5.3 Results of Engagement Analysis Method in Classroom-Wild Dataset

We collected a 20-min evaluation dataset from 30 students, designed to reflect a typical classroom size and to cover diverse instructional contexts within a single session. The dataset consists of three activities: a 5-min game-based warm up, a 9-min knowledge exploration, and a 6-min group discussion. This design was intentional to ensure diversity in instructional formats, thereby capturing both low-interaction and high-interaction learning contexts within a single session, thereby providing a compact yet representative testbed. The evaluation dataset is completely independent of the training set and is reserved solely for performance assessment. The preprocessing pipeline, including face detection, tracking, and the extraction of 15-frame sequences via a sliding window, follows the procedure described in Sect. 4.3. After applying the Video Swin Transformer to recognize emotions in each sliding-window segment and mapping them to four engagement levels using our proposed formula, we compared the automated predictions against learners' self-reported engagement data. The overall accuracy achieved is 78.83%.

Table 6 presents a minute-by-minute comparison between our predicted engagement levels and learners' self-reports for the first ten students in the evaluation dataset. To collect self-reports, after each classroom session the recorded video was replayed in one-minute segments, and students were asked to annotate their perceived engagement level for each segment. These annotations were then time-aligned with the corresponding video frames to serve as ground truth for evaluation. Most predictions match the self-reported Low (L) and Medium (M) levels, which correspond to the subtle "calm" and "sad" expressions that are often difficult to distinguish. This agreement highlights the reliability of our emotion-to-engagement mapping pipeline.

Table 6. Comparison of Predicted with Self-Reported Engagement (IDs 1–10)

ID	Type	Activity 1					Activity 2									Activity 3					
		1	2	3	4	5	6	7	8	9	10	11	12	13	14	15	16	17	18	19	20
1	Predicted	L	L	M	M	L	M	L	L	L	L	M	L	L	L	H	H	M	M	M	M
1	Self	L	L	M	M	L	M	L	L	L	L	M	L	H	L	M	M	M	M	H	M
2	Predicted	L	L	L	M	M	L	M	M	M	H	M	M	L	L	M	M	M	H	M	M
2	Self	L	L	L	H	H	L	M	M	M	H	M	M	L	L	M	H	M	H	H	M
3	Predicted	L	L	L	L	L	L	L	L	L	L	L	L	L	L				L	L	L
3	Self	L	L	L	L	L	L	M	M	L	L	L	L	L	L	M	H	H	L	M	L
4	Predicted	L	L	L	L	M	L	L	L	L	L	L	L	L	L				M	L	L
4	Self	L	L	L	L	M	M	M	M	L	L	L	L	L	L	H	L	L	M	L	L
5	Predicted	L	M	L	L	L	L	L	L	L	M	L	L	L	L	L	L	M	H	M	L
5	Self	L	M	L	L	L	L	L	L	L	M	L	L	L	L	M	M	H	H	M	L
6	Predicted	L	M	M	L	L	L	L	L	M	L	L	L	L	M	M	L	M	L	M	L
6	Self	L	M	M	L	L	L	L	L	L	L	L	L	L	H	H	H	L	L	M	L
7	Predicted	L	L	M	M	M	L	L	H	M	L	L	S	L	L	M	M	M	M	M	L
7	Self	L	L	L	M	M	L	L	M	M	L	L	H	L	L	H	H	H	M	M	L
8	Predicted	L	L	L	L	L	H	L	L	L	L	L	L	L	M	S	M	M	M	M	M
8	Self	L	L	L	L	L	L	L	L	L	L	L	L	L	M	S	M	H	M	M	L
9	Predicted	L	L	L	H	L	L	L	L	H	H	L	L	M	M	M	H	H	H	M	M
9	Self	L	L	L	M	L	L	L	L	H	H	L	L	M	H	H	H	H	H	M	L
10	Predicted	L	S	M	M	L	L	L	L	L	L	L	L	L	L				M	S	M
10	Self	L	L	M	M	L	L	L	L	L	L	L	L	M	M	H	H	M	H	H	M

Figure 6 analyzes how different classroom activities affect engagement patterns. During the 5-min warm-up (Activity 1), two cohorts (Figs. 6e, 6c) remain mostly at Low–Medium engagement, while the other two cohorts (Figs. 6b, 6d) fluctuate between Low and High engagement. These results indicate that interactive warm-up exercises effectively stimulate initial interest. In the 9-min knowledge exploration phase (Activity 2), the majority of learners return to Low–Medium engagement, reflecting a passive baseline when processing theoretical content. When learners are highly focused, their facial affect typically appears calm. For some learners, engagement rises when the content aligns with their preferences; others show a sudden increase after periods of low engagement, often coinciding with positive affect once the newly presented concept is understood. Finally, during the 6-min group discussion (Activity 3), engagement increases markedly across all cohorts and tends to exceed that observed in the other two activities, with levels peaking around minute 20. This increase confirms that highly interactive tasks drive stronger participation.

Although our model achieves high accuracy in engagement estimation, occasional gaps in the Table 6 and chart in Fig. 6 still occur. These interruptions

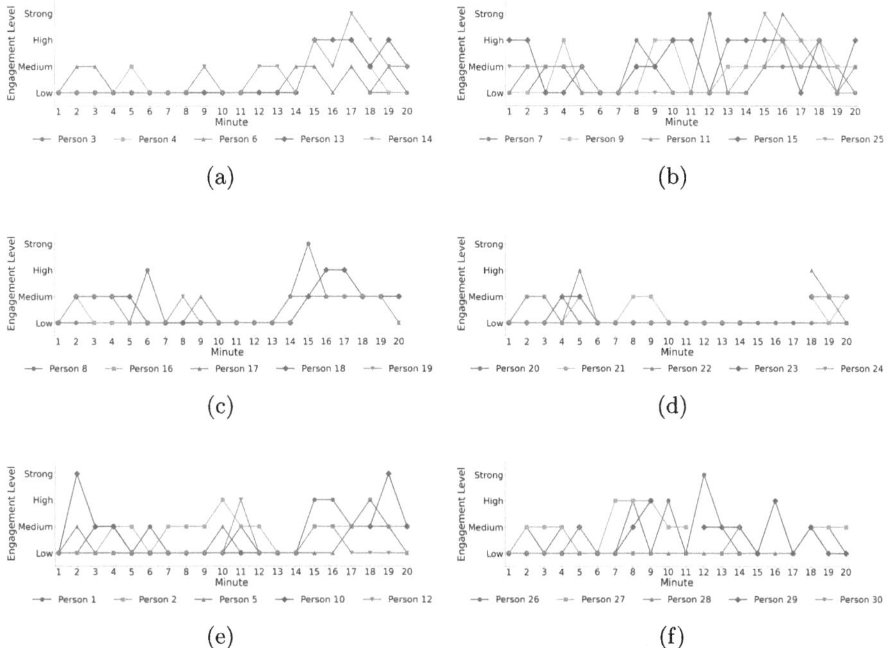

Fig. 6. Representation of Engagement Levels for ID Groups: (a) IDs 3, 4, 6, 13, 14; (b) IDs 7, 9, 11, 15, 25; (c) IDs 8, 16, 17, 18, 19; (d) IDs 20, 21, 22, 23, 24; (e) IDs 1, 2, 5, 10, 12; (f) IDs 26, 27, 28, 29, 30.

are primarily caused by failures in face detection and tracking when faces are occluded, rotated, or moving rapidly, resulting in the loss of critical expressive frames. Our model not only achieves high accuracy in predicting student engagement but also provides instructors with timely, objective insights into activity-specific participation patterns. In class, this helps teachers detect early signs of disengagement and adjust instruction, while after class, statistics such as median or peak engagement support evaluating effective activities and planning future lessons. For example, when engagement drops during a lecture, the teacher may pause for questions or introduce a short discussion; conversely, when engagement rises in group activities, more time can be allocated to such formats in subsequent sessions.

6   Conclusion

This research proposed a method for extracting facial expression information from students via live classroom videos. It utilized a Video Swin Transformer to classify seven basic emotions and introduced a novel mapping formula that integrates the temporal frequency of each emotion to estimate four levels of engagement. On the HCMUE-SEED dataset, the model achieved an emotion recognition accuracy of 83.41%. This study also contributed the HCMUE-SEED dataset, comprising videos recording seven natural expressions from 35 students in a real classroom setting, thus providing a foundation for future research. Although the obtained results are very promising, the current study still has some limitations that need to be addressed in subsequent versions. For instance, the HCMUE-SEED dataset was collected in a traditional classroom setting, which limits the model's generalizability when applied to interactive learning formats such as group work or discussions. Additionally, the current emotion recognition method primarily extracts holistic facial features but has not yet integrated micro-level information like Action Units or landmarks. Incorporating these elements will make the model more sensitive to small facial muscle movements and improve its ability to distinguish subtle emotional states. Future work will focus on expanding the dataset with more participants, activities, and longer recordings, as well as validating the model in live classroom environments. Beside, we therefore plan to implement class level engagement indices in live settings, providing timely insights that can support adaptive teaching.

Acknowledgment. This research was supported by The VNUHCM-University of Information Technology's Scientific Research Support Fund.

References

1. Fredricks, J.A., Blumenfeld, P.C., Paris, A.H.: School engagement: potential of the concept, state of the evidence. Rev. Educ. Res. **74**(1), 59–109 (2004)
2. Pantic, M., Rothkrantz, L.J.M.: Automatic analysis of facial expressions: The state of the art. IEEE Trans. Pattern Anal. Mach. Intell. **22**(12), 1424–1445 (2000)
3. Ministry of Education and Training. Circular no. 22/2021/tt-bgdđt: Regulations on assessment of lower secondary and upper secondary students. Circular (2021). https://vanban.chinhphu.vn/?pageid=27160&docid=203926. Accessed 20 June 2025
4. Ministry of Education and Training. Circular no. 27/2020/tt-bgdđt: Regulations on assessment of primary school students. Circular (2020). https://thuvienphapluat.vn/van-ban/Giao-duc/Thong-tu-27-2020-TT-BGDDT-quy-dinh-danh-gia-hoc-sinh-tieu-hoc-320659.aspx. Accessed 20 June 2025
5. Kruger, J., Dunning, D.: Unskilled and unaware of it: how difficulties in recognizing one's own incompetence lead to inflated self-assessments. J. Pers. Soc. Psychol. **77**(6), 1121 (1999)
6. Pennycook, G., Ross, R.M., Koehler, D.J., Fugelsang, J.A.: Dunning-Kruger effects in reasoning: theoretical implications of the failure to recognize incompetence. Psychon. Bull. Rev. **24**, 1774–1784 (2017)
7. Aly, M.: Revolutionizing online education: advanced facial expression recognition for real-time student progress tracking via deep learning model. Multimed. Tools Appl. **84**(13), 12575–12614 (2025)
8. Ly, D., Tran, N., Nguyen, H.Q., Nguyen, T., Nguyen, L., Nguyen, H.: A graph attention network-enhanced approach to facial expression recognition using hybrid pixel-geometry features. Int. J. Intell. Eng. Syst. **18**(5) (2025)

9. Bhardwaj, P., Gupta, P.K., Panwar, H., Siddiqui, M.K., Morales-Menendez, R., Bhaik, A.: Application of deep learning on student engagement in e-learning environments. Comput. Electr. Eng. **93**, 107277 (2021)
10. Altuwairqi, K., Jarraya, S.K., Allinjawi, A., Hammami, M.: A new emotion-based affective model to detect student's engagement. J. King Saud Univ.-Comput. Inf. Sci. **33**(1), 99–109 (2021)
11. Alkabbany, I., Ali, A.M., Foreman, C., Tretter, T., Hindy, N., Farag, A.: An experimental platform for real-time students engagement measurements from video in stem classrooms. Sensors **23**(3), 1614 (2023)
12. Tang, X., Gong, Y., Xiao, Y., Xiong, J., Bao, L.: Facial expression recognition for probing students' emotional engagement in science learning. J. Sci. Educ. Technol. **34**(1), 13–30 (2025)
13. Ashwin, T.S., Guddeti, R.M.R.: Automatic detection of students' affective states in classroom environment using hybrid convolutional neural networks. Educ. Inf. Technol. **25**(2), 1387–1415 (2020)
14. Pabba, C., Kumar, P.: A vision-based multi-cues approach for individual students' and overall class engagement monitoring in smart classroom environments. Multimed. Tools Appl. **83**(17), 52621–52652 (2024)
15. Mandia, S., Singh, K., Mitharwal, R.: Recognition of student engagement in classroom from affective states. Int. J. Multimed. Inf. Retrieval **12**(2), 18 (2023)
16. Selim, T., Elkabani, I., Abdou, M.A.: Students engagement level detection in online e-learning using hybrid efficientnetb7 together with TCN, LSTM, and bi-LSTM. IEEE Access **10**, 99573–99583 (2022)
17. Zhang, W.-L., Jia, R.-S., Wang, H., Che, C.-Y., Sun, H.-M.: A self-supervised learning network for student engagement recognition from facial expressions. IEEE Trans. Circuits Syst. Video Technol. (2024)
18. Lucey, P., Cohn, J.F., Kanade, T., Saragih, J., Ambadar, Z., Matthews, I.: The extended Cohn-Kanade dataset (CK+): a complete dataset for action unit and emotion-specified expression. In 2010 IEEE Computer Society Conference on Computer Vision and Pattern Recognition-Workshops, pp. 94–101. IEEE (2010)
19. Goodfellow, I.J., et al.: Challenges in representation learning: a report on three machine learning contests. In: Lee, M., Hirose, A., Hou, Z.-G., Kil, R.M. (eds.) ICONIP 2013. LNCS, vol. 8228, pp. 117–124. Springer, Heidelberg (2013). https://doi.org/10.1007/978-3-642-42051-1_16
20. Mollahosseini, A., Hasani, B., Mahoor, M.H.: AffectNet: a database for facial expression, valence, and arousal computing in the wild. IEEE Trans. Affect. Comput. **10**(1), 18–31 (2017)
21. Gupta, A., D'Cunha, A., Awasthi, K., Balasubramanian, V.: DAiSEE: towards user engagement recognition in the wild. arXiv preprint arXiv:1609.01885 (2016)
22. Singh, M., Hoque, X., Zeng, D., Wang, Y., Ikeda, K., Dhall, A.: Do i have your attention: a large scale engagement prediction dataset and baselines. In: Proceedings of the 25th International Conference on Multimodal Interaction, pp. 174–182 (2023)
23. C-H Wu, et al. CMOSE: comprehensive multi-modality online student engagement dataset with high-quality labels. In: Proceedings of the IEEE/CVF Conference on Computer Vision and Pattern Recognition, pp. 4636–4645 (2024)
24. Deng, J., Guo, J., Ververas, E., Kotsia, I., Zafeiriou, S.: RetinaFace: single-shot multi-level face localisation in the wild. In: Proceedings of the IEEE/CVF Conference on Computer Vision and Pattern Recognition, pp. 5203–5212 (2020)

25. Wojke, N., Bewley, A., Paulus, D.: Simple online and realtime tracking with a deep association metric. In: 2017 IEEE International Conference on Image Processing (ICIP), pages 3645–3649. IEEE (2017)
26. Liu, Z., et al.: Video swin transformer. In: Proceedings of the IEEE/CVF Conference on Computer Vision and Pattern Recognition, pp. 3202–3211 (2022)
27. Roy, A.K., Kathania, H.K., Sharma, A., Dey, A., Ansari, M.S.A.: ResEmoteNet: bridging accuracy and loss reduction in facial emotion recognition. IEEE Signal Process. Lett. **1**, 1 (2024)

ROPEVISION: A Computer Vision-Based System for Jump Rope Technique Recognition and Counting

Nguyen Hoang Pham[✉], Thanh Nguyen Thi Ngoc, and Thao Nguyen Pham

College of Information and Communication Technology, Can Tho University, Can Tho, Vietnam
pnhoang@ctu.edu.vn

Abstract. Amid the growing popularity of Computer Vision in Vietnam's technology landscape, we introduce a jump rope technique recognition application integrated with jump counting using MediaPipe. This application leverages MediaPipe Pose to extract skeletal features from video frames, then applies an LSTM model for real-time jump rope technique recognition. Additionally, the system incorporates a motion tracking algorithm to accurately count the number of jumps. Through experiments on a real-world collected dataset, the application achieves high accuracy in recognizing jump rope techniques and minimal error in jump counting. These results highlight the potential of computer vision in developing intelligent sports training support systems.

Keywords: Computer Vision in Sports · LSTM-based Action Classification · Human Pose Estimation

1 Introduction

During the rise of intelligent manifacturing and automation, applications of Information Technology and Artificial Intelligence (AI), particularly Computer Vision, have made significant advancements in the field of sports and fitness, with human pose estimation being a highly relevant topic. Jump rope is a sport that is accessible to everyone and causes minimal physical strain, making it an ideal candidate for technological integration. Accodingly, leveraging human pose estimation techniques for analyzing jump rope activities constitutes a promising and increasingly investigated research direction. In this study, we present the development of an application designed to assist jump rope training by recognizing jumping poses and accurately counting the number of jumps. This application can significantly reduce the complexity of manual tracking and recognition. In this context, MediaPipe has emerged as a pioneering tool, offering superior advantages over other human pose estimation methods. With its high-performance capabilities, precise keypoint detection, and extensive training on a large dataset, MediaPipe has been optimized for high accuracy, ensuring stability and reliability in various scenarios [1]. Therefore, this study explores the development of an application for jump rope technique recognition Jand jump counting using MediaPipe, which aligns with the current technological trends in sports and fitness.

N. Thai-Nghe et al. (Eds.): ISDS 2025, CCIS 2714, pp. 77–89, 2026.
https://doi.org/10.1007/978-981-95-3358-9_6

MediaPipe is a machine learning framework developed by Google, providing pre-trained models for human pose recognition. This framework can detect up to 33 key points on the body with high accuracy, including the head, neck, torso, arms, and legs. The key point detection is based on deep neural networks, enabling the processing of complex poses and movements [8]. In the paper "MediaPipe: A Framework for Building Perception Pipelines" [2] by Camillo Lugar et al., the authors describe how this framework manages execution, explain its working principles, architecture, and practical applications in processing and analyzing sensor data in real-world scenarios.

The use of MediaPipe for human pose estimation in the field of sports and fitness, as presented by Prof. Dr. Sunil Kale and colleagues in the study "Posture Detection and Comparison of Different Physical Exercises Based on Deep Learning Using MediaPipe, OpenCV" [6], is a notable application. Prof. Dr. Sunil Kale and his team developed a system capable of detecting and comparing workout postures in real-time. This system utilizes MediaPipe to track and analyze key body joints, extracting their coordinates to accurately recognize postures. Another notable application of MediaPipe was utilized in the study "Vision-Based Jump Rope Counter" [3] by Ayush Gupta. Ayush Gupta employed MediaPipe to track and analyze human body keypoints, focusing particularly on the ankle position throughout the video to count valid jump rope repetitions. Additionally, the angles formed by the wrist, ankle, knee, and hip were used to detect the presence of various jump rope techniques. Similarly, Prof. Anuja Garande and colleagues leveraged MediaPipe to track and extract keypoints for squat posture recognition. They developed a posture evaluation system based on established biomechanical principles by calculating critical joint angles such as knee flexion, hip flexion, and ankle dorsiflexion [9].

Video-based methods are the most common approaches for counting repetitions. Cutler tracked the subject and applied Fourier time-frequency analysis to detect periodic actions by calculating the self-similarity of the subject over time [4]. Xinxin Li converted video input into a sequence of frames, where each frame was processed using the Dense Optical Flow algorithm to extract motion information. Then, a ResNet model was used to predict the upward or downward movement type for each frame. Finally, a counting method was applied to tally and determine the number of repetitions [5].

Another study proposed a 3D human pose estimation system by combining MediaPipe with a convolutional neural network. The system employed a multi-stage approach to estimate poses and achieved high accuracy on public datasets [11–14].

MediaPipe has made significant advancements since its launch, continuously improving its features and supporting the developer community. One of the main reasons for its success can be attributed to its ecosystem of reusable computation graphs [2]. These innovations pave the way for new solutions across various domains, from recognition technologies in social applications to robotics, virtual reality, and especially sports.

Despite its remarkable progress and promising real-world applications, challenges remain in applying MediaPipe for behavior recognition and jump rope technique identification. One of the biggest challenges is ensuring robust recognition under varying environmental conditions, such as overexposure, low lighting, or partial body occlusions. In some cases, jumping actions may share visual similarities, making classification more difficult. Addressing these issues requires a combination of factors, including model design, fine-tuning, and algorithm optimization.

A promising approach to tackling this problem is using machine learning with an LSTM (Long Short-Term Memory) model. This approach involves training a model to recognize jump rope actions from skeleton data extracted from videos. To achieve high accuracy, a diverse dataset must be built, covering different lighting conditions, camera angles, and jumping speeds. Data collection and annotation play a crucial role in ensuring the model's success by enabling it to accurately recognize real-world scenarios.

LSTM is a specialized form of Recurrent Neural Network (RNN) that excels at processing and analyzing sequential data. It outperforms traditional RNNs by retaining long-term dependencies, making it well-suited for tasks such as time-series forecasting, natural language processing, video analysis, and motion tracking. LSTM networks rely on three primary gates: Forget Gate-discards unnecessary information, Input Gate-stores new relevant information and Output Gate-determines the current output. These gates function similarly to memory operations—writing, reading, and resetting memory cells [10]. By leveraging LSTM, the recognition model improves accuracy, significantly enhancing the application's performance.

The objective of this study is to design and develop an application for recognizing jump rope techniques and automatically counting repetitions, helping users track their workouts effectively and accurately. The system is designed to analyze different jump rope techniques and ensure precise detection and counting. This paper is structured into several sections. Section 2 presents the system model and its components, detailing the studied jump rope techniques, the process of building the recognition model, and the algorithm for counting jump repetitions. Section 3 describes the experimental setup and results, focusing on evaluating the application's performance and accuracy. Finally, Sect. 4 discusses the findings, challenges, and future work, particularly the integration of artificial intelligence to enhance the system's effectiveness.

2 Research Content

2.1 Dataset

The video data we collected came from various sources across online communities and forums. We selected and extracted 140 videos, each longer than one minute, which met our criteria in terms of lighting, camera angle, and action posture. Among them, 100 videos were used for training the model, and the remaining 40 videos were used as the testing dataset.

The collected videos were categorized into two groups based on the action sequence: JUMP and NON-JUMP. Within the jumping group, the data was further divided into six subcategories, each corresponding to a specific jump rope technique.

After classification, the training dataset underwent a preprocessing stage to remove noise and ensure consistency across the dataset. This process began with applying data augmentation techniques to the original frames to alter the coordinates of keypoints, thereby generating various data variations. This increases the diversity of the dataset and enables the model to generalize across different conditions. Additionally, we verified the number of collected frames, and if the count was below the minimum threshold of 1,000 frames, a missing data warning was issued to ensure the reliability of the training data.

Subsequently, all frames were resized uniformly to 720 × 720 pixels, and MediaPipe Pose was used to extract coordinate-based features. Keypoints that were successfully detected in a frame were retained, while undetected keypoints (in cases where a frame had fewer than 33 keypoints) were assigned a value of 0. After processing all videos in the dataset, the collected keypoint coordinates were recorded and exported into ".txt" files. Each file was named according to the corresponding action group: JUMP or NON-JUMP.

Following that, we used the data and previously saved keypoint information from the **JUMP** group to further classify the individual jump rope techniques. Each file in this step was named after the specific jump rope technique and saved with a ".txt" extension.

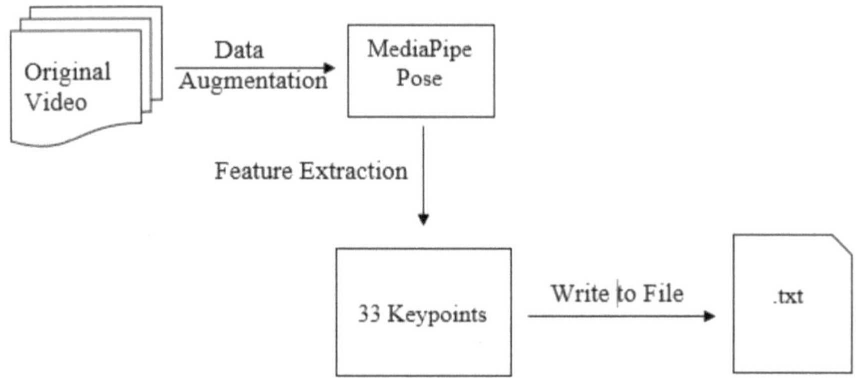

Fig. 1. Description of the data preprocessing pipeline.

2.2 System Model

In this paper, we propose an application, RopeVision, for recognizing jump rope techniques and counting repetitions from a sequence of video frames. The system consists of three main stages: detecting and extracting skeletal features, developing a jump repetition counting algorithm, and recognizing different jump rope techniques. In the skeletal feature detection and extraction stage, we utilize MediaPipe Pose, a module of MediaPipe, to estimate the 2D coordinates of human body joints in each frame. MediaPipe Pose constructs pipelines and processes perceptual data in video format using machine learning (ML). It employs BlazePose, a model capable of extracting 33 key landmarks on the human body, as illustrated in Fig. 1.

BlazePose is a lightweight machine learning architecture that achieves real-time performance on mobile phones and personal computers using only CPU inference [7]. When using normalized coordinates for pose estimation, it is necessary to inversely scale them with the pixel values on the y-axis [1]. We utilize 33 landmarks, as shown in Fig. 1, to estimate posture and movement in jump rope techniques. Each extracted landmark consists of four attributes: x, y, z, and visibility. However, we only use x, y, and z as input for the LSTM network. The landmarks extracted from video frames are fed into two models: one for action classification and the other for jump rope technique classification.

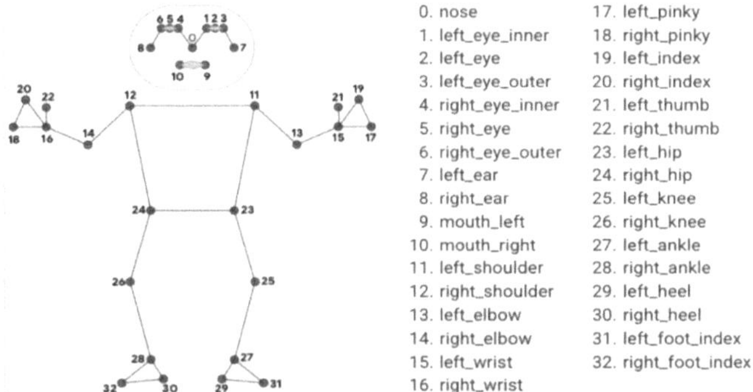

Fig. 2. Definition of Landmarks in MediaPipe Pose.

We use a sequence of frames from the video as input data, which is then fed into a binary classification model to determine the jumping action. Next, MediaPipe is used to extract skeletal coordinates, which are then processed for jump count detection and jump rope technique recognition. For jump count detection, we extract only the left hip coordinate to compare oscillation deviations between two consecutive frames. The initial state is set to "UP," and when the state changes to "DOWN," the counter is incremented by one. For jump rope technique recognition, we use all extracted skeletal coordinates and feed them into a multi-class classification model to identify the specific jump rope technique. The system's operation process is illustrated in Fig. 2 (Fig. 3).

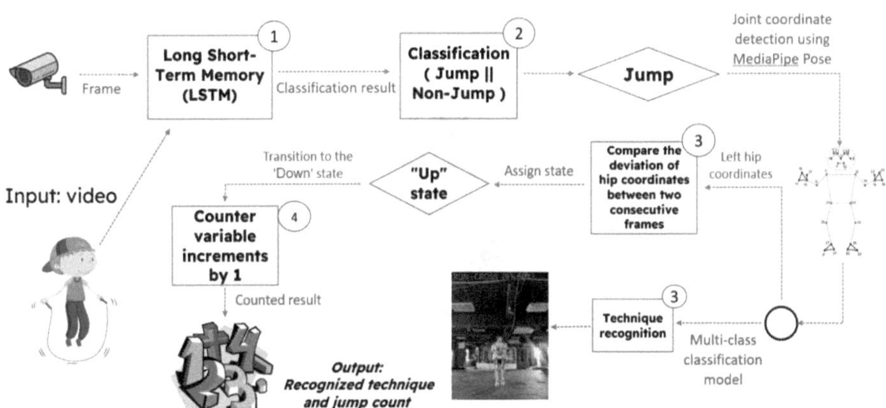

Fig. 3. System Activity Flowchart.

2.3 Developing the Counting Algorithm

The objective of this work is to study body movement during each jump cycle and how to recognize jumping actions through image processing and machine learning methods. The jump rope action is performed by jumping over a continuously rotating rope, causing the body's trajectory to move up and down along the vertical axis. A jump cycle begins with the "down" state, transitions to the "up" state, and finally ends the cycle back in the "down" state. Each jump cycle is considered one jump. We propose using a binary classification model based on the LSTM architecture to classify two actions: jumping (JUMP) and not jumping (NON-JUMP).

A counting algorithm is developed based on the trained binary classification model to assign states to the jumping action. To determine the jump state, we rely on the average amplitude of oscillation of the left hip as it moves up and down. If the coordinate of point (23) (left hip) is lower than the average value, the frame is assigned the "up" state; otherwise, if the left hip coordinate is higher than the average value, the frame is assigned the "down" state. Based on the assigned states, the system then displays the jump count result.

2.4 Recognition of Jump Rope Techniques

2.4.1 Jump Rope Techniques

The objective of this work is to study the characteristics of various jump rope techniques. Jump rope includes a wide range of techniques, from basic to advanced, along with several variations. In this paper, we limit the scope of our study to six techniques with the following characteristics:

Single Bounce: When the rope passes over the head and touches the ground, both feet remain together and jump once. Both feet will clear the rope simultaneously and land on the balls of the feet. The keypoints involved include the ankles, knees, and hips. These landmarks must be synchronized and lifted simultaneously, with a slight bounce motion upon landing. The arms maintain a consistent circular motion to rotate the rope, so the keypoints in the arms and wrists must move in coordination. The head and torso should remain stable, without significant movement, to support the jumping posture (Fig. 4).

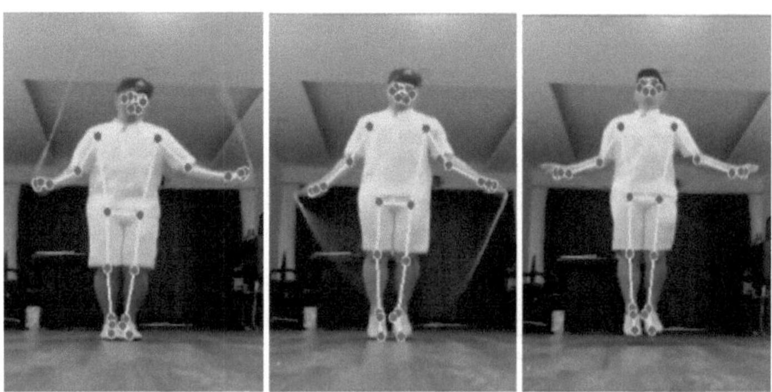

Fig. 4. Illustration of the Single Bounce technique.

Criss Cross: As the rope swings overhead, both hands cross in front of the chest, forming a crisscross loop. This movement affects keypoints at the shoulders and wrists, where the arms intersect. When the rope touches the ground, both feet should be together and jump over the rope, with keypoints at the knees and ankles synchronized and lifting simultaneously. The upper body posture should remain stable, while the arms adjust with each rope rotation (Fig. 5).

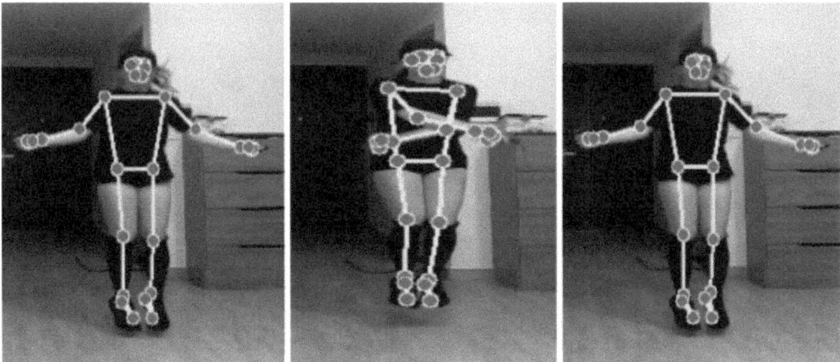

Fig. 5. Illustration of the Criss Cross technique.

Run: As the rope swings overhead and touches the ground, one foot lifts while the other jumps over the rope, alternating in a running motion. Keypoints such as the knees, ankles, and hips must adapt flexibly to each jump and leg movement. The arms maintain a steady and continuous motion to rotate the rope, affecting the wrists and shoulders (Fig. 6).

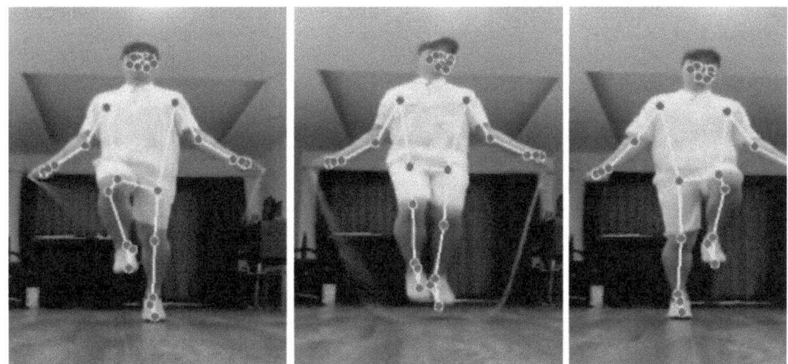

Fig. 6. Illustration of the Run technique.

Run Cross: Similar to the Run technique, but after a few jumps, the arms cross in front of the chest, forming a crossed loop with the rope. The leg movement continues alternating, lifting and jumping over the rope. Keypoints at the knees, ankles, and hips adjust dynamically with each step. When the arms cross in front of the chest, changes in the wrists and shoulders need to be tracked. As the rope swings overhead, the hands must uncross to return to the basic jumping rhythm (Fig. 7).

Fig. 7. Illustration of the Run Cross technique.

Side Straddle: On the first jump, as the rope swings overhead and touches the ground, both feet stay together and jump over the rope. On the second jump, when the rope touches the ground again, both feet spread apart to the sides. Keypoints include the knees, ankles, and hips. The legs must move in sync to execute both jump variations, while wrist and shoulder movements should remain controlled and not overly forceful (Fig. 8).

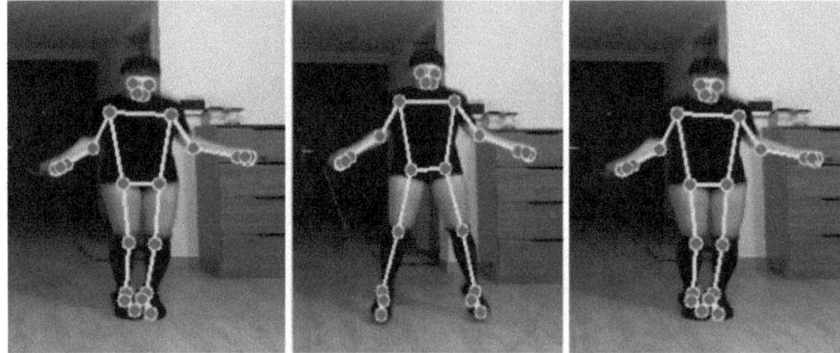

Fig. 8. Illustration of the Side Straddle technique.

Straddle Cross: On the first jump, both feet jump over the rope while spreading apart to the sides. On the second jump, as the rope swings overhead and touches the ground, both feet cross into an "X" shape upon landing. The keypoints at the knees, ankles, and hips undergo significant changes between the two jumps. The upper body posture must remain stable to support the complex leg movements, especially during the transitions between spreading and crossing the legs (Fig. 9).

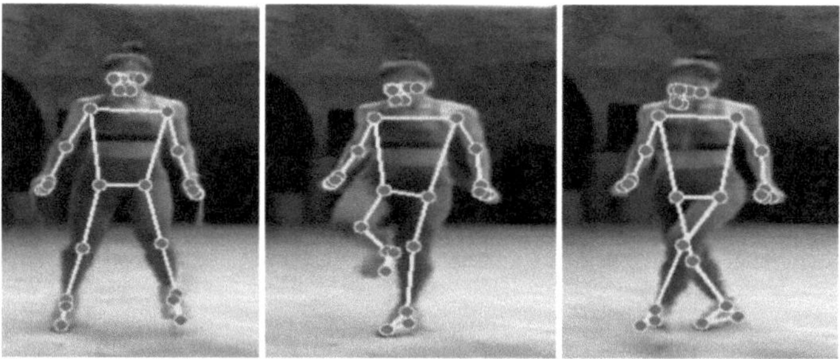

Fig. 9. Illustration of the Straddle Cross technique.

Studying the characteristics of each technique provides a deeper understanding of the movement dynamics involved in each jump rope technique. Analyzing the motions of these keypoints will aid in developing more accurate and efficient recognition methods.

2.4.2 Jump Rope Technique Recognition Model

The objective of this work is to develop a classification model for jump rope techniques based on sequential image frames from videos featuring human subjects. The model must achieve high accuracy in key areas such as the feet, ankles, knees, arms, and wrists. Additionally, upper-body landmarks, including the shoulders and hips, can help determine overall posture during jump rope exercises, aiding in distinguishing more complex techniques.

In this study, we propose using a multi-class classification model based on the LSTM architecture. The ability of LSTM to retain long-term information while discarding irrelevant data enhances its effectiveness in recognizing complex jump rope techniques, where the continuity of movements is crucial. The LSTM model will be trained on a large dataset of jump rope videos, with each video segmented into individual frames. Image features will be extracted from each frame using advanced image processing techniques. These features will then be fed into the LSTM model for analysis and classification of jump rope techniques (Fig. 10).

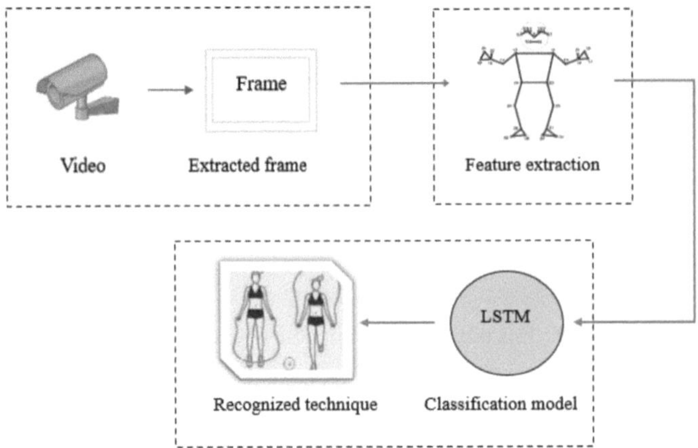

Fig. 10. The system model for recognizing jump rope techniques.

3   Experimental Results

Our experiments were carried out on a personal computer system equipped with an AMD Ryzen R5-4600H CPU, an NVIDIA GTX 1650Ti GPU, 8GB DDR4 RAM, CUDA 10.8, Python 3.10.8 and Windows 11 operating system.

3.1   Evaluation of the Counting Algorithm When Applying the Jump Action Classification Model

In the field of machine learning and classification tasks, evaluating the performance of a model is essential. Four commonly used metrics are Precision, Recall, F1-Score, and Accuracy. Each of these metrics provides different insights into how well a model performs, especially in the context of imbalanced datasets. In there, Precision is the ratio of correctly predicted positive observations to the total predicted positive observations. A high precision score means the classifier is accurate in its positive predictions, with few false positives; Recall (also known as Sensitivity or True Positive Rate) is the ratio of correctly predicted positive observations to all actual positive observations. It reflects the model's ability to detect true positives; The F1-Score is the harmonic mean of Precision and Recall, providing a single measure that balances both concerns. It is particularly useful when the cost of false positives and false negatives is high, offering a better perspective than accuracy alone in such cases; Accuracy is the ratio of correctly predicted observations (both positives and negatives) to the total number of observations. While intuitive and easy to understand, accuracy can be misleading in scenarios with class imbalance, as it does not differentiate between types of errors.

We evaluated the performance of the jump action classification model using the above-mentioned metrics: Precision, Recall, F1-score, and Accuracy on a dataset 40 test samples and computed the average value for each metric. After evaluation, we found that the average Precision reached 82.98%, indicating the proportion of correctly predicted JUMP labels among all predicted JUMP instances. The average Recall was 96.11%, demonstrating the model's ability to correctly identify nearly all actual JUMP labels.

Furthermore, the F1-score achieved 88.53%, reflecting a balance between Precision and Recall, suggesting a high overall performance. The average Accuracy was 83.96%, indicating that the model correctly classified the majority of the data.

Additionally, we assessed the performance of the jump counting algorithm using Mean Absolute Error (MAE) and Root Mean Square Error (RMSE). The discrepancy between the actual number of jumps and the counted jumps was minimal, with an MAE of 4 and an RMSE of 4.90. These results indicate that after integrating the JUMP and NON-JUMP classification model into the jump counting algorithm, the counting algorithm demonstrated superior performance, achieving an average accuracy of 92.77%.

Fig. 11. A sequence of frames in a jump rope cycle.

3.2   Evaluation of the Jump Rope Technique Recognition Model

We evaluate the performance of the jump rope technique recognition model using the following criteria: Precision, Recall, F1-score, and Accuracy. Next, we assess the model on the test set, achieving an Accuracy of 95.77%, with other evaluation metrics presented in Table 1. Additionally, we analyze the model using the confusion matrix shown in Fig. 11. The results indicate that the model performs well on labels 0 - Criss Cross, 1 - Run, 2 - Run Cross, and 4 - Single Bounce, with little to no misclassification. However, the model struggles with labels 3 - Side Straddle and 5 - Straddle Cross, which exhibit higher misclassification rates. This confusion may arise due to the similarities in movement patterns or keypoint characteristics between these classes (Fig. 12).

Table 1. Results of the evaluation metrics for the jump rope technique recognition model.

Label	Precision	Recall	F1-score
Criss Cross	0.87	1.00	0.93
Run	0.94	1.00	0.97

(*continued*)

Table 1. (*continued*)

Label	Precision	Recall	F1-score
Run Cross	1.00	1.00	1.00
Side Straddle	1.00	0.81	0.90
Single Bounce	0.99	1.00	0.99
Straddle Cross	1.00	0.92	0.96

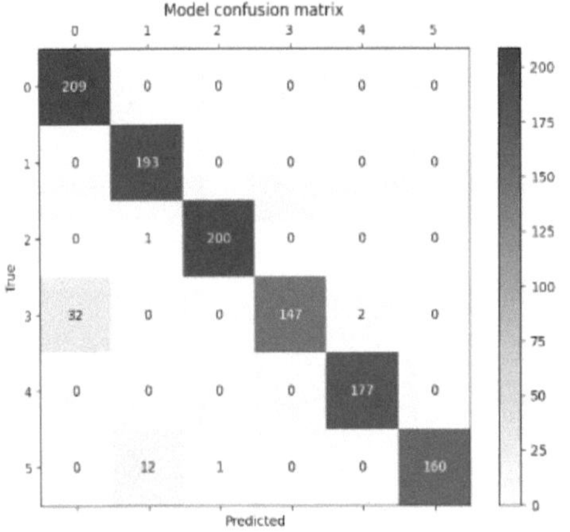

Jump technique	Label
Criss-cross	0
Run	1
Run-cross	2
Side-straddle	3
Single-bounce	4
Straddle-cross	5

Fig. 12. Confusion Matrix.

Additionally, we conducted real-world experiments and observed that environmental conditions such as lighting, resolution, and camera angles significantly impact the model's recognition performance. Each jump rope technique was tested with 10 samples of that technique under varying camera angles and lighting conditions. After testing, we obtained the following results: the Criss-Cross label achieved an accuracy of 97.75%, the Run label 74.87%, the Run-Cross label 86.95%, the Side-Straddle label 83.40%, the Single-Bounce label 91.88%, and the Straddle-Cross label 70.29%. Through the combined experimental results, we found that the model and application performed quite well, indicating that the development of the jump rope technique recognition application has been partially successful.

4 Discussion and Future Work

In this study, we developed an application utilizing computer vision to count jump repetitions and recognize jump rope techniques in real time. The application leverages deep learning models to detect and track body movements, using MediaPipe Pose to extract key body landmarks, which are then classified using an LSTM (Long Short-Term Memory) model. The system not only accurately counts jump repetitions but also identifies common jump rope techniques. The input data consists of video clips or real-time camera streams, allowing the application to analyze movements, determine body states, and provide visual feedback. The application performs stably under varying lighting conditions and can handle complex movements. However, due to hardware limitations and model processing time, the system has not yet achieved fully real-time performance in certain situations.

Future development directions include improving jump rope technique recognition through transfer learning and expanding the application to mobile devices. This application has great potential in assisting users with jump rope training, enhancing skills, and reducing the risk of injuries. Additionally, it can be integrated into professional sports training programs, physical education, and healthcare activities.

References

1. Sơn, D.T., et al.: Ước lượng tư thế người 3D trong video thể thao sử dụng MediaPipe. TẠP CHÍ KHOA HỌC ĐẠI HỌC TÂN TRÀO 9(3) (2023). ISSN: 2354 - 1431
2. Lugaresi, C., et al.: MediaPipe: a framework for building perception pipelines. arXiv:1906.08172v1 [cs.DC]. Accessed 14 June 2019
3. The University of Dublin, School of Computer Science and Statistics. Ayush Gupta. Vision Based Jump Rope Counter
4. Cutler, R., Davis, L.S.: Robust real-time periodic motion detection, analysis, and applications. IEEE Trans. Pattern Anal. Mach. Intell. 22(8), 781–796 (2002). https://doi.org/10.1109/34.868681
5. Wang, J., Li, X.: Video-Based Rope Skipping Repetition Counting with ResNet Model (2021). https://doi.org/10.18535/ijecs/v10i11.463
6. Kale, S., Kulkarni, N., Kumbhkarn, S., Khuspe, A., Kharde, S.: Posture detection and comparison of different physical exercises based on deep learning using media pipe, opencv. Int. J. Sci. Res. Eng. Manag. 7(04), 1–29 (2023). ISSN: 2582–3930
7. Bazarevsky, V., Grishchenko, I., Raceendran, K., Zhu, T., Zang, F., Grundmann, M.: BlazePose: on-device real-time body pose tracking. arXiv:2006.10204v1 [cs.CV]. Accessed 17 June 2020
8. Bhamidipati, V.S.P., Saxena, I., Saisanthiyavà, D., Retnadhas, M.: Robust Intelligent Posture Estimation for an AI Gym Trainer using Mediapipe and OpenCV (2023). https://doi.org/10.1109/ICNWC57852.2023.10127264
9. Garande, A., Patil, K., Deshmukh, R., Gurav, S., Yadav, C.: AI trainer: video-based squat analysis. In: Artificial Intelligence & Data Science Engineering, Zeal College of Engineering and Research, Pune, India (2024). https://doi.org/10.32628/IJSRSET2411221
10. Singh, D., et al.: Human active recognition using recurrent neural networks. In: Conference Paper in Lecture Notes in Computer Science (2017). https://doi.org/10.1007/978-3-319-66808-6_18
11. Kim, S.H., Chang, J.Y.: Single-shot 3d multi-person shape reconstruction from a single rgb image. Entropy 22(8), 806 (2020). https://doi.org/10.3390/e2080806
12. Li, X., Zhang, M., Gu, J., Zhang, Z.: Fitness action counting based on MediaPipe. In: 2022 15th International Congress on Image and Signal Processing, BioMedical Engineering and Informatics (CISP-BMEI), pp.1–7. IEEE (2022)
13. Zhang, S., Chen, W., Chen, C., Liu, Y.: Human deep squat detection method based on MediaPipe combined with Yolov5 network. In: 2022 41st Chinese Control Conference (CCC) (2022). https://doi.org/10.23919/CCC55666.2022.9902631
14. Palani, P., Panigrahi, S., Jammi, S.A., Thondiyath, A.: Real-time joint angle estimation using mediapipe framework and inertial sensors. In: 2022 IEEE 22nd International Conference on Bioinformatics and Bioengineering (BIBE), pp. 128–133. IEEE (2022)

CAHIM: Constructing Auto-generated HIstorical Mindmaps Based on Clustering and Ontology Structuring

Thanh Ma[✉], Hieu Nguyen, Phu-An Thai, Xuan Nguyen, Ky Nguyen, Thy Le, Le-Diem Bui[✉], Nguyen-Khang Pham[✉], and Won Ho[✉]

CICT, Can Tho University, Can Tho, Vietnam
{mtthanh,bldiem,pnkhang}@ctu.edu.vn, wohyho@gmail.com

Abstract. This study presents an innovative AI-driven system designed to generate interactive mind-maps tailored for history education. At its core, the system constructs a hierarchical ontology using a novel clustering algorithm that groups semantically related historical concepts. This structured ontology is then visualized as dynamic, interactive mind maps, enabling learners to comprehend and explore historical knowledge more intuitively and effectively. In addition, a RAG-based chatbot is integrated into the system, allowing users to ask questions and receive context-aware responses derived from the ontology. To support this functionality, the study introduces a framework called CAHIM, which automatically extracts and organizes knowledge from raw documents (e.g., PDFs) to construct the ontology used for both visualization and question answering. Experimental results demonstrate that this approach significantly improves both learning efficiency and user engagement when compared with conventional methods.

Keywords: Clustering · Mind-maps · Ontology · NLP · Historical Education

1 Introduction

History education plays a pivotal role in fostering critical thinking and cultural awareness, yet it often presents challenges due to the complexity and interconnectedness of historical events. Traditional pedagogical approaches, which typically rely on linear narratives, can hinder learners' ability to grasp the multifaceted relationships between historical concepts [7]. In response, interactive mindmaps have emerged as a promising tool for visualizing knowledge hierarchies, enabling learners to explore and comprehend historical content more intuitively [5]. However, the manual creation of such mindmaps is labor-intensive and impractical for large-scale educational use, necessitating automated solutions.

Recent advances in natural language processing (NLP), unsupervised machine learning, and knowledge representation have opened new avenues for

N. Thai-Nghe et al. (Eds.): ISDS 2025, CCIS 2714, pp. 90–104, 2026.
https://doi.org/10.1007/978-981-95-3358-9_7

automating the extraction and structuring of knowledge from unstructured text [6,12]. Knowledge graph construction, for instance, has seen significant progress through entity and relation extraction techniques [9,16]. Nonetheless, these methods often require extensive manual annotation or domain-specific rules, limiting their adaptability to diverse historical domains. Similarly, ontology learning approaches aim to automatically generate formal knowledge structures but frequently struggle to capture the nuanced hierarchies inherent in historical knowledge [2].

To address these challenges, we propose **CAHIM** (Constructing Auto-generated Historical Mindmaps), an innovative framework that leverages clustering and ontology structuring to transform unstructured historical documents into interactive, hierarchical mindmaps. At its core, CAHIM employs a novel clustering algorithm to group semantically related concepts, which are then organized into a hierarchical ontology. This ontology is visualized as a dynamic mindmap, allowing learners to navigate historical knowledge intuitively. Furthermore, CAHIM integrates a Retrieval-Augmented Generation (RAG)-based chatbot [1,11], enabling users to ask questions and receive context-aware responses grounded in the ontology [8].

Our main contributions are as follows:

- A novel clustering algorithm that automatically groups semantically related historical concepts from unstructured text.
- A method for constructing a hierarchical ontology by iteratively abstracting clustered concepts, providing a scalable solution for knowledge organization.
- An interactive mindmap visualization that enhances learner engagement and comprehension of historical content.
- A RAG-based chatbot that leverages the ontology to deliver accurate, contextually relevant responses, thereby enriching the educational experience.

The remainder of this paper is structured as follows: We briefly present a fundamental background in Sect. 2. Next, Sect. 3 describes the proposed framework. Then, Sect. 4 introduces an CAHIM algorithm. Section 5 presents the experiment and the results of the summary models. Finally, Sect. 6 shows the conclusions and future works.

2 Background

The architecture of our system is conceived as a multi-stage pipeline that transmutes unstructured textual corpora into a formal, hierarchical, and queryable knowledge structure. This process is underpinned by a synergistic application of principles from computer vision, unsupervised machine learning, and formal knowledge representation. The framework is logically partitioned into three foundational pillars: (1) Knowledge Fragment Extraction and Semantic Distillation, (2) Agglomerative Construction of Knowledge Hierarchies, and (3) Ontological Formalization and Conversational Knowledge Access.

2.1 Knowledge Fragment Extraction and Semantic Distillation

The initial stage of the pipeline addresses the critical challenge of converting a raw, unstructured document into a set of discrete, semantically potent units, which we define as Knowledge Fragments [13].

Document Layout Analysis: A document's visual presentation often encodes a logical structure that is lost during naive text extraction. To preserve this, the framework first employs document layout analysis [3]. A contemporary object detection model is tasked with parsing the two-dimensional page canvas, identifying and segmenting semantically distinct blocks such as titles, paragraphs, lists, and captions. This segmentation ensures that the textual content is processed within its intended logical context, preventing the erroneous concatenation of unrelated text blocks.

Semantic Distillation and Representation: Each extracted text block undergoes a semantic distillation process [10], managed by a state-of-the-art Large Language Model. For any given text block t, this process yields a concise summary s that captures its informational core and a salient keyword κ that encapsulates its central theme [15].

To facilitate computational analysis, each distilled summary s is projected into a high-dimensional vector space [4]. This is achieved using a transformer-based sentence embedding model, which functions as an encoder $E : \mathcal{T} \to \mathbb{R}^d$, where $\mathcal{T}$ represents the space of text and $\mathbb{R}^d$ is a d-dimensional Euclidean space. The resulting vector, or embedding, captures the nuanced semantics of the text.

A Knowledge Fragment $\mathcal{F}$ is thus formally defined as a tuple:

$$\mathcal{F} = (s, \kappa, v)$$

where s is the summary, κ is the keyword, and $v = E(s)$ is the corresponding semantic embedding. The set of all such fragments, $\{\mathcal{F}_i\}_{i=1}^{N}$, constitutes the foundational material for hierarchical construction.

2.2 Agglomerative Construction of Knowledge Hierarchies

The system autonomously organizes the discrete Knowledge Fragments into a coherent hierarchy that reflects the document's thematic structure. This is accomplished through an unsupervised, bottom-up agglomerative hierarchical clustering paradigm [14, 17]. The process is iterative, with each iteration abstracting a set of thematic concepts into a smaller set of higher-order concepts.

Let $\mathcal{D}^{(l)}$ be the set of conceptual nodes at hierarchy level l, where $\mathcal{D}^{(0)}$ is the initial set of N Knowledge Fragments. The transition from one level to the next is a function $\Phi : \mathcal{D}^{(l)} \to \mathcal{D}^{(l+1)}$ such that $|\mathcal{D}^{(l+1)}| < |\mathcal{D}^{(l)}|$. This function is realized through the following steps:

1. **Semantic Space Normalization**: All embedding vectors $\{v_i\}$ corresponding to the nodes in $\mathcal{D}^{(l)}$ are L2-normalized, $\hat{v}_i = v_i/\|v_i\|_2$. This projects the embeddings onto the surface of a unit hypersphere, making cosine similarity a direct measure of Euclidean distance, which is highly effective for clustering textual data.

2. **Adaptive Cluster Granularity**: At each level l, the optimal number of clusters, k^*, must be determined. The framework employs the Elbow Method, which analyzes the trade-off between the number of clusters k and the Within-Cluster Sum of Squares (WCSS). The point of maximum curvature on the WCSS-versus-k plot, or the "elbow" is selected as k^*. This allows the granularity of clustering to adapt dynamically to the thematic diversity of the concepts at that level.

3. **Spherical K-Means Clustering**: A `MiniBatchKMeans` algorithm partitions the normalized embeddings $\{\hat{v}_i\}$ into k^* clusters, $\{C_1, C_2, \ldots, C_{k^*}\}$. This partitioning effectively groups semantically related conceptual nodes.

4. **Medoid-based Cluster Representation**: A new, higher-level conceptual node must be created to represent each cluster C_j. The representation for this new node is not an artificial construct (like a geometric centroid) but is selected from the existing members of the cluster. The framework identifies the **medoid** of the cluster, i.e., the Knowledge Fragment whose embedding is most central, by finding the node that maximizes the average cosine similarity to all other nodes within the same cluster, or more efficiently, the one closest to the cluster's geometric centroid μ_j.

$$\mathcal{F}_{\text{rep}_j} = \arg \max_{\mathcal{F}_i \in C_j} \left(\frac{\hat{v}_i \cdot \mu_j}{\|\hat{v}_i\|\|\mu_j\|} \right)$$

The set of these k^* representative fragments $\{\mathcal{F}_{\text{rep}_j}\}_{j=1}^{k^*}$ forms the subsequent level of the hierarchy, $\mathcal{D}^{(l+1)}$.

This iterative process continues until a predefined termination condition is met, typically when the number of clusters converges to one, resulting in a single root node for the entire knowledge hierarchy.

2.3 Ontological Formalization and Conversational Knowledge Access

The final stage of the framework elevates the procedurally generated tree into a formal knowledge base and provides an intuitive, conversational interface for its exploration.

Ontology Generation: The computed hierarchy is translated into a Web Ontology Language (OWL) ontology, a W3C standard for knowledge representation. This formalization provides a robust, logical, and interoperable structure. The mapping is direct:

- Each node in the hierarchy is instantiated as an `owl:Class`.
- A parent-child relationship in the tree is expressed as an `rdfs:subClassOf` axiom, establishing a formal taxonomy.
- The semantic content of each Knowledge Fragment—its summary (s), keyword (κ), and embedding (v)—is attached to its corresponding `owl:Class` as `owl:AnnotationProperty` instances.

This act of formalization transforms the data structure into a true knowledge model, amenable to logical inference and standardized querying.

Conversational Access via Retrieval-Augmented Generation: To facilitate natural language querying, the system implements a Retrieval-Augmented Generation (RAG) architecture. This paradigm grounds the generative capabilities of an LLM in the factual content of the constructed ontology, significantly mitigating the risk of factual hallucination [8]. The process for answering a user query proceeds in three phases:

1. **Concept-driven Scoping**: The user's natural language query is first processed by an LLM that acts as a reasoning engine. It analyzes the query in the context of the ontology's class structure to identify a candidate set of relevant classes. This initial step effectively links the unstructured query to the formal concepts in the knowledge graph, dramatically narrowing the search space.
2. **Semantic Refinement**: Within the subgraph defined by the scoped classes and their descendants, a fine-grained retrieval process is initiated. The system leverages a high-performance vector search index (built with `faiss`) over the pre-computed `summary_embeddings`. It performs a similarity search to find the top-k text summaries that are most semantically aligned with the vector representation of the user's query. This step pinpoints the most direct textual evidence relevant to the query.
3. **Grounded Generation**: The retrieved text summaries are compiled into a single context block. This context is prepended to a final prompt, which instructs the LLM to synthesize a comprehensive and fluent answer based *exclusively* on the information provided. This final generation step ensures that the response is not only relevant but also factually tethered to the source document.

3   Proposal Framework

We present a novel, fully automated framework designed to transform unstructured documents into formally structured and conversationally accessible knowledge bases. The architecture operationalizes the theoretical principles of semantic distillation and agglomerative structuring within a robust, end-to-end pipeline. This system proceeds in two primary phases: (1) the construction of a hierarchical knowledge graph from the source document, and (2) the formalization of this graph into a queryable ontology equipped with a natural language interface. Figure 1 provides a high-level schematic of the entire process.

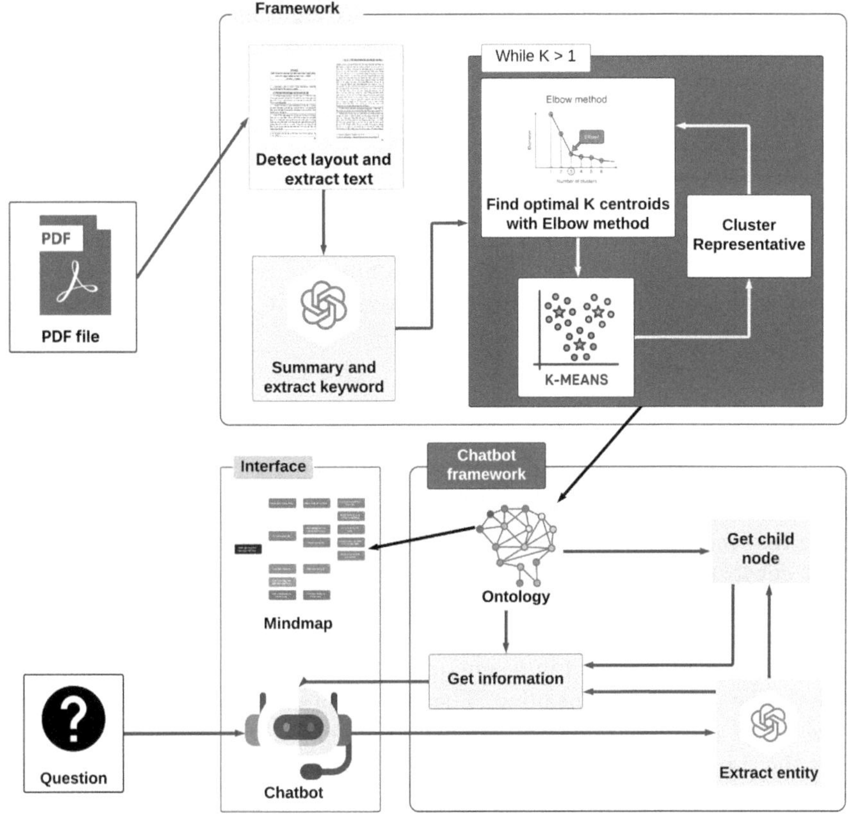

Fig. 1. The end-to-end architecture of proposed system. The pipeline ingests a source document (PDF) and executes a four-stage process: (1) It first performs vision-based layout analysis and semantic distillation to generate a set of foundational Knowledge Fragments. (2) It then iteratively clusters these fragments to construct a thematic hierarchy. (3) This hierarchy is formalized as a standard OWL ontology. (4) Finally, a Retrieval-Augmented Generation (RAG) module enables natural language querying, producing factually grounded responses.

3.1 Phase 1: Hierarchical Knowledge Graph Construction

The initial phase of CAHIM framework is dedicated to processing the raw source document and systematically organizing its semantic content into a coherent, hierarchical structure. This is accomplished through a sequence of three computational stages.

Stage 1: Vision-Based Document Segmentation. Traditional text extraction approaches that simply concatenate textual content fail to preserve the rich structural information encoded in document layout. Our framework begins with vision-based document segmentation, treating the document as a two-dimensional visual object rather than a linear text stream. We employ a contemporary object detection model to parse the document canvas, identifying and segmenting semantically distinct regions including headers, paragraphs, lists, captions, and tabular content. This approach ensures that the semantic context implicit in document structure—such as the relationship between a figure and its caption—is preserved during processing. The segmentation process produces an ordered sequence of logically coherent text blocks, each maintaining its original semantic context and document-level positioning. This structural preservation proves crucial for subsequent semantic analysis, as it prevents the erroneous merging of conceptually distinct content.

Stage 2: Knowledge Fragment Distillation. Each text block isolated in the previous stage undergoes a process of *Semantic Distillation.* This process is orchestrated by a state-of-the-art Large Language Model (LLM) which performs two tasks: first, it generates a concise, abstractive summary (s) that captures the core information of the block; second, it identifies a single, salient keyword (κ) that encapsulates the block's central theme.

To enable computational analysis, the generated summary (s) is then encoded into a high-dimensional vector space using a transformer-based sentence embedding model (E). This results in a semantic vector, $v = E(s)$, which numerically represents the summary's meaning. The successful execution of this stage produces a set of N foundational Knowledge Fragments, $\{\mathcal{F}_i\}_{i=1}^{N}$, where each fragment is the tuple $\mathcal{F}_i = (s_i, \kappa_i, v_i)$ as defined in theoretical basis. This set, denoted $\mathcal{D}^{(0)}$, forms the initial input for the hierarchy construction.

Stage 3: Agglomerative Hierarchy Construction. The framework then autonomously organizes the discrete Knowledge Fragments into a thematic hierarchy using an iterative, bottom-up clustering algorithm, detailed in Sect. 4. This process implements the transformation function $\Phi : \mathcal{D}^{(l)} \rightarrow \mathcal{D}^{(l+1)}$, progressively abstracting concepts level by level. A single iteration proceeds as follows:

1. **Vector Space Projection:** All embedding vectors $\{v_i\}$ corresponding to the nodes in the current level $\mathcal{D}^{(l)}$ are L2-normalized. This projects the embeddings onto the surface of a unit hypersphere, a standard practice that makes cosine similarity equivalent to Euclidean distance, thereby improving the performance of clustering algorithms on textual data.
2. **Dynamic Cluster Number Estimation:** To ensure that the granularity of clustering is appropriate for the data at each level of abstraction, the framework dynamically determines the optimal number of clusters, k^*. It automates the **Elbow Method** by computing the Within-Cluster Sum of Squares (WCSS) for a predefined range of k values and identifying the "elbow point" of maximum curvature.

3. **Clustering and Abstraction:** With k^* determined, a `MiniBatchKMeans` algorithm partitions the normalized embeddings into k^* clusters. This algorithm is chosen for its computational efficiency with high-dimensional data. To represent each newly formed cluster C_j, the framework selects a representative member rather than computing an artificial centroid. It identifies the cluster's **medoid**—the Knowledge Fragment $\mathcal{F}_i \in C_j$ whose embedding $\hat{v}_i$ is closest to the cluster's geometric centroid. These k^* medoids constitute the set of conceptual nodes for the next level of the hierarchy, $\mathcal{D}^{(l+1)}$.

This iterative process terminates when the number of clusters converges to one, yielding a single root node and a complete, multi-layered knowledge hierarchy.

3.2   Phase 2: Ontological Formalization and Conversational Access

In its final phase, the system elevates the generated tree into a formal, machine-readable knowledge base and equips it with an intuitive interface for human-in-the-loop exploration.

Stage 4: Automated Ontology Generation. The procedural hierarchy is systematically translated into the Web Ontology Language (OWL), a W3C standard for knowledge representation. This formalization provides a robust, logical structure amenable to standardized querying and logical inference. The translation follows a direct mapping:

- Each node in the hierarchy is instantiated as an `owl:Class`.
- A parent-child relationship in the tree is expressed as an `rdfs:subClassOf` axiom.
- The semantic content of each Knowledge Fragment (its summary s, keyword κ, and embedding v) is attached to its corresponding `owl:Class` as data annotations using custom `owl:AnnotationProperty` instances.

Stage 5: Retrieval-Augmented Query Answering. To facilitate natural language querying, the framework implements a Retrieval-Augmented Generation (RAG) architecture. This model grounds the generative capabilities of an LLM in the factual content of the newly created ontology, significantly mitigating the risk of factual hallucination. The query-answering process is executed in three steps:

1. **Ontology-Driven Scoping:** The user's query is first interpreted by an LLM configured as a reasoning agent. This agent navigates the `rdfs:subClassOf` taxonomy to identify a candidate set of `owl:Class`es that are conceptually relevant to the query. This step effectively uses the formal structure of the ontology to prune the search space to a highly relevant subgraph.

2. **Dense Vector Retrieval:** Within the scoped subgraph, a fine-grained retrieval process is initiated. The system leverages a pre-built `faiss` index of all summary embeddings (v) to perform a high-speed similarity search. It retrieves the top-k summaries from the Knowledge Fragments in the subgraph that are most semantically aligned with the vector representation of the user's query.

3. **Grounded Response Synthesis:** The retrieved summaries are compiled into a single context block. This context is prepended to a final prompt that instructs a generative LLM to synthesize a comprehensive and fluent answer based *exclusively* on the information provided. This final step ensures that the system's response is not only relevant but also factually tethered to the source document.

4   ASHI Algorithm

The Automated Semantic Hierarchy Induction (ASHI) algorithm, formally presented in Algorithm 1, constitutes the computational core of CAHIM framework. It delineates a deterministic procedure for transforming a flat set of semantically-encoded Knowledge Fragments into a multi-level conceptual hierarchy. This algorithm operationalizes the principles of agglomerative clustering and medoid-based abstraction to autonomously discover and structure the thematic hierarchy inherent in a source document.

Algorithm 1: Automated Semantic Hierarchy Induction (ASHI)

Input   : $\mathcal{D}^{(0)} = \{\mathcal{F}_i\}_{i=1}^{N}$: The initial set of N Knowledge Fragments, $\mathcal{F}_i = (s_i, \kappa_i, v_i)$
$\mathcal{E}_K$: Elbow Method analysis function to find optimal k
$\mathcal{K}$: MiniBatchKMeans clustering function
$\mathcal{T}$: Termination condition function (e.g., $|\mathcal{D}^{(l)}| \leq 1$)
Output: $\mathcal{H}$: A hierarchical tree structure of conceptual nodes

1   **begin**
2      $l \leftarrow 0$
3      $\mathcal{H} \leftarrow \text{InitializeTree}(\mathcal{D}^{(0)})$
4   **while** *not* $\mathcal{T}(\mathcal{D}^{(l)})$ **do**
5         $\{\hat{v}_i^{(l)}\}_{i=1}^{|\mathcal{D}^{(l)}|} \leftarrow \{v_i / \|v_i\|_2 \text{ for each node in } \mathcal{D}^{(l)}\}$
6         $k^* \leftarrow \mathcal{E}_K(\{\hat{v}_i^{(l)}\})$
7         $\{C_1, \ldots, C_{k^*}\} \leftarrow \mathcal{K}(\{\hat{v}_i^{(l)}\}, k^*)$
8         $\mathcal{D}^{(l+1)} \leftarrow \emptyset$
9   **for** $j \leftarrow 1$ **to** k^* **do**
10            $\mu_j \leftarrow \frac{1}{|C_j|} \sum_{\mathcal{F}_i \in C_j} \hat{v}_i$
11            $\mathcal{F}_{\text{rep}_j} \leftarrow \arg\max_{\mathcal{F}_i \in C_j} \left(\frac{\hat{v}_i \cdot \mu_j}{\|\hat{v}_i\| \|\mu_j\|} \right)$
12            $\mathcal{D}^{(l+1)} \leftarrow \mathcal{D}^{(l+1)} \cup \{\mathcal{F}_{\text{rep}_j}\}$
13            $\mathcal{H} \leftarrow \text{AddParentChildEdges}(\mathcal{H}, \mathcal{F}_{\text{rep}_j}, C_j)$
14         $l \leftarrow l + 1$
15   **return** $\mathcal{H}$

4.1   Procedural Description

The ASHI algorithm provides a structured, iterative methodology for constructing a knowledge hierarchy. Its execution is defined by a sequence of precise computational steps.

Inputs and Outputs: The algorithm requires as input: an initial set of N Knowledge Fragments, $\mathcal{D}^{(0)}$, where each fragment contains a summary, a keyword, and a semantic embedding; a function $\mathcal{E}_K$ for optimal cluster number determination; a clustering function $\mathcal{K}$; and a termination condition $\mathcal{T}$. Its final output is a tree data structure, $\mathcal{H}$, representing the complete knowledge hierarchy.

Algorithm Steps:

1. **Initialization (Lines 1–2):** The process begins at level $l = 0$. The hierarchy $\mathcal{H}$ is initialized as a flat structure containing all N Knowledge Fragments from the initial set $\mathcal{D}^{(0)}$ as its leaf nodes.
2. **Iterative Abstraction (Lines 3–16):** The core of the algorithm is a 'while' loop that continues as long as the termination condition $\mathcal{T}$ (e.g., the number of nodes at the current level is greater than 1) is not met. Each iteration of this loop constructs one higher level of the conceptual hierarchy.
3. **Semantic Space Normalization (Line 4):** Within each iteration, the embedding vectors v_i of all nodes at the current level $\mathcal{D}^{(l)}$ are L2-normalized. This step projects the vectors onto a unit hypersphere, ensuring that the subsequent clustering, based on Euclidean distance, effectively measures cosine similarity, which is highly suitable for semantic data.
4. **Adaptive Granularity Determination (Line 5):** To ensure the clustering is thematically meaningful, the algorithm dynamically computes the optimal number of clusters, k^*, for the current level. The function $\mathcal{E}_K$ implements the Elbow Method, analyzing the trade-off between the number of clusters and the within-cluster sum of squares to find the point of maximum curvature.
5. **Clustering and Partitioning (Line 6):** Using the determined k^*, the clustering function $\mathcal{K}$ (e.g., `MiniBatchKMeans`) partitions the normalized embeddings $\{\hat{v}_i^{(l)}\}$ into k^* distinct clusters, $\{C_1, \dots, C_{k^*}\}$. This operation groups semantically related conceptual nodes from level l.
6. **Medoid-Based Abstraction (Lines 8–14):** For each cluster C_j, a new, higher-level conceptual node must be formed. Instead of creating an artificial representative (like a geometric mean), the algorithm identifies the **medoid** of the cluster. It first computes the cluster's geometric centroid μ_j (Line 10). Then, it selects the existing Knowledge Fragment $\mathcal{F}_{\mathrm{rep}_j}$ from within the cluster whose embedding has the highest cosine similarity to this centroid (Line 11). This medoid becomes the representative node for the cluster in the next hierarchical level, $\mathcal{D}^{(l+1)}$. This set of k^* medoids forms the entirety of level $l + 1$. The hierarchy $\mathcal{H}$ is updated by establishing parent-child relationships between the new representative node $\mathcal{F}_{\mathrm{rep}_j}$ and all the nodes in its source cluster C_j (Line 14).
7. **Termination and Output (Line 17):** The loop continues, incrementing the level counter l, until the termination condition is met. This typically occurs when k^* converges to 1, leaving a single root node for the entire hierarchy. The final, fully-connected hierarchical tree $\mathcal{H}$ is then returned.

4.2 Computational Complexity Analysis

The time complexity of the ASHI algorithm is primarily driven by the operations inside its main iterative loop. Let L be the total number of levels in the final hierarchy, N_l be the number of nodes at level l (with $N_0 = N$), d be the dimensionality of the embeddings, and k_{max} be the maximum number of clusters evaluated by the Elbow Method.

- **Normalization:** The L2 normalization at each level requires a single pass over the nodes, resulting in a complexity of $O(N_l \cdot d)$.
- **Optimal k Determination:** The Elbow Method involves running the clustering algorithm for a range of k values (from 2 to k_{max}). Let the complexity of a single run of `MiniBatchKMeans` with i iterations be $O(i \cdot N_l \cdot d)$. The complexity for this step is thus $O(k_{max} \cdot i \cdot N_l \cdot d)$.
- **Clustering:** The final clustering run with the optimal k^* has a complexity of $O(i \cdot N_l \cdot d \cdot k^*)$. However, since k^* is determined in the previous step, we consider its dominant factor from that step.
- **Medoid Selection:** For each of the k^* clusters, calculating the centroid takes $O(|C_j| \cdot d)$, and finding the medoid requires comparing the centroid to each member, also $O(|C_j| \cdot d)$. Summing over all clusters, this step has a complexity of $\sum_{j=1}^{k^*} O(|C_j| \cdot d) = O(N_l \cdot d)$.

The total complexity is the sum of these costs over all levels L. The most computationally intensive part of each iteration is the determination of the optimal k. Therefore, the overall time complexity of the ASHI algorithm can be approximated as:

$$O\left(\sum_{l=0}^{L-1} (k_{max} \cdot i \cdot N_l \cdot d)\right)$$

Given that N_l decreases at each level (often exponentially), the complexity is dominated by the initial levels, particularly the first one ($l = 0$). Thus, the complexity is heavily influenced by the initial number of Knowledge Fragments, N: $O(L \cdot k_{max} \cdot i \cdot N \cdot d)$. This analysis highlights the algorithm's efficiency, as it scales linearly with the number of initial fragments, though it is modulated by the depth of the hierarchy and the search range for cluster granularity.

5 Experimental Result

We conducted a comprehensive evaluation of the CAHIM system, focusing on two key aspects: the performance of the RAG-based chatbot and the efficiency of the mindmap generation process.

5.1 Chatbot Performance Evaluation

To assess the effectiveness of the proposed chatbot (Method 1), which leverages entity extraction and ontology structuring, we compared it against a baseline

RAG approach (Method 2) that directly retrieves the top-k summaries based on similarity to the user's question. Both methods were evaluated on a set of 100 questions derived from historical content, using a suite of standard natural language processing metrics. These metrics include semantic similarity, improved accuracy (defined as the proportion of responses with semantic similarity exceeding 0.7), F1-score, ROUGE scores (ROUGE-1, ROUGE-2, ROUGE-L), BLEU score, and average response time. The results are summarized in Table 1.

Table 1. Comparison of chatbot performance between the proposed method (Method 1) and the baseline RAG approach (Method 2).

Metric	Method 1	Method 2
Semantic Similarity	0.5900	0.1383
Improved Accuracy	0.1200	0.0011
F1-Score	0.2307	0.0122
ROUGE-1	0.3559	0.2356
ROUGE-2	0.2129	0.0606
ROUGE-L	0.2309	0.1670
BLEU Score	0.0304	0.0001
Average Response Time (s)	9.966	0.0001

The proposed method (Method 1) demonstrates a significant improvement over the baseline in terms of semantic similarity (0.5900 vs. 0.1383), indicating that the responses are more contextually relevant to the user's questions. Additionally, Method 1 achieves an improved accuracy of 0.1200, meaning that 12% of the responses exceed the 0.7 similarity threshold, whereas the baseline fails to meet this threshold for any response. The F1-score, ROUGE metrics, and BLEU score also show superior performance for Method 1, reflecting enhanced precision and recall in the generated responses. However, the average response time for Method 1 is higher (9.966 s) compared to the near-instantaneous response of the baseline, likely due to the additional computational steps involved in entity extraction and ontology querying.

5.2   Mindmap Generation Efficiency

We also evaluated the efficiency of the mindmap generation process by measuring the time required for the clustering stage (PDF-to-Cluster) and the ontology creation stage across 58 historical documents with varying page counts (7 to 45 pages). The average times for these stages are presented in Table 2.

The results indicate that the clustering stage, which involves processing the document to extract and organize knowledge fragments, takes an average of 31.80 s per document. In contrast, the ontology creation stage is highly efficient, averaging only 0.86 s. These times demonstrate the system's capability to handle

Table 2. Average time required for mindmap generation stages.

Stage	Average Time (s)
PDF-to-Cluster	31.80
Ontology Creation	0.86

documents of varying sizes effectively, with the clustering time scaling reasonably with document complexity.

5.3 Application and Demonstration

To demonstrate the practical utility of the CAHIM system, we developed an interactive web-based application that enables users to explore historical knowledge through dynamically generated mindmaps and interact with the RAG-based chatbot. The application allows users to upload historical documents in PDF format, visualize the resulting hierarchical mindmaps, and query the system using natural language to retrieve context-aware responses grounded in the constructed ontology. The user interface, designed for intuitive navigation, displays the mindmap with expandable nodes representing clustered historical concepts and provides a conversational panel for querying. Screenshots of the application interface is shown in Fig. 2.

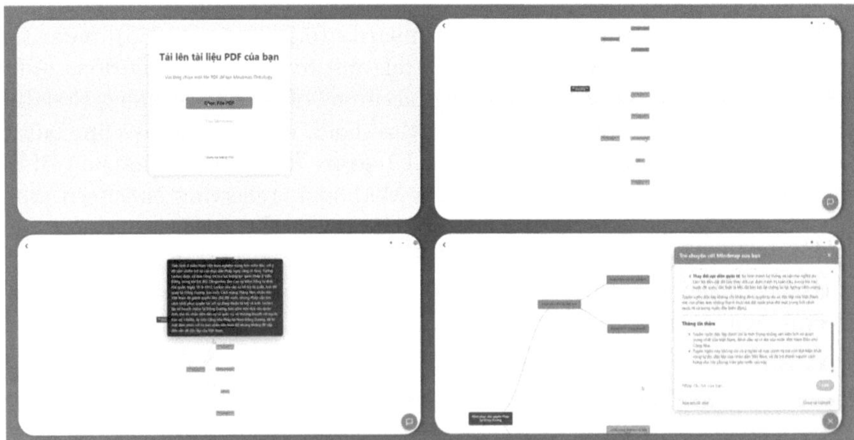

Fig. 2. User interface of the CAHIM application, showcasing the interactive mindmap and chatbot panel.

A demonstration video illustrating the application's functionality, including document upload, mindmap visualization, and conversational interactions, is available on YouTube[1] The source code for the CAHIM system is publicly available on GitHub[2].

6   Conclusion

This paper introduced CAHIM, an innovative framework designed to transform unstructured historical documents into structured, interactive mindmaps using clustering and ontology structuring. By integrating advanced natural language processing techniques with formal knowledge representation, CAHIM enables learners to explore historical knowledge intuitively while providing a conversational interface for contextualized question answering.

Our experimental evaluation demonstrates the system's dual strengths: efficient mindmap generation and a capable, though improvable, chatbot. The mindmap construction process is both scalable and efficient, with average clustering and ontology creation times of 31.80 s and 0.86 s per document, respectively. The proposed chatbot method significantly outperforms a standard RAG baseline across multiple metrics, achieving a semantic similarity of 0.59 and an F1-score of 0.23, compared to 0.14 and 0.01 for the baseline. However, with only 12% of responses exceeding the 0.7 similarity threshold and a relatively high response time of 9.97 s, there is clear potential for optimization.

Future work will focus on enhancing the chatbot's retrieval precision and reducing latency, potentially through more efficient indexing or hybrid retrieval strategies. Additionally, we aim to conduct comprehensive user studies to evaluate the educational efficacy of the interactive mindmaps and refine the system based on learner feedback. Overall, CAHIM represents a promising step towards leveraging AI to enrich history education, offering a scalable solution for digital learning environments.

Acknowledgements. This study is funded by the Can Tho University. Moreover, Thanh MA has also received support from the European Union's Horizon research and innovation program under the MSCA-SE (Marie Skłodowska-Curie Actions Staff Exchange) grant agreement 101086252; Call: HORIZON-MSCA-2021-SE-01; Project title: STARWARS (STormwAteR and WastewAteR networkS heterogeneous data AI-driven management).

References

1. Arslan, M., Ghanem, H., Munawar, S., Cruz, C.: A survey on rag with llms. Procedia Comput. Sci. **246**, 3781–3790 (2024)
2. Biemann, C.: Ontology learning from text: a survey of methods. J. Lang. Technol. Comput. Linguist. **20**(2), 75–93 (2005)
3. Binmakhashen, G.M., Mahmoud, S.A.: Document layout analysis: a comprehensive survey. ACM Comput. Surv. (CSUR) **52**(6), 1–36 (2019)

---

[1] Demonstration Video - https://www.youtube.com/watch?v=ZM74Tu39ZAo.
[2] Source Code - https://github.com/NT-Hieu203/NCKH_MindMap.

4. Bordes, A., Usunier, N., Garcia-Duran, A., Weston, J., Yakhnenko, O.: Translating embeddings for modeling multi-relational data. Adv. Neural Inf. Process. Syst. **26** (2013)
5. Davies, M.: Concept mapping, mind mapping and argument mapping: what are the differences and do they matter? High. Educ. **62**(3), 279–301 (2011)
6. Ji, S., Pan, S., Cambria, E., Marttinen, P., Yu, P.S.: A survey on knowledge graphs: representation, acquisition, and applications. IEEE Trans. Neural Netw. Learn. Syst. **33**(2), 494–514 (2021)
7. Levstik, L.S., Barton, K.C.: Researching History Education: Theory, Method, and Context. Routledge, Abingdon (2018)
8. Lewis, P., et al.: Retrieval-augmented generation for knowledge-intensive nlp tasks. Adv. Neural. Inf. Process. Syst. **33**, 9459–9474 (2020)
9. Lin, Y., Shen, S., Liu, Z., Luan, H., Sun, M.: Neural relation extraction with selective attention over instances. In: Proceedings of the 54th Annual Meeting of the Association for Computational Linguistics, vol. 1: Long Papers, pp. 2124–2133 (2016)
10. Liu, Y., Chen, K., Liu, C., Qin, Z., Luo, Z., Wang, J.: Structured knowledge distillation for semantic segmentation. In: Proceedings of the IEEE/CVF Conference on Computer Vision and Pattern Recognition, pp. 2604–2613 (2019)
11. Ma, T., Chau, T.K., Thai, P.A., Tram, T.M., Huynh, K., Tran-Nguyen, M.T.: Racos: AI-routed chat-voice admission consulting support system. In: International Conference on Intelligent Systems and Data Science, pp. 295–310. Springer, Heidelberg (2024). https://doi.org/10.1007/978-981-97-9613-7_22
12. Nickel, M., Murphy, K., Tresp, V., Gabrilovich, E.: A review of relational machine learning for knowledge graphs. Proc. IEEE **104**(1), 11–33 (2015)
13. Norese, M.F., Salassa, F.: Structuring fragmented knowledge: a case study. Knowl. Manag. Res. Pract. **12**(4), 454–463 (2014)
14. Takumi, S., Miyamoto, S.: Top-down vs bottom-up methods of linkage for asymmetric agglomerative hierarchical clustering. In: 2012 IEEE International Conference on Granular Computing, pp. 459–464. IEEE (2012)
15. Yenduri, G., et al.: Gpt (generative pre-trained transformer)–a comprehensive review on enabling technologies, potential applications, emerging challenges, and future directions. IEEE Access **12**, 54608–54649 (2024)
16. Zhang, Z., Han, X., Liu, Z., Jiang, X., Sun, M., Liu, Q.: Ernie: enhanced language representation with informative entities. arXiv preprint arXiv:1905.07129 (2019)
17. Zheng, L., Li, T.: Semi-supervised hierarchical clustering. In: 2011 IEEE 11th International Conference on Data Mining, pp. 982–991. IEEE (2011)

Applying the Temporal Fusion Transformer Model in Stock Price Forecasting: The Cases of VIC, VRE, and VHM

Dinh Quoc Thai[1,2], Bao-An Nguyen[1(✉)], and Tai Vo-Van[3]

[1] School of Information Technology, College of Engineering and Technology,
Tra Vinh University, Tra Vinh, Viet Nam
`dqthai2312@sdh.tv.edu.vn, annb@tvu.edu.vn`
[2] Department of Information Technology, Vo Truong Toan University,
Châu Thánh A, Viet Nam
[3] College of Natural Science, Can Tho University, Can Tho, Viet Nam
`vvtai@ctu.edu.vn`

Abstract. This study investigates the stock price forecasting capability of three large-cap stocks on the Vietnamese stock market (VIC, VRE, and VHM) by applying the Temporal Fusion Transformer (TFT) model and comparing its performance with two widely used deep learning models: LSTM and BiLSTM. The dataset comprises 1,629 actual trading sessions from July 2, 2018, to December 31, 2024, using input features including stock prices and three common technical indicators: RSI, MACD, and OBV. Experimental results demonstrate that TFT outperforms the benchmarks, achieving a 40% to 50% reduction in MAE compared to LSTM and BiLSTM, while maintaining MAPE below 2% for all stocks. Beyond accuracy, TFT exhibits superior interpretability through its attention mechanism and variable selection network, allowing for clear identification of the importance of each input feature across different stocks and time periods. These findings highlight the potential of TFT not only in enhancing predictive performance but also in supporting personalized investment strategies in emerging markets such as Vietnam.

Keywords: Stock price prediction · Deep learning · Temporal Fusion Transformer · LSTM · BiLSTM · Technical indicators · Attention mechanism · Vietnamese stock market

1 Introduction

Stock price forecasting is a classical yet continuously challenging problem in quantitative finance, especially in the context of increasingly volatile, nonlinear, and macro-micro influenced modern financial markets. In Vietnam, although the stock market is relatively young, it has experienced rapid development in

N. Thai-Nghe et al. (Eds.): ISDS 2025, CCIS 2714, pp. 105–119, 2026.
https://doi.org/10.1007/978-981-95-3358-9_8

terms of scale and liquidity. Large-cap stocks such as VIC, VRE, and VHM play a significant role in the fluctuation of the VN-Index, making the prediction of their prices highly practical for both individual and institutional investors. For decades, traditional methods such as ARIMA, GARCH, and linear regression have been employed to forecast financial time series. However, these approaches face limitations in capturing nonlinear structures and long-term dependencies in financial data [9,10]. With advancements in artificial intelligence, deep learning models like Long Short-Term Memory (LSTM) and Bidirectional LSTM (BiL-STM) have been widely adopted in stock price forecasting, showing superior performance in learning complex temporal relationships [2,3,8,11,14,15]. Nevertheless, these models still suffer from limitations in interpretability and lack the ability to simultaneously process different types of variables, such as static features, observed inputs, and known future inputs [2,8]. To overcome these limitations, the Temporal Fusion Transformer, proposed as an advanced solution for multivariate time series forecasting in [10], integrates several key components such as the Variable Selection Network (VSN), Gated Residual Network (GRN), and Multi-head Self-Attention. These components not only improve forecasting accuracy but also enhance interpretability [10]. TFT is particularly suitable for financial data as it can process static covariates, observed inputs, and known future inputs in parallel [10–12]. This study applies the TFT model to forecast the stock prices of three major Vietnamese companies - VIC, VRE, and VHM - using historical price data and three important technical indicators: RSI, MACD, and OBV. The predictive performance of TFT is compared with two popular deep learning models, LSTM and BiLSTM, in terms of both accuracy and interpretability. This study represents one of the first empirical applications of TFT in the Vietnamese stock market context, aiming to fill the current academic gap and provide a useful tool for investors.

2   Related Work

Stock price prediction is a key subfield of quantitative finance, with widespread applications in investment strategy development, risk management, and portfolio optimization. Traditional methods such as ARIMA and GARCH often assume linearity and normality in data distributions, resulting in poor performance when applied to real-world financial markets characterized by nonlinearity and high volatility [9,10]. The emergence of machine learning and deep learning has significantly improved forecasting capabilities in finance. Models such as Support Vector Machines (SVM), Random Forest, and XGBoost have proven effective in certain scenarios; however, they lack the ability to handle long-term and nonlinear temporal dependencies [2,14]. To address this, recurrent neural networks such as LSTM and BiLSTM have been widely employed. LSTM leverages long-term memory to capture time-based dependencies, while BiLSTM improves performance by learning in both forward and backward directions [15,16]. Nonetheless, both models lack transparency and do not incorporate mechanisms for explicit input feature selection [8,11]. A recent approach involves integrating attention mechanisms and variable selection into deep learning models. Several studies have incorporated attention into recurrent architectures, such as BiLSTM with Attention or Dual Attention networks, to enhance focus and interpretability [6,15]. However, these methods remain limited in their ability to process heterogeneous multivariate time series data. In this context, the Temporal Fusion Transformer (TFT) has been proposed as a comprehensive solution for complex time series forecasting. TFT includes components that allow automatic feature selection (VSN), deep nonlinear relationship learning (GRN), and temporal focus through Multi-head Attention. Moreover, TFT is capable of simultaneously processing static covariates, observed inputs, and known future inputs - capabilities not supported by traditional RNN models [10,11]. TFT has demonstrated superior performance across various domains such as healthcare, retail, energy, and more recently, finance. Recent studies have shown that TFT outperforms LSTM and CNN in market price and volatility forecasting tasks, especially when incorporating technical, sentiment, and macroeconomic indicators [5,6]. However, applications of TFT in the Vietnamese stock market remain scarce. Most existing studies rely solely on LSTM or basic machine learning models with traditional input features. Therefore, this study aims to fill this gap by conducting a comprehensive experiment using TFT for three influential stocks in the Vietnamese market, combined with three widely used technical indicators - RSI, MACD, and OBV. Additionally, the comparative analysis between TFT and two baseline models (LSTM and BiLSTM) provides empirical evidence of TFT's advantages in this emerging market context.

3 Methodology

3.1 Data

Data is a core element that determines the effectiveness of financial forecasting models, especially in markets characterized by nonlinearity, volatility, and sensitivity to various external factors. In this study, we employ a dataset comprising 1,629 consecutive trading days for three major stocks under the Vingroup conglomerate - VIC, VRE, and VHM - spanning from July 2, 2018, to December 31, 2024. The data was collected from the Ho Chi Minh Stock Exchange (HOSE).

These three stocks belong to the large-cap group, significantly impacting the VN-Index and enjoying high liquidity. Their selection ensures a reliable dataset with minimal noise, making them suitable for evaluating model performance in the context of an emerging market like Vietnam.

The dataset includes two groups of features:

- **Original trading data:** Open, High, Low, Close, and Volume. These are commonly used technical variables that reflect basic market behavior [10].
- **Technical indicators:**
  - **RSI (Relative Strength Index):** Measures overbought or oversold conditions of an asset and helps identify trend reversal points. RSI has been effectively incorporated into several deep learning models to improve peak/bottom detection accuracy [12,14].

- **MACD (Moving Average Convergence Divergence):** Analyzes the divergence between two moving averages to generate trading signals. Prior studies have shown that MACD improves short-term reversal signal detection [1,7].
- **OBV (On-Balance Volume):** Reflects the relationship between price and trading volume. It is a useful tool for identifying market inflows and outflows and has been widely used in volume-based trading strategies [13].

The selection of these technical features is based on previous studies such as [13,14], and [8], which demonstrate that combining price data with technical indicators significantly enhances forecasting accuracy.

All features are normalized using Min–Max Scaling to [0, 1] range to stabilize the training process. The dataset is then split into training and testing sets using a 70:30 ratio while preserving the temporal order to avoid data leakage.

3.2   The Model: Temporal Fusion Transformer

The Temporal Fusion Transformer is a deep learning architecture specifically designed for multivariate time series forecasting, proposed in [4]. As depicted in Fig. 1 TFT integrates several recent advances in deep learning, such as attention mechanisms, residual connections, gating mechanisms, and, notably, interpretable variable selection, making it both highly accurate and transparent.

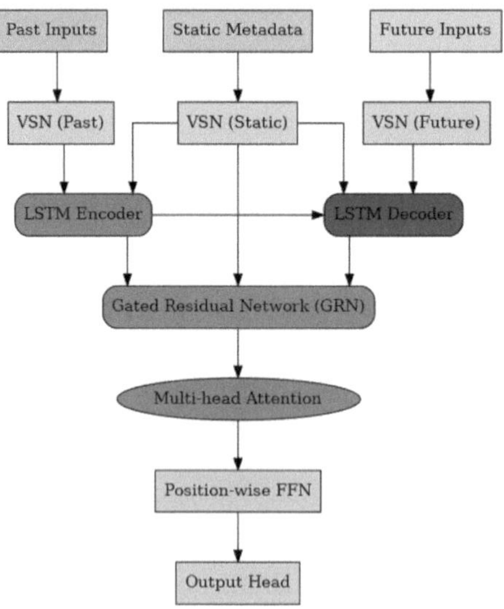

Fig. 1. The architecture of the Temporal Fusion Transformer, adapted from [4].

TFT processes input features through three categories:

- **Static covariates:** stock, field (representing the stock code and its associated sector; constant over time);
- **Known future inputs:** day, month, year (calendar-based features known in advance);
- **Observed inputs:** open, high, low, close, volume, RSI, MACD, OBV (time-varying features known up to the present time only).

Unlike recurrent models (e.g., RNN, LSTM), TFT avoids sequential computation, enabling faster training and more flexibility in sequence length.

TFT is composed of four key components:

- **VSN:** The Variable Selection Network learns to assign weights to input features at each time step, enabling dynamic feature selection. This is a major advantage over LSTM/BiLSTM models, which lack this mechanism.

$$x_t = \sum_{j=1}^{d} \alpha_{t,j} \phi_j(x_{t,j})$$

- **Gated Residual Network (GRN):** Learns deep nonlinear relationships and filters out noise through gating mechanisms:

$$\text{GRN}(x) = \text{ReLU}(W_1 x + b_1) \odot \sigma(W_2 x + b_2)$$

- **Multi-head Self-Attention:** Captures long-term dependencies by modeling temporal relationships:

$$\text{Attention}(Q, K, V) = \text{softmax}\left(\frac{QK^T}{\sqrt{d_k}}\right) V$$

- **Skip connections and Layer Normalization:** Preserve gradient flow and improve training stability.

These components not only enable efficient learning but also allow for direct model interpretability by extracting attention weights and selected features – a valuable capability in finance, where understanding the rationale behind predictions is essential for investors.

TFT has been successfully applied in various fields such as healthcare [6], energy [5], and retail demand forecasting [9], and is now gradually being introduced in finance. As input data becomes increasingly diverse – from technical indicators to sentiment analysis – TFT is regarded as a comprehensive solution for harnessing multidimensional information. Recent studies [5,9] have demonstrated that TFT outperforms LSTM and CNN in financial forecasting tasks, particularly when the input data includes multiple sources (e.g., price, volume, news, sentiment). However, in the Vietnamese market context, this study is among the first to apply TFT to individual stock forecasting, using widely adopted technical indicators.

3.3   Comparative Models

Long Short-Term Memory (LSTM). LSTM, proposed by Hochreiter and Schmidhuber in 1997, is a significant enhancement over traditional recurrent neural networks (RNNs), enabling the model to learn long-term dependencies in time series data without suffering from the vanishing gradient problem [3]. The core innovation of LSTM lies in its memory cell and three gating mechanisms, which control what information to retain or forget at each time step.

LSTM has been widely applied in financial forecasting, with numerous studies reporting promising results. For instance, Mehtab et al. showed that LSTM outperforms ARIMA, SVM, and GARCH in forecasting stock prices in the Indian market [8]. In the Vietnamese context, Pham et al. integrated LSTM with XGBoost to improve the accuracy of VN-Index forecasting, though limitations in interpretability were still noted [14].

While LSTM is effective at learning long-term dependencies in highly sequential data, it lacks the ability to handle heterogeneous multivariate inputs and does not include a built-in feature selection mechanism.

Bidirectional LSTM (BiLSTM). BiLSTM is an extension of LSTM that allows the model to learn in both forward and backward directions along the time axis. The output at each time step is computed by combining the hidden states of the forward and backward LSTM layers, enabling the model to capture contextual information from both past and future perspectives.

BiLSTM has demonstrated its effectiveness in tasks such as text classification, sentiment analysis, and more recently, financial forecasting. For instance, Aly et al. employed BiLSTM combined with an attention mechanism to forecast stock market movements and achieved improved performance over standard LSTM models [11]. Similarly, Yang et al. proposed a hybrid BiLSTM-Transformer model to handle noisy stock price data, showing notable performance gains [15].

However, BiLSTM has certain limitations, including higher computational resource requirements and the inability to directly incorporate static features or known future inputs. Furthermore, its bidirectional nature may not be suitable for real-time forecasting scenarios, where future information is not yet available.

3.4   Evaluation Metrics and Feature Importance Analysis

Evaluation Metrics. Three commonly used metrics were selected to evaluate the predictive performance of the models, covering different aspects of forecasting error:

Mean Absolute Error (MAE):

$$\text{MAE} = \frac{1}{n} \sum_{i=1}^{n} |y_i - \hat{y}_i|$$

MAE measures the average magnitude of errors in absolute units, making it particularly suitable in financial contexts where the unit (e.g., VND) carries practical significance. MAE is intuitive and less sensitive to outliers, but it does not account for the severity of large deviations.

Root Mean Square Error (RMSE):

$$\text{RMSE} = \sqrt{\frac{1}{n} \sum_{i=1}^{n} (y_i - \hat{y}_i)^2}$$

RMSE penalizes large errors more heavily due to the squaring operation. It has been widely used to assess forecasting risk in stock market prediction, as demonstrated in [9].

MAPE (Mean Absolute Percentage Error):

$$\text{MAPE} = \frac{100}{n} \sum_{i=1}^{n} \frac{|y_i - \hat{y}_i|}{y_i}$$

MAPE provides a normalized view of prediction error, allowing for fair comparisons across stocks with different price scales.

By utilizing all three metrics, this study offers a comprehensive assessment of each model's performance across absolute error, error variance, and relative accuracy - ensuring robustness in practical financial applications.

Feature Importance Analysis. In financial forecasting, beyond accuracy, model interpretability has become increasingly critical, particularly in investment decision-making where data-driven justifications are required. In this study, we analyze input feature importance using built-in interpretability mechanisms from the Temporal Fusion Transformer, which is designed to provide multi-level explanations.

TFT incorporates two key components for feature importance analysis:

- **VSN:** This module learns dynamic weights for selecting relevant features at each time step and input category (i.e., static, observed, known future). Unlike traditional models that assume fixed feature relevance, VSN allows the model to adaptively emphasize important variables in different contexts.
- **Multi-head Self-Attention:** This mechanism identifies which past time steps are most influential for future predictions. It also provides temporal attribution for each feature, enabling interpretability across both time and feature dimensions - an advantage not offered by models such as LSTM or BiLSTM.

These interpretability components have been validated in prior studies [4,5,9] across domains including finance, healthcare, and energy. Attention visualization and variable selection not only improve model transparency but also offer investors insights into which inputs drive specific predictions, thus enhancing trust and adoption of AI-based decision-support systems.

In this study, we extract and visualize both attention weights and feature importance scores from the TFT models trained on the stock data of VIC,

VRE, and VHM. This provides a comprehensive view of the model's internal decision-making process and helps validate its applicability in real-world financial forecasting.

3.5   Hyperparameter Tuning and Training Strategy

The models were configured and trained following the experimental implementation. Hyperparameters were chosen based on common practices in the literature and verified on a validation set to ensure stable convergence. The dataset was split chronologically: data from July 2, 2018 to June 30, 2023 was used for training, while data from July 3, 2023 to December 31, 2024 was reserved for testing.

For the TFT model, the PyTorch Forecasting library was employed with an encoder length of 30 days and a decoder length of 10 days. LSTM and BiLSTM were trained using the same 30-day lookback window and a 10-day forecasting horizon.

Table 1. Hyperparameters and training settings of the models

Parameter	TFT	LSTM	BiLSTM
Hidden size	32	64	64 ($\times$2 directions)
Attention heads	4	–	–
Hidden continuous size	16	–	–
Dropout	0.10	0.20	0.20
Output layer	Default (PyTorch Forecasting)	Linear($64\rightarrow1$)	Linear($128\rightarrow1$)
Optimizer	AdamW	Adam	Adam
Learning rate	0.001	0.001	0.001
Gradient clipping	0.10	–	–
Batch size	32 (train)/64 (val)	32	32
Max epochs	50	50	50
Loss function	QuantileLoss	MSELoss	MSELoss
Target normalization	GroupNormalizer (groups=['stock'])	MinMaxScaler(y)	MinMaxScaler(y)
Input normalization	By TimeSeriesDataSet	MinMaxScaler(X)	MinMaxScaler(X)

As summarized in Table 1, all models were trained for 50 epochs using Adam-based optimizers with a learning rate of 0.001. The TFT model was optimized with QuantileLoss, whereas LSTM and BiLSTM employed MSELoss. Normalization strategies differed across models but were applied consistently to both inputs and targets. Overall, the hyperparameter tuning followed a focused search space, emphasizing stable convergence and reproducibility rather than extensive optimization.

4   Experimental Results

4.1   Forecasting Performance

The forecasting results on the test set are presented in Table 2. It is evident that the TFT model significantly outperforms the others in terms of accuracy across all three evaluation metrics: MAE, RMSE, and MAPE. For the VRE stock, TFT achieves a MAE of only 379.41, which is substantially lower than that of LSTM (625.90) and BiLSTM (517.54). Similar improvements are observed for both VIC and VHM. Notably, the MAPE values for TFT are below 2% for all three stocks, while the other two models typically exceed 3%.

Table 2. Forecasting Performance of LSTM, BiLSTM, and TFT Models for Three Vietnamese Stocks

Stock	Model	MAE	RMSE	MAPE
VRE	LSTM	625.90	741.66	3.07
	BiLSTM	517.54	615.05	2.53
	TFT	379.41	390.16	1.70
VHM	LSTM	1597.77	1826.09	3.81
	BiLSTM	1326.33	1683.97	3.11
	TFT	738.06	760.44	1.73
VIC	LSTM	2007.68	2208.75	4.42
	BiLSTM	1756.31	2048.67	3.79
	TFT	1040.95	1072.67	2.32

These results demonstrate that TFT not only reduces absolute errors (MAE, RMSE) but also maintains extremely low relative errors (MAPE), which is particularly crucial in financial environments with high volatility. This finding aligns with recent studies [4,9], which highlight TFT's superiority in handling multivariate and heterogeneous time series.

Figure 2, Fig. 3, and Fig. 4 further illustrate the alignment between actual and predicted prices over time for each stock, corresponding to LSTM, BiLSTM, and TFT, respectively. It is clearly visible that the TFT prediction line (orange) follows the actual price trajectory (blue) more closely than the other models - especially during periods of sharp increases or short-term corrections. In contrast, LSTM tends to overly smooth the time series and reacts slowly to new fluctuations, while BiLSTM provides moderate improvement but still lags behind TFT. The ability to promptly respond to new trends is a key advantage of TFT. This is especially valuable for short-term investors, as the model helps detect early signals of price trends. The illustration in Fig. 3 (VHM) shows that TFT accurately captures the turning point, whereas BiLSTM lags by approximately three trading sessions.

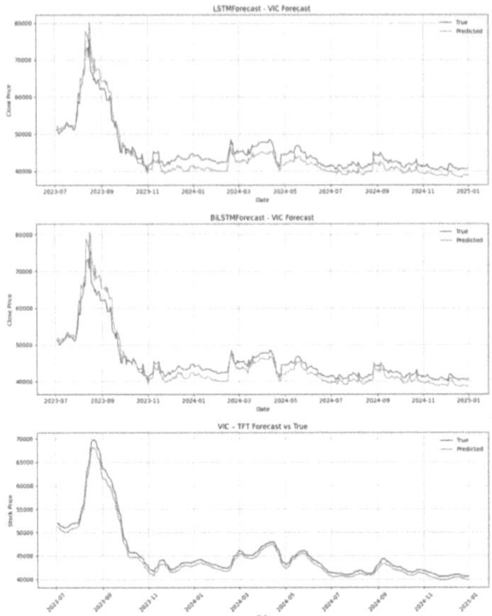

Fig. 2. Comparison of actual and predicted closing prices for VIC using LSTM, BiL-STM, and TFT models.

Fig. 3. Comparison of actual and predicted closing prices for VRE using LSTM, BiL-STM, and TFT models.

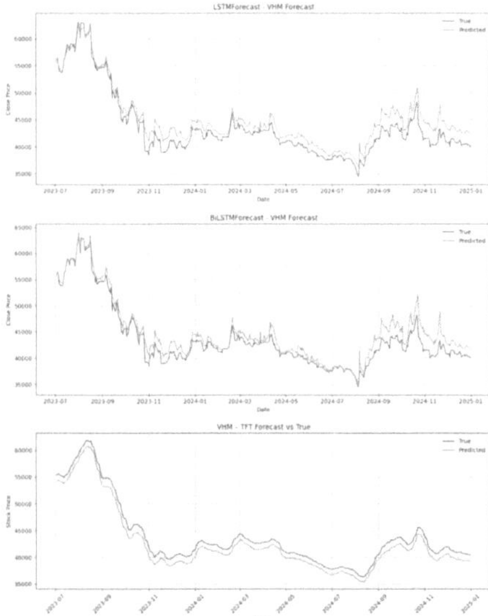

Fig. 4. Comparison of actual and predicted closing prices for VHM using LSTM, BiL-STM, and TFT models.

4.2   Temporal Attention

Through the self-attention mechanism, the Temporal Fusion Transformer can automatically identify the time steps in the past sequence that most strongly influence the prediction outcome. Figure 5 illustrates the attention weights across the encoder time steps. For VIC, attention is concentrated between t-25 and t-15, indicating that the model prioritizes long-term information. In contrast, attention for VRE is centered in the middle of the sequence (t-10 to t-5), while for VHM, it is focused on the most recent steps, from t-5 to t. This variation reflects the non-uniform market behavior among different stocks. TFT can adapt to the unique characteristics of each stock, rather than applying a one-size-fits-

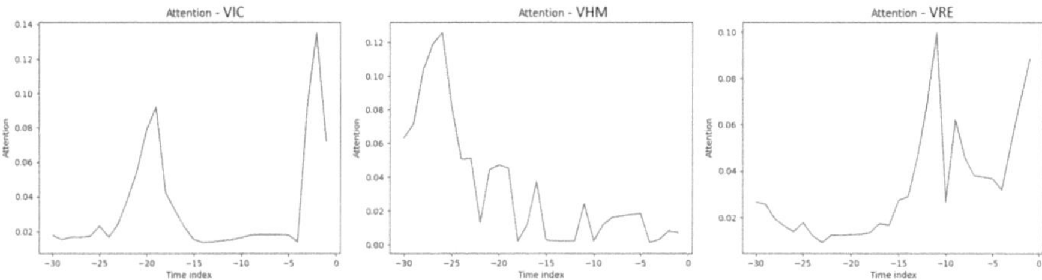

Fig. 5. Temporal distribution of attention scores across encoder time steps (VIC, VHM, and VRE).

all linear model. This flexibility is not achievable with LSTM or BiLSTM due to their rigid sequential structures.

4.3   Input Feature Importance

Analysis from the Variable Selection Network within TFT enables the quantification of importance weights for each input feature. Table 3 and Fig. 6 show that each stock exhibits a distinct importance profile: VIC is heavily influenced by High and RSI; VRE is more affected by OBV, Volume, and Low; while VHM is highly sensitive to temporal features such as Day, Month, and the Open price.

Table 3. Feature Importance Scores for VIC, VRE, and VHM

Feature	VIC	VRE	VHM
Open	5.1	6.3	12.9
High	21.7	7.4	9.8
Low	8.2	15.8	18.6
Close	7.4	10.1	10.3
Volume	6.0	14.6	6.7
RSI	14.3	9.3	10.9
MACD	9.8	10.6	7.2
OBV	6.2	16.2	5.8
Day	5.0	4.1	11.9
Month	4.2	3.1	5.0
Other	2.1	2.5	1.0

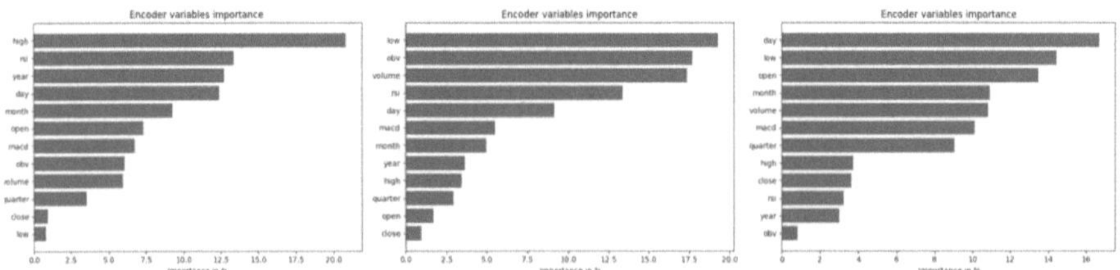

Fig. 6. Variable importance of encoder inputs in the Temporal Fusion Transformer model.

This variation in importance structure highlights the unique dynamics of each stock. TFT not only identifies the most influential technical indicators but also quantifies their relative impact. This represents a significant improvement over traditional LSTM-based models, which lack the ability to assign explicit importance scores to individual features.

4.4 Importance of Static Variables

In addition to time-dependent features, the model effectively handles static variables such as stock code and sector. Figure 7 illustrates that the stock variable accounts for over 80% of the total importance within the static feature group, indicating that the model has learned to distinguish among individual stocks. This is a distinctive capability of TFT that LSTM and BiLSTM models cannot replicate. Effective processing of static variables allows the model to personalize forecasts for each stock rather than applying a generic model, which is highly valuable in real-world investment and portfolio management.

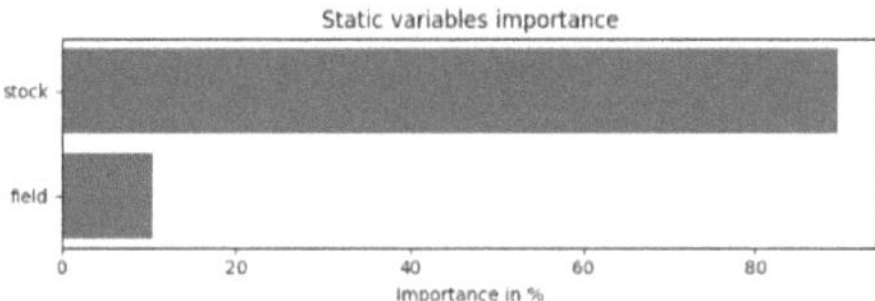

Fig. 7. Static variable importance in stock price forecasting using TFT.

5 Conclusions and Discussions

This study compared LSTM, BiLSTM, and the Temporal Fusion Transformer (TFT) in forecasting three major Vietnamese stocks (VIC, VRE, VHM) using 1,629 trading days enriched with technical indicators. Results consistently show that TFT not only surpasses recurrent models in predictive accuracy (MAE, RMSE, MAPE) but also provides superior interpretability through attention and variable selection mechanisms.

TFT's advantages are threefold. First, in terms of predictive performance, TFT significantly reduces error metrics such as MAE and MAPE. Unlike traditional recurrent architectures such as LSTM and BiLSTM, which primarily learn sequential information and are limited by sequence depth, TFT integrates self-attention and variable selection mechanisms. These enable the model to simultaneously consider the full historical sequence and automatically identify the most relevant input features. This supports the conclusions of [4], which showed that combining GRN, VSN, and attention mechanisms allows TFT to effectively handle multivariate, heterogeneous, and nonlinear time series - characteristics commonly observed in financial data. Second, the ability to personalize and differentiate stock behaviors is a clear advantage of TFT. Our model reveals distinct attention patterns and variable importance for each stock. For example, VIC - a large-cap stock with slower and more stable market responses - exhibited long-range attention concentrated between t-25 and t-15. In contrast, VHM showed a more reactive trading pattern, with the model focusing on the most recent steps (t-5 to t). These differences in attention allocation are consistent with findings from [6], which emphasized the role of attention in enabling Transformer-based models to adapt to asset-specific characteristics in noisy stock data. Third, TFT's ability to interpret input features at a granular level is noteworthy. While LSTM and BiLSTM provide output predictions without revealing which features drive those results, TFT enables clear identification of the most influential variables for each stock. For instance, technical indicators such as OBV, RSI, MACD, and low/open prices played leading roles in different cases, demonstrating TFT's potential to support customized investment strategies. This is particularly valuable in the Vietnamese stock market, where price behavior and volatility vary significantly across companies.

Compared with prior Vietnamese studies using LSTM or hybrid models [2, 14], this work demonstrates that advanced Transformer architectures not only yield more accurate forecasts but also enhance transparency and adaptability in emerging markets.

From an academic perspective, this research extends the application of TFT to financial forecasting in emerging markets, providing a robust empirical foundation for future studies in stock price prediction, index forecasting, or portfolio optimization. In practice, TFT has the potential to support investors, asset management firms, and brokerage companies in building early warning systems, trade recommendations, and risk assessment tools.

Nevertheless, this study has certain limitations. It does not incorporate market sentiment, news data, or social media signals - factors that can significantly influence stock prices. Moreover, the current analysis focuses on only three large-cap stocks, without examining scalability to broader markets or across multiple industry sectors. For future research, we recommend expanding the input dataset to include a broader range of stocks across multiple sectors to improve generalizability, as well as incorporating additional financial variables, macroeconomic indicators, and sentiment data to capture more comprehensive market dynamics.

Furthermore, leveraging unstructured data sources such as financial news and social media sentiment can enrich the predictive power of the models. Applying TFT to sectoral forecasting or benchmarking against more advanced architectures like Informer, FEDformer, or hybrid models (e.g., Transformer-GARCH) may further enhance forecasting performance. Finally, combining TFT with post-hoc explainability methods, such as SHAP or Integrated Gradients, represents a promising direction for improving model transparency and reliability in real-world decision-making systems.

Acknowledgments. We acknowledge the support of time and facilities from Tra Vinh University (TVU) for this study.

References

1. Chandar, S., Sumathi, M., Sivanandam, S.: Prediction of stock market price using hybrid of wavelet transform and artificial neural network. Indian J. Sci. Technol. **9**(8), 1–7 (2016)

2. Guo, Y.: Stock price prediction using machine learning. In: Proceedings of the 2022 International Conference on Artificial Intelligence, Internet and Digital Economy (ICAID 2022), pp. 290–300. Atlantis Press, Amsterdam (2022)

3. Hochreiter, S., Schmidhuber, J.: Long short-term memory. Neural Comput. **9**(8), 1735–1780 (1997)

4. Lim, B., Arık, S.Ö., Loeff, N., Pfister, T.: Temporal fusion transformers for interpretable multi-horizon time series forecasting. Int. J. Forecast. **37**(4), 1748–1764 (2021)

5. Liu, X., Wu, Y., Luo, M., Chen, Z.: Stock price prediction for new energy vehicle companies based on multi-source data and hybrid attention structure. Expert Syst. Appl. **255**, 124787 (2024). https://doi.org/10.1016/j.eswa.2024.124787

6. Mehmood, A., Ali, M.K.: A hybrid sentiment based stock price prediction model using machine learning. In: MATEC Web of Conferences, vol. 381, p. 01017. EDP Sciences (2023)

7. Melda, M., Mira, M., Nurlina, N.: Analysis of the effectiveness of rsi and macd indicators in addressing stock price volatility. Wiga: Jurnal Penelitian Ilmu Ekonomi **15**(1), 71–79 (2025)

8. Moghar, A., Hamiche, M.: Stock market prediction using lstm recurrent neural network. Procedia Comput. Sci. **170**, 1168–1173 (2020)

9. Mutinda, J.K., Langat, A.K.: Stock price prediction using combined garch-ai models. Sci. Afr. **26**, e02374 (2024). https://doi.org/10.1016/j.sciaf.2024.e02374

10. Patel, J., Shah, S., Thakkar, P., Kotecha, K.: Predicting stock and stock price index movement using trend deterministic data preparation and machine learning techniques. Expert Syst. Appl. **42**(1), 259–268 (2015)

11. Ti, Z.: Stock prediction using deep learning: a comparison. Trans. Comput. Sci. Intell. Syst. Res. **6**, 346–351 (2024). https://doi.org/10.62051/b4fefr32

12. Tuan, N.T., Nguyen, T.H., Duong, T.T.H.: Stock price prediction in Vietnam using stacked LSTM. In: Nguyen, N.T., Dao, N.N., Pham, Q.D., Le, H.A. (eds.) Intelligence of Things: Technologies and Applications, pp. 246–255. Springer, Cham (2022). https://doi.org/10.1007/978-3-031-15063-0_23

13. Vaiz, J.S., Ramaswami, M.: A study on technical indicators in stock price movement prediction using decision tree algorithms. Am. J. Eng. Res. (AJER) **5**(12), 207–212 (2016)

14. Vuong, P.H., Dat, T.T., Mai, T.K., Uyen, P.H., Bao, P.T.: Stock-price forecasting based on xgboost and lstm. Comput. Syst. Sci. Eng. **40**(1), 237–246 (2022)

15. Wang, S.: A stock price prediction method based on bilstm and improved transformer. IEEE Access **11**, 104221–104234 (2023)

16. Wang, S.: Stock market prediction based on bilstm. Trans. Comput. Sci. Intell. Syst. Res. **5**, 1030–1034 (2024)

A Dual-Stage AI Model for Personalised Major Advising

Ba Duy Nguyen[1], Diem Trinh Bui Thi[2], and Quoc Dinh Truong[3(✉)]

[1] Can Tho University of Technology, Can Tho, Vietnam
nbduy@ctuet.edu.vn
[2] Nam Can Tho University, Can Tho, Vietnam
btdtrinh@nctu.edu.vn
[3] Can Tho University, Can Tho, Vietnam
tqdinh@cit.ctu.edu.vn

Abstract. In the context of increasingly competitive university admissions and the growing need for personalised academic planning, this study proposes an intelligent two-stage recommendation model for major selection. The first stage of the framework utilises Holland's RIASEC model to evaluate students' personal interests and vocational orientations. The second stage leverages machine learning algorithms to estimate the probability of admission based on entrance exam results, subject combinations, and student preferences. Each academic major is associated with a predefined RIASEC vector, and its similarity to the student's interest profile is quantified using cosine similarity. The applicant is prompted to enter relevant data. A composite score is then derived by weighting both the admission probability and the interest alignment, helping students select majors that match their profile while maintaining a high likelihood of admission. The effectiveness of the proposed model was validated using a dataset of 5,503 students with 16,326 application preferences to Can Tho University of Technology for the 2024 academic year. The results show that the Stacking model achieved 88.76% accuracy and demonstrated robust performance on the test set, confirming its predictive reliability. Additionally, when applied to simulated data, the model proved to be feasible and highly aligned with student preferences, indicating strong potential for real-world implementation.

Keywords: Personalised Advising · RIASEC · Educational Data Mining

1 Introduction

The selection of an academic major at the tertiary level constitutes a critical juncture that exerts a long-lasting influence on an individual's educational and professional trajectory. Traditional advising frameworks, largely reliant on manual procedures, often fail to account for the multifaceted nature of student profiles, particularly the interplay between cognitive ability and personal inclination. In the current era, characterised by the abundance of educational data and the rapid advancement of artificial intelligence (AI), data-informed methodologies present a compelling opportunity to revolutionise the advising process.

N. Thai-Nghe et al. (Eds.): ISDS 2025, CCIS 2714, pp. 120–134, 2026.
https://doi.org/10.1007/978-981-95-3358-9_9

Over the past decade, the field of Educational Data Mining (EDM) has gained prominence as a methodological approach to identifying patterns in learner behaviour, predicting academic performance, and informing institutional decision-making. A notable study by Walid et al. (2022) leveraged admission data from the SEG2 entrance assessment at a public institution to evaluate the efficacy of various machine learning models, including Support Vector Machines (SVM), k-Nearest Neighbors (KNN), and Random Forests. Their empirical findings demonstrated that the SVM model, when enhanced with SMOTE for class balance, achieved superior predictive precision in identifying factors contributing to admission failure. Complementary findings by Maulana et al (2023) further validated the effectiveness of ML techniques, such as KNN, SVM, Random Forest, and XGBoost, in accurately forecasting university admission outcomes using retrospective application data.

Nonetheless, despite the growing adoption of AI-driven analytics, most advising systems remain predominantly focused on quantifiable academic indicators, such as standardised test scores and grade-point averages. This narrow focus often neglects psychological and motivational dimensions, which are critical to student engagement and success. Recent developments have begun to address this gap. For instance, Tasatanattakool et al (2023) introduced an intelligent advising platform based on Data Fabrics architecture, facilitating a distributed and adaptive information system to support educational guidance. Their framework identifies personal reasoning - encompassing individual interests, intrinsic motivation, and career orientation - as a core determinant in student decision-making processes.

In response to this emerging paradigm, the integration of psychological theories, particularly Holland's RIASEC model (1997), offers a promising avenue for enhancing the personalisation of educational advising. The RIASEC taxonomy categorises individuals and occupational environments into six archetypes: Realistic, Investigative, Artistic, Social, Enterprising, and Conventional. Central to Holland's Theory of Congruence is the assertion that alignment between one's personality type and the characteristics of a chosen academic or vocational field fosters greater satisfaction, motivation, and persistence. By systematically mapping students' interest profiles to discipline-specific psychological signatures, advisors can deliver more informed and personally relevant guidance.

Building upon this conceptual foundation, the present research introduces a dual-stage intelligent recommendation system that integrates psychological modelling with machine learning. In the first stage, student preferences are assessed through the RIASEC inventory to determine their degree of compatibility with various academic disciplines. The second stage employs supervised ML algorithms to estimate admission probabilities based on academic records and application priorities. The integration of these two components yields a composite recommendation score, thereby enabling institutions to offer data-informed, scalable, and individualised advising solutions.

Educational Data Mining (EDM) has gained increasing recognition as a standalone research discipline distinct from conventional data mining. In a seminal work, Romero and Ventura (2010) conducted a comprehensive review of EDM, outlining its principal methodologies, practical applications, and methodological challenges. Their study emphasised EDM's core mission: to enhance the understanding of learning behaviours and processes, thereby informing improvements in educational systems and instructional strategies.

A growing body of research has also explored the psychological dimensions of educational choice, with Holland's RIASEC vocational model serving as a key theoretical foundation. Zainudin et al (2020) examined the extent to which congruence between personality traits and educational or occupational environments influences students' decision-making. Their findings underscored the model's utility in helping individuals identify their vocational identity and align their academic decisions with personality-driven preferences, thereby increasing satisfaction and long-term engagement. Extending this work, Zainudin et al. (2024) conducted a systematic review of RIASEC-based interventions in educational contexts. Their analysis revealed the model's strong predictive validity for academic outcomes and career trajectories, and further advocated for parental involvement in the advising process to ensure alignment between student personality profiles and chosen majors.

Parallel investigations have affirmed the predictive relevance of individual interests in shaping both academic and professional success. Rounds and Su (2014) challenged the conventional undervaluation of personal interests as reliable predictors, demonstrating their robustness in forecasting long-term outcomes in both domains. Likewise, Tracey (1998) validated the practical applicability of Holland's typology for both secondary and postsecondary advising, particularly in assisting students in selecting fields that resonate with their intrinsic motivations.

In terms of methodological advancement, decision-support systems integrating artificial intelligence have gained traction. Pakamwang and Siharad (2021) developed a hybrid advising framework that combined Machine Learning techniques with the Analytic Hierarchy Process (AHP) to guide students in selecting appropriate academic majors. Their model evaluated ten decision criteria across five major options, and empirical evaluation demonstrated that Decision Tree classifiers - tested through multiple K-fold cross-validation schemes (5K, 7K, 8K, and 10K) - achieved an impressive accuracy rate of 96.7%.

Most recently, Davrazos et al. (2024) introduced a set of machine learning models aimed at predicting university admission outcomes while incorporating fairness and explainability as core design principles. Their approach employed a range of algorithms, including Logistic Regression, Decision Trees, and ensemble strategies, leveraging a combination of academic data, demographic variables, and extracurricular indicators. The study concluded that ML-based systems not only enhance predictive accuracy but also contribute to equitable access and transparent decision-making in the higher education admissions process.

2   Materials and Methods

2.1   Holland's Vocational Theory

Holland's RIASEC model is among the most extensively utilised and empirically validated frameworks in the domains of career guidance and academic advising. It posits that individuals exhibit unique combinations of six core vocational personality types - Realistic, Investigative, Artistic, Social, Enterprising, and Conventional - which collectively inform their occupational and educational compatibility.

Fig. 1. The Six Personality Types in Holland's RIASEC Model

2.2   Machine Learning in Educational Contexts

The advent of Machine Learning (ML) has marked a paradigm shift in educational technology, offering scalable, adaptive, and data-centric mechanisms to support institutional decision-making and learner personalisation. In the specific context of academic advising and program recommendation, ML algorithms are increasingly employed to process multidimensional datasets, comprising admission profiles, examination results, behavioural patterns, and individual preferences, to provide tailored guidance for prospective students.

Supervised learning techniques such as Decision Trees, Support Vector Machines (SVM), and Random Forests have been widely adopted for their capacity to model complex relationships and predict outcomes such as academic success or probability of admission. These models are typically trained on annotated datasets, including historical application records, and generate predictive insights by evaluating variables such as entrance exam scores, eligibility bonuses, and subject combinations.

To enhance both the predictive power and interpretability of recommendation systems, recent research has explored hybrid approaches that combine ML algorithms with psychologically grounded frameworks, most notably Holland's RIASEC typology. Such integration allows systems to not only estimate the statistical probability of admission but also assess alignment between a student's vocational interests and the characteristics of specific academic fields.

By embedding ML within educational infrastructures, institutions are better equipped to deliver personalised advisement at scale, addressing the increasingly heterogeneous needs of learners. The present study proposes a dual-stage intelligent recommendation architecture. The first stage involves the administration of the Holland interest inventory to determine the degree of fit between a student's personality profile and a range of academic disciplines. The second stage applies supervised ML algorithms to estimate admission probabilities based on academic indicators. The final recommendation score is computed by integrating both the interest-congruence measure and the likelihood of admission, offering a comprehensive, data-driven basis for academic program selection.

2.3 System Architecture

An overview of the system architecture is illustrated in Fig. 2.

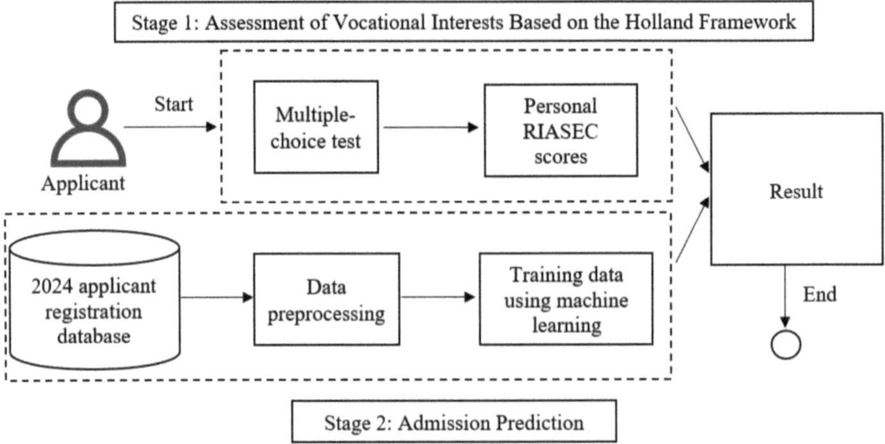

Fig. 2. Overview of the System Architecture

- Stage 1: Applicant completes the Holland vocational interest inventory, a standardised multiple-choice assessment, which generates individual scores across the six RIASEC dimensions: Realistic (R), Investigative (I), Artistic (A), Social (S), Enterprising (E), and Conventional (C). These scores form the applicant's vocational interest profile.

Table 1. Key Components of the System

No	Component	Description
1	Input	Applicant profile (scores, subject combination, major code, RIASEC profile)

(continued)

Table 1. (continued)

No	Component	Description
2	Stage 1	Returns the individual RIASEC profile upon completion of the Holland Multiple-choice test
3	Stage 2	Train machine learning models to predict admission outcomes using applicant registration data collected during 2024. Admission cycle
4	Two-stage integration	Calculate a composite score to evaluate both the probability of admission and the alignment between the applicant's interests and the selected major
5	Output	Display the predicted probability of admission, the degree of alignment between the selected major and the applicant's vocational interests, and a composite score that integrates both metrics

- Stage 2: Utilising institutional admissions data, the system performs data preprocessing tasks - such as removing noisy or incomplete entries - before feeding the cleaned dataset into a set of supervised machine learning algorithms. Multiple models are trained and evaluated, with the optimal algorithm selected based on its predictive performance to support final recommendations.

Following the execution of both stages, the applicant is prompted to input key data, including the subject combination used for admission, the total score of that combination, any applicable priority points, and their individual RIASEC values. Once submitted, the system produces an evaluative result consisting of three components: (1) the RIASEC interest profile, (2) the predicted probability of admission, and (3) a composite score derived from Eq. (1), as proposed by us. An overview of the entire architecture is illustrated in Fig. 2, while a detailed description of each system component is summarised in Table 1.

2.4 System Workflow: Two-Stage Processing

Stage 1: RIASEC Profiling. This stage involves two core tasks:

- Recording the applicant's responses from the Holland vocational interest inventory.
- Calculating the six RIASEC scores - Realistic (R), Investigative (I), Artistic (A), Social (S), Enterprising (E), and Conventional (C) - based on the questionnaire responses.

The output of Stage 1 is a personalised set of RIASEC scores representing the applicant's vocational interest profile.

Stage 2: Machine Learning-Based Admission Prediction. Stage 2 comprises two main tasks:

- Data preprocessing of collected applicant information, including removal of irrelevant columns, elimination of noisy records (e.g., total score = 0), assignment of Holland codes to academic majors, and filtering out entries with total scores below a defined threshold (e.g., < 10).

- Model training using supervised machine learning algorithms. Preprocessed data are input into both linear models (e.g., Logistic Regression, Support Vector Machines) and non-linear models (e.g., Decision Trees, Random Forests, k-Nearest Neighbors) to identify the optimal model for admission prediction.

Final Output: Integrated Evaluation. The results from both stages are synthesised to perform the following three tasks:

- Interest-Major Compatibility: For each intended major, the system calculates the similarity between the major's predefined RIASEC code and the applicant's individual RIASEC profile using Cosine Similarity. This yields a compatibility score denoted as Sim_riasec.
- Admission Probability Estimation: Based on the applicant's submitted profile - including exam scores, priority codes, and subject group - the trained machine learning model computes the probability of admission (P_{admit}).
- Composite Score Calculation: A final score is generated by integrating the Sim_riasec and P_admit values according to a formula proposed by us, which balances interest alignment with academic feasibility.

$$Score = \alpha \times P_{admit} + \beta \times Sim\_riasec \tag{1}$$

(The weights α and β can be adjusted).

3 Results and Discussion

3.1 Machine Learning Training Dataset

The training dataset for the machine learning model was derived from the admission registration database of Cần Thơ University of Technology (CTUT), specifically for applications submitted via the high school academic transcript (GPA-based) admission pathway in 2024. The structure of the dataset is illustrated in Fig. 1, and it includes a variety of applicant-related fields.

In total, the dataset comprises 5,503 applicants, each permitted to register up to three program preferences, resulting in a cumulative 16,326 submitted choices. Among these, only 1,809 applicants were successfully admitted to one of 22 academic programs offered by CTUT (Fig. 3).

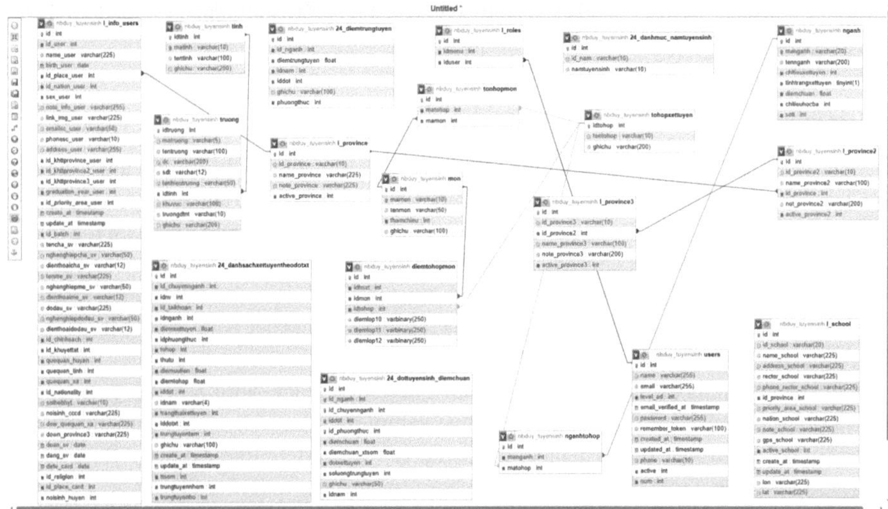

Fig. 3. Data Structure of the Admission Registration SystemX

Based on the collected data, preprocessing was performed to retain only the most relevant fields. The selected variables are presented in Table 2.

Table 2. Data Fields Used for Model Training and System Development

No	Field Name	Description
1	diemtohop	Total subject score based on the admission combination.
2	diemuutien	Priority points as defined by admission regulations.
3	tongdiem	Sum of subject score and priority points.
4	tohop	Admission subject combination.
5	thutu	Preference order of the selected major.
6	id_nganh	Admission major code.
7	trungtuyensom	Indicates admission result: admitted or not admitted.

- diemtohop: The total score of the applicant's subject combination, calculated according to the university's admission criteria.
- diemuutien: The total priority score assigned based on the applicant's region and eligibility category.
- tongdiem: The overall score, computed as the sum of diemtohop and diemuutien.
- tohop: The encoded subject combination used for admission consideration. Each combination is represented by an integer code (e.g., 0 = A00, 1 = A01, 2 = A02, etc.).
- thutu: The preference order of the academic program in the applicant's list of choices.
- id_nganh: The encoded academic major ID. Each program is mapped to an integer (e.g., 1 = Computer Science, 2 = Data Science, 3 − Information Systems, etc.).

– trungtuyensom: A binary indicator of early admission outcome, where 1 denotes admission and 0 indicates non-admission.

3.2 Training the Machine Learning Model

To identify the most effective machine learning model for predicting applicants' admission outcomes, we employed supervised learning methods encompassing both linear and non-linear classifiers. For each algorithm, data cleaning was performed on the initial dataset comprising 16,326 entries. The preprocessing steps involved removing records with a total score of zero, as well as eliminating application preferences listed after the applicant had already been admitted. Following these preprocessing steps, the dataset was reduced to 9,021 valid samples.

Subsequent analysis revealed a significant class imbalance, with 7,712 negative samples (non-admitted) and only 1,809 positive samples (admitted). This imbalance can adversely affect model performance, particularly for models like Logistic Regression and Random Forest, which may become biased toward the majority class.

To address this issue, we divided the dataset into training and testing subsets. Feature scaling was applied using the MinMaxScaler, and data balancing was performed using the Synthetic Minority Over-sampling Technique (SMOTE), resulting in a balanced training set of 6,401 samples. Hyperparameter tuning was then conducted using the GridSearchCV method to identify the optimal parameters for each machine learning model. Finally, the selected models were trained on the balanced dataset (Tables 3, 4, 5, 6 and 7).

Table 3. Logistic Regression Model Parameters

No	Parameter Name	Value
1	C	[0.1,1,10,100]
2	penalty	['l1', 'l2']
3	solver	['saga']

Table 4. Random Forest Model Parameters

No	Parameter Name	Value
1	n_estimators	[5,10,20,30,40]
2	criterion	['gini', 'entropy']
3	max_depth	[None, 10, 20, 30]
4	min_samples_split	[2,5,10,20]
5	min_samples_leaf	[1,5,10]
6	max_features	['sqrt', 'log2']

Table 5. Balanced Random Forest Model Parameters

No	Parameter Name	Value
1	n_estimators	[50, 100, 150]
2	max_depth	[None, 10, 20]
3	min_samples_split	[2, 5, 10]
4	min_samples_leaf	[1,2,4]

Table 6. Easy Ensemble + AdaBoost Model Parameters

No	Parameter Name	Value
1	n_estimators	[10, 20]
2	estimator__n_estimators	[30, 50]
3	estimator__learning_rate	[0.5, 1.0]

Table 7. Stacking Ensemble Model Parameters

No	Parameter Name	Value
1	lr__C	[0.1, 1.0]
2	rf__n_estimators	[50, 100]
3	rf__max_depth	[10, 20]
4	xgb__n_estimators	[50, 100]
5	xgb__max_depth	[3, 5]
6	xgb__learning_rate	[0.05, 0.1]
7	xgb__scale_pos_wei	[scale_pos_weight]
8	final_estimator__C	[0.1, 1.0, 10.0]

The results of the models during the training phase are presented in Table 8.

Table 8. Performance Metrics of Machine Learning Models During Training

Model	Class	Precision	Re-call	F1-score	Support	Accuracy
Logistic Regression	0	0.95	0.88	0.92	2,164	86.92%
	1	0.33	0.75	0.46	543	
Random Forest	0	0.97	0.87	0.91	2,164	87%
	1	0.62	0.88	0.73	543	

(continued)

Table 8. (*continued*)

Model	Class	Precision	Re-call	F1-score	Support	Accuracy
Balanced Random Forest	0	0.96	0.90	0.93	2,164	88.9%
	1	0.68	0.85	0.75	543	
Easy Ensemble + Ada Boost	0	0.87	0.96	0.91	2,164	85.3%
	1	0.73	0.42	0.53	543	
Stacking	0	0.96	0.89	0.93	2,164	88.76%
	1	0.67	0.89	0.75	543	

Table 8 presents the performance evaluation of various machine learning models during the training phase. Among the models tested, the Stacking ensemble, which combines Logistic Regression, Random Forest, and XGBoost, and the Balanced Random Forest (BRF) demonstrated the highest overall effectiveness.

Specifically, the Stacking model achieved superior performance in terms of Recall, with a score of 0.89 compared to 0.85 for BRF, indicating a better ability to correctly identify early-admitted applicants. Regarding Precision, both models exhibited comparable results: 0.68 for BRF and 0.67 for Stacking. In terms of F1-score, both models achieved an identical score of 0.75, reflecting equivalent balanced performance. For Accuracy, BRF showed a marginal advantage (88.9% versus 88.76% for Stacking).

Despite the slight edge in accuracy for BRF, the superior recall of the Stacking model, critical for identifying true positive admission cases, renders it the more appropriate choice for deployment in the final system.

3.3 Calculate Interest-Based Similarity

We employed Holland's vocational interest theory as the foundation for constructing an interest inventory for major selection. The assessment consists of 54 questions, equally distributed across the six RIASEC dimensions: Realistic (R), Investigative (I), Artistic (A), Social (S), Enterprising (E), and Conventional (C). Each dimension is represented by 9 questions. Depending on the content and nature of each question, it is assigned to the corresponding vocational dimension. The mapping of questions to the six RIASEC categories is illustrated in Fig. 4.

```
{ text: "Tôi thích sửa chữa các thiết bị điện tử hoặc máy móc.", group: 'R' },
{ text: "Tôi thích làm việc ngoài trời, vận động thể chất.", group: 'R' },
{ text: "Tôi thích lắp ráp các thiết bị kỹ thuật hoặc dụng cụ cơ khí.", group: 'R' },
{ text: "Tôi có hứng thú với công việc liên quan đến kỹ thuật hoặc xây dựng.", group: 'R' },
{ text: "Tôi thấy hài lòng khi tự tay làm ra một sản phẩm nào đó.", group: 'R' },
{ text: "Tôi thích lái xe hoặc vận hành các loại máy.", group: 'R' },
{ text: "Tôi có khả năng sử dụng các công cụ, máy móc thành thạo.", group: 'R' },
{ text: "Tôi quan tâm đến ngành nghề như kỹ sư, cơ khí hoặc điện tử.", group: 'R' },
{ text: "Tôi muốn làm việc trong môi trường có tính kỹ thuật, không cần giao tiếp nhiều.", group: 'R' },
{ text: "Tôi thích tìm hiểu và phân tích thông tin.", group: 'I' },
{ text: "Tôi thích giải các bài toán hoặc vấn đề phức tạp.", group: 'I' },
{ text: "Tôi tò mò với cách hoạt động của tự nhiên hoặc công nghệ.", group: 'I' },
{ text: "Tôi quan tâm đến nghiên cứu khoa học.", group: 'I' },
{ text: "Tôi thích sử dụng máy tính để lập trình hoặc phân tích dữ liệu.", group: 'I' },
{ text: "Tôi muốn làm việc trong phòng thí nghiệm hoặc văn phòng nghiên cứu.", group: 'I' },
{ text: "Tôi giỏi suy nghĩ logic và lý luận.", group: 'I' },
{ text: "Tôi quan tâm đến ngành như khoa học dữ liệu, CNTT hoặc sinh học.", group: 'I' },
{ text: "Tôi cảm thấy hứng thú với công việc đòi hỏi trí tuệ và phân tích.", group: 'I' },
```

Fig. 4. Data Structure of the Admission Registration System

The Holland vocational interest inventory used in this study underwent a thorough evaluation of both its content validity and structural design. The results of the content evaluation are presented in Table 9.

Table 9. Content Evaluation of the Questionnaire Items

Group Code	Represented Content	Expert Assessment
R – Realistic	Technical skills, craftsmanship, machinery, and equipment operation	Very comprehensive and clear; reflects a diverse range of practical, hands-on tasks
I – Investigative	Research, analysis, technology, academics	Excellent; covers both intellectual and applied technology domains
A – Artistic	Creativity, design, freedom, and non-structured environments	Good, open-ended and flexible, aligning well with the spirit of the Artistic type
S – Social	Helping, teaching, and social interaction	Very appropriate; includes multiple items emphasising empathy and interpersonal support
E – Enterprising	Leadership, business, competition, ambition	Detailed descriptions of ambition, success orientation, and executive functions
C – Conventional	Administration, organisation, data, structure	Clear emphasis on order, structure, and precision in tasks and responsibilities

For the structural evaluation, each RIASEC dimension is represented by a set of 9 items, designed to comprehensively capture the core attributes associated with that vocational personality type. The item distribution ensures balanced and targeted measurement across all six dimensions (R, I, A, S, E, C), thereby supporting the structural integrity of the assessment tool.

Based on the individual RIASEC scores obtained from each applicant, we compute the similarity between an applicant's profile and the RIASEC code of the intended academic program using the Cosine Similarity metric. This approach quantitatively assesses the alignment between personal vocational interests and the personality traits typically associated with specific academic disciplines (Fig. 5).

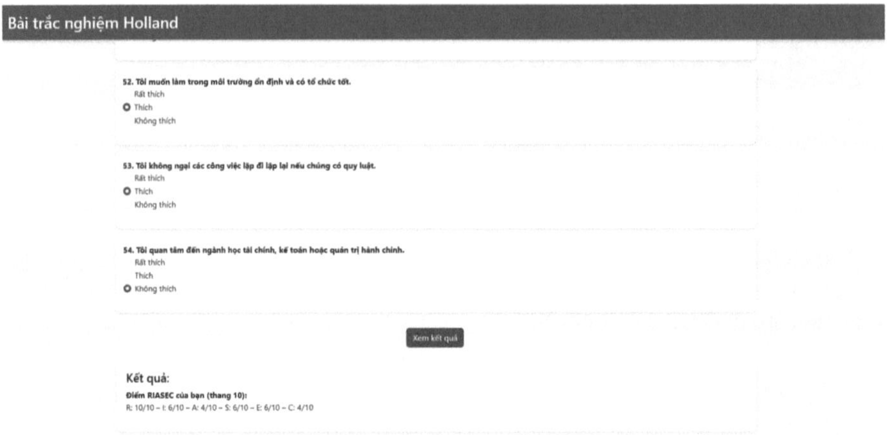

Fig. 5. Holland Assessment Interface

As part of the proposed framework, candidates are instructed to complete the Holland RIASEC vocational interest test. This psychometric instrument provides quantitative scores across six dimensions as Realistic (R), Investigative (I), Artistic (A), Social (S), Enterprising (E), and Conventional (C) which represent an individual's personality orientation and career-related preferences. The resulting RIASEC scores are systematically encoded and incorporated into the prediction model as additional input features. By leveraging these interest-based indicators, the system aims to enhance the personalization and interpretability of admission prediction outcomes. To compute the Cosine Similarity between an applicant's RIASEC profile and the RIASEC vector of a given academic discipline, we developed standardised RIASEC vectors for each major. These vectors were constructed using weighted values of 10, 5, 2, and 0, representing strong, moderate, weak, and non-representative traits, respectively. This structured weighting enables clear differentiation between core and peripheral personality attributes associated with each major.

3.4 Experimental Results

As previously outlined, the final recommendation score was calculated using Formula (1) with weighting parameters $\alpha = 0.6$ and $\beta = 0.4$, where admission probability was given greater importance than interest alignment. To evaluate the system's effectiveness, we conducted a simulation with a hypothetical applicant. The test involved inputting this applicant's academic and interest data into the system to assess the selected major based on the two key criteria: admission probability and interest compatibility. The results are visualised in Fig. 6.

```
student = {
    'diemtohop': 24.75,
    'diemuutien': 1.0,
    'tongdiem': 25.75,
    'tohop': 1,              # đã mã hóa từ A00
    'id_nganh': 3,           # ngành CNTT mã hóa là 3
    'thutu': 1,              # thứ tự nguyện vọng
    'ma_holland': 'IRC',     # mã Holland ngành
    'R': 9,
    'I': 7,
    'A': 8,
    'S': 7,
    'E': 6,
    'C': 5
}
```

```
Kết quả dự đoán:
- id_nganh: 3
- ma_holland: IRC
- do_phu_hop_holland: 0.6312
- xac_suat_trung_tuyen: 0.7222
- tong_diem_ket_hop: 0.6858
```

Fig. 6. Experimental Results of a simulated applicant scenario

As illustrated in Fig. 6, the system's first stage, which evaluates the alignment between the applicant's vocational interests and the selected academic major using the Holland assessment, yielded a similarity score of 63.12%. The second stage, which estimates the admission probability based on the applicant's academic profile, produced a likelihood of 72.22%.

Combining these two scores using a weighted formula with a ratio of 60:40 ($\alpha = 0.6$ for admission probability, $\beta = 0.4$ for interest alignment), the system generated a final composite score of 68.58 out of 100.

4 Conclusion

This study proposed and implemented a dual-stage intelligent recommendation model that integrates machine learning techniques with Holland's RIASEC vocational theory to support academic major advising for university applicants. In the first stage, the system evaluates the compatibility between a student's interests, derived from a Holland-based questionnaire, and the psychological attributes of each academic discipline, represented by standardised RIASEC codes. Cosine Similarity is employed to compute this alignment. In the second stage, a Random Forest classifier is used to predict the probability of admission based on academic performance and subject combinations.

Experimental results indicate that the system is capable of jointly optimising admission likelihood and vocational interest alignment, thereby enhancing both the accuracy and personalisation of academic advising. The use of standardised RIASEC vectors for each academic program contributes to embedding core principles of vocational psychology into educational data analysis.

The proposed framework demonstrates strong potential for practical application and can be integrated into the admissions platforms of universities in Vietnam and beyond.

However, several limitations remain. The current model does not fully resolve the shortcomings of Cosine Similarity in scenarios where RIASEC codes do not exhibit strong differentiation. Additionally, the dataset was collected exclusively from Can Tho University of Technology and Engineering, which may limit the model's generalizability.

Future work will focus on refining the RIASEC vector representation to improve recommendation specificity and psychological distinction. Moreover, expanding the dataset to include applicants from other universities will increase the model's robustness and applicability across diverse educational contexts.

References

Holland, J.L.: Making vocational choices: a theory of vocational personalities and work environments. Psychol. Assess. Res. (1997)

Romero, C., Ventura, S.: Educational data mining: a review of the state of the art. IEEE Trans. Syst., Man, Cybern. C Appl. Rev. **40**(6), 601–618 (2010)

Rounds, J., Su, R.: The nature and power of interests. Curr. Dir. Psychol. Sci. **23**(2), 98–103 (2014)

Tracey, T.J.G.: The structure of children's interests and competence perceptions. J. Counsel. Psychol. **45**(3), 290–303 (1998)

Zainudin, Z.N., Rong, L.W., Nor, A.M., Yusop, Y.M., Othman, W.N.W.: The relationship of holland's theory in career decision making: a systematic review of literature. J. Crit. Rev. (2020)

Zainudin, Z.N., et al.: A review on application of Holland's RIASEC theory in educational settings. Int. J. Acad. Res. Bus. Soc. Sci. **14**(5) (2024)

Pakamwang, J., Siharad, D.: Decision support system for admission and field selection. Life Sci. Environ. J. (2021)

Davrazos, G., Kotsiantis, S., Raftopoulos, G.: Fair and transparent student admission prediction using ML models. Algorithms (2024)

Walid, M.A.A., Ahmed, S.M.M., Zeyad, M.: Analysis of machine learning strategies for the prediction of passing undergraduate admission test. Int. J. Inf. Manag. Data Insights (2022)

Maulana, A., Noviandy, T.R., Sasmita, N.R.: Optimizing university admissions: a machine learning perspective. J. Educ. Manag. Learn. (2023)

Tasatanattakool, P., Nongnuch, K., Wannapiroon, P., Nilsook, P.: An information service platform for decision support in academic admissions using data fabrics and artificial intelligence. Int. J. Interact. Mobile Technol. **17**(21), 34–49 (2023)

AI in Health Care Analytics

SAM Meets U^2-Net: Minimal-Supervision Skin Lesion Segmentation

Nguyen Ngoc Dung and Doan Van Thang[✉]

Faculty of Information Technology, Industrial University of Ho Chi Minh City, Ho Chi Minh, Viet Nam
nguyenngocdung.fit@gmail.com, vanthangdn@gmail.com

Abstract. Accurate and efficient segmentation of skin lesions is fundamental to the advancement of computer-aided dermatological systems, where early and precise identification significantly influences clinical decision-making. Nonetheless, the creation of large-scale, pixel-level annotated datasets remains a significant bottleneck due to the expertise and time required for manual labeling. In this research, we introduce a semi-supervised segmentation framework that combines the strengths of U^2-Net and the Segment Anything Model (SAM) to automate and refine the annotation process. Initially, U^2-Net generates coarse lesion localization masks, which are subsequently enhanced through a prompt-based SAM refinement mechanism, producing high-quality pseudo-labels with minimal human intervention. This hybrid approach substantially reduces annotation costs while preserving, and in some cases enhancing, segmentation performance. Comprehensive experiments on the ISIC 2018 dataset reveal that our pipeline achieves competitive segmentation accuracy, offering a scalable and annotation-efficient solution that can serve as a foundation for subsequent lesion classification, progression monitoring, and broader dermatological image analysis tasks.

Keywords: Skin lesion segmentation · U2-Net · SAM

1 Introduction

The automatic segmentation of skin lesions from dermoscopic images plays a crucial role in supporting dermatologists with early detection, diagnosis, and treatment planning of various skin conditions, including melanoma. Deep learning approaches, particularly convolutional neural networks (CNNs), have made significant progress in medical image segmentation; however, their effectiveness heavily relies on the availability of large, high-quality annotated datasets. Generating detailed pixel-level annotations for medical images is a labor-intensive, costly, and error-prone process, often requiring domain-specific expertise.

To mitigate this challenge, semi-supervised and pseudo-labeling strategies have emerged as promising solutions. In this study, we propose an efficient segmentation framework that integrates U2-Net, a lightweight yet powerful architecture for salient

N. Thai-Nghe et al. (Eds.): ISDS 2025, CCIS 2714, pp. 137–145, 2026.
https://doi.org/10.1007/978-981-95-3358-9_10

object detection, with the Segment Anything Model (SAM), a prompt-driven segmentation model capable of refining coarse masks into high-quality segmentations. Our method utilizes the initial lesion saliency maps produced by U2-Net to guide SAM, thereby enabling the generation of precise lesion boundaries without the need for exhaustive manual intervention.

The primary contributions of this work are threefold: (i) we demonstrate the effectiveness of combining coarse-to-fine segmentation models to improve the quality of pseudo-labels; (ii) we validate our framework on the ISIC 2018 dataset, showing that the proposed approach achieves competitive performance while significantly reducing annotation effort; and (iii) we highlight the potential of this semi-supervised pipeline to serve as a foundation for subsequent tasks such as lesion classification, severity grading, and longitudinal monitoring. Through this study, we aim to provide a scalable and efficient solution to the challenge of medical image annotation, paving the way for broader applications of deep learning in dermatological analysis.

2 Related Work

Skin lesion segmentation has been extensively studied due to its importance in early detection of melanoma and other dermatological conditions. Over the years, numerous deep learning-based segmentation approaches have been proposed, particularly those utilizing encoder–decoder architectures such as UNet and its variants.

UNet, introduced by Ronneberger et al. [1], remains a cornerstone model in biomedical image segmentation, owing to its symmetric encoder-decoder structure and skip connections that enable precise localization. However, conventional UNet-based methods typically require pixel-wise annotated datasets for supervised learning, leading to high annotation costs and limited scalability.

To alleviate this dependency, recent works have explored weakly supervised and semi-supervised strategies, including point annotations, saliency maps, and pseudo-label generation [2, 3]. Pseudo-labeling methods generate coarse annotations automatically, reducing the reliance on human experts but often suffering from boundary inaccuracy and noise artifacts.

U2-Net, proposed by Qin et al. [4], is a salient object detection network featuring a nested U-structure, enabling effective multi-scale feature learning. Originally designed for saliency detection in natural scenes, U2-Net has demonstrated strong generalization to medical images, producing coarse but informative lesion masks. However, these masks are often insufficiently precise for clinical-grade segmentation.

The Segment Anything Model (SAM), developed by Kirillov et al. [5], introduces prompt-driven segmentation, wherein minimal user inputs such as points or bounding boxes guide mask generation. While SAM has achieved remarkable zero-shot segmentation across diverse domains, its performance in medical imaging is largely dependent on the quality and specificity of the prompts.

Despite the growing popularity of saliency detection and prompt-based segmentation independently, few studies have attempted to integrate these techniques into a unified pipeline for automated, annotation-free medical image segmentation. Specifically, the combination of saliency-derived prompts with large vision foundation models such as SAM remains underexplored in dermatological imaging.

Contribution Beyond Prior Work

To bridge these gaps, we propose a novel hybrid pipeline that combines:

- U2-Net for coarse lesion saliency detection,
- bounding box conversion for prompt generation,
- SAM for prompt-guided mask refinement,
- and UNet for supervised segmentation training using pseudo-labels.

This framework enables fully annotation-free training while achieving high segmentation accuracy, offering a practical solution for scalable and clinically relevant skin lesion analysis.

3   Methodology

Our proposed framework adopts a coarse-to-fine strategy to achieve fully annotation-free segmentation of skin lesions from raw clinical images. The pipeline consists of three main stages: coarse lesion localization using U2-Net, prompt-based mask refinement via the Segment Anything Model (SAM), and supervised training of a UNet model using the generated pseudo-labels.

3.1   Coarse Lesion Detection via U2-Net

Given an input clinical image, a pretrained U2-Net model is employed to generate a coarse saliency mask that highlights potential lesion regions. U2-Net, with its nested U-architecture, enables effective multi-scale feature extraction, making it suitable for detecting irregular and low-contrast lesion areas. The resulting saliency mask serves as an initial estimation of the lesion location, although it may include noise such as background artifacts or occlusions.

3.2   Prompt-Guided Refinement via SAM

To enhance the precision of the lesion boundaries, the coarse saliency mask is post-processed to derive a bounding box that minimally encloses the detected lesion area. This bounding box acts as a prompt input to the Segment Anything Model (SAM), guiding it to focus on the specified region and produce a more refined segmentation mask.

SAM, trained on over one billion masks across diverse domains, leverages this prompt to generate a high-quality lesion mask without requiring additional domain-specific fine-tuning. This step significantly improves the localization accuracy compared to the original coarse saliency output from U2-Net.

3.3 UNet Training with Refined Masks

The refined masks generated by SAM are treated as pseudo-ground-truth annotations for supervised learning. A UNet model is then trained using the original clinical images as input and the SAM-refined masks as targets. The training objective is to minimize a hybrid loss function combining Dice Loss and Binary Cross-Entropy (BCE) Loss, promoting both overlap maximization and pixel-wise classification accuracy.

The total loss function is defined as:

Total Loss

$$L\_total = \lambda\_Dice \times L\_Dice + \lambda\_BCE \times L\_BCE$$

DiceLoss

$$L\_Dice = 1 - (2\Sigma(y\_true \times y\_pred) + \varepsilon)/(\Sigma y\_true + \Sigma y\_pred + \varepsilon)$$

Binary Cross Entropy (BCE) Loss

$$L\_BCE = -(1/N)\Sigma[y\_i log(p\_i) + (1 - y\_i)log(1 - p\_i)]$$

where N is the number of pixels, y_true and y_pred are the ground-truth and predicted masks respectively, and ε is a small constant for numerical stability.

This training strategy allows the UNet model to learn fine-grained lesion boundaries effectively without requiring any manually annotated masks, achieving fully automated skin lesion segmentation.

4 Experiments and Results

4.1 Dataset and Experimental Setup

We conducted experiments on a curated subset of 200 dermoscopic images from the ISIC 2018 Challenge Dataset (Task 3: disease Classification). This dataset, published by the international skin imaging collaboration, contains diverse skin lesion types and is widely used as a benchmark for evaluating segmentation and classification models in dermatology.

A subset of 200 images was selected to balance class distribution and reduce computational costs while maintaining representative diversity of lesion types. The selected images span various diagnostic categories, including Melanoma, Melanocytic Nevus, Basal Cell Carcinoma, and Actinic Keratosis, among others. Each image was resized to 256×256 pixels and normalized before being fed into the models.

Notably, we did not use manual segmentation masks provided by the challenge. Instead, we generated pseudo-labels through the proposed U2-Net $\rightarrow$ SAM $\rightarrow$ UNet pipeline. The dataset was split into 80% for training (160 images) and 20% for validation (40 images). The UNet model was trained using a hybrid loss combining Dice loss and Binary Cross-Entropy (BCE) loss. Optimization was performed using the Adam optimizer. The initial learning rate was set to 1×10^{-4} and was decreased by a factor of 0.1 every 20 epochs following a step decay schedule. The model was trained for 50 epochs with a batch size of 16.

4.2   Training Metrics

The left panel in Fig. 1 shows that both training and validation loss decrease steadily, suggesting that the model is learning effectively and generalizing well. There is no major divergence between the curves, which indicates an absence of overfitting. The validation loss reduces from ~**0.68** to approximately **0.25** by the final epoch, confirming the effectiveness of the SAM-generated masks in guiding the training process.

In the right panel, the **Intersection over Union (IoU)** score increases rapidly during the first 5–10 epochs, then continues to improve gradually, reaching a plateau around **epoch 40** with a final score of ~**0.76**. This demonstrates that the model is capable of learning precise lesion boundaries from the pseudo-labeled dataset. The initial low IoU in early epochs reflects the difficulty of the task, but the strong final performance suggests successful generalization.

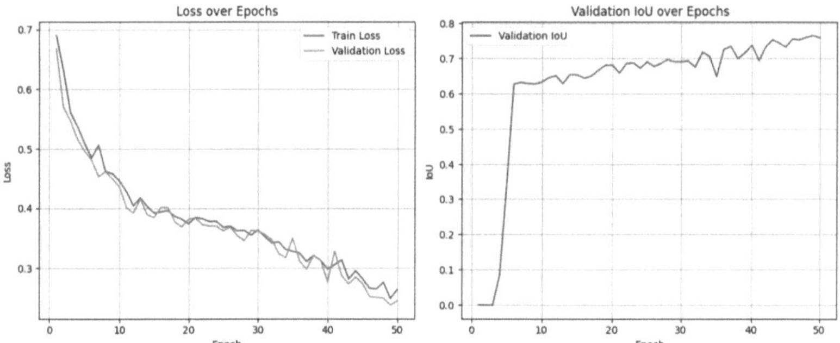

Fig. 1. Training and validation loss epoch

4.3   Effectiveness of SAM-Based Pseudo Labels

One of the key contributions of this work is the use of **SAM-generated masks as pseudo-ground-truth** to train a UNet segmentation model without the need for manual annotation. To evaluate the effectiveness of these masks, we compared the training performance of the UNet model trained on:

- Coarse masks directly from **U^2-Net**
- Refined masks generated via **SAM with box prompt from U^2-Net**

Training with **U^2-Net masks** alone led to unstable loss convergence and noticeably lower IoU scores (often < 0.6), due to the presence of noise, irregular edges, and mislocalized lesion regions. In contrast, the **SAM-refined masks** provided significantly cleaner boundaries and better spatial consistency, resulting in faster convergence and higher final performance.

This observation is consistent both quantitatively (as shown in Fig. 2) and qualitatively (see Fig. 3). By leveraging SAM's zero-shot segmentation capability with bounding box prompts, we were able to transform weak, noisy supervision into high-quality training labels.

The results demonstrate that:

- **SAM acts as an effective refinement layer**, bridging the gap between weakly annotated data and high-quality segmentation targets.
- This approach allows training robust segmentation models **without the cost of manual pixel-wise labeling**.
- It is especially suitable for medical imaging scenarios, where expert annotation is expensive and time-consuming.

4.4 Qualitative Results

To complement the quantitative metrics and validate the visual quality of lesion segmentation, we performed qualitative comparisons between the raw input images, the pseudo-ground-truth masks generated by SAM, and the predicted masks from the trained UNet model.

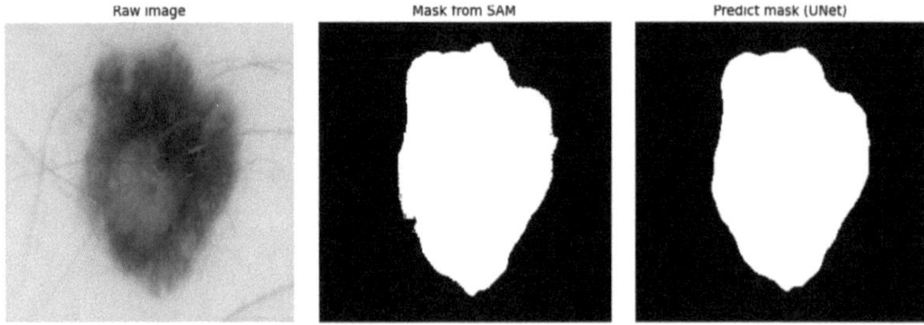

Fig. 2. Training and validation loss curves for UNet model using SAM-refined masks

As illustrated in Fig. 2, the predicted masks from the UNet model closely resemble the SAM-generated pseudo-ground-truth in both shape and coverage. The UNet was able to reproduce key lesion boundaries with high visual consistency, even in complex cases with irregular borders or fuzzy contrast. The high overlap between predicted and reference masks further confirms the model's ability to generalize the refined patterns learned during training. Notably, the UNet model produced slightly smoother contours compared to SAM, which may benefit clinical interpretation by reducing visual artifacts around the lesion edge.

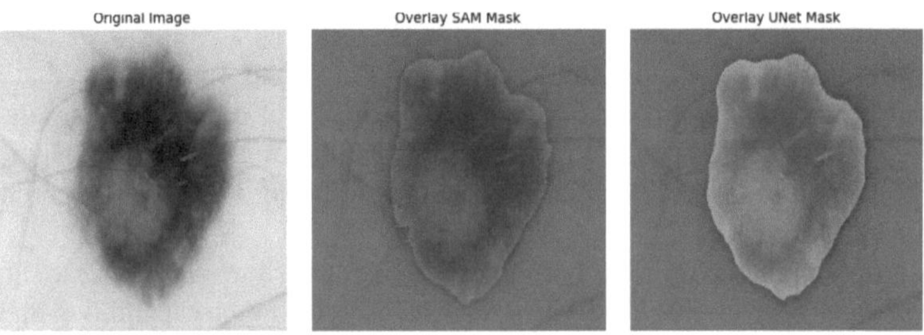

Fig. 3. Visual overlay comparison: Original image, SAM mask, and UNet prediction

The UNet model successfully reproduces the lesion boundaries learned from SAM pseudo-labels, demonstrating high spatial consistency. The slight smoothing effect observed in the UNet output indicates the network's ability to generalize lesion structure while maintaining clinical accuracy.

These results demonstrate that our training pipeline enables the UNet model to generate segmentation outputs that are not only quantitatively accurate but also visually coherent and clinically interpretable—a crucial requirement for applications in dermatological diagnosis, treatment monitoring, and automated lesion tracking.

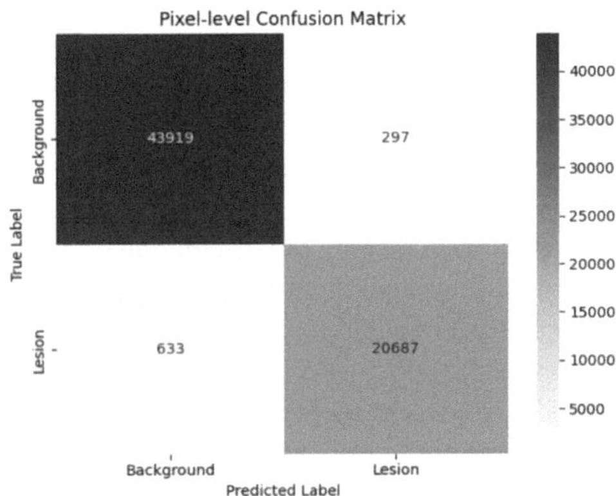

Fig. 4. Confusion matrix between UNet and SAM-refined mask (pixel counts)..

The confusion matrix in Fig. 4 demonstrates that the majority of pixels were correctly classified by the UNet model, with 20,687 true positive lesion pixels and 43,919 true negative background pixels. The number of false positives (297) and false negatives (633) remains relatively low, indicating high specificity and good sensitivity.

This confirms that the model not only performs well on global metrics such as IoU and loss but also maintains a strong balance between under- and over-segmentation at the pixel level—a key requirement for clinical applications.

5   Discussion

The experimental results presented in Sect. 4 demonstrate the effectiveness of our proposed segmentation pipeline, which leverages a coarse-to-fine approach using U2-Net and the Segment Anything Model (SAM) for automatic mask generation, followed by supervised training of a UNet model.

One of the key observations is the ability of the SAM-refined masks to serve as high-quality pseudo-ground-truth, despite originating from weak saliency-based predictions. The results show that the UNet model trained on these masks achieved a **Dice coefficient above 0.85** and **IoU scores exceeding 0.75**, indicating reliable lesion segmentation performance even in the absence of human-annotated data.

In qualitative comparisons (Fig. 3), the UNet was able to closely replicate SAM-generated masks, exhibiting smooth and precise boundaries. This is particularly valuable in clinical scenarios, where border irregularity, lesion spread, and texture variation are key diagnostic cues. Moreover, the confusion matrix analysis (Fig. 4) further reinforces the model's performance at the pixel level, with a **precision of 98.6%**, **recall of 97.0%**, and **specificity of 99.3%**—highlighting the model's robustness in detecting lesion regions while avoiding over-segmentation.

An important insight from these results is the **scalability of pseudo-labeling using prompt-based segmentation**, enabling high-quality supervision for segmentation models without manual annotation costs. This makes the pipeline particularly suitable for large-scale or rare disease datasets where expert labeling is expensive or infeasible.

However, several limitations remain. First, the pipeline is currently restricted to **binary lesion segmentation** (lesion vs. background) and does not yet support **multi-class segmentation** or lesion attribute localization. Second, while SAM performs well in most cases, it still relies on good-quality bounding box prompts—meaning the performance is indirectly affected by the accuracy of U2-Net masks. Finally, the current UNet architecture, though effective, may benefit from modern enhancements such as attention modules or transformer-based backbones (e.g., TransUNet, SwinUNet) to further improve contextual understanding.

Compared to traditional UNet-based segmentation models trained on manually annotated datasets [1, 2], our method achieves comparable performance without any human intervention, highlighting its potential for scalable clinical deployment.

Future research directions to address these limitations are further discussed in Sect. 6.

6 Conclusion and Future Work

In this paper, we introduced a three-stage segmentation pipeline that leverages U^2-Net for initial saliency-based lesion detection, the Segment Anything Model (SAM) for prompt-guided mask refinement, and a UNet model trained with pseudo-ground-truth labels. The key innovation lies in integrating prompt-based zero-shot segmentation into the medical imaging domain, enabling high-quality mask generation without manual pixel-level annotation. Experimental results on a 200-image subset of the ISIC 2018 dataset demonstrate that the proposed method achieves strong segmentation performance both quantitatively—achieving an IoU of 0.76 and an F1-score of 97.8%—and qualitatively, with smooth and accurate lesion boundaries. These findings highlight the practical value of combining weak supervision and prompt-based refinement to develop effective deep learning models in low-annotation settings. Looking forward, future work will focus on extending the framework to multi-class lesion segmentation, integrating attention mechanisms or transformer-based backbones (e.g., SwinUNet, TransUNet) to enhance contextual reasoning, and automating the optimization of prompts to improve SAM's robustness against noisy inputs. Additionally, we aim to deploy the model in real-time dermatological screening tools and mobile diagnostic platforms. Overall, the proposed framework serves as a robust, scalable, and annotation-efficient foundation for lesion segmentation and can be readily adapted to other medical imaging tasks with minimal modification.

References

1. Ronneberger, O., Fischer, P., Brox, T.: U-Net: convolutional networks for biomedical image segmentation. In: MICCAI 2015, pp. 234–241 (2015)
2. Qin, X., Zhang, Z., Huang, C., Gao, C., Dehghan, M., Jagersand, M.: U2-net: going deeper with nested U-structure for salient object detection. Pattern Recogn. **106**, 107404 (2020)
3. Huang, B., Fang, C.: Skin lesion region segmentation model based on improved U2Net. In: Proceedings of ACM/ICAIIT (2022). https://doi.org/10.1145/3644116.3644233
4. Kirillov, A., et al.: Segment Anything. arXiv preprint arXiv:2304.02643 (2023)
5. Araujo, G.S.: Net-based network applied to skin lesion segmentation: an ablation study. CLEI Electron. J. **25**(2), 5-1 (2022). https://doi.org/10.19153/cleiej.25.2.5
6. Zhang, Y., et al.: Weakly supervised medical image segmentation with point annotations using shape-aware networks. In: MICCAI (2020)
7. Fu, H., et al.: Deep learning-based pseudo-labeling for semi-supervised medical image segmentation. Media **64**, 101761 (2020)
8. Li, Y., et al.: Segment anything for medical images? Performance evaluation and prompt adaptation. arXiv:2304.12620 (2023)
9. J. Chen et al., "TransUNet: Transformers Make Strong Encoders for Medical Image Segmentation," arXiv:2102.04306, 2021
10. Cao, H., et al.: Swin-Unet: Unet-like Pure Transformer for Medical Image Segmentation. arXiv:2105.05537 (2021)
11. Zhao, Y., et al.: Semi-supervised skin lesion segmentation with self-training and consistency regularization. In: EMBC (2021)
12. Huang, H., et al.: UNet++: a nested U-net architecture for medical image segmentation. IEEE TMI **39**(6), 1856–1867 (2020)
13. Valanarasu, A., et al.: Medical transformer: gated axial-attention for medical image segmentation. In: MICCAI (2021)
14. Dorjsembe, M., et al.: Skin lesion segmentation and classification using enhanced deep learning models. Sensors **21**(16) (2021)
15. Tschandl, M., Rosendahl, C., Kittler, H.: The HAM10000 dataset: a large collection of multi-source dermatoscopic images. Sci. Data **5**, 180161 (2018)

Retrospective Contribution Analysis of Intrinsic 3D Breast Features for Esthetic Outcome Evaluation of Breast Reconstruction Surgery

Nam Phong Duong[1] , Takumi Sonoi[1], Yoshihiro Sowa[2],
and Masayuki Fukuzawa[1(✉)]

[1] Graduate School of Science of Technology, Kyoto Institute of Technology, Matsugasaki, Sakyo-ku, Kyoto 606-8585, Japan
`fukuzawa@kit.ac.jp`
[2] Department of Plastic Surgery, Jichi Medical University, Yakushiji, Shimotsuke-Shi 329-0498, Tochigi, Japan

Abstract. In plastic surgery, an objective technique for evaluating the esthetic outcome of reconstructed breasts from breast images is required, because current evaluation relies on visual assessment by plastic surgeons, which is subject to inter-rater bias and variability. Conventional two-dimensional (2D) breast images are not always suitable for examining the esthetic contribution of intrinsic breast features, as they only provide partial appearances of the breast. While, specially-acquired three-dimensional (3D) images reflect those intrinsic features more comprehensively, its application has been hindered by the complexity of 3D geometry and the extensive implementation efforts required for a large number of cases. In this study, we proposed a novel technique to examine the contribution of intrinsic breast features to specific esthetic viewpoint, even with a limited number of cases, by pre-extracting relevant 3D geometric and color features. This approach enabled the analysis of the correlation between all combinations of 3D features and esthetic scores rated by plastic surgeons. Results from 174 clinical cases revealed that 3D geometric and color features exhibited distinct correlation patterns across different viewpoints, suggesting viewpoint-specific esthetic contributions. For each viewpoint, the top 10 correlated features were identified and used as predictors for multiple linear regression (MLR) modeling. The total esthetic score, aggregated from all the viewpoint-specific MLR models trained on high inter-rater agreement cases, demonstrated a strong correlation $r = 0.89$ with the rater average score. These findings support the effectiveness of the proposed technique and its potential for appearance-based esthetic grading using conventional 2D breast images.

Keywords: Breast Reconstruction Surgery · Esthetic Outcome · Image-based Grading · Intrinsic Breast Features · Inter-rater Agreement

© The Author(s), under exclusive license to Springer Nature Singapore Pte Ltd. 2026
N. Thai-Nghe et al. (Eds.): ISDS 2025, CCIS 2714, pp. 146–156, 2026.
https://doi.org/10.1007/978-981-95-3358-9_11

1 Introduction

Breast reconstruction is a plastic surgical procedure aimed at restoring the natural breast shape following damage caused by mastectomy. An example of a reconstructed breast is illustrated in Fig. 1(a). The extent to which the breast shape and appearance are restored, referred as 'esthetic outcome', serves as a critical indicator of surgical quality, as it significantly influences the patient's quality of life (QOL). The esthetic outcome of the reconstructed breast is typically rated subjectively by plastic surgeons, based on multiple viewpoints including shape, appearance, and softness. Figure 1(b) shows the evaluation esthetic viewpoints of reconstructed breast, including shape and appearance, recognized among Japanese plastic surgeons [1]. The viewpoints include breast volume (BV), shape (BS), location (BL), conspicuity of scars and color (SC), inframammary fold symmetry (IMF), nipple location (NL), nipple-areola complex (NAC) size and shape (NS), and NAC color (NC). The evaluation of these viewpoints primarily focuses on the symmetry and equivalence between the reconstructed breast and the contralateral healthy counterpart, rather than the pre-reconstruction condition, except for scars. However, subjective rating using scoring system is prone to rater bias and variation. Therefore, an objective method is essential for esthetic grading of the reconstructed breasts.

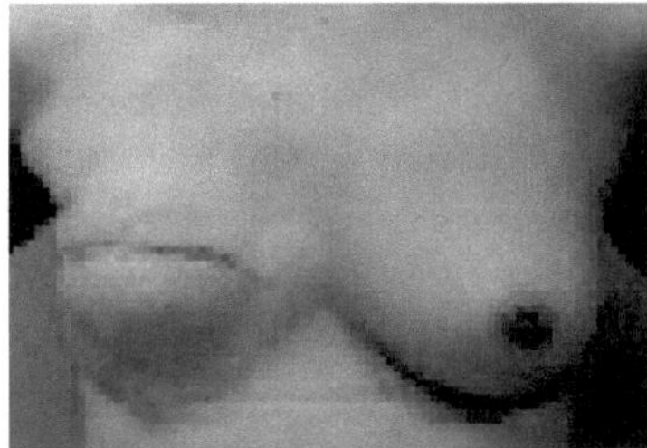

(a) Example of breast reconstruction surgery

Abbreviation	Evaluation viewpoint	Evaluation score (Scoring criteria)
BV	Breast volume	2 (equivalent), 1 (slightly different), 0 (considerably different)
BS	Breast shape	2 (equivalent), 1 (slightly different), 0 (considerably different)
BL	Breast location	2 (equivalent), 1 (slightly different), 0 (considerably different)
SC	Scars and color (Non-symmetric)	2 (inconspicuous), 1 (slightly conspicuous), 0 (conspicuous)
IMF	Inframammary fold symmetry (height)	1 (less than 2cm), 0 (2cm and over)
NL	Nipple location (nipple-to-sternal notch distance)	1 (less than 2cm), 0 (2cm and over)
NS	Nipple-Areola Complex (NAC) size and shape	1 (equivalent), 0 (different)
NC	NAC color	1 (equivalent), 0 (different)

(b) Evaluation viewpoint of esthetic outcome recognized among Japanese plastic surgeons

Fig. 1. (a) Example of reconstructed breast, and (b) evaluation viewpoint of esthetic outcome recognized among Japanese plastic surgeons [1]

Several attempts have been made to facilitate esthetic grading of reconstructed breasts from breast imaging. Among these, the software BCCT.core was developed to assist esthetic grading by interactively analyzing conventional two-dimensional (2D) breast images, and has shown significant correlation with subjective measure based on the Harris scale [2–4]. However, conventional 2D breast images capture only partial appearances of the breast and do not fully represent its intrinsic features. Consequently, esthetic grading based on these images is limited to partial appearances and may not be suitable for evaluating the esthetic contribution of intrinsic breast features.

Three-dimensional (3D) imaging offers an approach to examine intrinsic breast features and their contribution to esthetic outcome. 3D images can be obtained by integrating data captured using a handheld depth camera from multiple positions and orientations. These images consist of approximated curved surfaces formed by numerous triangular meshes with shared edges and textures to be projected on the surfaces, allowing geometric and color features of the breast to be extracted in a relatively straightforward manner. However, due to the geometric complexity of 3D images and the extensive implementation efforts required for a large number of cases, it has been difficult to utilize this imaging technique to examine the esthetic contribution.

Recently, Harada et al. developed an image processing system applicable to both 2D and 3D breast images to reduce efforts required for case collection [5]. This system facilitates anonymization, annotation support, and feature extraction of 3D breast images, enabling effective collection of reconstructed breast cases in clinical settings.

In this study, we propose a novel technique to examine the contribution of intrinsic breast features to each esthetic viewpoint, even with a limited number of cases, by extending on Harada's system to pre-extract relevant 3D geometric and color features, which took roughly 16 min for each case [5]. Using this technique, we conducted correlation analysis, for the first time, between all combinations of objectively extracted geometric and color features and the esthetic scores subjectively rated by plastic surgeons across multiple viewpoints. For each viewpoint, top correlated features were then selected for multiple linear regression (MLR) modeling. Finally, total esthetic score was calculated by aggregating the predictions from all the viewpoint-specific MLR models, allowing us to assess the effectiveness of the proposed technique.

The remainder of this paper is organized as follows. Section 2 describes the methodology in detail, including data preparation, feature extraction, and model construction. Section 3 presents the experimental results and provides an in-depth analysis. Finally, Sect. 4 summarizes the findings and concludes the paper with discussions on limitations and future research directions.

2 Methodology

2.1 Process Flow of Contribution Analysis of Intrinsic 3D Breast Features

Previous studies have introduced a case-collection process with high applicability for clinical settings [5, 6]. Upon completion of each collection cycle, cases were progressively accumulated, eventually becoming sufficient for use in the subsequent step of contribution analysis of this study. The process flow illustrated in Fig. 2 outlines our proposed analyses performed on the latest case collection, which comprises intrinsic

3D breast features and corresponding esthetic scores rated by different plastic surgeons. These analyses include correlation analysis and regression analysis, which were realized by extending Harada's proposed system.

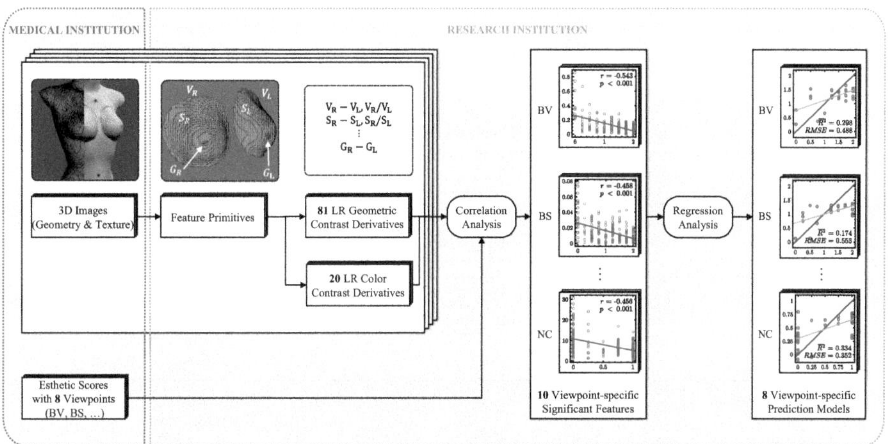

Fig. 2. Process flow of the contribution analysis of intrinsic 3D breast features

Feature primitives represent fundamental geometric and color measurement quantities extracted independently from the left and right breasts, such as 'breast volume' or 'nipple color'. These primitives are designed to comprehensively characterize various aspects and regions of each side of the breast. To capture a detailed and holistic representation, a wide range of feature primitives were extracted. Once corresponding pairs of left and right feature primitives were obtained, left-right (LR) contrast derivatives were calculated to quantify the contrast between the two sides. For example, a LR-ratio was calculated for 'breast volume'. The procedure of feature extraction and contrast calculation forms the basis of the naming convention for LR-contrast derivatives shown in Table 1.

Since LR-contrast derivatives were designed to serve as key indicators of various aspects of breast asymmetry, different subsets of them are expected to yield esthetic contributions to different viewpoints scoring, exhibiting linear relationships. Thus, following the contribution analysis, a regression analysis was conducted. For each esthetic viewpoint, a preliminary correlation analysis was performed to identify the top 10 features with the highest correlation coefficients, which were then selected as predictor variables. These predictors were used to construct MLR models to evaluate their combined influence on the esthetic viewpoint scores. After establishing the viewpoint-specific MLR models, the total esthetic score for each case was predicted by aggregating the predicted scores from all individual models. The predictive performance was then evaluated by comparing these predicted total scores with the actual esthetic scores.

Table 1. Naming convention for LR-contrast derivatives

{FeatureType}_{ContrastType}_LR_{ROI}_{FeaturePrimitive}

Order of Operation	Component	Definition	Symbol	Symbol Meaning
↑	{FeatureType}	Type of feature	G C	Geometric Color
	{ContrastType}	Mathematical operation applied between L and R breasts primitive features to obtain the contrast measurement	d r ED CIED	Difference Ratio Euclidean Distance CIEDE2000 Color Distance
	LR	Indicating the completion of feature primitive extraction independently on L and R breasts		
	{ROI}	Region of interest where feature primitive is measured.	Breast Nipple IMF NAC	Whole breast region Nipple specific region IMF specific region NAC specific region
	{FeaturePrimitive}	Identifier of feature primitive	V S z 3S Lab	Volume Surface Area z-coordinate Summation of 3 sides CIELAB Color Space

2.2 3D Breast-Image Acquisition and Esthetic Scoring

The 3D images used in this study were specially acquired from May 2019 to December 2022 at the previous affiliated medical institutions (Kyoto Prefectural University of Medicine and Kyoto University) of the co-author Sowa, Y. under the approval of their ethics committee. The final dataset includes 174 patient cases.

The process of image acquisition is as follow: as a pretreatment, the plastic surgeon places two types of markers on the patient's body surface before image acquisition. The first type consists of two cross marks, one at the center of the clavicle and the other 25 cm below along the body axis, used for scaling purposes. The second type involves outlining the breast region, which the plastic surgeon places directly on the patient's body surface. If this direct placement was not easy to achieve, the surgeon may instead draw the outline retrospectively on the 3D image post-acquisition.

The 3D images were captured using a hand-held depth camera Intel RealSense L515 [7], and multiple viewpoints were merged using ImFusion RecFusion software [8]. For each case, cross-annotation was performed by four plastic surgeons, following the Harada proposed medical-image processing system [5].

The evaluation of the reconstructed breast is a highly subjective task, where inconsistency and variability are unavoidable. Figure 3 presents inter-rater variation of esthetic score among four plastic surgeons as bubble plots, accompanied by the calculated coefficient of variation (CV) as a quantitative measure.

High inter-rater variability was consistently observed across all evaluation perspectives, particularly in cases with lower esthetic score. To investigate the impact of this variability on the contribution of intrinsic breast features, we prepared two datasets: 1)

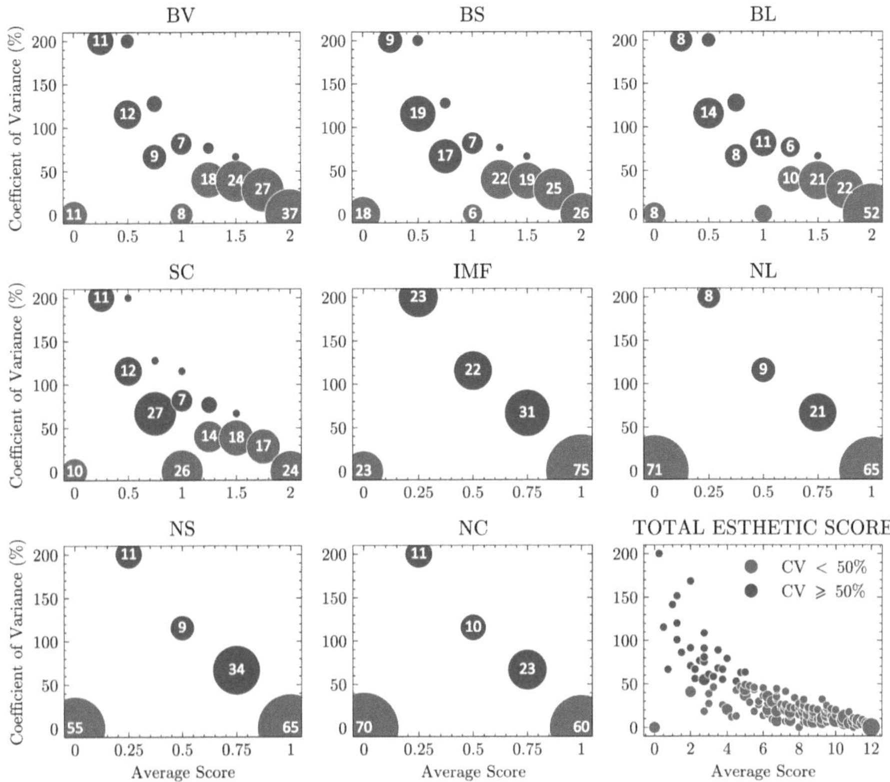

Fig. 3. Inter-rater variations of esthetic scores among four plastic surgeons

the Agreement Dataset, consisting of selected cases with high inter-rater agreement (CV < 50%) and 2) the Full Dataset, comprising all 174 cases. Both datasets were used for esthetic contribution analysis and regression modeling. The number of available cases in the Agreement Dataset varied across viewpoints, ranging from 98 to 130, corresponding to 56% to 74% of the Full Dataset.

3   Experimental Results

3.1   Esthetic Contribution of Intrinsic 3D Breast Features

The contribution of intrinsic 3D breast features to esthetic scoring was examined through correlation analysis with scores assigned to each esthetic viewpoint. Viewpoints sharing similar esthetic meanings were grouped into categories. The Overall Appearance category includes BV, BS, BL, and IMF. The Detailed Appearance category comprises NL, NS, and NC. The remaining viewpoint, SC, represents the esthetic imperfections of the reconstructed breast and is categorized separately.

Figure 4 shows the highest correlated LR-contrast derivatives for eight esthetic viewpoints across two datasets: (a) the Agreement Dataset and (b) the Full Dataset. Although

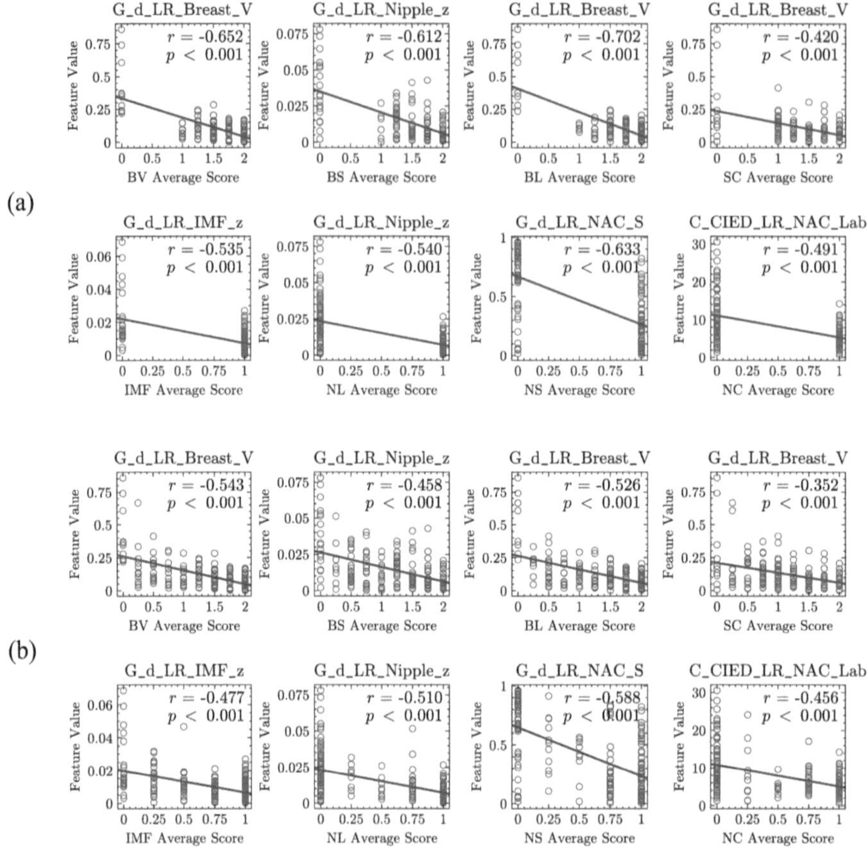

Fig. 4. The highest correlated LR-contrast derivatives for eight esthetic viewpoints in two datasets (a) the Agreement Dataset and (b) the Full Dataset.

the value of r was not excellent, highest at around 0.7, relatively stronger correlations were observed for viewpoints categorized under the Overall Appearance category, typically exceeding 0.45 in the Full Dataset and 0.53 in the Agreement Dataset. In the Agreement Dataset, all viewpoints showed an increase in the correlation coefficient for the same LR-contrast derivatives compared with the Full Dataset. However, the degree of improvement varied across viewpoints. The most significant increase was particularly observed in viewpoints belonged to the Overall Appearance category, with an average of approximately 25%.

Table 2 presents the top 10 LR-contrast derivatives exhibiting the highest correlations across two datasets. In both datasets, the combination of highly correlated features varied across viewpoints. However, derivatives such as f7 and f45, were consistently found across almost all viewpoints, each occurring in at least six instances, suggesting their common significance as intrinsic features.

Additionally, a few LR-contrast derivatives were commonly included in all the viewpoints of a certain category. For example, the derivative f1 and f82 were included in all

Table 2. Top 10 LR-contrast derivatives for eight viewpoints in two datasets: (a) the Agreement Dataset, and (b) the Full Dataset.

(a)

#		BV		BS		BL		SC
1	f7	G_d_LR_Breast_V	f45	G_d_LR_Nipple_z	f7	G_d_LR_Breast_V	f7	G_d_LR_Breast_V
2	f82	G_r_LR_BreastSkin_CurveAvg	f7	G_d_LR_Breast_V	f1	G_d_LR_BreastSkin_S	f49	G_d_LR_IMF_z
3	f1	G_d_LR_BreastSkin_S	f49	G_d_LR_IMF_z	f82	G_r_LR_BreastSkin_CurveAvg	f20	G_r_LR_Nipple_DroopTan
4	f80	G_r_LR_BreastSkin_Curve3sr	f1	G_d_LR_BreastSkin_S	f2	G_r_LR_BreastSkin_S	f45	G_d_LR_Nipple_z
5	f2	G_r_LR_BreastSkin_S	f18	G_r_LR_Nipple_DroopA	f80	G_r_LR_BreastSkin_Curve3sr	f17	G_d_LR_Nipple_DroopA
6	f81	G_d_LR_BreastSkin_CurveAvg	f48	G_d_LR_VolumeCentroid_z	f84	G_r_LR_BreastSkin_CurveMax	f48	G_d_LR_VolumeCentroid_z
7	f79	G_d_LR_BreastSkin_Curve3sr	f2	G_r_LR_BreastSkin_S	f20	G_r_LR_Nipple_DroopTan	f18	G_r_LR_Nipple_DroopA
8	f84	G_r_LR_BreastSkin_CurveMax	f82	G_r_LR_BreastSkin_CurveAvg	f81	G_d_LR_BreastSkin_CurveAvg	f8	G_r_LR_Breast_V
9	f45	G_d_LR_Nipple_z	f17	G_d_LR_Nipple_DroopA	f17	G_d_LR_Nipple_DroopA	f47	G_d_LR_VertexCentroid_z
10	f12	G_r_LR_BreastBBox_3S	f12	G_r_LR_BreastBBox_3S	f83	G_d_LR_BreastSkin_CurveMax	f4	G_r_LR_d_FrontBackSkin_S
#		IMF		NL		NS		NC
1	f49	G_d_LR_IMF_z	f45	G_d_LR_Nipple_z	f50	G_d_LR_NAC_S	f66	C_CIED_LR_NAC_Lab
2	f45	G_d_LR_Nipple_z	f50	G_d_LR_NAC_S	f51	G_r_LR_NAC_S	f50	G_d_LR_NAC_S
3	f7	G_d_LR_Breast_V	f21	G_d_LR_r_Nipple_DroopAlt1	f66	C_CIED_LR_NAC_Lab	f65	C_WED_LR_NAC_RGB
4	f5	G_d_LR_r_FrontBackSkin_S	f51	G_r_LR_NAC_S	f63	C_ED_LR_NAC_HSV	f62	C_ED_LR_NAC_RGB
5	f6	G_r_LR_r_FrontBackSkin_S	f23	G_d_LR_r_Nipple_DroopAlt2	f65	C_WED_LR_NAC_RGB	f63	C_ED_LR_NAC_HSV
6	f3	G_d_LR_d_FrontBackSkin_S	f22	G_r_LR_r_Nipple_DroopAlt1	f49	G_d_LR_IMF_z	f64	C_ED_LR_NAC_Lab
7	f17	G_d_LR_Nipple_DroopA	f24	G_r_LR_r_Nipple_DroopAlt2	f62	C_ED_LR_NAC_RGB	f51	G_r_LR_NAC_S
8	f18	G_r_LR_Nipple_DroopA	f49	G_d_LR_IMF_z	f45	G_d_LR_Nipple_z	f72	G_r_LR_Nipple_x
9	f1	G_d_LR_BreastSkin_S	f66	C_CIED_LR_NAC_Lab	f64	C_ED_LR_NAC_Lab	f58	C_ED_LR_Nipple_HSV
10	f82	G_r_LR_BreastSkin_CurveAvg	f7	G_d_LR_Breast_V	f7	G_d_LR_Breast_V	f7	G_d_LR_Breast_V

(b)

#		BV		BS		BL		SC
1	f7	G_d_LR_Breast_V	f45	G_d_LR_Nipple_z	f7	G_d_LR_Breast_V	f7	G_d_LR_Breast_V
2	f1	G_d_LR_BreastSkin_S	f7	G_d_LR_Breast_V	f49	G_d_LR_IMF_z	f45	G_d_LR_Nipple_z
3	f2	G_r_LR_BreastSkin_S	f1	G_d_LR_BreastSkin_S	f1	G_d_LR_BreastSkin_S	f49	G_d_LR_IMF_z
4	f45	G_d_LR_Nipple_z	f49	G_d_LR_IMF_z	f45	G_d_LR_Nipple_z	f17	G_d_LR_Nipple_DroopA
5	f79	G_d_LR_BreastSkin_Curve3sr	f2	G_r_LR_BreastSkin_S	f2	G_r_LR_BreastSkin_S	f8	G_r_LR_Breast_V
6	f81	G_d_LR_BreastSkin_CurveAvg	f11	G_d_LR_BreastBBox_3S	f3	G_d_LR_d_FrontBackSkin_S	f2	G_r_LR_BreastSkin_S
7	f49	G_d_LR_IMF_z	f3	G_d_LR_d_FrontBackSkin_S	f5	G_d_LR_r_FrontBackSkin_S	f3	G_d_LR_d_FrontBackSkin_S
8	f82	G_r_LR_BreastSkin_CurveAvg	f12	G_r_LR_BreastBBox_3S	f6	G_r_LR_r_FrontBackSkin_S	f1	G_d_LR_BreastSkin_S
9	f80	G_r_LR_BreastSkin_Curve3sr	f31	G_d_LR_BreastBBox_S	f82	G_r_LR_BreastSkin_CurveAvg	f18	G_r_LR_Nipple_DroopA
10	f3	G_d_LR_d_FrontBackSkin_S	f5	G_d_LR_r_FrontBackSkin_S	f48	G_d_LR_VolumeCentroid_z	f48	G_d_LR_VolumeCentroid_z
#		IMF		NL		NS		NC
1	f49	G_d_LR_IMF_z	f45	G_d_LR_Nipple_z	f50	G_d_LR_NAC_S	f66	C_CIED_LR_NAC_Lab
2	f7	G_d_LR_Breast_V	f50	G_d_LR_NAC_S	f51	G_r_LR_NAC_S	f50	G_d_LR_NAC_S
3	f1	G_d_LR_BreastSkin_S	f51	G_r_LR_NAC_S	f66	C_CIED_LR_NAC_Lab	f65	C_WED_LR_NAC_RGB
4	f5	G_d_LR_r_FrontBackSkin_S	f21	G_d_LR_r_Nipple_DroopAlt1	f65	C_WED_LR_NAC_RGB	f63	C_ED_LR_NAC_HSV
5	f6	G_r_LR_r_FrontBackSkin_S	f49	G_d_LR_IMF_z	f63	C_ED_LR_NAC_HSV	f62	C_ED_LR_NAC_RGB
6	f45	G_d_LR_Nipple_z	f23	G_d_LR_r_Nipple_DroopAlt2	f62	C_ED_LR_NAC_RGB	f51	G_r_LR_NAC_S
7	f2	G_r_LR_BreastSkin_S	f66	C_CIED_LR_NAC_Lab	f45	G_d_LR_Nipple_z	f64	C_ED_LR_NAC_Lab
8	f3	G_d_LR_d_FrontBackSkin_S	f22	G_r_LR_r_Nipple_DroopAlt1	f64	C_ED_LR_NAC_Lab	f72	G_r_LR_Nipple_x
9	f11	G_d_LR_BreastBBox_3S	f7	G_d_LR_Breast_V	f49	G_d_LR_IMF_z	f58	C_ED_LR_Nipple_HSV
10	f12	G_r_LR_BreastBBox_3S	f18	G_r_LR_Nipple_DroopA	f48	G_d_LR_VolumeCentroid_z	f7	G_d_LR_Breast_V

the viewpoints in the Overall Appearance category. Such viewpoint-category-specific LR-derivatives were also found in other categories. This result is an important finding regarding the contributing factors to esthetic scoring, as it indicates a viewpoint-specific property of intrinsic breast features.

Through the correlation analysis of LR-contrast feature derivatives, promising key factors that significantly contribute to individual viewpoints and the overall esthetic score were successfully identified. These LR-contrast derivatives are expected to encapsulate various aspects of the reconstructed breast related to esthetic evaluation and are suitable for use as predictor inputs in developing basic regression models for each viewpoint.

3.2 Total Esthetic Score Prediction from Viewpoint-Specific MLR-Models

Since there were two datasets, two types of test sets were prepared with different data partitioning strategies. The Agreement Test Set was obtained by filtering down to an exclusive set of overlapping 30 cases which are shared across all viewpoints' Agreement Datasets. In contrast, the Full Test Set was determined by random selection of 30 cases from the Full Dataset to minimize inter-rater variation bias. The training and validation sets were prepared accordingly using the remaining cases after test set splitting, and the number of cases in these sets may vary for each viewpoint.

When evaluating the test sets, each case was first evaluated independently using viewpoint-specific MLR models to obtain eight predicted viewpoint scores. These scores were then aggregated to produce the final prediction of the total esthetic score. Figure 5 presents a comparison between the predicted and actual total esthetic scores across different data partitioning strategies. The evaluation metrics include correlation coefficient (r), coefficient of determination (R^2), mean absolute error (MAE), root mean square error (RMSE), and root mean square percentage error (RMSPE), which collectively offer a comprehensive assessment of both the accuracy and robustness of the predictions.

When the model is evaluated on the Full Test Set, a moderate predictive capability is observed, with noticeable degree of variability across training condition. Relatively high RMSPE value is observed at 84.9 and 107.7% respectively, emphasizing the influence of cases with low inter-rater agreement on the model's accuracy.

In contrast, the model predictions on the Agreement Test Set under both training conditions reveal stronger linear relationship between predicted and actual scores. This is supported by higher correlation coefficients ($r = 0.92$ and 0.89, respectively) as well as reduced prediction errors. This indicates improved consistency when low-variability cases are selected. Notably, when the viewpoint-specific MLR models are trained on the Agreement Dataset, the evaluation metrics show a marked and consistent improvement across all measures. In this setting, the models achieve the best predictive performance, with the RMSPE reduced to approximately 17%, which represents the lowest error rate observed.

To our best knowledge, this is the first study to achieve esthetic grading of reconstructed breasts based on intrinsic 3D features by aggregating the predicted scores from multiple viewpoint-specific MLR models. While this approach establishes a methodological foundation, the predictive performance of the models remains insufficient to support practical use in clinical decision-making, highlighting the need for further refinement in both data quality as well as evaluation method.

A key challenge lies in improving the quality of annotation and enhancing inter-rater agreement, particularly in cases that receive low-score evaluations. Several strategies are currently under consideration to mitigate these challenges. One promising direction is the introduction of a comparative scoring framework that evaluates cases within a shared context, enabling more objective and reliable agreement among raters.

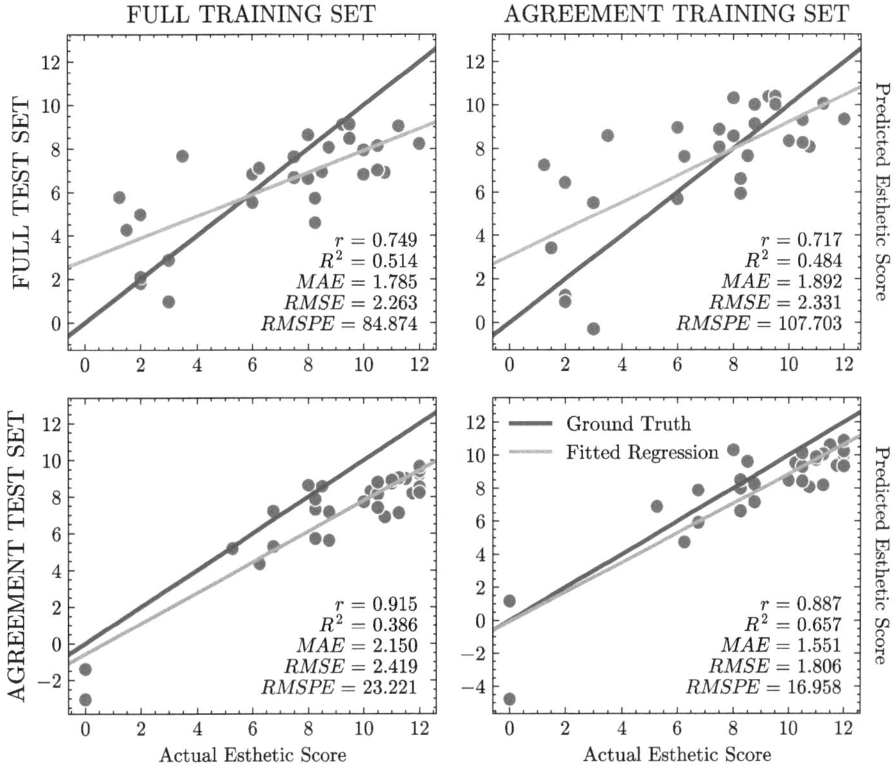

Fig. 5. Predicted vs. actual esthetic scores across under different data partitioning strategies

4 Conclusion

In this study, we proposed a novel technique to investigate the contribution of intrinsic breast features to each esthetic perspective, even with a limited number of images, by pre-extracting geometric and color features from 3D breast images. This technique allowed correlation analysis between any combination of these features and esthetic scores subjectively rated by plastic surgeons.

Analysis from 174 clinical cases showed that these intrinsic features correlated differently depending on the viewpoint, indicating unique esthetic contributions specific to each viewpoint. Top 10 correlated features were selected and used to obtain MLR models for each viewpoint. The overall grading, aggregated from viewpoint-specific MLR models trained on cases with high inter-rater agreement, showed a strong correlation $r = 0.89$ with the surgeons' evaluation score and a low RMSPE of 17%. These results suggest the effectiveness of the proposed technique and its potential for appearance-based esthetic grading using conventional 2D breast images.

On the other hand, the subjective evaluation scores of plastic surgeons tended to show low inter-rater agreement at low evaluation scores, making correlation analysis of these particular cases difficult. Additionally, the introduced MLR models showed strong correlations only when trained on cases with high inter-rater agreement. Therefore, studies on the low-scoring cases, including how to improve the quality of annotation and increase their inter-rater agreement, remain a challenge for the future.

Acknowledgments. The authors extend their gratitude to Mr. Kazuya Koyanagi, Ms. Kotori Harada and Mr. Takahiro Yoshimoto for their invaluable contribution during the early development stage of this study. This work was supported by JST SPRING, Grant Number JPMJSP2107 and by JSPS Core-to-Core Program, Grant Number JPJSCCB20230005.

Disclosure of Interests. The authors have no competing interests to declare that are relevant to the content of this article.

References

1. Yano, K.: Cosmetic results after breast reconstruction (in Japanese). Jpn. J. Clin. Med. **65**(6), 465–468 (2007)
2. Cardoso, M.J., et al.: Turning subjective into objective: the BCCT.core software for evaluation of cosmetic results in breast cancer conservative treatment. Breast **16**(5), 456–461 (2007). https://doi.org/10.1016/j.breast.2007.05.002
3. Heil, J., Carolus, A., Dahlkamp, J., Golatta, M., Domschke, C., Schuetz, F., et al.: Objective assessment of aesthetic outcome after breast conserving therapy: subjective third party panel rating and objective BCCT.core software evaluation. Breast **21**(1), 61–65 (2012). https://doi.org/10.1016/j.breast.2011.07.013
4. Preuss, J., Lester, L., Saunders, C.: BCCT.core – can a computer program be used for the assessment of aesthetic outcome after breast reconstructive surgery? Breast **21**(4), 597–600 (2012). https://doi.org/10.1016/j.breast.2012.05.012
5. Harada, K., Yoshimoto, T., Duong, N.P., Nguyen, M.N., Sowa, Y., Fukuzawa, M.: A new integrated medical-image processing system with high clinical applicability for effective dataset preparation in ML-based diagnosis. In: Thai-Nghe, N., Do, T. N., Haddawy, P. (eds.) ISDS 2023. CCIS, vol. 1950, pp. 41–50. Springer, Singapore (2023). https://doi.org/10.1007/978-981-99-7666-9_4
6. Nguyen, M.N., Harada, K., Yoshimoto, T., Duong N.P., Sowa, Y., Fukuzawa, M.: Integrated dataset-preparation system for ML-based medical image diagnosis with high clinical applicability in various modalities and diagnoses. SN Comput. Sci. **5**(676) (2024). https://doi.org/10.1007/s42979-024-03025-7
7. Intel Realsense L515. https://www.intelrealsense.com/lidar-camera-l515. Accessed 23 June 2025
8. ImFusion RecFusion. https://www.recfusion.net/. Accessed 23 June 2025
9. Harada, K.: Study on Dataset Preparation and Feature Analysis of Breast Images for Esthetic Outcome Evaluation. Master Thesis of Kyoto Institute of Technology (2024)
10. Koyanagi, K.: Extraction of Geometric Features from 3D Camera Images for Esthetic Outcome Evaluation on Breast Reconstruction Surgery. Master Thesis of Kyoto Institute of Technology (2023)
11. Stern, C., Kim, L.N., Plotsker, E., Boyce, L., Dayan, J., Nelson, J.A.: An updated systematic review of esthetic grading tools in postmastectomy breast reconstruction. J. Surg. Oncol. **127**(5), 782–790 (2023). https://doi.org/10.1002/jso.27186

Gestational Diabetes Prediction Using Classification Methods

Al Maruf Hassan[1], The-Phi Pham[2(✉)], Md. Maruf Hassan[3],
Abdul Kadar Muhammad Masum[3], and Dewan Md. Farid[3]

[1] Department of Electrical and Computer Engineering, North South University, Plot:
15, Block: B, Bashundhara, Dhaka 1229, Bangladesh
[2] College of Information and Communication Technology, Can Tho University,
3/2 Street, Can Tho City, Vietnam
`ptphi@ctu.edu.vn`
[3] Department of Computer Science and Engineering, Southeast University, 252,
Tejgaon Industrial Area, Dhaka 1208, Bangladesh

Abstract. Gestational diabetes (GD) is a form of diabetes that is first identified during pregnancy. It is a rapidly emerging condition among pregnant women and has become widespread in various populations around the world. Although its impact is considerable, there is currently no definitive cure; only symptomatic treatment is possible. This study aims to assess the risk of GD in women using modern data mining techniques for diagnostic purposes. The dataset was sourced from two well-known private hospitals in Dhaka, Bangladesh. We applied six classification algorithms: Support Vector Machine (SVM), Logistic Regression (LR), Decision Tree (DT), Gaussian Naive Bayes (GaussianNB), k-Nearest Neighbors (KNN), and Random Forest (RF) - both before and after implementing a feature selection step (selecting the top seven features), along with 5-fold cross-validation. In addition, a custom ensemble approach and two ensemble methods, bagging and boosting, were used. Our analysis revealed that the custom ensemble technique paired with the RF classifier achieved the highest accuracy of 88.24%. The findings of this research demonstrate strong potential to aid in the early detection and prevention of GD.

Keywords: Gestational Diabetes (GD) · Classification · Feature Selection · Ensemble Techniques

1 Introduction

Gestational Diabetes (GD), like other types of diabetes, affects how the body's cells process glucose. It occurs when blood sugar levels rise during pregnancy, which can negatively impact the mother's health and the baby's development. Managing this condition is especially critical due to the complications it may cause during pregnancy. However, GD can often be controlled through a healthy diet, regular physical activity, and, if necessary, medical guidance and treatment.

© The Author(s), under exclusive license to Springer Nature Singapore Pte Ltd. 2026
N. Thai-Nghe et al. (Eds.): ISDS 2025, CCIS 2714, pp. 157–175, 2026.
https://doi.org/10.1007/978-981-95-3358-9_12

By closely monitoring blood sugar levels, the mother can reduce health risks for herself and her baby, helping to prevent complications during delivery. As a precaution, frequent testing for blood sugar fluctuations is essential. According to the International Diabetes Federation (IDF), in 2019, approximately 16% of pregnancies were affected by hyperglycemia, with 84% of those cases being attributed to gestational diabetes [1] reported GDM. If left unmanaged, Gestational Diabetes can lead to serious health issues for both the mother and the baby. With the growing availability of medical data, a vast amount of real and practical patient information is now stored in electronic health records. The complications associated with diabetes are complex and influenced by numerous factors. In healthcare, preventing chronic illnesses is often considered more crucial than relying solely on medication. Since a permanent cure for diabetes has not yet been discovered, current approaches focus on effective management. Gestational Diabetes Mellitus (GDM) is a common pregnancy complication and is linked to various adverse maternal and neonatal outcomes, particularly in women with obesity [7]. It is also related to the developed danger of both short-term and long-term complexity in mothers and babies [10]. High-sensitivity C-reactive protein (Hs-CRP) and sex hormone-binding globulin (SHBG) are critical early indicators of Gestational Diabetes Mellitus (GDM). Incorporating Hs-CRP and SHBG into the analysis enhances the specificity and improves the overall accuracy of the diagnostic results [14].

Therefore, with the support of data mining technology, the most influential factors can be extracted from diverse patient data, aiding in symptom management and enhancing the quality of healthcare services. This study applied various data mining methods to analyze and predict the occurrence of gestational diabetes among the Bangladeshi population. The key contributions of this research are outlined below:

- Data is collected from two reputed hospitals of the Dhaka region, containing 605 records and 27 features
- Six classification algorithms—k-Nearest Neighbors, Decision Tree, Logistic Regression, Random Forest, Support Vector Machine, and Gaussian naive Bayes (GaussianNB)—were applied to the dataset. A feature selection approach was implemented to identify the top seven most relevant features. Additionally, a 5-fold cross-validation technique was employed. The performance results were evaluated before and after applying feature selection and cross-validation.
- This study utilized two ensemble learning techniques: bagging and boosting. For the bagging approach, a classifier is implemented where six distinct classification algorithms serve as base learners, evaluated using 5-fold cross-validation. In addition, two boosting methods—AdaBoost and Gradient Boosting Classifier—were employed to enhance predictive performance.
- A custom ensemble method was implemented on the dataset. In this approach, the AdaBoost classifier was used with three different classification algorithms serving as base learners, replacing the default base estimator.

Limited research has been conducted to identify the key factors contributing to Gestational Diabetes among Bangladeshi women. Even fewer studies have explored the prediction of this condition using data mining approaches. In this study, a new dataset was gathered from a hospital in Bangladesh, encompassing various symptoms and risk factors associated with gestational diabetes. Several data mining techniques are then applied to this dataset to achieve accurate and meaningful results.

The rest of the paper is structured as follows. Section 2 contains a review of related work. Section 3 describes the methodology. Section 4 discusses the experimental analysis with a description of the datasets. Finally, the conclusion and future work are presented in Sect. 5 (Table 1).

2   Related Works

Artzi et al. [2] demonstrated the potential for highly accurate prediction of Gestational Diabetes Mellitus (GDM) during the early stages of pregnancy. Their investigation encompassed approximately 588,600 pregnancies from 368,300 women between 2010 and 2017. Of these, 451,402 pregnancies were allocated to the training cohort. Model performance was subsequently assessed using three distinct validation strategies: (i) a **feature validation set** comprising $\sim$82,500 pregnancies occurring in 2017 and thereafter, (ii) a **geographical validation set** including $\sim$46,000 pregnancies from women residing in Jerusalem, and (iii) a **geo-temporal validation set** of 8,540 pregnancies meeting both temporal and geographical inclusion criteria. Gradient boosting was employed to estimate the probability of GDM. Predictive performance was evaluated across multiple scenarios, including women identified as high risk, those in early pregnancy, and those with prior glucose challenge test (GCT) results. Feature importance was quantified using the Shapley value attribution framework, which revealed that pregnancy history and laboratory test variables contributed most substantially to model performance relative to other features.

Gnanadass [9] investigated the causes of Gestational Diabetes Mellitus (GDM) using several machine learning (ML) classifiers, including Support Vector Machines (SVM), Logistic Regression (LR), and Random Forests (RF), based on the Polyisocyanurate Insulation Manufacturers Association (PIMA) dataset. The dataset contains approximately eight clinical features and 800 instances, comprising about 250 positive and 450 negative cases of diabetes, with origins linked to the Indian population. Similarly, Mazumder et al. [15] examined the prevalence and risk factors of GDM in the context of Bangladesh. Their study utilized secondary data from around 250 pregnant women collected through the Bangladesh Demographic and Health Survey (BDHS) 2017âĂŞ2018. The dataset included features such as sleep-related breathing disturbances, obesity, family history of diabetes, hypothyroidism, and polycystic ovary syndrome. Both bivariate and multivariate statistical analyses were employed to identify factors associated with GDM, and multicollinearity was specifically examined between residence and wealth indices.

Table 1. Summary of the literature review.

Authors	No. of Dataset	No. of Record	Data Type	Data Source	Class	Feature	Algorithm	Evaluation
Artzi et al. [2]	3	(82678,46002,8540)	Clinical	Nature Medicine	Geo-graphical and temporal	20	Gradient Boosting Model (GBM)	auROC, precision-recall-(auPR) curve
Gnanadass [9]	1	768	Clinical	Poly-isocyanurate Insulation Manufacturers Association (PIMA)	Numerical	8	NB, LR, RF, AdaBoost, XGBoost, SVM and GBM	Confusion matrix, ROC, AUC
Mazumder et al. [15]	1	272	Survey	Bangladesh Demographic and Health Survey (BDHS)	Numerical	9	Statistical Analysis, LR, Stata 14	Crude, AOR and CI
Qiu et al. [19]	1	4,378	Clinical	West China Second Hospital	Numerical	50	LR, BN, ANN, SVM, CHAID Trees, Ensemble Model	AUC, TPR, FPR, ROC
Xiong et al. [22]	2	(215,275)	Clinical	West China Second University Hospital	Numerical	43	SVM, LightGBM	TPR, FPR, AUC(ROC)
Zhang and Wang [23]	1	1,000	Clinical	Beijing Qingwutong Health Technology Company	Mixed	85	CatBoost, LightGBM XGBoost	AUC, Recall, F1 score, Precision
Yan Ting [16]	1	16,819	Clinical	International Peace Maternity and Child Health Hospital	Mixed	73	LR, BN, ANN, SVM, CHAID Trees, Ensemble Model	AUC, TPR, FPR, ROC
Our research	1	605	Clinical	Two Bangladeshi hospital	Numerical	27	LR, DTR, RFC, SVM, NBCGaussian, knN, Ensemble Model (Bagging, AdaBoost, and GB)	Accuracy, Precision, Recall, F1-score

Qiu et al. [19] developed a cost-sensitive hybrid model (CSHM) to construct a machine learning framework for predicting the risk of Gestational Diabetes Mellitus (GDM) during early pregnancy using electronic health records (EHRs). Their study analyzed data from approximately 34,000 pregnant women in China between 2013 and 2016, including ~4,500 confirmed GDM cases. Following data preprocessing, around 50 features were retained for modeling. The prediction framework was based on supervised classification algorithms, including Support Vector Machines (SVM), Chi-squared Automatic Interaction Detector (CHAID) trees, Logistic Regression (LR), Bayesian Networks (BN), and Artificial Neural Networks (ANN). Similarly, Xiong et al. [22] constructed a risk prediction model for GDM during the first 19 weeks of pregnancy, incorporating predictors of renal, metabolic, and liver function. Their case-control study involved ~500 pregnant women, comprising ~200 GDM cases and ~300 controls. Prediction models were built using SVM and the Light Gradient Boosting Machine (LightGBM) to identify potential associations with the risk of GDM.

Zhang and Wang [23] proposed an ensemble-based machine learning framework for the early prediction of Gestational Diabetes Mellitus (GDM), utilizing models such as XGBoost, LightGBM, and CatBoost. Their experimental dataset comprised approximately 1,000 training samples with 85-dimensional features, including 30 anatomical markers such as weight, cholesterol, age, height, and body mass index (BMI). In a related study, Hou et al. [10] employed the Light-GBM classifier and compared its performance with Random Forest (RF) and XGBoost for genetic risk prediction of GDM. Their findings indicated that pregnant women with high insulin resistance (VAR00007) and advanced maternal age exhibited an elevated risk of developing GDM. Similarly, Kumar [11] analyzed a diabetes-related dataset using multiple classifiers, including Support Vector Machines (SVM), Random Forest (RF), k-Nearest Neighbors (KNN), Classification and Regression Trees (CART), and Linear Discriminant Analysis (LDA). This dataset comprised approximately 600 instances and 15 categorical and numerical features related to GDM patients across various age groups. The data were collected from a clinical laboratory in Warangal, India.

Liu et al. [13] developed a machine learningâ ŞŞbased predictive model for Gestational Diabetes Mellitus (GDM) during early pregnancy in a cohort of Chinese women. Their study analyzed data from approximately 19,000 pregnant women in Tianjin, China, collected between late 2010 and the end of 2012. The authors employed XGBoost as the primary classifier and compared its performance against Logistic Regression (LR). Key risk factors considered in the model included maternal age, fasting plasma glucose, and pre-pregnancy body mass index (BMI), among others. In a related study, Raveendra et al. [20] proposed an approach for predicting GDM using association rule mining in combination with other data mining techniques. Their methodology incorporated Random Tree (RT), Naïve Bayes (NB), Multilayer Perceptron (MLP), Random Forest (RF), and J48 classifiers to diagnose GDM. The authors emphasized the identification of critical risk indicators and abnormalities in women with GDM before, during, and after delivery.

Donovan et al. [6] developed a machine learningâ ŞŞbased model for early prediction of Gestational Diabetes Mellitus (GDM) using clinical risk indicators. Their study was conducted in California, USA, from 2007 to 2012. The prediction framework incorporated five primary risk factors: genetic predisposition, pre-pregnancy body mass index (BMI), and maternal age at delivery. In another contribution, Christobel and Kamalakannan [4] proposed a methodology for the early prediction of diabetes (Type 1, Type 2, and GDM) by employing a Hybrid Convolutional Neural Networkâ ŞŞLong Short-Term Memory (HCNN-LSTM) architecture. Their approach leveraged Big Data technologies with storage and processing implemented through the Hadoop Distributed File System (HDFS).

Lee et al. [12] developed an early prediction model for Gestational Diabetes Mellitus (GDM) using machine learning approaches, including Logistic Regression (LR), Random Forest (RF), Support Vector Machines (SVM), and deep Artificial Neural Networks (ANN). Their study analyzed 1,443 women, of whom 86 (6.0%) were diagnosed with GDM. The authors reported that including non-alcoholic fatty liver disease (NAFLD)âĂŞrelated variables significantly improved predictive performance. Similarly, Wu et al. [21] constructed a predictive framework for early GDM detection in both Chinese and multi-ethnic populations. Their dataset comprised approximately 17,000 training cases and 15,000 test cases. Using 73 variables, the deep ANN model demonstrated strong discriminative ability, and their findings highlighted that machine learningâĂŞbased models can achieve high accuracy in forecasting GDM during early pregnancy.

Gao et al. [8] identified a set of risk indicators for predicting Gestational Diabetes Mellitus (GDM). Correa et al. [5] examined existing hypotheses regarding the pathophysiology of GDM, evaluated current screening approaches, and provided recommendations for clinical management and treatment strategies. Bogdanet et al. [3] explored novel protein-based biomarkers with potential diagnostic utility for GDM, suggesting that these biomarkers may serve as alternatives to the oral glucose tolerance test (OGTT), which is currently the standard screening protocol. Pustozerov et al. [18] proposed a data-driven approach using a decision treeâĂŞgradient boosting (DT-GB) classifier to model postprandial glycemic responses for GDM prediction. In another study, Naser et al. [17] investigated first-trimester serum magnesium and high-sensitivity C-reactive protein (hs-CRP) levels, concluding that these biomarkers were not significantly associated with the development of GDM.

3   Methodology

In this section, the process of data collection, dataset cleaning, data labeling, and a statistical analysis of the data has been demonstrated exhaustive manner. Figure 1 shows the overview of the steps that we have applied in our study to classify gestational Diabetes.

Implementation of our research work, named, gestational diabetes prediction using classificationÂămethods, can be accessed from the *GitHub page*[1]

---

[1] https://github.com/marufgreat/GestationalDiabetesPredictionUsingClassificationMethods. git.

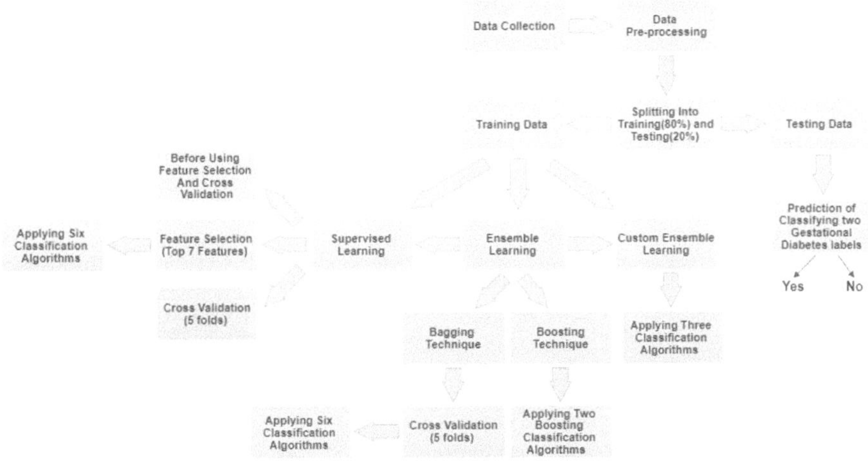

Fig. 1. Working sequence of the proposed work.

3.1   Data Collection

The dataset for this study was collected from several hospitals in Bangladesh specializing in gynecological care. It comprises records of 605 pregnant women across different age groups. In the initial stage, consultations were held with gynecologists to identify the key symptoms associated with Gestational Diabetes Mellitus (GDM). Based on this expert input, a structured data table was developed. The attributes employed in this study, along with their descriptions and types, are presented in Table 2.

3.2   Pre-processing

Following the data collection phase, 605 records were obtained, of which 337 were retained after data cleaning. Feature names were initially standardized for clarity. Data quality was ensured by addressing missing values, duplicate entries, and outliers. The **overweight** feature contained many missing values and was consequently removed along with the associated records. Additionally, 269 duplicate records were identified and excluded. The **occupation** attribute was manually encoded, with **housewife** assigned a value of 0, and **job holder, teacher, job holder, and doctor** assigned a value of 1. All remaining categorical features were encoded using a label encoder. Finally, numerical features, including age, pregnancy duration, height, and weight, were normalized using the MinMaxScaler.

Table 2. All attributes with descriptions and types

Attribute	Description	Type
occupation	patient line of work	Numerical
age	age of patient	Numerical
pregnancy-Duration	postnatal period	Numerical
height	patient height	Numerical
weight	patient Weight	Numerical
pregnant-Before	having child before	Numerical
having-Baby(9 pounds)	this is a medical complication where the newborn weighs 9 pounds, which is larger than normal. This complication is called fetal macrosomia	Numerical
abortion or miscarriage	reason behind the death of an embryo or fetus before it can survive independently	Numerical
pre-Diabetes	before pregnancy having diabetes or not	Numerical
pre-GDM	pre-GDM outside of pregnancy	Numerical
high-Cholesterol	cholesterol concentration in the body	Numerical
high-BP	pressure Diastolic blood Pressure	Numerical
heart disease	patient heart-related issues during Pregnancy	Numerical
PCOS	women's hormones are out of balance	Numerical
taken-Steroid	any kind of steroid taken during pregnancy or before pregnancy	Numerical
exercise	having regular exercise or not	Numerical
dry mouth	feeling thirsty more than usual time	Numerical
frequent Urination	needing to urinate more often than usual	Numerical
excessive-Sweating	sweating more often than usual	Numerical
Hypertension	is the patient having a blood pressure issue?	Numerical
Congenital Anomalies	structural or functional anomalies that occur during intrauterine life	Numerical
macrosomia	a newborn with an excessive birth weight	Numerical
pre-Elampsia	serious blood pressure condition that develops during pregnancy	Numerical
neonatal-Loss	a baby dies within 28 days after they are born	Numerical
OGTT (**class label**)	utilized to detect or identify diabetes in individuals whose fasting blood glucose levels are over 125 mg/dL, which is higher than usual yet not too high to qualify as having the disease	Numerical

Final Data Set Description. After pre-processing, the data set contains in total of 337 records and 26 features, which are shown in Table 3.

Table 3. Total records in each class

Classes	Total Records
Yes	234
No	103

3.3   Feature Selection

In our study, *SelectKBest* class is used in Univariate Selection as Feature selection technique, which finds out the top useful features on the k highest scores, where k was set as 7 and the score function was set as Chi-square. Chi-Square Test method selects highly dependent features which has a higher Chi-Square value. The formula to calculate chi-square is shown in (1).

$$X_c^2 = \sum \frac{(O_i - E_i)^2}{E_i} \tag{1}$$

where, O = observed count, E = expected count, and C = degree of freedom. Hence, 7 features out of 27 were found to be optimal. The features are:

1. Taken steroids
2. Congenital anomalies
3. Excessive sweating
4. Macrosomia
5. Frequent urination
6. High BP
7. Pre-elampsia

3.4   Applications of Algorithms

After pre-processing the data, a total of 337 records were split for training(80%) and testing(20%) purpose using a model selection method named *train_test_split* from the python package named *sklearn*[2]. The data that were split for training and testing were applied to fit and evaluate six different models. Finally, Supervised learning and ensemble learning e.g., 1. Bagging and 2. Boosting were applied. Apart from that, Custom ensemble techniques were used to compare the performance.

Supervised Learning. Supervised learning involves training algorithms on labeled datasets to enable accurate data classification. Since our study had input features and corresponding outputs, it is based on supervised learning. Initially, six classification algorithms were trained on the dataset without applying feature selection, and their performance was evaluated. Afterwards, a feature selection technique was employed to identify the top seven features, which were then used to train the same six algorithms, followed by performance analysis. A 5-fold cross-validation method was implemented to train and evaluate the six classifiers consistently. The six supervised learning algorithms applied include:

1. *Logistic Regression:* Logistic Regression (LR) is one of the most commonly used and straightforward classification algorithms among supervised learning methods. It estimates the probability of a particular class based on the given input features. Mathematically, for an input X, it predicts the probability P(Y=1). This method determines the output class by applying a suitable threshold to the calculated probabilities of multiple input variables. In our study, LR was used on our data set, keeping the default hyperparameter values of the library called *scikit learn.*

---

[2] https://scikit-learn.org.

2. *Decision Tree:* A Decision Tree (DT) is a supervised learning algorithm that divides data based on specific criteria. It makes predictions by following a series of decision rules. The tree continues to split the data until it reaches pure leaf nodes, where all samples belong to the same class, although the splitting depth can be limited. Our study used the training data to fit a DT model, and its performance was evaluated using the default hyperparameter values of *scikit learn.*

3. *Random Forest:* Random Forest (RF) generates multiple DT through the bagging method to improve prediction reliability. It determines the final output by selecting the class with the majority of votes from these trees. Typically, the process involves four steps: selecting random samples, building a DT for each sample, applying a voting mechanism, and choosing the class with the most votes. In our research, the hyperparameter, i.e., *n-estimators* and *random state*, which controls the randomness of the model, were set as default values of *scikit learn.* Then, the training data was fitted, and the performance of this model was evaluated.

4. *Support Vector Machine:* Support Vector Machine (SVM) classifies data by leveraging key concepts such as the hyperplane, margin, and kernel. It represents data points in an n-dimensional space, where n corresponds to the number of features. The main objective of SVM is to identify the optimal hyperplane that best separates the classes. In this study, a linear kernel was used. Other hyperparameters used the default values of the *scikit learn* library.

5. *Naive Bayes (Gaussian):* Gaussian Naive Bayes (GaussianNB) classifiers use Bayes' Theorem and the Gaussian normal distribution to determine the output.
   Here, the attributes of each pair to be classified are not dependent on each other. The equation to calculate the likelihood of the features is shown as:

$$P(x_i \mid y) = \frac{1}{\sqrt{2\pi\sigma_y^2}} \exp\left(-\frac{(x_i - \mu_y)^2}{2\sigma_y^2}\right) \qquad (2)$$

   where μ and σ represent the mean and standard deviation of X. It is assumed that, variance is independent of Y (i.e., σ_i), or independent of X_i (i.e., σ_k) or both (i.e., σ).

6. *K Nearest Neighbors:* K-Nearest Neighbour (K-NN) is a supervised learning algorithm for classification and regression tasks. The parameter 'K' represents the number of nearest neighbors considered for classification. The algorithm follows several steps: selecting the value of K, calculating the Euclidean distance between data points, identifying the K closest neighbors, counting the data points in each category, and finally assigning the new data point to the most common category among its neighbors. The training data were fitted in this model using the default hyperparameter values of *scikit learn,* and the performance of that model was analyzed after that.

Ensemble Learning. Ensemble techniques combine individual models to improve stability and better result of the model. In our research, two types of ensemble techniques were used, including:

1. *Bagging technique:* The bagging technique creates multiple base learners from the original model, each trained on a different sample drawn from the dataset. During prediction, every model produces an output for the new data, and the final decision is made based on the majority vote among these outputs. In our study, the bagging classifier model was trained on our data for the six different classification models, and these models were set as the *base_estimator*, keeping *n_estimators* and *max_samples* as 50 and 0.8, respectively. This was achieved by using 5-fold cross validation as well.

2. *Boosting technique:* Boosting techniques build base learners one after another in sequence. Initially, a set of random records is used to train the first base learner, and subsequent learners are trained on records passed along in sequence. In our study, two different ensemble algorithms based on boosting were implemented. e.g.

    (a) *AdaBoost Classifier:* AdaBoost is a boosting technique where each data record is initially assigned an equal weight. The first base learner is then created sequentially, typically using a Decision Tree (DT) with a depth of one, known as a stump. These stumps are generated for each feature based on the initial Decision Tree (DT) learner. The AdaBoost model is trained in this way, and during prediction, the overall error is computed by summing the weights of all misclassified records. In our research, the AdaBoost model was fitted with the training data, keeping the default hyperparameter values of *scikit learn*, and the performance of that model was analyzed after then.

    (b) *Gradient Boosting Classifier:* Gradient Boosting is an ensemble method applicable to both continuous and categorical target variables. It improves upon initial weak assumptions by iteratively refining the model to reduce bias error. When addressing classification problems, the mean squared error (MSE) is used as the cost function. In this study, the Gradient Boosting Classifier was trained on the dataset using default hyperparameter settings, and its performance was subsequently evaluated.

Custom Ensemble Learning. Custom ensemble methods were employed to achieve improved results. In this approach, the AdaBoost classifier is utilized. Still, instead of relying on the default base learner, it was modified to incorporate three different classification algorithms described in detail earlier. The *n_estimators* and *learning_rate* are set as 100 and 1, respectively. The classification algorithms are:

1. Logistic Regression (LR)
2. Decision Tree (DT)
3. Gaussian Naive Bayes (GaussianNB)

4   Experimental Analysis

Following the training of the classifiers, their performance is thoroughly evaluated and analyzed. **Accuracy** is the primary metric to assess overall model performance. In addition, **precision, recall**, and **F1-score** are calculated. Precision quantifies the proportion of correctly predicted positive instances among all predicted positives, whereas recall measures the proportion of true positives that are correctly identified. The F1-score, representing the harmonic mean of precision and recall, provides a balanced assessment of both metrics.

The equations to calculate those are show in (3), (4), and (5).

$$Precision = \frac{TP}{TP + FP} \tag{3}$$

$$Recall = \frac{TP}{TP + FN} \tag{4}$$

$$F1 - score = 2 \times \frac{precision \times recall}{precision + recall} \tag{5}$$

where, TP = True positive rate, FP = False positive rate, FN = False negative rate.

4.1   Performance of Different Algorithms Using Supervised Learning

Table 4 presents the performance of six classifiers applied to the dataset. The Random Forest (RF) algorithm achieved the highest overall accuracy at 85.29%, corresponding recall and F1-score values of 0.842 and 0.762, respectively. This indicates that the RF model produced fewer false positives and false negatives while maximizing true positives, fulfilling the desired predictive criteria. The Decision Tree (DT) classifier exhibited the highest precision at 0.782. In contrast, the Gaussian Naive Bayes classifier demonstrated the lowest performance across all evaluated metrics, including accuracy, precision, recall, and F1-score.

Table 4. Performance matrices for different algorithms.

Algorithms	Accuracy (%)	Precision	Recall	F1-score
LR	57.35	0.261	0.333	0.292
DT	79.41	**0.782**	0.667	0.720
RF	**85.29**	0.696	**0.842**	**0.762**
SVM	72.06	0.348	0.667	0.458
GaussianNB	53.90	0.100	0.378	0.548
KNN	77.94	0.566	0.722	0.634

Table 5 summarizes the performance of six classifiers when trained using the top seven features. All models achieved an accuracy above 65%. The k-Nearest

Neighbors (KNN) classifier attained the highest accuracy and recall, with recall reaching 87.5%. In contrast, Logistic Regression (LR) exhibited the lowest performance among the models, with an accuracy of 63.24%, precision of 11.54%, and recall of 19.35%.

Table 5. Performance matrices of different algorithms using the Top 7 Features.

Algorithms	Accuracy (%)	Precision	Recall	F1-score
LR	63.24	0.116	0.600	0.193
DT	67.65	0.192	0.833	0.312
RF	66.18	0.192	0.714	0.303
SVM	67.65	0.192	0.833	0.312
GaussianNB	60.29	**0.461**	0.480	**0.480**
KNN	**70.59**	0.270	**0.876**	0.411

Table 6 presents the performance of six classifiers evaluated using 5-fold cross-validation on all features. The Random Forest (RF) classifier achieved the highest accuracy and precision among the models, and it also attained the highest recall, indicating robust and balanced predictive performance across all evaluation metrics.

The Random Forest (RF) classifier achieved an accuracy of 86.91% and a precision of 89.34%, with a recall of 90.29%. These results indicate that the model produced fewer false positives and false negatives while maximizing true positives, fulfilling the desired predictive criteria. In addition, the k-Nearest Neighbors (KNN) and Decision Tree (DT) classifiers also demonstrated strong performance, achieving accuracies of 81.81% and 79.78%, respectively. In contrast, the Gaussian Naive Bayes classifier exhibited the lowest performance across all metrics, with an accuracy of 43.36%, precision of 37.52%, and recall of 27.52%, highlighting its limited effectiveness for this dataset.

Table 6. Performance matrices for different algorithms using Cross-Validation.

Algorithms	Accuracy (%)	Precision	Recall	F1-score
LR	71.76	0.754	0.870	0.810
DT	79.78	0.864	0.860	0.860
RF	**86.91**	**0.893**	0.919	**0.902**
SVM	73.56	0.747	**0.936**	0.830
Gaussian	48.36	0.376	0.273	0.418
KNN	81.81	0.857	0.880	0.867

4.2 Performance of Different Algorithms Using Ensemble Learning

Table 7 presents the performance of six classifiers when combined with a bagging ensemble and evaluated using 5-fold cross-validation. The Random Forest (RF) classifier achieved the highest accuracy and F1-score, with 86.29% and 90.78%, respectively. The Decision Tree (DT) demonstrated the second-highest accuracy at 86.01% and the highest precision of 89.31%, closely matching the performance of RF. The k-Nearest Neighbors (KNN) classifier also exhibited strong performance. In contrast, the Gaussian Naive Bayes classifier performed poorly, with an accuracy of 46.94%, a recall of 25.18%, and an F1-score of 38.40%, indicating limited predictive capability for this dataset.

Table 7. Performance matrices for different algorithms using Ensemble Bagging techniques.

Algorithms	Accuracy (%)	Precision	Recall	F1-score
LR	69.10	0.732	0.871	0.791
DT	86.01	0.893	0.910	0.900
RF	**86.29**	0.877	0.944	**0.908**
SVM	73.87	0.744	**0.949**	0.833
GaussianNB	46.29	**0.924**	0.251	0.384
KNN	80.94	0.851	0.884	0.867

Table 8 presents the performance of two boosting classifiers. Among them, the Gradient Boosting classifier achieved superior results across all evaluation metrics, with an accuracy of 86.76% and a precision of 84.3%. Recall and F1-score were 82.98% and 81.02%, respectively, indicating that this model produced fewer false positives and negatives while maximizing true positives. The AdaBoost classifier also demonstrated strong performance, though slightly lower than Gradient Boosting.

Table 8. Performance matrices for different algorithms using Ensemble Boosting techniques.

Algorithms	Accuracy (%)	Precision	Recall	F1-score
AdaBoost	70.59	0.566	0.566	0.566
Gradient Boosting Classifier	**86.76**	**0.844**	**0.830**	**0.810**

4.3 Performance of Different Algorithms Using Custom Ensemble Learning

According to Table IX, RF gave the highest accuracy, which is 88. 24%, and the rest of them gave low accuracies, which are 57.35% (LR) & 33.82% (NB)

Table 9 presents the performance of three classifiers evaluated using a custom AdaBoost ensemble. In this scenario, the Random Forest (RF) classifier emerged as the best base estimator for AdaBoost. Furthermore, this configuration achieved the highest performance among all models evaluated in this study, attaining the top scores in accuracy and across all other metrics, including precision, recall, and F1-score.

For this model, the Random Forest (RF) classifier within the custom AdaBoost framework achieved an accuracy of 88.24% and a precision of 78.26%. Recall and F1-score were 85.71% and 81.82%, respectively, indicating that the model produced fewer false positives and false negatives while maximizing true positives. In contrast, the Gaussian Naive Bayes classifier performed poorly in this setting, achieving an accuracy of 33.82% and a precision of 59%, with a recall and F1-score of 33.82% and 50.55%, respectively.

Table 9. Performance matrices for different algorithms using a Custom AdaBoost.

Algorithms	Accuracy (%)	Precision	Recall	F1-score
LR	57.30	0.260	0.333	0.292
RF	**88.24**	**0.782**	**0.858**	**0.819**
GaussianNB	33.82	0.590	0.337	0.506

Figure 2 compares classifier accuracy before and after feature selection and cross-validation. Before applying feature selection and cross-validation, the highest accuracy of 85.29% was achieved by the Random Forest (RF) classifier. After

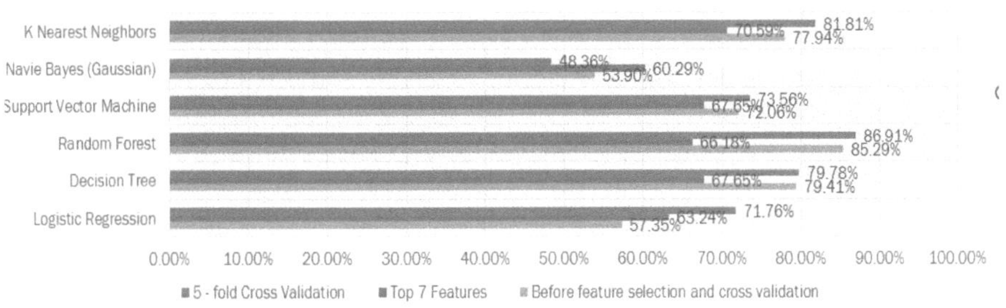

Fig. 2. Analyzing accuracy variations before and after feature selection combined with Cross-Validation.

selecting the top seven features, accuracy decreased for all models except Gaussian Naive Bayes. k-Nearest Neighbors (KNN) achieved the highest accuracy in this scenario at 70.59%, indicating a notable reduction in predictive performance. Conversely, when a 5-fold cross-validation approach was applied, the accuracy of all models improved, with KNN attaining the highest accuracy of 81.81%.

Figure 3 presents the accuracy of different classifiers evaluated using cross-validation and the Bagging ensemble. The Random Forest (RF) classifier achieved the highest accuracy in both scenarios. The corresponding accuracies for all models are plotted for comparison, indicating that k-Nearest Neighbors (KNN) attained the second-highest accuracy across both evaluation methods.

Figure 3 compares the accuracy of six classifiers using cross-validation and the Bagging ensemble technique. The Random Forest (RF) classifier achieved the highest accuracy in both cases, with 86.91% and 86.29%, respectively. The highest accuracy slightly decreased when cross-validation was applied to the Bagging classifier. In contrast, the accuracy of Logistic Regression (LR), Decision Tree (DT), and Support Vector Machine (SVM) improved under this approach, reaching 69.10%, 86.10%, and 73.87%, respectively.

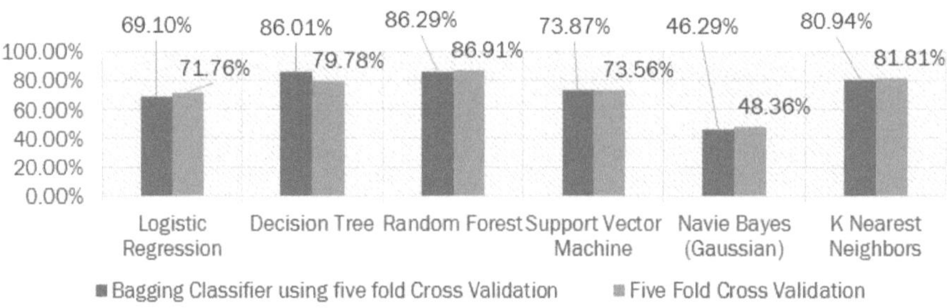

Fig. 3. Analyzing accuracy across multiple algorithms via Cross-Validation and Bagging approaches.

Figure 4 compares accuracy, precision, recall, and F1-score for two ensemble boosting classifiers. The Gradient Boosting classifier outperformed AdaBoost across all metrics, achieving an accuracy of 86.76% and a precision of 84.3%, with a recall and F1-score of 82.98% and 81.02%, respectively. Nonetheless, AdaBoost also demonstrated satisfactory performance, albeit lower than Gradient Boosting.

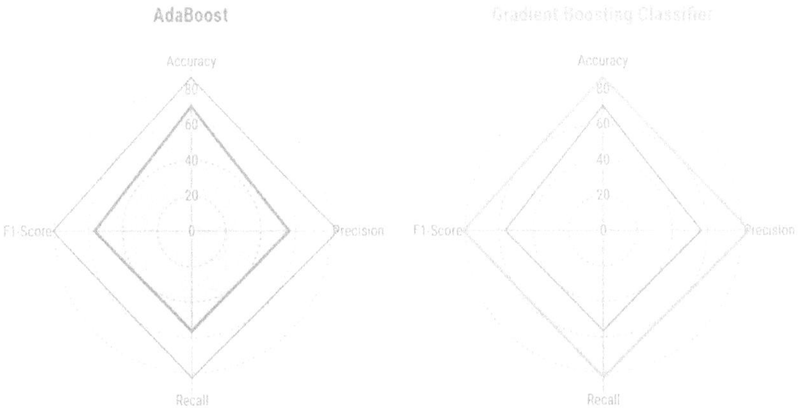

Fig. 4. Analyzing performance metrics of multiple algorithms through Ensemble Boosting approaches.

Figure 5 compares accuracy, precision, recall, and F1-score for three classifiers evaluated using a custom AdaBoost ensemble. The Random Forest (RF) classifier achieved the best performance across all metrics, with an accuracy of 88.24%, precision of 78.26%, recall of 85.71%, and F1-score of 81.82%. Logistic Regression (LR) achieved moderate accuracy at 57.35%, whereas Gaussian Naive Bayes demonstrated the lowest accuracy at 33.82%. Notably, while Naive Bayes performed poorly in accuracy, it achieved somewhat higher recall and F1-score values, with a recall of 50.55% and an F1-score of 33.82%, indicating a relatively better ability to identify positive cases despite overall low accuracy.

Fig. 5. Analyzing performance metrics of multiple algorithms through a Custom AdaBoost.

In our experiment, the accuracy decreases significantly after applying feature selection for most models for several reasons (i.e., the dataset is imbalanced (about 70% positive cases) and small in size, as well as complex interactions between features).

5 Conclusion and Future Works

This study presents a data mining-based approach for predicting Gestational Diabetes among Bangladeshi women. A comparative analysis uses six classification algorithms: Logistic Regression (LR), Random Forest (RF), Support Vector Machine (SVM), Decision Tree (DT, Gaussian Naive Bayes (GaussianNB), and k-Nearest Neighbors (KNN)—before and after applying feature selection (top 7 features) and 5-fold cross-validation. Additionally, the performance of the six classifiers is compared using a bagging ensemble method with 5-fold cross-validation. Boosting techniques and a custom ensemble method were also developed and evaluated. Model performance was measured using accuracy, precision, recall, and F1 score. Among the supervised learning models, Random Forest (RF) achieved the highest accuracy of 86.29% with cross-validation. Gradient Boosting yielded the highest accuracy within the ensemble techniques at 86.76%. The custom ensemble method combined with the Random Forest (RF) classifier achieved the best overall result with an accuracy of 88.24%. Moreover, Random Forest (RF) outperformed other models regarding recall, precision, and F1 score for Logistic Regression (LR). In the future, we plan to collect additional data to identify more relevant factors associated with this condition and address the issue of the significant decrease in accuracy after applying feature selection. We also aim to explore and analyze the emerging field of Explainable Artificial Intelligence (XAI).

References

1. Afsana, F., et al.: Current practices in diagnosis and management of gestational diabetes: a Bangladesh study. J. Diabetol. **12**(5), 79 (2021)
2. Artzi, N.S., et al.: Prediction of gestational diabetes based on nationwide electronic health records. Nat. Med. **26**(1), 71–76 (2020)
3. Bogdanet, D., et al.: Emerging protein biomarkers for the diagnosis or prediction of gestational diabetes–a scoping review. J. Clin. Med. **10**(7), 1533 (2021)
4. Christobel, T.P., Kamalakannan, T.: Predictive analysis in gestational diabetic mellitus (GDM) using HCNN-LSTM/DPNN (big data). In: 2020 3rd International Conference on Intelligent Sustainable Systems (ICISS), pp. 407–413. IEEE (2020)
5. Correa, P.J., Vargas, J.F., Sen, S., Illanes, S.E.: Prediction of gestational diabetes early in pregnancy: targeting the long-term complications. Gynecol. Obstet. Invest. **77**(3), 145–149 (2014)
6. Donovan, B.M., et al.: Development and validation of a clinical model for preconception and early pregnancy risk prediction of gestational diabetes mellitus in nulliparous women. PLoS ONE **14**(4), e0215173 (2019)
7. Du, Y., Rafferty, A.R., McAuliffe, F.M., Wei, L., Mooney, C.: An explainable machine learning-based clinical decision support system for prediction of gestational diabetes mellitus. Sci. Rep. **12**(1), 1–14 (2022)

8. Gao, S., et al.: Development and validation of an early pregnancy risk score for the prediction of gestational diabetes mellitus in Chinese pregnant women. BMJ Open Diabetes Res. Care **8**(1), e000909 (2020)

9. Gnanadass, I.: Prediction of gestational diabetes by machine learning algorithms. IEEE Potentials **39**(6), 32–37 (2020)

10. Hou, F., Cheng, Z., Kang, L., Zheng, W.: Prediction of gestational diabetes based on LightGBM. In: Proceedings of the 2020 Conference on Artificial Intelligence and Healthcare, pp. 161–165 (2020)

11. Kumar, P.S., Pranavi, S.: Performance analysis of machine learning algorithms on diabetes dataset using big data analytics. In: 2017 International Conference on Infocom Technologies and Unmanned Systems (trends and future directions)(ICTUS), pp. 508–513. IEEE (2017)

12. Lee, S.M., et al.: Nonalcoholic fatty liver disease and early prediction of gestational diabetes mellitus using machine learning methods. Clin. Mol. Hepatol. **28**(1), 105 (2022)

13. Liu, H., et al.: Machine learning risk score for prediction of gestational diabetes in early pregnancy in Tianjin, china. Diabetes Metab. Res. Rev. **37**(5), e3397 (2021)

14. Maged, A.M., Moety, G.A.F., Mostafa, W.A., Hamed, D.A.: Comparative study between different biomarkers for early prediction of gestational diabetes mellitus. J. Matern.-Fetal Neonatal Med. **27**(11), 1108–1112 (2014)

15. Mazumder, T., Akter, E., Rahman, S.M., Islam, M.T., Talukder, M.R.: Prevalence and risk factors of gestational diabetes mellitus in Bangladesh: findings from demographic health survey 2017–2018. Int. J. Environ. Res. Public Health **19**(5), 2583 (2022)

16. Moreira, M.W., Rodrigues, J.J., Kumar, N., Al-Muhtadi, J., Korotaev, V.: Evolutionary radial basis function network for gestational diabetes data analytics. J. Comput. Sci. **27**, 410–417 (2018)

17. Naser, W., Adam, I., Rayis, D.A., Ahmed, M.A., Hamdan, H.Z.: Serum magnesium and high-sensitivity C-reactive protein as a predictor for gestational diabetes mellitus in Sudanese pregnant women. BMC Pregnancy Childbirth **19**(1), 1–5 (2019)

18. Pustozerov, E.A., et al.: Machine learning approach for postprandial blood glucose prediction in gestational diabetes mellitus. IEEE Access **8**, 219308–219321 (2020)

19. Qiu, H., et al.: Electronic health record driven prediction for gestational diabetes mellitus in early pregnancy. Sci. Rep. **7**(1), 1–13 (2017)

20. Raveendra, C., Thiyagarajan, M., Thulasi, P., Priya, S.K.: Role of association rules in medical examination records of gestational diabetes mellitus. In: 2017 International Conference on Computing, Communication and Automation (ICCCA), pp. 78–81. IEEE (2017)

21. Wu, Y.T., et al.: Early prediction of gestational diabetes mellitus in the Chinese population via advanced machine learning. J. Clin. Endocrinol. Metab. **106**(3), e1191–e1205 (2021)

22. Xiong, Y., et al.: Prediction of gestational diabetes mellitus in the first 19 weeks of pregnancy using machine learning techniques. J. Matern.-Fetal Neonatal Med. **35**(13), 2457–2463 (2022)

23. Zhang, J., Wang, F.: Prediction of gestational diabetes mellitus under cascade and ensemble learning algorithm. Comput. Intell. Neurosci. **2022** (2022)

GeKAN: Enhancing High-Dimensional Gene Expression Classification with Feature-Selected Kolmogorov-Arnold Networks

Bich-Chung Phan, Thanh Ma, and Huu-Hoa Nguyen[✉]

College of Information and Communication Technology, Can Tho University, 3/2 Street, Can Tho City, Vietnam
{pbchung,mtthanh,nhhoa}@ctu.edu.vn

Abstract. The classification of high-dimensional gene expression data presents formidable computational and accuracy challenges, primarily stemming from inherent dimensionality, noise, and feature redundancy. To address these limitations, this study introduces GeKAN, a novel framework that synergistically integrates advanced feature selection methodologies (Boruta and mRMR) with Kolmogorov-Arnold Network (KAN) variants to optimise nonlinear classification. Our approach demonstrably enhances both predictive precision and computational efficiency. The experiment confirms that our proposal achieves substantial improvements in classification accuracy (elevating performance by 8–10%) while reducing computational time by approximately threefold compared to conventional methods. These results underscore the framework's efficacy in mitigating dimensionality-related constraints, establishing it as a robust solution for high-throughput genomic data analysis.

Keywords: Gene expression classification · mRMR · Boruta · KAN · Feature Selection

1 Introduction

Gene expression classification [1,8–11] has become an indispensable approach in bioinformatics, facilitating disease subtype identification, outcome prediction, and the discovery of therapeutic targets. Recent advancements in high-throughput sequencing technologies have enabled researchers to profile thousands of genes simultaneously. Despite these advancements, the analysis of gene expression data remains challenging due to the pervasive issue of high dimensionality, often referred to as the *curse of dimensionality* [2,15]. Typically, the number of genes (p) vastly surpasses the number of available samples (n), creating significant hurdles for traditional machine learning classifiers. Such imbalanced scenarios frequently result in severe overfitting, diminished predictive performance, and increased computational overhead. To mitigate these issues, feature selection strategies [17,20] have emerged as crucial preprocessing steps.

N. Thai-Nghe et al. (Eds.): ISDS 2025, CCIS 2714, pp. 176–190, 2026.
https://doi.org/10.1007/978-981-95-3358-9_13

By retaining only the most informative and relevant genes, these methods significantly enhance the accuracy, interpretability, and efficiency of subsequent predictive models [8,12]. Notably, gene expression data inherently contains biological redundancy and noise, reinforcing the importance of robust feature selection techniques. Conventional approaches include filter-based [5], wrapper-based [3,13,21,26], and embedded methods [24]. Nevertheless, no single method consistently excels across diverse datasets, leading to increased interest in hybrid and integrative approaches such as Boruta [16,27] and mRMR (Minimum Redundancy Maximum Relevance)[23].

In parallel with advances in feature selection, neural network architectures have evolved substantially, providing increasingly powerful nonlinear modeling capabilities. Recently, Kolmogorov–Arnold Networks (KAN) [19] and their variants, including Efficient-kan [6], FastKAN [18] and FasterKAN have emerged as effective solutions to capture complex nonlinear relationships in high-dimensional spaces. KAN architectures leverage spline-based nonlinear transformations, offering superior performance compared to traditional multilayer perceptrons (MLP) [25] in various applications. These methods are particularly advantageous for gene expression classification tasks due to their inherent capability to manage nonlinear interactions among genes.

Motivated by these insights, this study proposes GeKAN, a novel hybrid algorithm that integrates advanced feature selection methods (Boruta and mRMR) with the robust nonlinear classification capabilities of KAN variants. GeKAN systematically exploits the strengths of *global feature* selectors as Boruta and redundancy-minimizing strategies as mRMR to distill the feature space prior to classification effectively. Subsequently, the selected feature subsets are input into various KAN-based classifiers, effectively addressing the complexities associated with high-dimensional gene expression data. The combination of feature-selected subsets with KAN variants (i.e., FasterKAN) is anticipated to improve predictive accuracy and computational efficiency.

The remainder of this paper is structured as follows: Sect. 2 provides a fundamental background of the methods employed. Section 3 details the proposed GeKAN algorithmic framework. Section 4 describes the experimental settings and evaluates the performance of GeKAN alongside comparative analyses. Finally, Sect. 5 concludes the study and outlines future research directions.

2   Background

In this section, we describe the foundation of our proposed method, which integrates feature selection techniques such as Boruta and mRMR with advanced KANs. Additionally, we introduce the ML models utilized for classifying gene expression data.

2.1   Gene Expression Classification with DL Models

Gene expression classification [8,9,11] lies at the forefront of biomedical research, offering profound insights into the molecular mechanisms underlying various diseases. Deep Learning (DL) models have become indispensable in this domain, as they can uncover complex patterns within vast and high-dimensional gene expression datasets. However, these datasets often contain numerous features, many of which are redundant or irrelevant, potentially obscuring the most critical biological signals and leading to overfitting. Consequently, feature selection becomes imperative—it refines the dataset by isolating the most informative genes, thereby enhancing model accuracy and computational efficiency. By focusing solely on pivotal biomarkers, this research achieves more reliable predictive outcomes. In this paper, we investigate and evaluate classification using nonlinear DL techniques, specifically KAN variants [19], including Efficient-KAN [6], FastKAN [18], and FasterKAN.

Definition 1 (Classification). *Let $D = (X, y)$ be a dataset where $X \subseteq \mathbb{R}^n$ is the feature space and $y \in \mathcal{Y} = \{1, 2, \ldots, k\}$ represents the class labels. A classifier is a function $f : X \to \mathcal{Y}$, assigning a predicted label $\hat{y} = f(x)$ to each input $x \in X$. The function f is learned from labeled instances: $D = \{(x_i, y_i) \mid x_i \in X, \ y_i \in \mathcal{Y}, \ i = 1, \ldots, N\}$, by minimizing a loss function $\ell : \mathcal{Y} \times \mathcal{Y} \to \mathbb{R}_{\geq 0}$ that quantifies the error between the predicted and true labels. Once trained, f is used to classify new, unseen inputs.*

2.2   Leveraging Boruta for Robust Feature Extraction

Boruta [16,27] is a powerful wrapper-based feature selection algorithm designed to identify all truly relevant variables in a dataset. By comparing the importance of actual features with that of randomly generated *"shadow"* features, Boruta systematically filters out irrelevant variables while preserving essential predictors. This rigorous selection process is particularly valuable in high-dimensional applications, such as gene expression classification, where capturing meaningful signals is crucial. For clarity, we formally define Boruta as follows:

Definition 2 (Boruta feature selection). *Let $D = (X, y)$ be a dataset with features $X = \{x_1, x_2, \ldots, x_p\}$ and target y. The Boruta algorithm identifies all relevant features in X as follows: (1) Shadow Feature Generation: For each $x_i \in X$, create a shadow feature x_i^{shadow} by randomly permuting its values, forming the set X^{shadow}; (2) Importance Estimation: Train a classifier (e.g., Random Forest) on the combined set $X \cup X^{shadow}$ and compute the importance score $I(z)$ for each z; (3) Feature Comparison: For each x_i, define $I_{\max}^{shadow} = \max_{z \in X^{shadow}} I(z)$.   Then classify x_i as* relevant *if $I(x_i)$ is significantly greater than $I_{\max}^{shadow}$,* irrelevant *if significantly lower, or* tentative *otherwise; Iteration: (4) Remove irrelevant and tentative features and repeat until all features are decisively classified. The final selected subset $X^* \subseteq X$ comprises all features deemed relevant.*

After applying the Boruta algorithm, we retain only the relevant features (confirmed) and excluded the tentative and irrelevant features (rejected). To further enhance the selection of features in X^*, we employed the AI explanation technique outlined in the following section.

2.3   Leveraging mRMR for Discriminative Feature Extraction

Minimum Redundancy Maximum Relevance (mRMR) [7,22] is a widely adopted filter-based feature selection technique that aims to select features with the highest relevance to the target variable while minimizing redundancy among the selected features. In high-dimensional gene expression data, mRMR is particularly advantageous as it identifies compact subsets of features that are both informative and non-redundant, improving the efficiency and performance of downstream classifiers. For clarity, we formally define mRMR as follows:

Definition 3 (mRMR feature selection). *Let $D = (X, y)$ be a dataset with features $X = \{x_1, x_2, \ldots, x_p\}$ and target y. The mRMR algorithm selects a subset $S^* \subseteq X$ of m features by maximizing the following criterion:*

$$\max_{S \subseteq X,\ |S|=m} \left[\frac{1}{|S|} \sum_{x_i \in S} \mathrm{Rel}(x_i, y) - \frac{1}{|S|^2} \sum_{x_i, x_j \in S} \mathrm{Red}(x_i, x_j) \right]$$

where $\mathrm{Rel}(x_i, y)$ denotes the relevance of feature x_i to the target y (e.g., mutual information $I(x_i; y)$), and $\mathrm{Red}(x_i, x_j)$ measures the redundancy between features x_i and x_j (e.g., mutual information $I(x_i; x_j)$).

1. ***Initialization:*** *Start with an empty set $S = \emptyset$.*
2. ***Selection:*** *At each step, add the feature $x_j \in XS$ that maximizes $\mathrm{Score}(x_j) = \mathrm{Rel}(x_j, y) - \frac{1}{|S|} \sum_{x_i \in S} \mathrm{Red}(x_j, x_i)$.*
3. ***Iteration:*** *Repeat until $|S| = m$ or another stopping criterion is met.*

The final subset S^* contains the features that achieve maximum relevance to y and minimum redundancy among themselves.

After applying the mRMR algorithm, we obtain a reduced feature set S^*, which is expected to enhance classification performance by removing irrelevant and redundant features while preserving those most informative for the classification.

2.4   Gene Expression Classification with KAN Variants

Kolmogorov–Arnold Networks (KAN) [19] represent a novel neural network architecture inspired by the Kolmogorov–Arnold representation theorem [14]. This theorem states that any continuous multivariate function $f : [0,1]^n \to \mathbb{R}$ can be represented as a composition of continuous single-variable functions:

$$f(x_1, \ldots, x_n) = \sum_{q=1}^{2n+1} \Phi_q \left(\sum_{p=1}^{n} \phi_{q,p}(x_p) \right). \tag{1}$$

KANs replace fixed activation functions with learnable univariate functions, typically parameterized as B-splines. This feature significantly improves flexibility compared to traditional Multi-Layer Perceptrons (MLPs), where activations are fixed and the complexity relies predominantly on linear weight combinations.

In KANs, the learnable univariate functions $\phi_{q,p}$ are expressed using B-splines: $\text{spline}(x) = \sum_i c_i B_i(x)$, where c_i are trainable coefficients and $B_i(x)$ are B-spline basis functions. Moreover, KANs utilize residual activation functions combining a base function $b(x)$ with B-spline approximations: $\phi(x) = w_b b(x) + w_s \text{spline}(x)$, where $b(x)$ is a sigmoid linear unit (SiLU), and w_b, w_s are trainable weights. KAN models are trained using backpropagation, compatible with standard optimization techniques such as Differentially Private SGD (DP-SGD). Recent advancements in KAN architectures have focused on improving computational efficiency, memory usage, and flexibility:

Efficient-kan [6][1] significantly reduces the number of spline parameters by employing sparser grids or lower-order splines, coupled with optimized spline evaluation techniques. This approach greatly enhances computational and memory efficiency, making it suitable for high-dimensional data applications.

FastKAN [18][2] further improves computational speed by substituting traditional third-order B-spline bases with Gaussian radial basis functions (RBFs). Formally, a FastKAN univariate function is represented as: $\text{fast\_spline}(x) = \sum_i \alpha_i \tilde{B}_i(x)$, where $\tilde{B}_i(x)$ are Gaussian RBFs and α_i are trainable coefficients. Additionally, FastKAN employs automatic input normalization techniques, ensuring that input values remain within the effective domain of the RBFs, thus stabilizing training and maintaining accuracy. Its simplified architecture facilitates parallel computation, significantly accelerating training and inference without sacrificing performance.

FasterKAN[3] introduces the Reflectional Switch Activation Function (RSWAF) as an advanced basis function, defined as: $b_i(u) = 1 - \left(\tanh\left(\frac{u-u_i}{h}\right)\right)^2$, where u_i represents the basis center and h controls the width of the function. This novel basis effectively approximates traditional B-splines while being computationally efficient. Moreover, FasterKAN incorporates automatic scaling of inputs to spline grids through learnable parameters, eliminating manual adjustments and enhancing model adaptability via trainable width parameters h.

Collectively, these innovations in Efficient-kan, FastKAN, and FasterKAN expand the applicability of KAN architectures to large-scale, high-dimensional gene expression datasets, achieving remarkable improvements in computational efficiency and predictive accuracy. For variants of KAN, we propose an algorithm that takes advantage of the power of KAN to improve the classification result. Our main proposal in this paper will be presented in the following section.

3   GeKAN Algorithm

Our methodology systematically addresses the high-dimensionality challenge of gene expression datasets through advanced feature selection. We leverage two

---

[1] https://github.com/Blealtan/efficient-kan.
[2] https://github.com/Blealtan/FastKAN.
[3] https://github.com/AthanasiosDelis/faster-kan.

complementary feature selection methods, Boruta and mRMR, to effectively identify informative gene subsets. Boruta uses random forests to globally select relevant genes, whereas mRMR prioritizes genes that are highly informative but minimally redundant. By integrating and exploring these approaches, we ensure robustness in the selected features, significantly enhancing the interpretability and accuracy of subsequent classifiers.

Namely, this study proposes a feature-aware classification framework that integrates state-of-the-art feature selection techniques with nonlinear KANs. The primary motivation stems from the need to reduce noise, improve model interpretability, and improve classification accuracy in datasets where the number of features vastly exceeds the number of samples. The proposed algorithm (Algorithm 1) starts by applying one of several feature selection strategies - Borta, mRMR, or a hybrid of both through union or intersection - to isolate the most informative gene features. This step mitigates the risk of overfitting while preserving biologically relevant signals. The reduced feature set is then used to train multiple KAN-based classifiers, including FastKAN, FasterKAN, and EfficientKAN, each offering varying trade-offs between speed and expressiveness.

To ensure robustness, the algorithm evaluates all trained models on a validation set using a chosen performance metric, such as accuracy or F1-score. The classifier with the highest validation score is selected as the final model. This approach allows for adaptive model selection while leveraging the strengths of feature selection and advanced non-linear modeling. Hence, it offers a scalable and effective solution for classifying complex, high-dimensional biological data.

Algorithm 1: Feature Selection and Model Training using Boruta, mRMR, and KAN Variations

---

$\quad$ **Input** $\quad$: $D = (X, y)$: Dataset where $X \in \mathbb{R}^{m \times n}$ and $y \in \mathbb{R}^m$ are features and class labels.
$\qquad\qquad$ $method \in \{\text{Boruta}, \text{mRMR}, \text{Intersection}, \text{Union}\}$: Feature selection strategy.
$\qquad\qquad$ $\mathcal{M} \in \{\text{Accuracy}, \text{F1-Score}\}$
$\quad$ **Output**: f_{opt}: Optimal classifier after feature selection and model training.

1   **begin**
2   **if** $method = Boruta$ **then**
3         $X_{\text{selected}} \leftarrow \text{Boruta}(X, y)$
4   **else if** $method = mRMR$ **then**
5         $X_{\text{selected}} \leftarrow \text{mRMR}(X, y)$
6   **else if** $method = Intersection$ **then**
7         $X_{\text{selected}} \leftarrow \text{Boruta}(X, y) \cap \text{mRMR}(X, y)$
8   **else**
9         $X_{\text{selected}} \leftarrow \text{Boruta}(X, y) \cup \text{mRMR}(X, y)$
10      $\oplus \leftarrow \emptyset$
11   **foreach** $model \in \{FastKAN, FasterKAN, EfficientKAN\}$ **do**
12         $f_{\text{model}} \leftarrow \text{Train}(X_{\text{selected}}, y, \text{model})$
13         $\oplus \leftarrow \oplus \cup f_{\text{model}}$
14      $best\_score \leftarrow -\infty$
15      $f_{\text{opt}} \leftarrow \text{None}$
16   **foreach** $f \in \oplus$ **do**
17   score $\leftarrow \mathcal{M}(f(X_{\text{val}}), y_{\text{val}})$ `// Evaluate each` $f \in \oplus$ `on validation set` $(X_{\text{val}}, y_{\text{val}})$
$\qquad\qquad\qquad$ `using metric` $\mathcal{M}$
18   **if** $score > best\_score$ **then**
19            $best\_score \leftarrow score$
20            $f_{\text{opt}} \leftarrow f$
21   **return** f_{opt}

Computational Complexity. The overall time complexity of Algorithm 1 consists of three main components: feature selection, model training, and evaluation. For feature selection, Boruta operates at $O(n \cdot m \cdot \log m \cdot T)$ due to its reliance on random forests, while mRMR typically incurs $O(n^2)$ complexity for pairwise mutual information computation. In the worst case, the union or intersection strategy combines both, doubling this cost. Each KAN variant is trained on the reduced feature set $X_{\text{selected}} \in \mathbb{R}^{m \times n'}$, with training cost denoted as $O(C(n', m))$ per model, resulting in a total of $3 \cdot O(C(n', m))$. The evaluation phase contributes $O(3 \cdot m_{\text{val}} \cdot n')$ for validation scoring. Therefore, the end-to-end complexity can be summarized as: $O(n \cdot m \cdot \log m + n^2 + 3 \cdot C(n', m))$, where $n' \ll n$ after feature selection. This design effectively balances computational cost and classification performance on high-dimensional gene expression data.

To evaluate our proposal, an experiment will provide in the following section.

4   Experiment and Results

This section offers a brief description of the gene expression datasets while delivering a detailed comparative analysis of the classification models. Furthermore, we also provide the results of Boruta. Our source code has been made publicly accessible on GitHub[4].

4.1   Dataset and Configurations

Table 1. Datasets and feature selection results (Boruta[col6], mRMR[col7])

ID	Datasets	#Datapoints	#Dimensions	#Classes	#Boruta	#mRMR
1	E-GEOD-20685	327	54,627	6	545	1,878
2	E-GEOD-20711	90	54,675	5	111	701
3	E-GEOD-21050	310	54,613	4	72	3,028
4	E-GEOD-21122	158	22,283	7	271	856
5	E-GEOD-29354	53	22,215	3	28	237
6	E-GEOD-30784	229	54,675	3	171	732
7	E-GEOD-31312	498	54,630	3	213	3,301
8	E-GEOD-31552	111	33,297	3	79	592
9	E-GEOD-32537	217	11,950	7	96	1,444
10	E-GEOD-33315	575	22,283	10	483	3,132
11	E-GEOD-36895	76	54,675	14	39	768
12	E-GEOD-37364	94	54,675	4	59	514
13	E-GEOD-39582	566	54,675	6	640	4,251
14	E-GEOD-39716	53	33,297	3	124	311
15	E-GEOD-44077	226	33,252	4	227	210
16	E-GEOD-44760	582	15,261	10	135	3,339
17	E-GEOD-63270	104	18,989	9	80	879
18	E-GEOD-63885	101	54,675	4	38	916
19	E-GEOD-65106	59	33,297	2	3	328
20	E-GEOD-65329	327	22,645	5	87	1,012
21	E-GEOD-66533	58	54,675	4	130	156
22	E-GEOD-66848	390	22,283	6	278	1,035
23	E-GEOD-68606	274	22,283	16	700	651
24	E-GEOD-7307	677	54,675	12	3	4,958
25	E-GEOD-73685	183	33,297	8	343	1,087

---

[4] https://github.com/Pbchung75/GeKAN/.

We carry out a series of experiments using 25 gene expression datasets, all meticulously sourced from the reputable ArrayExpress repository [4]. The gene expression datasets [8] summarized in Table 1 exemplify the inherent challenges of high-dimensional biomedical data. The Datapoints column indicates the number of samples, ranging from 53 to 677. The Dimensions column specifies the number of features, spanning from 11,950 to 54,675, reflecting typical high-dimensional and low-sample scenarios commonly encountered in gene expression data analysis. The Classes column shows the number of classification labels, varying between 3 and 16, reflecting biological diversity or disease grouping variability among the studied samples. This high dimensionality, coupled with limited sample sizes, accentuates the risk of overfitting and underscores the critical need for effective feature selection.

In this study, our feature selection strategy is specifically designed to address these challenges, ensuring that only the most relevant features are retained for subsequent classification tasks. The experimental results summarized in Table 1 show that Boruta tends to select a relatively smaller number of features, ranging from 3 to 700 across different datasets, thus effectively reducing dimensionality and noise. On the other hand, mRMR selects a broader range of features, from 156 to 4958, reflecting its approach of balancing relevance and redundancy. These results illustrate distinct strategies and strengths of the two feature selection methods, allowing us to optimize dataset characteristics and enhance classification performance.

To evaluate our proposed approach and training model, we conducted experiments on a computer configured as follows: *Intel Core i5-12400 CPU at 2.50 GHz, 32 GB RAM, Windows 11 Pro OS, and NVIDIA GeForce RTX™ 4060 Ti GPU.* The implementation was performed in Python using several open-source libraries: *scikit-learn (version 1.5.2)* for traditional modeling and evaluation; *featurewiz (version 0.6.1)* for feature selection; *pandas (version 2.2.3)* and *numpy (version 2.2.2)* for data processing and organization; and *matplotlib (version 3.10.3)* and *seaborn (version 0.13.2)* for result visualization.

4.2   Classification Results

In this section, we provide a comparative analysis of KAN before and after feature selection. We further benchmark the performance of feature-selected KANs against conventional machine learning models, including Support Vector Machines (SVM) and Random Forests. Finally, we investigate the impact of combining feature selection methods using two aggregation strategies (i.e., Union and Intersection) to assess their effectiveness of the classification.

Figure 1 and Table 2 provide a comprehensive comparison of classification accuracy for EfficientKAN, FastKAN, FasterKAN, SVM, and Random Forest across multiple gene expression datasets, evaluated with three feature selection strategies: Boruta, mRMR, and the original (all dimension) feature set. The results clearly indicate that feature selection substantially enhances both accuracy and stability for all models. Notably, **FasterKAN consistently outperforms all other classifiers**, achieving the highest mean accuracy

Table 2. Summary of mean accuracy (± std) and training time (seconds, ± std) for each classifier and feature selection method across all datasets.

Feature Selection	Models	Accuracy	Training Time (s)
Boruta	EfficientKAN	0.856 ± 0.085	212.6 ± 305.0
	FastKAN	0.870 ± 0.084	98.7 ± 143.8
	FasterKAN	0.869 ± 0.079	87.0 ± 131.8
	SVM	0.839 ± 0.102	0.510 ± 1.01
	Random forest	0.833 ± 0.089	13.5 ± 16.7
mRMR	EfficientKAN	0.873 ± 0.091	197.2 ± 318.7
	FastKAN	0.881 ± 0.085	92.6 ± 136.5
	FasterKAN	0.884 ± 0.086	80.8 ± 115.5
	SVM	0.849 ± 0.107	0.946 ± 1.19
	Random forest	0.836 ± 0.094	14.6 ± 15.1
Original (Full Dim)	EfficientKAN	0.832 ± 0.110	725.4 ± 898.3
	FastKAN	0.829 ± 0.113	404.6 ± 505.2
	FasterKAN	0.834 ± 0.116	379.8 ± 484.3
	SVM	0.809 ± 0.133	337.4 ± 486.9
	Random forest	0.805 ± 0.123	84.2 ± 119.1

across most datasets and settings. For example, under the mRMR scheme, FasterKAN reaches a mean accuracy of **0.909** ± 0.069, compared to 0.899 ± 0.068 for FastKAN, 0.894 ± 0.077 for EfficientKAN, 0.855 ± 0.073 for SVM, and 0.856 ± 0.071 for Random Forest. With Boruta, all KAN variants achieve mean accuracies around 0.87–0.88, again with FasterKAN slightly ahead.

Feature selection improves average accuracy and reduces variability, as evidenced by lower standard deviations and more uniform accuracy trends across datasets. In contrast, using the original feature set results in a clear drop in performance: mean accuracies for all models fall to the 0.76–0.80 range, with increased standard deviations, highlighting the challenges posed by high-dimensional gene expression data without dimensionality reduction. In terms of computational efficiency, both Boruta and mRMR dramatically reduce the average training time for all models. For instance, FasterKAN achieves mean training times below 3 s with Boruta or mRMR, while training with the original feature set can take more than 60 s on some datasets. This substantial improvement in speed, along with top-tier accuracy, further highlights the practical advantages of FasterKAN. Full results for all datasets and parameter settings are reported in Supplementary Table 4 and Table 5.

4.3 Discussion

The experimental results clearly demonstrate that integrating advanced feature selection methods, i.e., Boruta and mRMR, with modern KAN variants yields substantial benefits for gene expression classification. Feature selection aims to enhance predictive accuracy by identifying the most informative and non-redundant genes but also dramatically reduces training time and computational variance, making large-scale analysis more feasible (Table 3).

Table 3. FasterKAN performance with different feature selection methods.

Datasets ID	Boruta (Bo)		mRMR (mR)		Union (Bo,mR)		Intersection (Bo,mR)	
	Time	Acc	Time	Acc	Time	Acc	Time	Acc
1	16.803	0.930	15.228	0.963	18.165	0.966	16.526	0.930
2	41.451	0.822	38.876	0.900	42.135	0.911	42.984	0.822
3	15.489	0.748	19.835	0.777	20.935	0.781	17.382	0.726
4	114.401	0.911	107.663	0.918	119.377	0.924	120.982	0.937
5	17.770	0.849	16.274	0.868	19.083	0.906	17.694	0.849
6	252.480	0.948	218.765	0.952	265.927	0.952	247.498	0.948
7	23.370	0.894	33.326	0.964	31.491	0.946	23.737	0.902
8	58.696	0.883	54.370	0.910	58.502	0.910	59.050	0.874
9	221.193	0.802	197.620	0.816	224.347	0.811	224.171	0.797
10	32.243	0.911	35.137	0.908	38.077	0.906	30.762	0.906
11	31.027	0.697	27.447	0.697	30.475	0.711	31.114	0.671
12	44.957	0.872	39.200	0.904	43.820	0.883	44.237	0.830
13	30.629	0.862	47.944	0.906	38.908	0.896	27.912	0.869
14	18.090	0.943	15.838	0.981	17.544	0.962	17.569	0.981
15	348.227	0.991	210.796	0.996	243.142	0.996	253.799	0.996
16	54.838	0.807	43.578	0.840	43.496	0.833	36.278	0.797
17	89.522	0.702	53.386	0.817	59.491	0.837	60.969	0.750
18	86.034	0.713	50.252	0.901	55.610	0.881	59.417	0.663
19	29.576	0.576	18.389	0.932	20.268	0.949	19.268	0.901
20	23.453	0.911	15.966	0.920	17.632	0.914	16.386	0.911
21	29.606	0.983	17.150	0.983	18.744	0.983	19.166	0.983
22	28.285	0.977	16.132	0.972	18.211	0.979	16.724	0.982
23	545.237	0.949	324.709	0.985	366.009	0.982	370.540	0.974
24	40.935	0.829	34.010	0.829	37.883	0.829	27.767	0.829
25	240.602	0.858	136.922	0.891	160.503	0.891	156.674	0.863

Among the evaluated models, **FasterKAN** stands out for its consistently superior accuracy and robustness across various datasets and selection strategies. When combined with Boruta or mRMR, FasterKAN achieves both the highest mean accuracy and the lowest standard deviation, indicating stable and reliable performance. Importantly, FasterKAN also exhibits the lowest average training time with feature selection, underscoring its computational efficiency despite its increased model complexity. FastKAN remains competitive in accuracy and training time, but generally trails FasterKAN in both consistency and peak performance. In contrast, EfficientKAN, although effective, shows lower accuracy and greater variability, particularly when handling the original, unreduced feature sets. It is worth noting, however, that FasterKAN's expressive power and additional learnable parameters may increase the risk of overfitting, especially for smaller or noisier datasets, and often necessitate careful hyperparameter tuning. Nevertheless, the empirical results highlight that rigorous feature selection—particularly mRMR—combined with FasterKAN provides a powerful and scalable framework for high-dimensional gene expression analysis. This approach successfully balances accuracy, robustness, and computational efficiency, making it well-suited for practical applications in complex biological data domains.

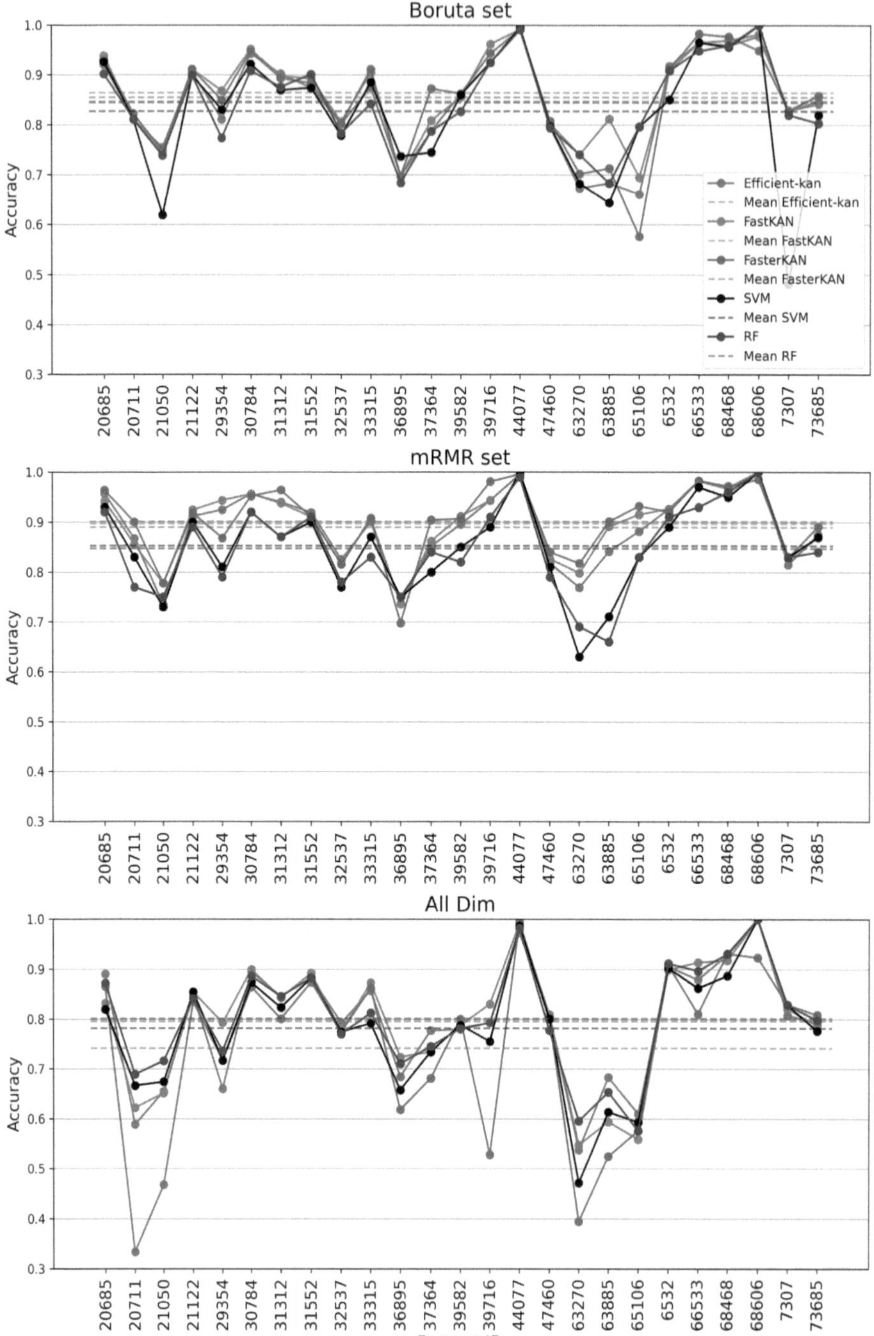

Fig. 1. Accuracy using mRMR, Boruta feature selection, and All Dim.

Table 4. Accuracy of all classifiers across all datasets and feature selection methods.

Dataset	Boruta					mRMR					Original				
	Efficient	Fast	Faster	SVM	RF	Efficient	Fast	Faster	SVM	RF	Efficient	Fast	Faster	SVM	RF
20685	0.921	0.939	0.930	0.927	0.902	0.957	0.942	0.963	0.930	0.920	0.890	0.832	0.865	0.820	0.871
20711	0.811	0.811	0.822	0.811	0.811	0.867	0.856	0.900	0.830	0.770	0.589	0.622	0.333	0.667	0.689
21050	0.755	0.745	0.748	0.619	0.739	0.735	0.777	0.777	0.730	0.750	0.655	0.652	0.468	0.674	0.716
21122	0.905	0.911	0.911	0.899	0.899	0.911	0.924	0.918	0.900	0.890	0.848	0.854	0.835	0.854	0.842
29354	0.811	0.868	0.849	0.830	0.774	0.925	0.943	0.868	0.810	0.790	0.660	0.792	0.717	0.717	0.736
30784	0.943	0.952	0.948	0.921	0.908	0.956	0.956	0.952	0.920	0.920	0.895	0.900	0.865	0.873	0.886
31312	0.902	0.898	0.894	0.869	0.876	0.940	0.938	0.964	0.870	0.870	0.845	0.843	0.801	0.823	0.845
31552	0.874	0.892	0.883	0.874	0.901	0.919	0.910	0.910	0.900	0.910	0.883	0.892	0.874	0.883	0.883
32537	0.802	0.806	0.802	0.779	0.783	0.825	0.816	0.816	0.770	0.780	0.793	0.774	0.774	0.774	0.770
33315	0.875	0.904	0.911	0.885	0.842	0.901	0.906	0.908	0.870	0.830	0.857	0.873	0.862	0.791	0.812
36895	0.697	0.697	0.697	0.737	0.684	0.737	0.737	0.697	0.750	0.750	0.684	0.724	0.618	0.658	0.711
37364	0.787	0.809	0.872	0.745	0.787	0.851	0.862	0.904	0.800	0.840	0.777	0.734	0.681	0.734	0.745
39582	0.862	0.855	0.862	0.860	0.827	0.896	0.911	0.906	0.850	0.820	0.779	0.788	0.800	0.788	0.781
39716	0.925	0.962	0.943	0.925	0.925	0.943	0.943	0.981	0.890	0.910	0.792	0.830	0.528	0.755	0.792
44077	0.991	0.991	0.991	0.996	0.991	0.996	0.996	0.996	1.000	0.990	0.973	0.991	0.996	0.987	0.982
47460	0.796	0.802	0.807	0.797	0.794	0.818	0.826	0.840	0.810	0.790	0.807	0.809	0.801	0.801	0.777
63270	0.673	0.740	0.702	0.683	0.740	0.769	0.798	0.817	0.630	0.690	0.538	0.548	0.394	0.471	0.596
63885	0.683	0.812	0.713	0.644	0.683	0.842	0.891	0.901	0.710	0.660	0.683	0.594	0.525	0.614	0.653
65106	0.661	0.695	0.576	0.797	0.797	0.881	0.915	0.932	0.830	0.830	0.610	0.559	0.576	0.593	0.576
6532	0.917	0.908	0.911	0.850	0.908	0.926	0.926	0.920	0.890	0.910	0.905	0.899	0.908	0.902	0.911
66533	0.966	0.966	0.983	0.966	0.948	0.983	0.983	0.983	0.970	0.930	0.879	0.914	0.810	0.862	0.897
68468	0.959	0.969	0.977	0.956	0.959	0.969	0.967	0.972	0.950	0.960	0.928	0.918	0.931	0.887	0.931
68606	0.978	0.982	0.949	1.000	1.000	1.000	1.000	0.985	1.000	1.000	1.000	1.000	0.923	1.000	1.000
7307	0.829	0.829	0.829	0.481	0.820	0.815	0.827	0.829	0.830	0.830	0.808	0.821	0.829	0.829	0.829
73685	0.842	0.847	0.858	0.820	0.803	0.874	0.869	0.891	0.870	0.840	0.792	0.787	0.809	0.776	0.798

Table 5. Training time (seconds) of all classifiers across all datasets and feature selection methods.

Dataset	Boruta					mRMR					Original				
	Efficient	Fast	Faster	SVM	RF	Efficient	Fast	Faster	SVM	RF	Efficient	Fast	Faster	SVM	RF
20685	32.754	18.011	16.803	0.135	3.069	36.491	18.909	15.228	0.500	4.435	466.146	256.319	306.082	149.100	26.414
20711	88.430	46.855	41.451	0.156	8.156	86.999	47.860	38.876	0.290	12.023	1,151.738	625.278	561.464	318.549	50.255
21050	38.326	18.363	15.489	0.081	1.618	41.840	21.753	19.835	0.764	5.387	436.189	243.448	215.198	121.285	21.315
21122	246.087	129.065	114.401	0.636	22.559	247.394	133.525	107.663	1.198	30.565	1,280.964	697.517	700.475	681.914	120.315
29354	36.322	20.153	17.770	0.055	3.809	38.497	21.226	16.274	0.079	4.958	127.973	73.318	63.009	20.241	10.077
30784	546.007	283.719	252.480	0.569	34.817	516.324	276.334	218.765	1.309	51.883	4,868.835	2,559.787	2,421.035	1,421.496	357.964
31312	52.406	26.456	23.370	0.174	3.264	75.024	38.546	33.326	1.948	9.724	439.154	281.419	359.917	268.249	45.014
31552	125.037	68.557	58.696	0.199	9.769	131.237	70.333	54.370	0.305	14.865	707.286	372.772	327.188	164.686	52.059
32537	473.649	244.827	221.193	0.648	31.786	481.678	253.062	197.620	4.410	75.351	1,127.862	631.947	559.815	1,936.004	214.188
33315	66.499	34.309	32.243	0.340	6.554	78.014	40.559	35.137	3.032	13.031	240.849	136.324	174.095	538.892	38.746
36895	63.731	34.911	31.027	0.117	5.689	61.436	33.517	27.447	0.392	9.454	547.148	290.724	257.489	192.646	34.409
37364	91.826	50.028	44.957	0.173	7.472	88.828	49.113	39.200	0.259	11.484	815.112	432.104	402.871	120.015	43.673
39582	62.012	31.755	30.629	0.341	6.616	101.015	60.829	47.944	3.807	13.707	570.803	311.193	387.531	416.846	55.088
39716	38.722	21.066	18.090	0.072	4.017	36.859	20.624	15.838	0.082	5.041	189.928	104.657	91.522	19.010	11.038
44077	915.848	374.048	348.227	0.458	27.080	497.237	266.266	210.796	0.519	29.102	2,977.453	1,577.511	1,429.752	936.077	156.802
47460	139.203	57.574	54.838	0.180	4.170	83.114	42.865	43.578	3.653	16.603	157.234	104.157	126.687	426.928	34.284
63270	248.162	92.496	89.522	0.161	9.808	117.102	63.422	53.386	0.548	16.749	384.330	230.312	216.799	155.147	41.695
63885	228.446	91.921	86.034	0.161	9.028	112.456	61.157	50.252	0.376	16.336	1,029.619	553.126	487.848	128.299	64.685
65106	80.841	32.547	29.576	0.117	3.960	39.801	22.043	18.389	0.090	6.015	221.504	117.891	102.096	21.729	13.760
6532	67.263	26.343	23.453	0.033	1.448	36.967	18.785	15.966	0.112	2.831	119.771	69.021	56.267	48.839	11.780
66533	80.532	32.006	29.606	0.076	4.243	38.668	21.095	17.150	0.081	5.155	334.449	176.884	161.246	21.482	13.413
68468	74.025	29.252	28.285	0.080	2.858	37.004	19.764	16.132	0.374	4.644	138.038	85.531	101.663	153.617	21.731
68606	1,467.975	546.115	545.237	2.969	77.513	759.450	397.915	324.709	3.801	70.474	2,910.387	1,686.508	1,696.306	2,155.449	313.977
7307	114.811	44.741	40.935	5.663	1.353	100.764	48.868	34.010	15.397	74.562	632.663	317.458	260.394	1,372.892	307.474
73685	634.907	257.165	240.602	0.758	32.440	343.967	177.514	136.922	2.111	42.493	1,914.710	1,041.852	987.225	768.838	183.292

5   Conclusion and Future Work

This study introduces a systematic framework for gene expression classification, integrating advanced feature selection techniques with state-of-the-art KAN variants. A key contribution is the implementation of comprehensive grid search for hyperparameter optimization in FasterKAN, which enables the identification of optimal model configurations for each dataset. This approach consistently delivers high accuracy and robust performance across diverse gene expression profiles, highlighting the essential role of tailored hyperparameter tuning in harnessing the full potential of expressive neural network architectures for high-dimensional biomedical data.

For future work, we will focus on developing and integrating adaptive hyperparameter optimization strategies to further enhance both the efficiency and scalability of the framework. By automating the tuning process and enabling dynamic adjustment to specific dataset characteristics, such adaptive methods are expected to reduce computational costs and improve model generalizability. Additionally, exploring extensions to multi-omics data integration and advancing model interpretability will be valuable for expanding the practical impact of this approach in biomedical and clinical research.

Acknowledgements. Thanh MA and Thanh-Nghi DO are received support from the European Union's Horizon research and innovation program under the MSCA-SE (Marie Skłodowska-Curie Actions Staff Exchange) grant agreement 101086252; Call: HORIZON-MSCA-2021-SE-01; Project title: STARWARS.

References

1. Ahmed, O., Brifcani, A.: Gene expression classification based on deep learning. In: SICN'19, pp. 145–149. IEEE (2019)
2. Bach, F.: Breaking the curse of dimensionality with convex neural networks. J. Mach. Learn. Res. **18**(19), 1–53 (2017)
3. Bajer, D., Dudjak, M., Zorić, B.: Wrapper-based feature selection: how important is the wrapped classifier? In: SST'20, pp. 97–105. IEEE (2020)
4. Brazma, A., et al.: ArrayExpress–a public repository for microarray gene expression data at the EBI. Nucleic Acids Res. **31**(1), 68–71 (2003)
5. Cervante, L., Xue, B., Zhang, M., Shang, L.: Binary particle swarm optimisation for feature selection: a filter based approach. In: 2012 IEEE Congress on Evolutionary Computation, pp. 1–8. IEEE (2012)
6. Chen, Z., Zhang, X.: LSS-SKAN: efficient Kolmogorov-Arnold networks based on single-parameterized function. arXiv preprint: arXiv:2410.14951 (2024)
7. Ding, C., Peng, H.: Minimum redundancy feature selection from microarray gene expression data. J. Bioinform. Comput. Biol. **3**(02), 185–205 (2005)
8. Do, T.N.: Enhancing gene expression classification through explainable machine learning models. SN Comput. Sci. **5**(5), 1–16 (2024)
9. Do, T.N., Tran-Nguyen, M.T.: Ensemble learning with SVM for high-dimensional gene expression data. In: ISDS'23, pp. 29–40. Springer (2023)

10. Huynh, P.H., Nguyen, V.H., Do, T.N.: Random ensemble oblique decision stumps for classifying gene expression data. In: proceedings of the 9th International Symposium on information and Communication Technology, pp. 137–144 (2018)
11. Huynh, P.H., Nguyen, V.H., Do, T.N.: Novel hybrid DCNN-SVM model for classifying RNA-sequencing gene expression data. J. Inf. Telecommun. **3**(4), 533–547 (2019)
12. Karim, M.R., Cochez, M., Beyan, O., Decker, S., Lange, C.: OncoNetExplainer: explainable predictions of cancer types based on gene expression data. In: BIBE'19, pp. 415–422. IEEE (2019)
13. Kasongo, S.M., Sun, Y.: A deep learning method with wrapper based feature extraction for wireless intrusion detection system. Comput. Secur. **92**, 101752 (2020)
14. Kolmogorov, A.N.: On the representation of continuous functions of several variables by superpositions of continuous functions of a smaller number of variables. Am. Math. Soc. (1961)
15. Köppen, M.: The curse of dimensionality. In: 5th online world conference on soft computing in industrial applications (WSC5), vol. 1, pp. 4–8 (2000)
16. Kursa, M.B., Jankowski, A., Rudnicki, W.R.: Boruta-a system for feature selection. Fund. Inform. **101**(4), 271–285 (2010)
17. Li, J., et al.: Feature selection: a data perspective. ACM Comput. Surv. (CSUR) **50**(6), 1–45 (2017)
18. Li, Z.: Kolmogorov-Arnold networks are radial basis function networks. arXiv preprint: arXiv:2405.06721 (2024)
19. Liu, Z., et al.: KAN: Kolmogorov-Arnold networks (2024). arXiv preprint: arXiv:2404.19756
20. Miao, J., Niu, L.: A survey on feature selection. Procedia Comput. Sci. **91**, 919–926 (2016)
21. Mufassirin, M.M., Ragel, R.G.: A novel filter-wrapper based feature selection approach for cancer data classification. In: 2018 IEEE International Conference on Information and Automation for Sustainability (ICIAfS), pp. 1–6. IEEE (2018)
22. Peng, H., Long, F., Ding, C.: Feature selection based on mutual information criteria of max-dependency, max-relevance, and min-redundancy. IEEE Trans. Pattern Anal. Mach. Intell. **27**(8), 1226–1238 (2005)
23. Radovic, M., Ghalwash, M., Filipovic, N., Obradovic, Z.: Minimum redundancy maximum relevance feature selection approach for temporal gene expression data. BMC Bioinformatics **18**, 1–14 (2017)
24. Stańczyk, U.: Feature evaluation by filter, wrapper, and embedded approaches. Feature Sel. Data Pattern Recogn., 29–44 (2015)
25. Taud, H., Mas, J.F.: Multilayer perceptron (MLP). In: Geomatic approaches for modeling land change scenarios, pp. 451–455. Springer (2017)
26. Wang, H., Khoshgoftaar, T.M., Napolitano, A.: Stability of filter-and wrapper-based software metric selection techniques. In: Proceedings of IEEE IRI 2014, pp. 309–314. IEEE (2014)
27. Zhou, H., Xin, Y., Li, S.: A diabetes prediction model based on Boruta feature selection and ensemble learning. BMC Bioinformatics **24**(1), 224 (2023)

Fuzzy-EDTrans: A Fuzzy Ensemble Learning Approach for Brain Tumor Classification

Anh-Cang Phan[✉], Chan-Khoa Ho[✉], and Khac-Tuong Nguyen[✉]

Vinh Long University of Technology Education, Vĩnh Long, Vietnam
{cangpa,khoahc,tuongnk}@vlute.edu.vn

Abstract. Brain tumors are among the most critical neurological disorders, significantly affecting patients' health and prognosis. Early detection and accurate classification play a vital role in the treatment process. Although Magnetic Resonance Imaging (MRI) is a widely used noninvasive imaging modality, classification accuracy remains limited due to the morphological similarities among tumor types and inconsistencies in image quality. Deep learning models have shown remarkable success in medical image analysis; however, individual models often lack robustness and may suffer performance degradation when dealing with complex or atypical data. This paper proposes Fuzzy-EDTrans, an ensemble deep learning model that integrates three advanced architectures: EfficientNet, Vision Transformer, and DenseNet, using the Fuzzy Sugeno Integral to optimize classification accuracy for brain tumors. This approach not only leverages the individual strengths of each architecture but also enhances the aggregation of information and improves decisionmaking precision. Moreover, it increases the model's stability when processing complex medical images. Experiments conducted on the public BrTC2020 dataset demonstrate that the proposed model achieves a classification accuracy of 97.91%, outperforming standalone models. These results confirm the potential of the proposed approach in supporting AI-based medi-cal image diagnosis, especially in the development of intelligent healthcare systems.

Keywords: Brain Tumor · Ensemble Learning · Deep Learning

1 General Appearance

1.1 Practical Basis

Brain and central nervous system tumors are considered a serious global health issue, with approximately 347,992 new cases and 246,253 deaths reported annually worldwide [1]. In Vietnam, brain cancer ranks 14th among the most common cancer types, with an estimated 2,829 new cases per year and a high mortality rate of 48%, primarily due to late diagnosis [2]. Although brain tumors account for only about 1.6% of all cancer cases, they exhibit an abnormally high fatality

N. Thai-Nghe et al. (Eds.): ISDS 2025, CCIS 2714, pp. 191–203, 2026.
https://doi.org/10.1007/978-981-95-3358-9_14

rate, with only 13% of patients surviving five years after diagnosis in the United Kingdom [1]. Notably, 38.9% of cases in the UK are detected through emergency admissions, highlighting limitations in early detection [1]. Brain tumors are also among the leading causes of cancer-related deaths in children and young adults in many countries, including Vietnam. Magnetic Resonance Imaging (MRI) is currently the primary diagnostic tool due to its ability to provide detailed anatomical imaging [3]; however, interpretation of MRI scans remains heavily reliant on expert knowledge and time-consuming analysis. To improve classification accuracy and model stability in brain tumor diagnosis, this paper proposes Fuzzy-EDTransâĂŤa deep learning model that integrates EfficientNet, DenseNet, and Transformer architectures using the Fuzzy Sugeno Integral.

1.2 Related Works

In recent years, the application of deep learning to brain tumor classification using MRI images has become a prominent research focus, driven by the increasing incidence and complexity of tumor cases. Many studies have aimed to improve classification accuracy and streamline diagnostic workflows. J. Cheng et al. [4] developed a region of interest (ROI) enhancement technique based on dilation and ring-shaped partitioning, incorporating intensity histograms, gray-level co-occurrence matrix (GLCM), and Bag-of-Words (BoW) features. Their approach achieved an accuracy of 91.28% on a dataset comprising 3,064 MRI T1 slices from 233 patients. In another study, V. Sajja and Kalluri [5] utilized the Fuzzy C-means algorithm for brain tumor segmentation and applied a VGG16-based classifier on the BRATS dataset, attaining an accuracy of 96.70% and sensitivity of 97.05%. Sameh Samir [6] proposed a brain tumor classification approach based on a pretrained EfficientNet model. This highlights the model's capability in handling multiclass tumor classification using transfer learning techniques, achieving an accuracy of 97.98%. Medina and Sanchez [7] investigated the effectiveness of the EfficientNet-B0 model for brain tumor classification using MRI images. The model achieved an accuracy of 97.5%, highlighting the potential of transfer learning for accurate brain tumor diagnosis with limited data. Sharma and Shukla [8] applied transfer learning using Efficient-Net-B0 and B7 architectures to classify brain tumors from MRI images. Their method was evaluated on the Kaggle SARTAJ dataset containing 3,264 images and achieved an accuracy of 98%. Meanwhile, Muhannad Alanazi et al. [9] leveraged transfer learning using a 22-layer deep neural network, trained on the Figshare brain tumor dataset, and reported an accuracy of 96.89%. W. Chen et al. [10] designed a hybrid feature fusion framework by combining deep features from ResNet101, DenseNet121, and EfficientNetB0, achieving classification accuracies of 99.18% and 97.24% on two different datasets. These works demonstrate the effectiveness of transfer learning and architectural optimization. However, most rely on individual models, lacking ensemble strategies that could further enhance classification robustness—motivating the development of our proposed method. Jaeyong Kang et al. [11] proposed an ensemble of deep features extracted from DenseNet-

169, ShuffleNet V2, and MnasNet, which were then classified using a Support Vector Machine (SVM), resulting in an accuracy of 93.72% on an MRI dataset.

2   Background

2.1   Brain Tumor Types and MRI Diagnosis

Common symptoms of brain tumors include headaches, nausea, vision problems, seizures, and cognitive decline [12,13]. Morning headaches and unexplained vomiting may signal increased intracranial pressure. Early detection is essential for treatment. Figure 1a–1c show typical tumor types: glioma (aggressive, poor prognosis), meningioma (mostly benign), and pituitary tumor (hormone-related). Figure 1d shows a normal brain MRI for comparison.

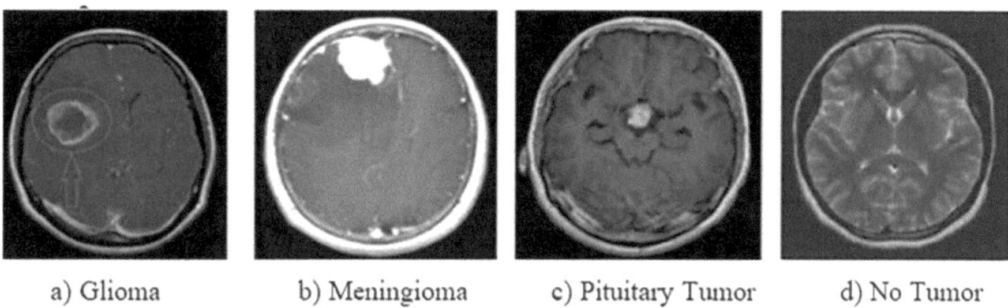

Fig. 1. MRI images tumor types and no tumors were obtained from the BrTC dataset [6–8].

2.2   Transfer Learning

Transfer learning [14] utilizes pretrained models to handle new tasks with limited data, making it well-suited for medical imaging. This study employs three architectures: EfficientNet [15], Vision Transformer [16], and DenseNet [17] for brain tumor classification.

2.3   Ensemble Learning

Soft Voting
Soft Voting [18] is a probabilistic ensemble method that averages the predicted class probabilities from multiple base models. For each class c, let pi denote the probability assigned to class c by the i. The final Voting score for class c, denoted as P voting(c) is computed as:

$$P_{voting}(c) = \frac{1}{n} \sum_{i=1}^{n} p_i^{(c)} \tag{1}$$

Hard Voting Hard Voting [19] is a decision-level ensemble method that selects the final class based on majority voting among models. Each model contributes one vote to the class it assigns the highest confidence to. For a given class c, the P voting(c) score is computed as:

$$P_{voting}(c) = \frac{1}{n} \sum_{i=1}^{n} p_i^{(c)} \tag{2}$$

Fuzzy Voting

Fuzzy Voting [20] applies fuzzy logic to combine the outputs of multiple models while accounting for uncertainty and interactions. Instead of treating models independently, this approach uses fuzzy measures to represent how individual models or their combinations contribute to the overall decision. Among fuzzy ensemble strategies, we adopt the additive form of the Fuzzy Sugeno Integral [12], which aggregates prediction scores based on a predefined fuzzy measure. This measure reflects the behavior of subsets of models during the voting process, enabling more flexible and context-aware integration without assuming equal influence across classifiers. We adopt the Sugeno integral with additive fuzzy measure to compute the final voting score for each class. Let the set of models be $X = x_1, x_2, ..., x_n$, where each model x_i outputs a prediction value $p(x_i) \in [0, 1]$. The fuzzy measure μ satisfies:

$$\mu(X) = \sum_{x_i \in A} \mu(x_i) \tag{3}$$

Then, the Fuzzy Sugeno Integral computes the P_{voting} score for class c as follows:

$$P_{voting}(c) = \max_i \left[\min \left(P(x_{\sigma(i)}), \sum_{j=i}^{n} \mu(x_{\sigma(j)}) \right) \right] \tag{4}$$

Here, σ is a permutation of indices such that: $p(x_{\sigma(1)}) \leq p(x_{\sigma(2)}) \leq \cdots \leq p(x_{\sigma(n)})$, and the corresponding fuzzy measure values $\mu(x_{\sigma(j)})$ are ordered accordingly.

3   Proposed Method

Figure 2 illustrates the proposed brain tumor classification framework that integrates deep learning with ensemble decision fusion through three voting strategies: Soft Voting, Hard Voting, and Fuzzy Voting. The pipeline consists of four main stages: data preprocessing, individual model training, Voting-based fusion, and final decision. MRI images are first standardized and divided into training and testing subsets. Three transfer learning models - EfficientNet-B0, DenseNet-169, and Vision Transformer (ViT) - are trained independently to perform four-class classification: Glioma, Meningioma, Pituitary Tumor, and No Tumor. For

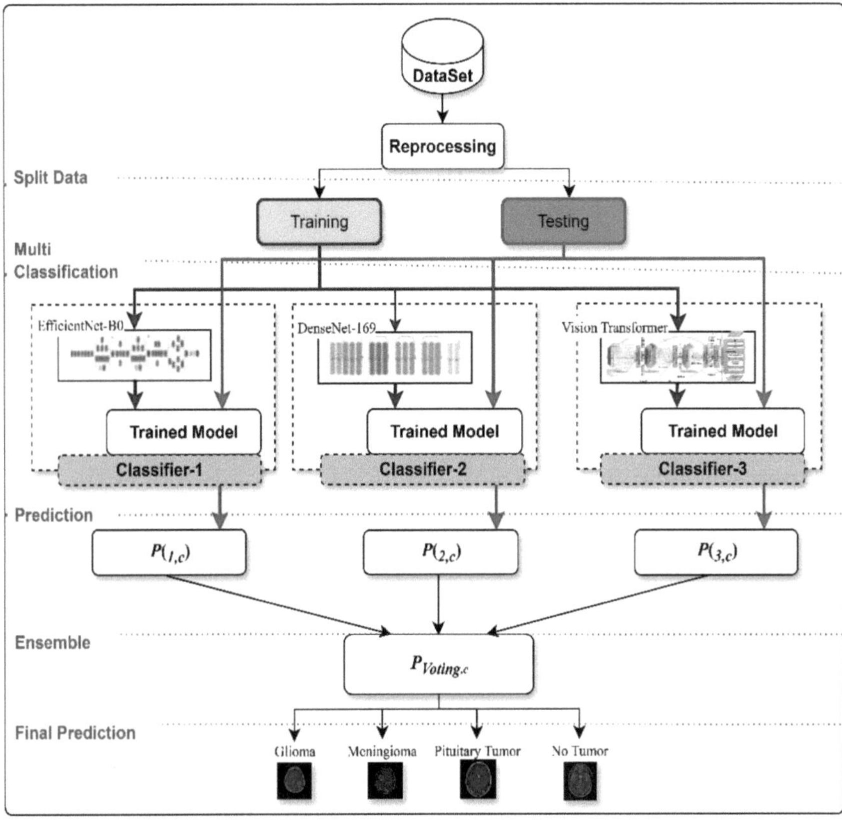

Fig. 2. Overall workflow of the proposed approach.

each test sample, all predicted class probabilities from the three models are retained. These outputs are aggregated using three distinct Voting strategies. In Soft Voting, the class probabilities are averaged across models (Eq. 1). In Hard Voting, each model casts a vote for its top predicted class, and the majority label is selected (Eq. 2). In Fuzzy Voting, the predicted scores are combined using the Fuzzy Sugeno Integral (Eq. 4), with fuzzy measure values ðİIJĞ(i) chosen through iterative experimentation. The final decision is made by selecting the class with the highest aggregated score for each voting method. Comparative experiments are conducted to evaluate and analyze the effectiveness of each strategy.

Algorithm 1 outlines the computation process of ensemble learning using the Fuzzy Sugeno Integral [21]. For each class, prediction scores from all models are collected and sorted in ascending order. These scores are then combined with their corresponding fuzzy measures. The aggregation is performed by computing the maximum of the minimum values between each sorted score and the cumulative fuzzy measure. The class with the highest aggregated value is selected as the final prediction.

Algorithm 1. Pseudo code for Ensemble Learning using Fuzzy Sugeno Integral

1: **Input:** Predicted probabilities: p
2:             Number of models: n
3:             Number of classes: c
4:             Fuzzy measures: μ
5:
6: **Output:** Final predicted label: y

7: **Initialize:** predictions $\leftarrow$ empty list of size c
8: **for** each class index $j \in \{0, 1, \ldots, c-1\}$ **do**
9:     $p\_r \leftarrow [p_1^{(j)}, p_2^{(j)}, \ldots, p_n^{(j)}]$
10:     Sort p_r in ascending order $\rightarrow p\_sorted$
11:     Reorder μ according to the sort order $\rightarrow \mu\_sorted$
12:     $fuzzy\_pred \leftarrow 0$
13:     **for** $i \in \{1, \ldots, n\}$ **do**
14:         $sum\_mu \leftarrow \sum(\mu\_sorted[i \text{ to } n])$
15:         $min\_val \leftarrow \min(p\_sorted[i], sum\_mu)$
16:         $fuzzy\_pred \leftarrow \max(fuzzy\_pred, min\_val)$
17:     **end for**
18:     $predictions[j] \leftarrow fuzzy\_pred$
19: **end for**
20: $y \leftarrow \arg\max(predictions)$
21: **return** y

---

4   Experiments and Results

4.1   Dataset

This study uses the Brain Tumor Classification MRI dataset (BrTC2020) [6–8], which includes 3,264 brain MRI images across four classes: glioma, meningioma, pituitary tumor, and no tumor. The data were manually annotated and split into 70% training, 15% validation, and 15% testing. Table 1 shows the class-wise distribution used for model development and evaluation.

Table 1. Detailed overview of the dataset

No.	Class	Training	Validation	Testing
1	Glioma	648	138	140
2	No Tumor	277	59	60
3	Meningioma	655	140	142
4	Pituitary	630	135	136

4.2   Experimental Scenarios

We used the Brain Tumor Classification MRI dataset (BrTC2020) [7,8], which includes 3,264 brain MRI images across four classes: glioma, meningioma, pituitary tumor, and no tumor. The data were manually annotated and split into 70% training, 15% validation, and 15% testing. Table 2 summarizes their configurations. Scenarios 1 to 3 are trained separately with the same settings, while Scenarios 4 to 6 apply Soft, Hard, and Fuzzy Voting during testing.

Table 2. Summary of experimental scenarios and training configurations

Scenario	Backbone Model	Batch Size	Num Classes	Learning Rate	Epochs
1	EfficientNet-B0	32	4	1e-4	100
2	Vision Transformer	32	4	1e-4	100
3	DenseNet-169	32	4	1e-4	100
4	Soft-EDTrans	–	4	–	–
5	Hard-EDTrans	–	4	–	–
6	Fuzzy-EDTrans	–	4	–	–

4.3   Training Results

Loss Curves of Training Scenarios. Figure 3 shows the training and validation loss for the three scenarios. Scenario 2 converged fastest and most stably, with validation loss stabilizing at 0.0391. Scenario 1 also showed consistent performance with final validation loss of 0.0386. Scenar-io 3 recorded higher validation loss at 0.0989, indicating weaker generalization. Overall, Scenarios 1 and 2 demonstrated better convergence than Scenario 3

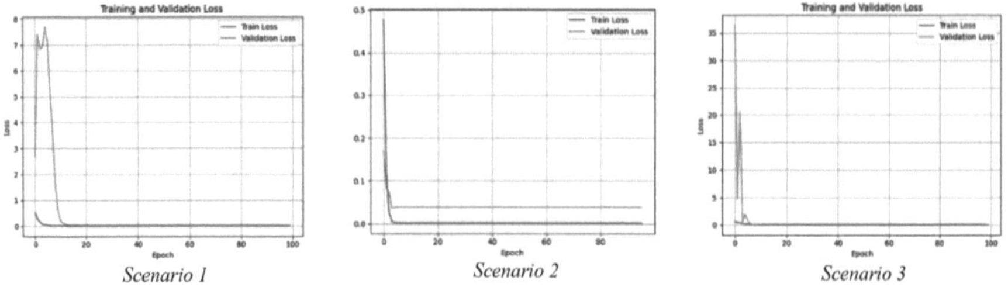

Scenario 1   *Scenario 2*   *Scenario 3*

Fig. 3. Training and validation loss curves for all scenarios.

Accuracy Curves of Training Scenarios. Figure 4 presents training and validation accuracy on the BrTC dataset. Scenarios 1 and 2 showed high and stable performance with validation accuracy around 0.9873, while Scenario 2 reached peak training accuracy at 0.9995. Scenario 3 achieved lower validation accuracy at 0.9661 despite high training performance, indicating signs of overfitting.

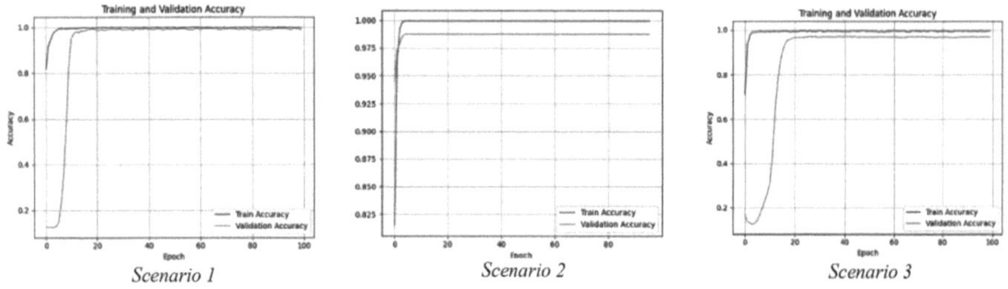

Scenario 1   *Scenario 2*   *Scenario 3*

Fig. 4. Training and validation accuracy curves for all scenarios.

4.4   Testing Results

To apply the Ensemble Learning approach using the Fuzzy Sugeno Integral, we needed to determine the fuzzy measure values (μ) for each model. On the BrTC dataset, the optimal values were set as [$\mu_1 = 0.7; \mu_2 = 0.9; \mu_3 = 0.6$], based on the highest classification accuracy achieved through multiple experimental trials. Figure 5 shows the confusion matrices for all scenarios. Scenario 1 misclassified several glioma cases as meningioma, while Scenario 2 reduced this but introduced minor confusion with no tumor. Scenario 3 showed the highest misclassification, particularly between glioma and meningioma. Ensemble strategies in Scenarios 4 to 6 clearly improved class separability. Scenario 4 reduced errors across all classes, Scenario 5 further minimized confusion, and Scenario 6 achieved the most balanced distribution, including perfect classification for no tumor. These results highlight the effectiveness of ensemble learning, with fuzzy aggregation in Scenario 6 offering the most robust performance.

Table 3 summarizes the evaluation results of all six scenarios. Scenarios 1 and 2 showed strong performance among individual models, while Scenario 3 had the lowest accuracy. Ensemble methods in Scenarios 4 to 6 performed better overall, with Scenario 6 achieving the highest accuracy of 0.9791 and strong F1-scores across all classes. These results highlight the effectiveness of fuzzy-based fusion in enhancing consistency and reducing misclassification.

Table 3. Evaluation metrics of all proposed scenarios on the test data

Scenario	Class	Precision	Recall	F1-score	Accuracy
1	Glioma	0.9571	0.9571	0.9571	0.9749
	No tumor	1.000	1.000	1.000	
	Meningioma	0.9580	0.9647	0.9614	
	Pituitary	1.000	0.992	0.9963	
2	Glioma	0.9432	0.9500	0.9466	0.9707
	No tumor	1.000	0.9666	0.9830	
	Meningioma	0.9366	0.9366	0.9366	
	Pituitary	0.9854	0.9926	0.9890	
3	Glioma	0.9770	0.9142	0.9446	0.9603
	No tumor	0.9833	0.9833	0.9833	
	Meningioma	0.9183	0.9507	0.9342	
	Pituitary	0.9642	0.9926	0.9782	
4	Glioma	0.9643	0.9643	0.9643	0.9749
	No tumor	1.000	1.000	1.000	
	Meningioma	0.9650	0.9718	0.9684	
	Pituitary	1.000	0.9926	0.9963	
5	Glioma	0.9571	0.9571	0.9571	0.9770
	No tumor	1.000	1.000	1.000	
	Meningioma	0.9580	0.9647	0.9614	
	Pituitary	1.000	0.992	0.9963	
Proposed method	Glioma	0.9583	0.9857	0.9718	0.9791
	No tumor	1.000	1.000	1.000	
	Meningioma	0.9854	0.9507	0.9677	
	Pituitary	0.9854	0.9926	0.9890	

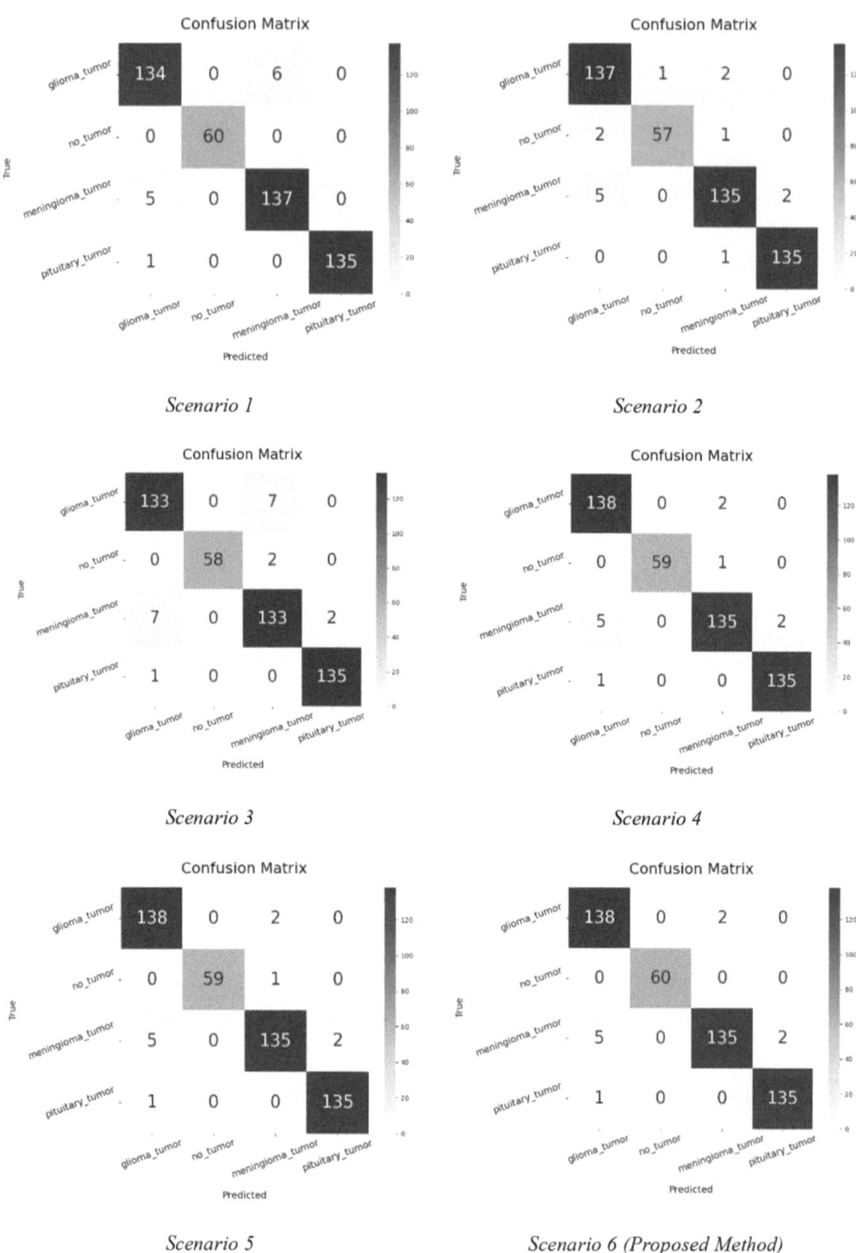

Fig. 5. Confusion matrices of all experimental scenarios on the BrTC test set.

Table 4. Sample prediction results of the scenarios on test images

	True label	Scenario 1	Scenario 2	Scenario 3	Proposed Method
	Glioma	Glioma: 0.9775	Glioma: 0.9979	Meningioma: 0.9742	Glioma: 0.9775
	Glioma	Meningioma: 0.6355	Glioma: 0.7291	Meningioma: 0.7086	Glioma: 0.7
	Meningioma	Meningioma: 0.7851	Glioma: 0.6213	Meningioma: 0.7086	Meningioma: 0.8370

4.5   Comparison and Discussion

Table 5 presents a comparison of classification accuracy between the proposed method and several recent deep learning approaches for brain tumor classification. Previous studies employed various architectures, including single models as well as ensemble methods, across different MRI datasets. Reported accuracies range from approximately 97.7% to 98.9%. The proposed Fuzzy-EDTrans method achieved an accuracy of 97.91%, which is comparable to the results of prior studies. However, the remaining differences may stem from factors such as the number of classification classes, the scale and diversity of the training datasets, and the variations in feature extraction strategies used by different approaches (Table 4).

Table 5. Comparison of Brain Tumor Classification Methods

Authors	Method/ Architecture	Dataset	Ensemble Method	Class	Accuracy (%)
Sameh Samir [6]	EfficientNet (pre-trained)	BraTS (2487 images)	None	3	97.98
Medina et al. [7]	EfficientNet (Fine-tuned)	Kaggle (2872 images)	None	4	97.75
Preeti Sharma et al. [8]	EfficientNetB0 & B7	Figshare BrTC (3264 images)	None	4	98
T. Mahmud et al. [22]	CNN + VGG19	Figshare + SARTAJ + Br35H (7028 images)	Weighted Average Ensemble	4	98.01
S. Patil et al. [23]	Shallow CNN, VGG16	Figshare Brain Tumor (3064 images)	EDCNN	3	97.77
G.S. Tandel et al. [24]	AlexNet, VGG16, ResNet18, GoogleNet, ResNet50	Molecular Brain Tumor Data (3567 images)	Majority Voting	2	98.88
Proposed Method	EfficientNetB0+ VisionTransformer+ DenseNet169 (Fuzzy-EDTrans)	Figshare BrTC (3264 images)	Fuzzy Voting	4	97.91

5   Conclusion

This study introduced Fuzzy-EDTrans, a novel ensemble deep learning framework for brain tumor classification using MRI images. By integrating EfficientNet-B0, DenseNet-169, and Vision Transformer through the Fuzzy Sugeno Integral, the proposed method effectively combines the strengths of each architecture to enhance classification accuracy and stability. Experimental results on the BrTC dataset demonstrated that Fuzzy-EDTrans outperforms most existing single-model and en-semble-based approaches, achieving an accuracy of 97.91%. The comparative analysis further confirmed the robustness of fuzzy-based aggregation strategies in reducing misclassification, especially in cases with overlapping tumor morphology. These findings suggest that the proposed approach holds significant potential for supporting intelligent diagnostic systems in clinical practice. Future work will focus on extending the method to multi-modal datasets and incorporating explainable AI techniques to enhance interpretability for medical professionals.

References

1. Charity. Brain Tumour Statistics. United Kingdom (2023)
2. International Agency for Research on Cancer, Viet Nam Cancer Fact Sheet (2022)
3. M. C. Center, Brain Tumor Diagnosis - MRI. https://www.moffitt.org/cancers/brain-tumor/diagnosis/mri/. Accessed 15 May 2025

4. J. Cheng, W., et al.: Enhanced performance of brain tumor classification via tumor region augmentation and partition. PLOS ONE **10**(10), e0140381 (2015)

5. Sajja, V.R., Kalluri, H.K.: Classification of brain tumors using fuzzy c-means and VGG16. Turk. J. Comput. Math. Educ. **12**(9), 2103–2113 (2021)

6. Samir, R.S.: EfficientNet algorithm for classification of different types of cancer. arXiv preprint: arXiv:2304.08715 (2023)

7. Medina, J.M., Sanchez, J.: High accuracy brain tumor classification with EfficientNet and magnetic resonance images. In: Proceedings of the 5th International Conference on Advances in Signal Processing and Artificial Intelligence (ASPAI), Tenerife, Spain, pp. 1–6 (2023)

8. Sharma, P., Shukla, A.: Transfer learning based brain tumor detection using EfficientNet-B0 and B7. Int. J. Comput. Appl. **184**(35), 25–30 (2022)

9. Alanazi, M.F., et al: Brain tumor/mass classification framework using magnetic-resonance-imaging-based isolated and developed transfer deep-learning model. Sensors **22**(1), 372 (2022)

10. Chen, W.: A robust approach for multi-type classification of brain tumor using deep feature fusion. Front. Neurosci. **18** (2024)

11. Kang, J., Ullah, Z., Gwak, J.: MRI-based brain tumor classification using ensemble of deep features and machine learning classifiers. Sensors **21**, 202

12. Symptoms, Brain Tumor Symptoms. https://virtualtrials.org/symptoms.pdf. Accessed 31 May 2025

13. Symptoms, Brain Tumour Research. https://braintumourresearch.org/pages/information-brain-tumour-symptoms. Accessed 31 May 2025

14. Pan, S.J., Yang, Q.: A survey on transfer learning. IEEE Trans. Knowl. Data Eng. **22**(10), 1345–1359 (2010)

15. Tan, M., Le, Q.V.: EfficientNet: rethinking model scaling for convolutional neural networks. In: Proc. 36th Int. Conf. Machine Learning (ICML), Long Beach, CA, USA, pp. 6105–6114 (2019)

16. Dosovitskiy, A., et al.: An image is worth 16x16 words: transformers for image recognition at scale. In: Proc. Int. Conf, Learning Representations (ICLR) (2021)

17. Huang, G., Liu, Z., van der Maaten, L., Weinberger, K.Q.: Densely connected convolutional networks. In: Proc. IEEE Conf. Computer Vision and Pattern Recognition (CVPR), Honolulu, HI, USA, pp. 2261–2269 (2017)

18. Dietterich, T.G.: Ensemble methods in machine learning. In: Kittler, J., Roli, F. (eds.) MCS 2000. LNCS, vol. 1857, pp. 1–15. Springer, Heidelberg (2000). https://doi.org/10.1007/3-540-45014-9_1

19. Rokach, L.: Ensemble-based classifiers. Artif. Intell. Rev. **33**(1), 1–39 (2010)

20. Sugeno, M.: Fuzzy Measures and Fuzzy Integrals—A Survey. Elsevier, pp. 251–257 (1993)

21. Kundu, R., Basak, H., Kollala, A., Chattopadhyay, S., Chakraborty, O., Das, N.: Ensemble of CNN classifiers using Sugeno fuzzy integral technique for cervical cytology image classification (2021)

22. Mahmud, T., Barua, A., Barua, K., Basnin, N., Monju, M., Sharmen, N.: Anik Barua 1, Koushick Barua 1, Nanziba Basnin 2. https://www.researchgate.net/publication/379431047

23. Patil, S., Kirange, D.: Ensemble of deep learning models for brain tumor detection. Procedia Comput. Sci., 2468–2479 (2022)

24. Tandel, G.S., Tiwari, A., Kakde, O.G., Gupta, N., Saba, L., Suri, J.S.: Role of ensemble deep learning for brain tumor classification in multiple magnetic resonance imaging sequence data. Diagnostics **13**(3) (2023)

A Machine Learning-Based Tool for Autism Screening Using Psychological Medical Records

Hao Nguyen Thi Bich, Thanh Nguyen Van Quoc, Thuan Nguyen Dinh[(✉)], and Nhut Nguyen Minh

Faculty of Information Systems, University of Information Technology - Vietnam National University, Ho Chi Minh City, Vietnam
{21521447,21522049}@gm.uit.edu.vn, thuannd@uit.edu.vn,
nhutnm.17@grad.uit.edu.vn

Abstract. Autism Spectrum Disorder (ASD) is a neurodevelopmental condition that impacts social interaction and behavior, where early detection is crucial for optimizing intervention outcomes.

This study proposes a machine learning-based screening tool for Autism Spectrum Disorder (ASD) using psychological electronic medical records (EMR) for children aged 1–8 years. Two datasets were utilized: (1) a public dataset with approximately 300 records and (2) a clinical EMR dataset with approximately 600 records. Both were labeled by psychology experts based on 18 behavioral evaluation criteria, including eye contact, pointing, and imitation.

Ten significant features were selected as inputs for training four machine learning models: Decision Tree, XGBoost, CatBoost, and Overall Local Accuracy (OLA). OLA demonstrated superior adaptability but faced challenges with class imbalance (90% ASD) and feature bias due to high-accuracy EMR data.

To address these, SMOTE-IPF was applied to balance the data, and OLA was enhanced with dynamic weighting. The proposed system shows high feasibility for ASD screening.

Keywords: Autism Spectrum Disorder · Machine Learning · Psychological Medical Records · SMOTE-IPF · OLA · Screening Tool

1 Introduction

Autism Spectrum Disorder (ASD) is a neurodevelopmental condition marked by difficulties in social interaction, communication, and repetitive behaviors [1]. Early detection is vital for effective interventions, yet diagnosis becomes more challenging with age due to symptom overlap with other disorders. In the U.S., ASD prevalence has increased from 1 in 150 (2000) to 1 in 36 (2023) [2]; in Vietnam, estimates suggest 1 in 68 children, though official data remains limited.

Conventional tools like M-CHAT-R/F and ADOS require expert administration and are less accessible in low-resource settings such as Vietnam [3]. The

N. Thai-Nghe et al. (Eds.): ISDS 2025, CCIS 2714, pp. 204–216, 2026.
https://doi.org/10.1007/978-981-95-3358-9_15

absence of definitive medical tests further complicates diagnosis, relying on subjective behavioral assessments. Recent advances in AI and ML enable automated ASD screening using electronic medical records (EMRs), though challenges like imbalanced data and feature bias remain [4]. While models like Decision Trees, XGBoost, and CatBoost perform well in controlled environments, they often lack generalizability in real-world data.

This study introduces an ML-based ASD screening system for children aged 18, using a combined dataset of 300 public and 600 clinical EMRs. We tackle class imbalance with SMOTE-IPF and improve prediction fairness via an enhanced Overall Local Accuracy (OLA) framework. Key contributions include: (1) a curated dataset with expert-annotated features; (2) a bias-reduced OLA model; and (3) a scalable, cost-effective solution for under-resourced environments.

The paper is structured as follows: Sect. 2 reviews related work; Sect. 3 describes the data and preprocessing; Sect. 4 outlines the methodology; Sect. 5 presents results; and Sect. 6 discusses contributions, limitations, and future directions.

2   Related Works

Harshita Chandrappa [5] developed a hybrid ML framework for ASD prediction across all age groups, integrating classifiers like Decision Tree, SVM, and Random Forest into a web-based diagnostic tool. SVM and Random Forest achieved the highest accuracy. The system also includes a chatbot interface to support early ASD awareness and symptom understanding.

Jose A. Saez [6] proposed SMOTE-IDF, a hybrid method combining SMOTE and IDF weighting to address class imbalance in text classification. Using a DNN, the approach preserved rare term significance and outperformed traditional methods on datasets like 20 Newsgroups in multiple evaluation metrics.

Youngkyu Hong [9] proposed a dual-branch learning framework to address bias in image classification. It uses Bias-Contrastive Learning (BCL) to capture spurious correlations in a biased branch and Bias-Balanced Learning (BBL) to focus on robust features in an unbiased branch. By minimizing biased branch loss and maximizing unbiased branch agreement with true labels, the method reduces reliance on dataset biases, outperforming traditional debiasing techniques on datasets like Biased-MNIST and ImageNet-9.

Y. Zeng [10] introduces Ola, an advanced omni-modal language model capable of understanding images, audio, and video. It uses progressive modality alignment-gradually learning from image-text, audio, then video-to reduce training costs. Ola supports streaming speech generation and demonstrates state-of-the-art performance among open-source omni-modal models, rivaling specialized systems across diverse tasks.

Conventional screening for Autism Spectrum Disorder (ASD) typically relies on standardized instruments such as the Modified Checklist for Autism in Toddlers, Revised with Follow-Up (M-CHAT-R/F) and the Autism Diagnostic Observation Schedule (ADOS). These tools are widely used and clinically validated, but they are time-consuming, require trained professionals, and remain less accessible in low-resource environments. Such limitations have motivated the integration of machine learning (ML) methods to automate and scale ASD screening.

Duda et al. [11] investigated the use of logistic regression and random forest classifiers on behavioral questionnaire data, demonstrating that ML could replicate clinical judgments with high reliability. Thabtah [12] later proposed a rule-based ML system for ASD detection in toddlers, highlighting how decision trees and rule induction can provide interpretable predictions suitable for non-specialist use. More recently, Abbas et al. [13] applied deep learning architectures such as convolutional neural networks (CNNs) to multimodal behavioral and medical features, achieving promising performance in early ASD prediction.

Other researchers have also advanced this area. Alshammari et al. [14] developed an ensemble framework combining SVM, Random Forest, and k-NN classifiers to analyze ASD screening questionnaires, reporting improved accuracy compared to single models. Similarly, Thabtah and Peebles [15] evaluated ML classifiers on both adult and child ASD datasets, emphasizing the importance of dataset characteristics and feature selection in clinical applicability.

Compared with these works, our study contributes by integrating both public survey data and real clinical EMRs from Vietnamese children, applying SMOTE-IPF for data balancing, and enhancing the Overall Local Accuracy (OLA) framework to mitigate feature dominance (notably *Eye Contact*). In addition, we emphasize recall as the primary evaluation metric, reflecting the clinical priority of minimizing false negatives in ASD screening.

3   Dataset

This study employs two datasets to develop and evaluate the proposed ASD detection approach. The first is a public dataset obtained from Kaggle [22], comprising 292 records with a balanced class distribution (48.3% ASD). Each entry includes binary responses to 10 behavioral screening questions together with demographic and ancillary information (20 attributes in total), aligned with expert-defined evaluation criteria and later used to support consistent labeling of the private dataset. The second dataset consists of 594 anonymized medical records collected from a pediatric psychology clinic. Derived from an initial pool of more than 1,200 developmental assessment files, the records were curated through filtering, exclusion of incomplete or low-quality entries, and expert annotation of key behavioral features such as eye contact, pointing, response to name, joint attention, imitation, and symbolic play. The resulting dataset contains 36 structured features (raw and expert-labeled), though it remains highly imbalanced, with approximately 93% of cases diagnosed with or monitored for ASD. Together, these datasets provide a robust foundation for training and validating machine learning models for early ASD screening in children aged 1–8 years.

- **Step 1:** Raw records were extracted from PDF documents and categorized by diagnosis, resulting in over 800 cases relevant to ASD or non-clinical controls.

- **Step 2:** Incomplete or low-quality entries were excluded, yielding 600 usable samples. Key behavioral features—such as eye contact [16], pointing [17], response to name [18], joint attention [19], imitation [20], and symbolic play [21]—were annotated with the support of domain experts. All personally identifiable information was securely encoded to ensure confidentiality.
- **Step 3:** The final dataset consisted of 594 structured records with 36 features (raw and expert-labeled). The distribution is highly imbalanced, with approximately 93

Together, these data sets provide a comprehensive foundation for building and validating a machine learning model for early detection of ASD in children aged 1 to 8 years. These data sets have 18 labeled features that are used for analysis and prediction. These labeled features are listed below (Table 1).

Table 1. List of labeled features in the dataset

Feature 1	Feature 2	Feature 3
Medical History	Task Requesting	Symbolic Play
Speech Delay	Response to Name in Infants	Repetitive/Stereotyped Play
Behavioral Rigidity	Toe Walking after 3 years	Social Communication Skills
Behavioral Isolation	Making a Point	Turn-Taking Play
Sensorimotor Play	Joint Attention	Imitation
Functional Play	Eye Contact	Combinatorial Play

4 Proposed Methodology

This section presents the proposed methodology for developing an automated screening tool for Autism Spectrum Disorder (ASD). It outlines the overall workflow, the underlying theoretical foundations, and the implementation procedures adopted in this study. By combining advanced machine learning models with tailored data preprocessing strategies, the methodology is designed to address the unique challenges of psychological medical records, including class imbalance, feature dominance, and data heterogeneity. A structured pipeline is introduced to provide a clear understanding of how each component contributes to building a robust and clinically relevant ASD screening system. The detailed implementation of each step is illustrated below (Fig. 1).

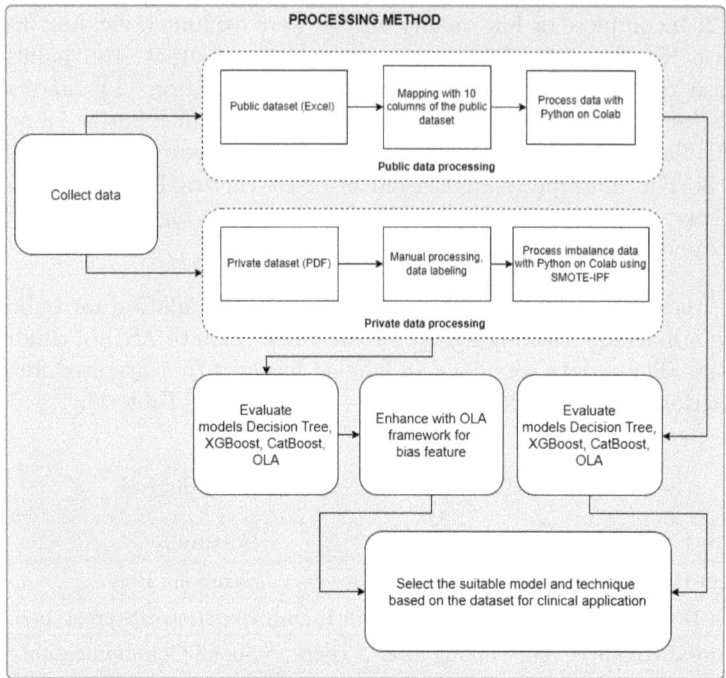

Fig. 1. Overview of implementation methods.

4.1   Data Preprocessing

Effective data preprocessing is crucial to ensure the quality and usability of electronic medical records (EMR) for developing an Autism Spectrum Disorder (ASD) screening tool. The pipeline comprised five sequential stages, each with a clear *objective* and *method*, as detailed below.

Step 1: Anonymization All personally-identifiable information was either encrypted or removed before model training.

Step 2: Redundant Column Removal Columns lacking analytical value-such as purely administrative metadata-were eliminated.

Step 3: Missing Value Imputation Numerical attributes were imputed with the median, while categorical attributes were filled with the mode; domain-expert psychology rules guided these choices where appropriate.

Step 4: Feature Encoding

- Target variable (ASD diagnosis) was encoded as binary - 0 (non-ASD) and 1 (ASD).
- Behavioral features were encoded as binary (0/1) or ordinal values (0, 0.5, 1).

Step 5: SMOTE-IPF Balancing

SMOTE-IPF [6] was selected as the main balancing method due to its effectiveness in handling class imbalance and filtering out noisy data. Unlike SMOTE-KNN [7], which imputes missing values but lacks noise filtering, SMOTE-IPF reduces the risk of overfitting.

To improve model training, unsafe ASD samples were further resampled using KNN, resulting in three datasets:

- A0: After SMOTE-IPF.
- A1: Fully amplified unsafe ASD samples.
- A2: Selective amplification based on neighbor voting.

These datasets, exported for training, showed that SMOTE-IPF offered the best trade-off between balance and robustness.

4.2 Evaluate Models

a, Models

Decision Tree. A hierarchical structure of binary decisions based on feature thresholds to classify samples as ASD or non-ASD. Its simplicity and interpretability make it suitable for clinical applications. However, Decision Trees are prone to overfitting, especially on imbalanced datasets, and may overly rely on dominant features, such as specific behavioral indicators, leading to biased predictions. The model was implemented with default parameters, using Gini impurity as the splitting criterion, and evaluated on the preprocessed dataset. Additionally, its performance benefits from pruning techniques to reduce complexity, though it struggles with noisy data and lacks robustness against feature interactions, necessitating careful preprocessing to enhance accuracy in ASD classification.

XGBoost. An ensemble learning method based on gradient boosting, XGBoost integrates multiple weak learners (decision trees) to enhance predictive performance. It effectively manages complex feature interactions and employs regularization to reduce overfitting. However, in imbalanced datasets, XGBoost may favor dominant features, potentially reducing its generalizability in pediatric ASD classification. Additionally, its efficiency stems from parallel processing and optimized tree pruning, though it requires careful hyperparameter tuning to address bias and improve adaptability across diverse data distributions.

CatBoost. CatBoost is an advanced Gradient Boosting Decision Trees (GBDT) algorithm that builds an ensemble of symmetric (oblivious) decision trees to enhance predictive performance and computational efficiency. Unlike traditional Decision Trees, CatBoost combines multiple weak learners with ordered boosting and regularization to reduce overfitting, achieving superior generalization on complex datasets. It was configured with a depth of 6 and 100 iterations, leveraging its ability to process ordinal features (e.g., poor-moderate-good encodings) directly.

However, CatBoost may still struggle with bias toward dominant features in imbalanced datasets, as applied in this study.

OLA (Overall Local Accuracy) [23]. Overall Local Accuracy (OLA) [24] is a dynamic classifier selection method that chooses the most accurate base classifier for each test sample based on its performance within the sample's local neighborhood. For a given input, it finds the k-nearest neighbors in the training set and selects the classifier with the highest local accuracy to make the prediction.

In this study, OLA was applied to an ensemble of diverse classifiers (XGBoost, CatBoost and Random Forest). By adapting to local data patterns, OLA enhances prediction fairness and accuracy in ASD classification.

b, Evaluation Metrics and Validation

Evaluation Metrics. Recall, Specificity, Precision, F1-score Prioritizing the Recall (sensitivity) metric is driven by the primary goal of maximizing the detection of ASD cases, even if it results in some false negatives. This is particularly critical because missing a diagnosis of ASD in a child can lead to a lack of early intervention, potentially causing severe long-term developmental impacts. Recall measures the proportion of actual ASD cases correctly identified by the model, making it a preferred metric over Precision, F1-score. Even the low Specificity but high Recall is still rated as a suitable model.

Validation Techniques

Cross-Validation. A 5-fold stratified cross-validation was conducted.

Process: The dataset was divided into 5 folds with preserved class distribution. Each fold was used once as test while others served as training.

Feature Importance Analysis. Evaluates the influence of original features on the model's predictions, providing insights into feature relevance for the ASD classification task.

4.3   Enhance with OLA Framework

Framework with OLA. This subsection illustrates the enhanced OLA pipeline designed to mitigate feature bias, particularly caused by dominant features such as `TiepXucMat`

- **Step 1:** Data Preprocessing
    - Standardization with StandardScaler: Transform data to a standard normal distribution (mean $= 0$, variance $= 1$) to reduce the influence of features with large values (e.g.: TiepXucMat), ensuring efficient model performance.
    - Dimensionality Reduction with PCA (Principal Component Analysis): Reduce data dimensions, removing noise and bias from dominant features while lowering computational cost.

- **Step 2:** Choosing internal model
  - Model Group Creation: XGBClassifier, CatBoostClassifier, and RandomForestClassifier trained on PCA data, focusing on key features to minimize noise.
  - This diversity enhances flexibility, allowing OLA to select the most suitable model for each test sample.
- **Step 3:** Deploy OLA Model
  Implement with the above models on the PCA-processed dataset to provide predictions for each data sample.

5 Results and Discussion

5.1 Experimental Results

Public Dataset

a. Cross-Validation Metrics
On the public dataset, the Decision Tree model showed the lowest performance across all metrics, reflecting its limited robustness. CatBoost and XGBoost achieved strong and comparable results, with XGBoost slightly outperforming CatBoost (mean Recall 0.979 vs. 0.968). The OLA model consistently outperformed all baselines, reaching a Recall of 0.989 and an F1-score of 0.985. These results demonstrate OLA's superior adaptability and robustness in handling balanced public data (Table 2).

Table 2. Cross-validation performance on the public dataset

Model	Recall	Specificity	Precision	F1-Score
Decision Tree	0.820	0.813	0.757	0.783
XGBoost	0.979	0.950	0.944	0.954
CatBoost	0.968	0.935	0.916	0.939
OLA	**0.989**	**0.986**	**0.979**	**0.985**

b. Feature Importance
The feature importance charts indicate a relatively uniform distribution across features, suggesting minimal column bias in the public dataset. CatBoost and OLA achieve the most balanced recognition, followed by XGBoost and Decision Tree (Fig. 2).

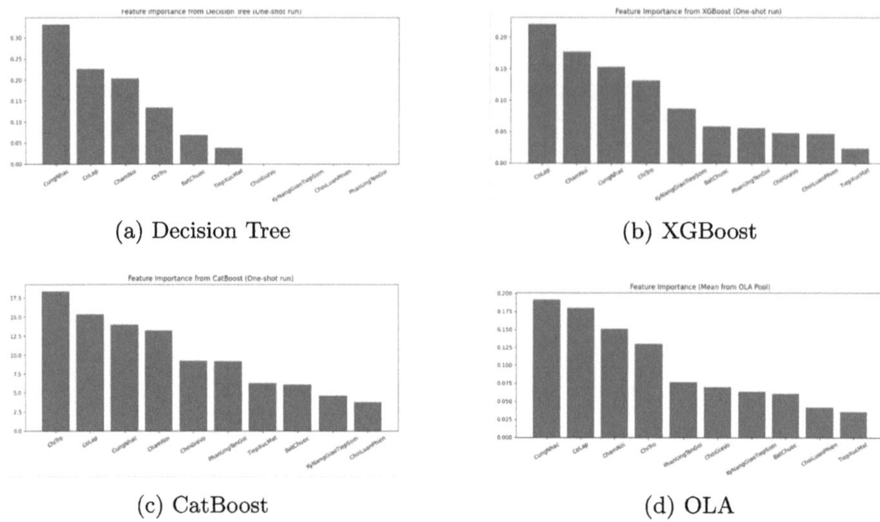

Fig. 2. Feature importance analysis on the public dataset across four models.

Private Dataset

a. Cross-Validation Metrics

On the private clinical dataset, performance dropped across all models due to noise and imbalance. Decision Tree remained the weakest (Recall 0.828), while XGBoost and CatBoost both achieved Recall close to 0.90. OLA again delivered the highest Recall (0.964) and F1-score (0.979), and the enhanced OLA framework further improved balance with Recall of 0.967. This demonstrates that OLA is more resilient to unstable and biased clinical data compared to individual models (Table 3).

Table 3. Cross-validation performance on the private dataset

Model	Recall	Specificity	Precision	F1-Score
Decision Tree	0.828	0.836	0.839	0.830
XGBoost	0.904	0.896	0.898	0.900
CatBoost	0.896	0.892	0.896	0.895
OLA	**0.964**	**0.996**	**0.996**	**0.979**
Enhanced OLA	**0.967**	0.984	0.988	0.977

b. Feature Importance

Unlike the public dataset, the private dataset exhibited a pronounced bias toward the *Eye Contact (TiepXucMat)* feature, which inflated evaluation metrics and undermined robustness. This bias limited the generalizability of all baseline models—although CatBoost handled it slightly better, none provided a reliable solution. Leveraging its adaptability, OLA emerged as the most promising approach but still required refinement. To overcome this limitation, the Enhanced OLA framework was developed, effectively reducing the dominance of *Eye Contact*, diversifying feature contributions, and producing more balanced and stable predictions (Fig. 3).

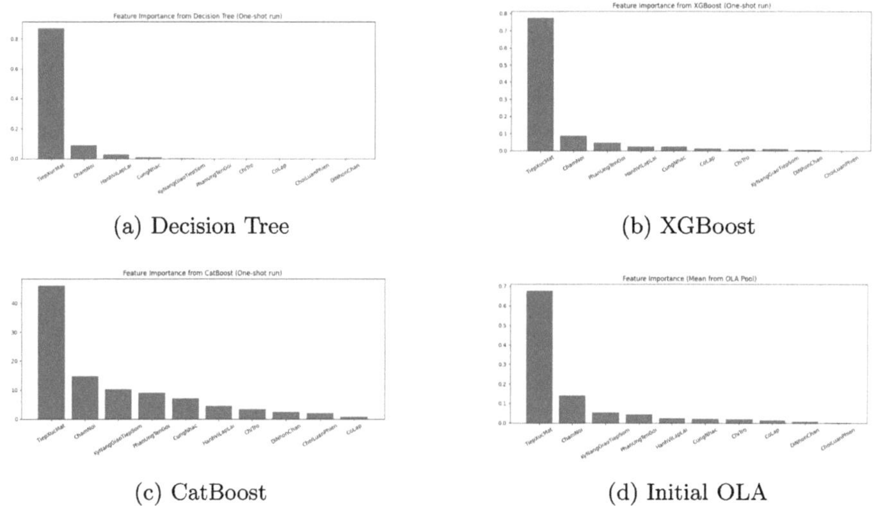

(a) Decision Tree

(b) XGBoost

(c) CatBoost

(d) Initial OLA

Fig. 3. Feature importance distribution on the private dataset across four models.

The Enhanced OLA framework successfully reduced the dominance of *TiepX-ucMat*, leading to more balanced feature utilization and improved robustness for ASD classification (Fig. 4).

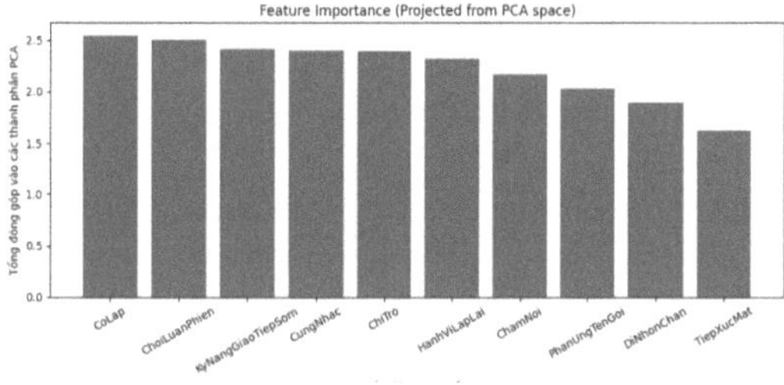

Fig. 4. Feature importance after applying Enhanced OLA on the private dataset.

5.2 Discussion

Overall, OLA and its enhanced version consistently outperformed Decision Tree, XGBoost, and CatBoost in both datasets, particularly in recall—the most critical metric for screening, where missing true ASD cases has serious consequences. The drop in performance from public to private datasets highlights the challenges of real-world variability and dataset bias.

Compared with prior works, our approach achieves higher recall while maintaining computational efficiency, making it deployable in low-resource clinical settings. However, the results rely heavily on a limited dataset (594 clinical records) and lack external validation from other hospitals. Interpretability also remains an open challenge: future integration of SHAP or LIME will allow clinicians to better understand model predictions and feature contributions.

Clinically, the tool is envisioned as a decision-support module that can operate either as a standalone screening app or integrated into electronic medical record (EMR) systems. Ethically, false positives may cause unnecessary concern for families and must be followed by expert confirmation, while strict anonymization and privacy safeguards are required for sensitive child health data.

6 Conclusion

This paper presented a machine learning-based screening tool for Autism Spectrum Disorder (ASD) in children, combining data balancing (SMOTE-IPF) with an enhanced Overall Local Accuracy (OLA) framework. Across both public and private datasets, OLA achieved superior recall compared to Decision Tree, XGBoost, and CatBoost, demonstrating strong potential for real-world screening.

The study makes three contributions: (i) a curated ASD dataset combining public and clinical records, (ii) an enhanced OLA framework that mitigates feature bias (notably *Eye Contact*), and (iii) a lightweight model suitable for clinics with limited computational resources.

Nevertheless, limitations remain. The private dataset is small and biased, raising concerns about generalizability. No external validation has yet been performed, and interpretability is still limited. Future work will focus on validating the tool with data from multiple hospitals, integrating interpretability methods (e.g., SHAP, LIME) for clinical decision support, and extending the framework to screen other developmental conditions.

From a clinical perspective, this tool is not intended to replace professional diagnosis but to support early identification and referral. Ethical considerations—particularly false positives and privacy—must be carefully managed. With these refinements, the proposed approach has the potential to become a scalable, cost-effective, and clinically relevant tool for early ASD screening in resource-constrained environments.

Acknowledgment. We would like to express our sincere gratitude to the University of Information Technology, Vietnam National University Ho Chi Minh City (UIT, VNU-HCM), for their support and for providing the necessary resources throughout this research. This research was supported by The VNUHCM-University of Information Technology's Scientific Research Support Fund.

We also would like to thank experts in the field of psychological health for providing young medical record data and professional documents, who supported the research team to authenticate and consolidate professional knowledge so that the group can achieve the best research results.

References

1. Hodges, H., Fealko, C., Soares, N.: Autism spectrum disorder: definition, epidemiology, causes, and clinical evaluation (2020). https://doi.org/10.21037/tp.2019.09.09

2. Le, B.: Ty le tre tu ky o Viet Nam va The gioi: Nhung con so bao dong (2025). https://tretuky.edu.vn/ty-le-tre-tu-ky-o-viet-nam-va-the-gioi/

3. dos Santos, C.L., Barreto, I.I., Floriano, I., Tristao, L.S., Silvinato, A., Bernardo, W.M.: Screening and diagnostic tools for autism spectrum disorder: systematic review and meta-analysis (2024). https://doi.org/10.1016/j.clinsp.2023.100323

4. Olubudo, P.: Statistical bias correction in imbalanced datasets (2025). https://www.researchgate.net/publication/390805682

5. Chandrappa, H., Swaathi, B.R., Pahuja, S.: Prediction of autism spectrum disorder based on machine learning approach. Int. Res. Appl. Sci. Eng. Technol. (IJRASET) **11**(V), 3521–3526 (2023). https://doi.org/10.22214/ijraset.2023.52417 https://doi.org/10.22214/ijraset.2023.52417

6. Sáez, J.A., Luengo, J., Stefanowski, J., Herrera, F.: A novel technique for text classification using SMOTE-IPF based oversampling and deep neural network. Int. J. Adv. Comput. Sci. Appl. **12**(10) (2014). https://doi.org/10.14569/IJACSA.2021.051

7. Karamti, H., et al.: SMOTE-KNN: an improved algorithm based on SMOTE and K-nearest neighbors for imbalanced data. IEEE Access **9**, 95690–95701 (2023). https://doi.org/10.1109/ACCESS.2023.5174412

8. Hung, B.D., Thoa, V.V., Dang, X.T.: KSI - a combined clustering and resampling method with noise filtering algorithm for imbalanced data classification. J. Inf. Commun. Technol. **1**(1) (2019)

9. Hong, Y., Yang, E.: Unbiased classification through bias-contrastive and bias-balanced learning. Adv. Neural Inf. Process. Syst. (NeurIPS) **34**, 20153–20165 (2021)

10. Liu, Z., et al.: Ola: pushing the frontiers of omni-modal language model with progressive modality alignment. arXiv preprint arXiv:2502.04328 (2025)

11. Duda, M., Ma, R., Haber, N., Wall, D.P.: Use of machine learning for behavioral distinction of autism and ADHD. Transl. Psychiatry **6**(2), e732–e732 (2016). https://doi.org/10.1038/tp.2015.221

12. Thabtah, F.: Machine learning in autistic spectrum disorder behavioral research: a review and ways forward. Inform. Health Soc. Care **44**(3), 278–297 (2019). https://doi.org/10.1080/17538157.2017.1399132

13. Abbas, H., Alsheikh, M., Sodhro, A.H., Pirbhulal, S., Al-Bander, B.: ASD–autism spectrum disorder detection using deep learning. Comput. Biol. Med. **145**, 105495 (2022). https://doi.org/10.1016/j.compbiomed.2022.105495

14. Alshammari, T., Sagayam, K., Alzahrani, A., Alshammari, F.: Autism spectrum disorder detection using ensemble machine learning approaches. Comput. Mater. Continua **66**(3), 2937–2952 (2021). https://doi.org/10.32604/cmc.2021.013731
15. Thabtah, F., Peebles, D.: Autism spectrum disorder: machine learning prediction models and feature significance evaluation. J. Inf. Knowl. Manage. **19**(1), 2040009 (2020). https://doi.org/10.1142/S0219649220400098
16. Moriuchi, J.M., Klin, A., Jones, W.: Mechanisms of diminished attention to eyes in autism. Am. J. Psychiatry **173**(8), 825–833 (2016). https://doi.org/10.1176/appi.ajp.2016.15081034
17. Ramos-Cabo, S., Vulchanov, V., Vulchanova, M.: Different ways of making a point: a study of gestural communication in typical and atypical early development. Autism Res. **14**(5), 984–996 (2021). https://doi.org/10.1002/aur.2474
18. Miller, M., et al.: Response to name in infants developing autism spectrum disorder: a prospective study. J. Pediatrics **165**(2), 332–337 (2014). https://doi.org/10.1016/j.jpeds.2014.04.017
19. Montagut-Asunción, M., et al.: Joint attention and its relationship with autism risk markers at 18 months of age. Children **9**(4) (2022). https://doi.org/10.3390/children9040480.
20. Ingersoll, B.: The social role of imitation in autism: implications for the treatment of imitation deficits. Infants Young Child. **21**(2), 107–119 (2008). https://doi.org/10.1097/01.IYC.0000314482.24087.14
21. Lam, Y.G., Yeung, S.S.: Symbolic play in children with autism. In: Comprehensive Guide to Autism, pp. 491–508. Springer (2014). https://doi.org/10.1007/978-1-4614-4788-7_26
22. Tabtah, F.F.: "Autism spectrum disorder screening" dataset about list 10 question to predict ASD in Kaggle. In: Proceedings of the 1st International Conference on Medical and Health Informatics, Taichung City, Taiwan. ACM (2017). http://fadifayez.com/wp-content/uploads/2017/11/Autism-Spectrum-Disorder-Screening-Machine-Learning.pdf
23. Overall Local Accuracy. https://deslib.readthedocs.io/en/latest/modules/dcs/ola.html
24. Cruz, R.M.O., Sabourin, R., Cavalcanti, G.D.C.: DESlib: a dynamic ensemble selection library in Python (2018). https://github.com/scikit-learn-contrib/DESlib

A Deep Learning Framework with Squeeze-and-Excitation Attention for Guava Disease Classification

Tuong Le[1(✉)], Duc-Tan Nguyen[2], Hoang-Duy Nguyen[3], and Thang Cap[3]

[1] Faculty of Information Technology, HUTECH University, Ho Chi Minh City, Vietnam
`lc.tuong@hutech.edu.vn`
[2] University of Management and Technology, Ho Chi Minh City, Vietnam
`tan.nguyen@umt.edu.vn`
[3] University of Information Technology, Vietnam National University, Ho Chi Minh City, Vietnam
`22520328@gm.uit.edu.vn`, `thangcpd@uit.edu.vn`

Abstract. Timely and accurate detection of plant diseases is essential for safeguarding crop yields and ensuring sustainable agricultural practices. This study presents a deep learning-based framework for guava leaf disease classification that leverages transfer learning and attention mechanisms to enhance model performance. Five state-of-the-art convolutional neural networks (CNNs)—DenseNet201, EfficientNetV2B0, InceptionV3, MobileNetV2, and Xception—were evaluated on the Guava Foliar Lesion Dataset (GFLD) to benchmark their feature extraction capabilities. While DenseNet201 achieved the highest baseline accuracy (85.99%), its recall and F1-score remained relatively low. To address this, we propose DAGuavaNet, an enhanced architecture that integrates a squeeze-and-excitation (SE) attention module into DenseNet201. DAGuavaNet maintained the same accuracy (85.99%) but significantly improved recall (from 79.89% to 85.99%) and F1-score (from 80.22% to 85.81%), indicating more balanced and consistent performance across disease classes. These results demonstrate that attention-enhanced CNNs can provide a robust and effective solution for automated guava leaf disease detection.

Keywords: Guava Disease Classification · Deep Learning · DenseNet201 · Attention Mechanism · Transfer Learning

1 Introduction

In recent decades, artificial neural networks, especially within computer vision, have revolutionized how machines interpret and interact with visual data. The field of computer vision has rapidly evolved from basic rule-based automation to intelligent systems empowered by deep learning models. Among various domains

N. Thai-Nghe et al. (Eds.): ISDS 2025, CCIS 2714, pp. 217–227, 2026.
https://doi.org/10.1007/978-981-95-3358-9_16

of artificial intelligence, computer vision stands out as a highly dynamic field that enables machines to extract meaningful features from images and videos through techniques such as pattern classification, image processing, and machine learning. Real-world applications of computer vision are extensive and continually expanding: in healthcare, it enhances diagnostic accuracy in medical imaging, notably during the COVID-19 pandemic [1–3]; in the automotive sector, it plays a critical role in self-driving systems through object detection [4,5], lane tracking [6], and traffic sign recognition [7]; and in agriculture, it enhances precision farming by enabling crop monitoring [8], disease detection [9–12], and yield estimation [13,14].

Agriculture is undergoing a technological transformation, with computer vision playing an increasingly vital role in applications such as yield prediction [13,14] and plant health monitoring [8]. For tropical crops like guava, early detection of leaf and fruit diseases—such as leaf spot, fruit rot, and powdery mold—is critical to maintaining product quality and minimizing yield losses [10]. Traditionally, disease identification relies on manual inspection by farmers or specialists, a process that is time-consuming, labor-intensive, and prone to human subjectivity [9]. These limitations are particularly pronounced in large-scale farms or remote areas with limited access to expert support. To address these challenges, this study proposes a deep learning-based computer vision system for the automatic detection and classification of guava diseases, aiming to support farmers with timely, accurate diagnostics and improve overall agricultural productivity.

This study presents a comprehensive evaluation of transfer learning techniques for guava leaf disease classification, emphasizing the effectiveness of state-of-the-art convolutional neural networks (CNNs) on the GFLD dataset [15]. We begin by benchmarking five well-established architectures—DenseNet201 [16], EfficientNetV2B0 [17], InceptionV3 [18], MobileNetV2 [19], and Xception [20]—to determine the most suitable backbone for the task. Among these, Dense-Net201 achieved the highest classification accuracy (85.99%), confirming its superior feature extraction capabilities in the context of guava disease recognition.

Building on this observation, we propose DAGuavaNet, an enhanced model that integrates a squeeze-and-excitation (SE) attention mechanism into the Dense-Net201 architecture to improve the network's focus on disease-relevant features. This modification enables more effective channel-wise feature recalibration, leading to improved representation of subtle visual patterns associated with guava leaf diseases. Experimental results show that DAGuavaNet outperforms the baseline DenseNet201 in key performance metrics: the F1-score increases from 80.22% to 85.81%, and recall improves from 79.89% to 85.99%, while maintaining the same overall accuracy of 85.99%. These improvements demonstrate the effectiveness of incorporating attention mechanisms into CNN architectures for achieving more balanced, accurate, and reliable guava disease classification.

The remainder of this paper is organized as follows. Section 2 provides a detailed description of the dataset, data preprocessing techniques, and the deep learning architectures employed in this study. Section 3 presents the experimental design, including training configurations, evaluation metrics, and comparative analysis of model performance. Finally, Sect. 4 summarizes the key findings and discusses the practical implications of the proposed approach for real-world applications in agricultural disease diagnosis.

2   Materials and Methods

2.1   The GFLD Dataset

The GFLD dataset, introduced by Shihab et al. [15] in 2025, contains 3,049 high-resolution images of healthy and diseased guava fruits and leaves, captured using mobile devices across various locations in Bangladesh. It covers fruit diseases such as anthracnose, scab, and stylar end rot, as well as leaf diseases including rust, dot, anthracnose, and canker. All images were carefully verified and labeled with input from agricultural experts.

Figure 1 shows the number of images per class, revealing a clear imbalance. The Healthy (Leaf) class dominates with 1,248 images. In the fruit category, Healthy (Fruit) has 342 images, followed by Anthracnose (Fruit) with 263, Stylar end rot (Fruit) with 262, and Scab (Fruit) with 119. For leaf diseases, Anthracnose (Leaf) has 237 images, Dot (Leaf) has 219, Canker (Leaf) has 192, and Rust (Leaf) has 167. This distribution indicates a strong bias toward healthy leaf samples.

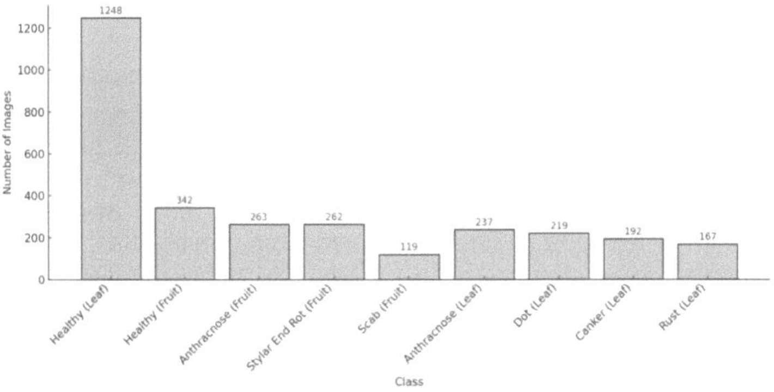

Fig. 1. Class distribution of guava images in the GFLD dataset.

The dataset was partitioned into training, validation, and test subsets using an approximate 8:1:1 ratio, resulting in 2435, 300, and 314 images, respectively. To address class imbalance in the training set, targeted augmentation was applied to underrepresented classes by replicating existing samples. Specifically, the following classes were augmented: Anthracnose (Fruit) from 210 to 500 images, Anthracnose (Leaf) from 189 to 500, Canker (Leaf) from 153 to 500, Dot (Leaf) from 175 to 800, Healthy (Fruit) from 273 to 500, Rust (Leaf) from 133 to 500, Scab (Fruit) from 95 to 500, and Stylar End Rot (Fruit) from 209 to 500. The Healthy (Leaf) class, with 998 original samples, remained unchanged.

This selective augmentation strategy improved class balance, enhanced training stability, and supported more robust model performance across disease categories. Care was taken to avoid excessive duplication that might lead to overfitting, ensuring that the augmentation process remained effective without compromising generalization.

2.2 System Architecture

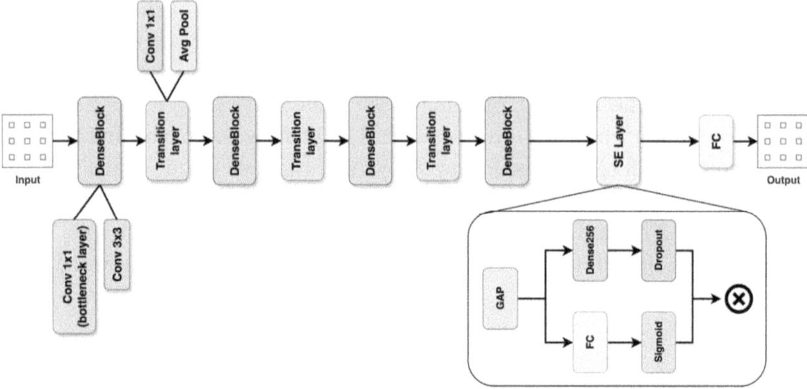

Fig. 2. System Architecture of DAGuavaNet for Disease Classification in Guava.

The proposed DAGuavaNet architecture for guava disease classification is depicted in Fig. 2. The system is composed of three major stages: data preprocessing, feature extraction using a DenseNet-based backbone, and attention-enhanced classification via Squeeze-and-Excitation (SE) modules.

Preprocessing Pipeline. Prior to training, all images from the GFLD dataset are passed through a preprocessing pipeline designed to ensure input consistency and enhance model generalization. Each image is first resized to a fixed input size to maintain uniformity across the dataset. Pixel values are then normalized, either by scaling them to the range [0,1] or by standardizing them based on the dataset's mean and variance. To increase data variability and reduce overfitting, various data augmentation techniques are applied, including random rotations, horizontal and vertical flipping, brightness and contrast adjustments, as well as zoom transformations.

Feature Extraction Backbone (DenseNet201). DAGuavaNet utilizes a modified version of DenseNet201 as its core feature extractor. DenseNet introduces dense connectivity, where each layer receives feature maps from all previous layers, formally represented as:

$$x_l = H_l\left([x_0, x_1, \ldots, x_{l-1}]\right), \tag{1}$$

Here, x_l denotes the output of the $l - th$ layer, $H_l(.)$ is a composite function consisting of Batch Normalization (BN), ReLU activation, and convolution, and $[x_0, x_1, \ldots, x_{l-1}]$ is the concatenation of all feature maps from preceding layers. This architecture improves feature reuse and enables better gradient flow during training.

Each DenseBlock comprises multiple composite layers arranged in the form of bottleneck blocks. Each bottleneck block typically consists of a 1×1 convolutional layer that performs dimensionality reduction, effectively decreasing the number of input feature maps. This is followed by a 3×3 convolutional layer that extracts local spatial features from the reduced representation. This structure not only improves computational efficiency but also preserves salient features through compact representations.

Between DenseBlocks, Transition Layers are incorporated to control model complexity and compress the feature maps. Each Transition Layer consists of a 1×1 convolutional layer to reduce the number of feature channels, followed by a 2×2 average pooling operation that downsamples the spatial dimensions by a factor of two. This design helps maintain a balance between network depth and computational efficiency while avoiding overfitting.

Channel Attention with Squeeze-and-Excitation (SE) Module. To improve the model's sensitivity to important feature channels, DAGuavaNet incorporates a SE block after the final DenseBlock. This mechanism allows the network to adaptively recalibrate channel-wise feature responses by modeling interdependencies between channels. The SE block operates in three steps:

(1) Squeeze: Global Information Embedding. For each channel c, global average pooling is applied to condense spatial information into a single scalar:

$$z_c = \frac{1}{H \times W} \sum_{i=1}^{H} \sum_{j=1}^{W} x_c(i,j) \tag{2}$$

where $x_c(i,j)$ is the activation at location (i,j) for channel c, and z_c is the global descriptor.

(2) Excitation: Channel-wise Dependencies Modeling. The aggregated descriptors are passed through a bottleneck structure of two fully connected (FC) layers with non-linearities:

$$s = \sigma(W_2 \cdot \delta(W_1 \cdot z)) \tag{3}$$

where:

- $z \in \mathbb{R}^C$ is the squeezed vector for C channels,
- $W_1 \in \mathbb{R}^{\frac{C}{r} \times C}, W_2 \in \mathbb{R}^{C \times \frac{C}{r}}$ are the learnable weights of the FC layers,
- $\delta(\cdot)$ is the ReLU activation function,
- $\sigma(\cdot)$ is the sigmoid activation function,
- r is the reduction ratio (commonly 16 or 32) to reduce parameter overhead.

(3) Recalibration: Feature Reweighting. The resulting attention vector s is then used to recalibrate the original feature maps:

$$\tilde{x}_c = s_c \cdot x_c \tag{4}$$

This operation enhances important feature channels while suppressing irrelevant ones, helping the model focus on disease-relevant patterns.

Classification Layer. After attention-based refinement, the output feature maps are passed through a fully connected (FC) layer. A softmax activation function is used at the final layer to compute the probability distribution over K disease classes:

$$\hat{y}_k = \frac{\exp(z_k)}{\sum_{j=1}^{K} \exp(z_j)} \tag{5}$$

where z_k is the logit for class k, and $\hat{y}_k$ is the predicted probability.

3 Experiments

3.1 Experimental Setting

We conducted a series of experiments on the GFLD dataset to evaluate and compare the performance of five state-of-the-art pretrained convolutional neural network architectures—DenseNet201, EfficientNetV2B0, InceptionV3, MobileNetV2, and Xception—for the task of guava disease classification. Beyond these baseline models, we propose a novel architecture, **DAGuavaNet**, which enhances the DenseNet201 backbone by integrating Squeeze-and-Excitation (SE) attention modules to improve channel-wise feature recalibration and boost discriminative learning capacity.

For DAGuavaNet, input images were resized to 224×224 pixels. The model was trained using the Adam optimizer with an initial learning rate of 1×10^{-4} and a batch size of 64. Training was conducted for a maximum of 50 epochs, employing an early stopping strategy with a patience of 10 epochs to prevent overfitting. Furthermore, a learning rate reduction strategy was applied, lowering the rate to 1×10^{-6} when the validation performance plateaued, allowing for finer convergence during later training stages.

All models were trained and fine-tuned using Google Colab, a widely adopted cloud-based development environment for machine learning research. The platform offers free access to NVIDIA Tesla T4 GPUs, enabling efficient and accelerated training of deep CNN models. This setup allowed us to experiment with computationally intensive architectures while maintaining reproducibility and accessibility.

To objectively evaluate the performance of the models, we employed four standard classification metrics: **Accuracy**, **Precision**, **Recall**, and **F1-Score**. These metrics were computed on both the validation and independent test sets. The definitions are as follows:

$$\text{Accuracy} = \frac{TP + TN}{TP + TN + FP + FN} \tag{6}$$

$$\text{Precision} = \frac{TP}{TP + FP} \tag{7}$$

$$\text{Recall} = \frac{TP}{TP + FN} \tag{8}$$

$$\text{F1-Score} = \frac{2 \cdot \text{Precision} \cdot \text{Recall}}{\text{Precision} + \text{Recall}} \tag{9}$$

where:

- **TP (True Positives)**: correctly predicted positive cases.
- **TN (True Negatives)**: correctly predicted negative cases.
- **FP (False Positives)**: negative cases incorrectly predicted as positive.
- **FN (False Negatives)**: positive cases incorrectly predicted as negative.

3.2   Model Convergence and Learning Dynamics

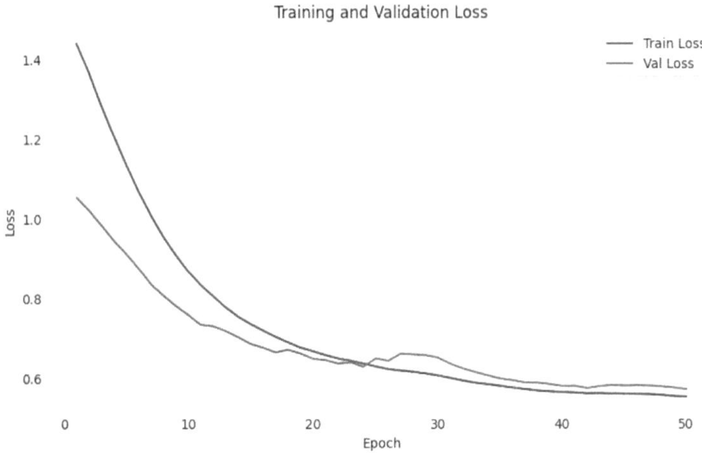

Fig. 3. Training and validation loss curves of DAGuavaNet.

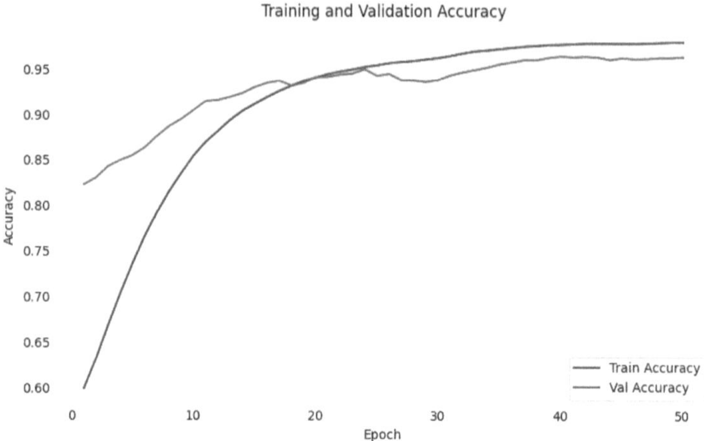

Fig. 4. Training and validation accuracy curves of DAGuavaNet.

Figures 3 and 4 illustrate the training and validation accuracy and loss curves of DAGuavaNet over the course of model training. As shown, the model demonstrates consistent convergence behavior, with both training and validation losses steadily decreasing as the number of epochs increases. Similarly, the accuracy curves exhibit a smooth upward trend, indicating progressive improvement in the model's predictive performance.

The gap between training and validation curves remains minimal throughout the training process, suggesting that DAGuavaNet generalizes well and is not prone to overfitting, despite the inherent class imbalance in the dataset. This stability can be attributed to several factors: the use of data augmentation to mitigate class imbalance, the feature reuse facilitated by the DenseNet architecture, and the channel-wise recalibration capability provided by the squeeze-and-excitation (SE) module.

Overall, the training dynamics indicate that DAGuavaNet achieves a good balance between learning capacity and generalization, which supports the reliability of the subsequent evaluation results reported in Sect. 3.3.

3.3   Experimental Results

Table 1 presents a comprehensive comparison of the performance of several deep learning architectures on the guava leaf disease classification task. Among the evaluated models, DAGuavaNet achieved the best overall performance, recording an accuracy of 85.99%, an F1-score of 85.81%, a recall of 85.99%, and a precision of 86.18%. These results demonstrate DAGuavaNet's superior ability to balance sensitivity and specificity, making it the most reliable model for this classification task.

Although DAGuavaNet and DenseNet201 achieved the same overall accuracy of 85.99%, this metric alone does not fully capture the model's effectiveness

Table 1. Classification Performance of Various Models on the GFLD Dataset.

No	Models	Accuracy (%)	F1-Score (%)	Recall (%)	Precision (%)
1	DenseNet201 [16]	85.99	80.22	79.89	81.00
2	EfficientNetV2B0 [17]	82.80	76.97	77.33	77.13
3	InceptionV3 [18]	81.21	73.28	72.45	75.02
4	MobileNetV2 [19]	82.48	76.49	76.00	79.11
5	Xception [20]	81.53	76.00	75.00	79.00
6	DAGuavaNet (ours)	85.99	85.81	85.99	86.18

in dealing with class imbalance. Given the skewed distribution of the GFLD dataset, where the Healthy (Leaf) class dominates, accuracy may give a misleading impression of model reliability. DAGuavaNet's significant improvements in recall and F1-score—over 6% higher than DenseNet201—indicate its enhanced capacity to correctly classify minority disease classes. This balanced performance is particularly critical for real-world agricultural deployment, where missing rare disease cases could lead to substantial crop damage.

In contrast, other baseline models such as InceptionV3, MobileNetV2, and Xception exhibited noticeably lower scores across all metrics. For instance, InceptionV3 achieved only 81.21% accuracy and a relatively low F1-score of 73.28%, indicating its limitations in effectively capturing the nuanced features necessary for precise disease classification.

Overall, the results clearly validate the effectiveness of DAGuavaNet as a robust and well-balanced model for guava leaf disease detection. Its consistent performance across all key metrics underscores its potential for practical deployment in real-world agricultural diagnostics, especially where accurate and dependable disease identification is critical.

4   Conclusion

This study proposed a deep learning-based framework for guava leaf disease classification by evaluating several state-of-the-art convolutional neural networks (CNNs) and introducing DAGuavaNet, an enhanced model built upon DenseNet-201 with an integrated squeeze-and-excitation (SE) attention mechanism. While DenseNet201 achieved the highest baseline accuracy (85.99%) among all tested models, its relatively lower recall and F1-score highlighted the need for improved feature discrimination. DAGuavaNet effectively addressed this by enhancing the model's ability to focus on disease-relevant features, leading to significant gains in recall (85.99%) and F1-score (85.81%), while maintaining the same accuracy. These improvements indicate that incorporating attention mechanisms into deep CNN architectures can provide more balanced and reliable performance for plant disease classification tasks.

Looking ahead, several research directions will be pursued to further enhance the proposed approach. First, the GFLD dataset will be expanded to include a wider range of guava diseases, growth stages, and environmental conditions to improve model generalization. Second, explainable AI (XAI) techniques, such as Grad-CAM and SHAP, will be integrated to provide visual and semantic insights into model decisions, thereby improving transparency and usability for farmers and agronomists. Third, we aim to optimize DAGuavaNet for real-time deployment on mobile and edge devices, enabling in-field disease diagnosis and decision support. Finally, we plan to explore multimodal learning approaches that combine visual data with contextual information (e.g., weather, soil conditions, and cultivation practices) to further boost predictive performance and contribute to precision agriculture initiatives.

References

1. Shi, F., et al.: Review of artificial intelligence techniques in imaging data acquisition, segmentation and diagnosis for COVID-19. IEEE Rev. Biomed. Eng. **14**, 4–15 (2021)
2. Vo, M.T., Vo, A.H., Le, T.: A robust framework for shoulder implant X-ray image classification. Data Technol. Appl. **56**(3), 447–460 (2022)
3. Luong, H.H., Nguyen, H.T., Thai-Nghe, N.: Toward supporting breast cancer diagnosis based on captioning mammogram and ultrasound images. In: Proc. Int. Conf. on Intelligent Systems and Data Science, pp. 71–85. Springer Nature Singapore, Singapore (2024)
4. Mao, J., Shi, S., Wang, X., Li, H.: 3D object detection for autonomous driving: a comprehensive survey. Int. J. Comput. Vision **131**(8), 1909–1963 (2023)
5. Almazrouei, K., Kamel, I., Rabie, T.: Dynamic obstacle avoidance and path planning through reinforcement learning. Appl. Sci. **13**(14), 8174 (2023)
6. Sultana, S., Ahmed, B., Paul, M., Islam, M.R., Ahmad, S.: Vision-based robust lane detection and tracking in challenging conditions. IEEE Access **11**, 67938–67955 (2023)
7. Grigorescu, S., Trasnea, B., Cocias, T., Macesanu, G.: A survey of deep learning techniques for autonomous driving. J. Field Robot. **37**(3), 362–386 (2020)
8. d'Andrimont, R., Yordanov, M., Martinez-Sanchez, L., Van der Velde, M.: Monitoring crop phenology with street-level imagery using computer vision. Comput. Electr. Agric. **196**, 106866 (2022)
9. Dang, M., et al.: Computer vision for plant disease recognition: a comprehensive review. Bot. Rev. **90**(3), 251–311 (2024)
10. Rashid, J., Khan, I., Ali, G., Alturise, F., Alkhalifah, T.: Real-time multiple guava leaf disease detection from a single leaf using hybrid deep learning technique. Comput. Mater. Continua **74**, 1 (2023)
11. Pham, T.C., Le, C.H., Packianather, M., Hoang, V.D.: Artificial intelligence-based solutions for coffee leaf disease classification. In: IOP Conf. Ser.: Earth and Environmental Science, vol. 1278, no. 1, art. no. 012004 (2023)
12. Thanh, T.N., Nguyen, L.X., Cap, T., Le, T.: A durian leaf image dataset of common diseases in Vietnam for agricultural diagnosis. Data Brief, art no. 111845 (2025)
13. Talaviya, T., Shah, D., Patel, N., Yagnik, H., Shah, M.: Implementation of artificial intelligence in agriculture for optimisation of irrigation and application of pesticides and herbicides. Artif. Intell. Agric. **4** (2020)

14. Pham, H.T., Awange, J., Kuhn, M., Nguyen, B.V., Bui, L.K.: Enhancing crop yield prediction utilizing machine learning on satellite-based vegetation health indices. Sensors **22**(3), 719 (2022)
15. Shihab, M.R., Saim, N.I., Mojumdar, M.U., Raza, D.M., Siddiquee, S.M.T., Noori, S.R.H., Chakraborty, N.R.: Image dataset for classification of diseases in guava fruits and leaves. Data Brief **111378** (2025)
16. Huang, G., Liu, Z., van der Maaten, L., Weinberger, K.Q.: Densely connected convolutional networks. In: Proc. IEEE Conf. Computer Vision and Pattern Recognition (CVPR), Honolulu, HI, pp. 2261–2269 (2017)
17. Tan, M., Le, Q.V.: EfficientNetV2: smaller models and faster training. In: Proc. 38th Int. Conf. Machine Learning, vol. 139, pp. 10096–10106 (2021)
18. Szegedy, C., Vanhoucke, V., Ioffe, S., Shlens, J., Wojna, Z.: Rethinking the inception architecture for computer vision. In: Proc. IEEE Conf. Computer Vision and Pattern Recognition (CVPR), Las Vegas, NV, pp. 2818–2826 (2016)
19. Sandler, M., Howard, A., Zhu, M., Zhmoginov, A., Chen, L.C.: MobileNetV2: inverted residuals and linear bottlenecks. In: Proc. IEEE/CVF Conf. Computer Vision and Pattern Recognition (CVPR), Salt Lake City, UT, pp. 4510–4520 (2018)
20. Chollet, F.: Xception: Deep learning with depthwise separable convolutions. In: Proc. IEEE Conf. Computer Vision and Pattern Recognition (CVPR), Honolulu, HI, pp. 1251–1258 (2017)

Big Data, IoT, and Cloud Computing

Towards End-to-End Traffic Congestion Estimation Using Learned ROI and Vehicle Object Dynamics

Khang Le[1,2] , Anh Duc Tran[1,2] , Thi Minh Tam Tran[3] ,
and Quang Tran Minh[1,2(✉)]

[1] Faculty of Computer Science and Engineering, Ho Chi Minh City University of
Technology (HCMUT), 268 Ly Thuong Kiet, Dien Hong Ward, Ho Chi Minh City,
Vietnam
{khang.le2710,duc.trananh0502,quangtran}@hcmut.edu.vn
[2] Viet Nam National University Ho Chi Minh City, Linh Xuan Ward,
Ho Chi Minh City, Vietnam
[3] Nguyen Thuong Hien High School, Tan Son Nhat Ward,
Ho Chi Minh City, Vietnam

Abstract. Traffic congestion estimation has long been a critical aspect in intelligent transportation systems and the need for such an automatic and precise congestion estimation system is increasingly demanding. However, existing approaches often overlook the unique traffic patterns in motorbike-centric cities like Ho Chi Minh City, Vietnam. To address this gap, we propose a novel traffic flow estimation framework, combining vehicle detection and an enhanced road segmentation workflow. The proposed system integrates a lightweight instance segmentation module with squeeze-and-excitation and depthwise-separable convolution blocks, enabling automatic mapping of region-of-interest (ROI) with minimal computational cost. This segmentation component achieves 75.4% Dice Score and 73.9% mIoU. For vehicle detection and tracking, we incorporate an object localization framework augmented through custom training on motorbike-rich scenes, achieving 83.1% mAP@50 and delivering consistently high F1-scores across all congestion levels. We also introduce a new benchmark dataset designed to capture motorbike-rich traffic scenes. This framework offers a balanced trade-off between accuracy and efficiency, enhancing congestion analysis in dense, dynamic urban environments.

Keywords: Computer Vision · Speed Estimation · Deep Learning · Intelligent Transportation System · Traffic Congestion Analysis · Road Markings Detection

1 Introduction

As urban areas expand, efficient traffic management has become important for sustainable development and quality of life. Traditional traffic monitoring methods, mostly manual surveillance, are labor-intensive and insufficient to handle

© The Author(s), under exclusive license to Springer Nature Singapore Pte Ltd. 2026
N. Thai-Nghe et al. (Eds.): ISDS 2025, CCIS 2714, pp. 231–245, 2026.
https://doi.org/10.1007/978-981-95-3358-9_17

the complexity and scale of modern urban traffic. Therefore, the shift towards intelligent traffic surveillance systems represents a critical step forward. These intelligent systems have achieved significant milestones in developed countries, providing integrated solutions for complex traffic congestion monitoring. However, developing countries, particularly in densely populated cities such as Ho Chi Minh City (HCMC), Vietnam, face unique challenges in deploying state-of-the-art traffic analysis systems. One main challenge is the traffic flow characteristics, where motorbikes dominate as the primary mode of transportation. Existing public datasets, such as COCO [25] and KITTI [12], contain mostly cars, making them less effective for motorbike-heavy environments. While some domestic datasets, like UIT-VinaDeveS22 [41], include motorbike instances, many remain inaccessible for public research. Furthermore, traffic monitoring infrastructure in HCMC often operates with restricted processing and memory capabilities (e.g., in surveillance cameras), which fosters the development of lightweight models.

Traffic congestion is not solely determined by vehicle density. In practice, even high-density traffic can flow smoothly if vehicles maintain adequate speed. Therefore, congestion estimation systems must consider not only object counts, but also vehicle speed to reflect real-world traffic conditions. Moreover, congestion mapping results from density and vehicle speed must follow a determined and widely acceptable standard, which in this study, we use LoS for urban streets proposed in Highway Capacity Manual [40]. This protocol categorizes traffic conditions into six levels, from free flow (LoS A) to heavily congested (LoS F), making it a reliable quantitative and qualitative identifier. Additionally, even though recent frameworks of vehicle counting and speed estimation have proven high capability in real-time detection [34,35], they still rely on manually defined Regions of Interest (ROI). This poses a significant problem in practical deployments, as new camera installations may require additional human resources to reconfigure the ROI.

To address these critical limitations, this paper introduces a novel, lightweight deep learning framework for traffic congestion classification. The main contributions of this paper are summarized as follows.

1. An automatic traffic analysis pipeline, including lightweight road markings detection model using DeepLabV3 with modified ResNet34 with squeeze-and-excitation blocks and depthwise-separable convolution for automatic ROI lines generation and per-lane vehicle detection and speed estimation using a fine-tuned YOLO11s model
2. The collection of new dataset to reflect traffic conditions in Ho Chi Minh City, which can be open for further research in this field.
3. The framework is validated on real-world traffic, and succeeds to demonstrate strong segmentation, detection and congestion classification performance.

The rest of the paper is organized as follows: A review of existing techniques and their limitations in lane detection and traffic estimation is provided in Sect. 2. The problem statement, along with the overall architecture and design objectives, is introduced in Sect. 3. Section 4 outlines the proposed automatic congestion estimation pipeline, including road segmentation, object detection, and speed analysis. Implementation details and experimental evaluations using real-world data are presented in Sect. 5, accompanied by a discussion of future directions. Finally, Sect. 6 concludes the paper.

2   Related Works

Traditional lane detection methods employed handcrafted features like Hough transforms and edge detectors, which prove effective under ideal conditions but unreliable in real-world urban scenarios [23]. Recent studies adopted deep learning methods with various models such as DeepLabV3 [5], UnetEdge [10] to further improve segmentation power and inference speed. To support the development of these deep learning approaches, several annotated datasets-namely CULane [31], Ceymo [21], and RLMD [18] has emerged as standard benchmarks. Pan et al. employs a Spatial CNN (SCNN) to model strong spatial relationship of large objects [31] with an F_1-score of 80.9% at kernel 9×9 on their collected dataset CULane. Some studies have also implemented ROI extraction from segmentation masks [4,16], but few put a step forward on using them as references for traffic congestion classification.

Vehicle detection also plays a pivotal role in mapping congestion levels using the Level of Service (LoS) framework. While transformer-based detectors [9,36,38] offer high precision, real-time and edge deployment demands fast architectures such as the YOLO family [24]. Alif et al. demonstrate YOLO11 outperforms YOLOv8 and YOLOv10 by 2.9% and 2.3% respectively on mAP@50 [1]. Accurate detection underpins Multi-Object Tracking (MOT), which enables extraction of dynamic traffic metrics. Azimjonov et al. combine YOLO with a CNN-based reclassifier and Kalman filtering, achieving 95.45% detection accuracy [2]. Quang et al. propose a speed estimation method based on reference object calibration, reporting 7.4% and 7.47% error on their dataset and BrnoCompSpeed, respectively [29].

Building upon accurate detection and tracking, congestion analysis plays a vital role in urban planning and the development of comprehensive intelligent transportation systems. Traditional methods often depend on costly physical infrastructure, such as speed enforcement cameras and embedded sensors, as well as flow-based analytical models. However, these approaches lack scalability to dynamic urban environments. To address this, Quang et al. employ the LoS standard to accurately classify congestion levels from A to F, offering a structured interpretation of traffic states [39]. Recent studies leverage CNNs to embed traffic metrics directly into detection pipelines, enhancing efficiency and eliminating complex post-processing [11]. Accordingly, robust object counting and speed estimation, aligned with YOLO11's nature and an automatic ROI detection mechanism, is vital for deployment in diverse and changing environments.

Despite advances in road marking segmentation, vehicles detection and traffic congestion analysis, existing approaches often lack a unified, scalable framework suitable for motorbike-dominant and resource-constrained urban environments. While some studies address automatic ROI extraction from segmentation outputs, they rarely integrate this into an end-to-end pipeline for classifying congestion levels based on standards such as LoS. Therefore, this research is motivated by the need for a unified system that not only automates the entire analysis workflow from ROI generation to per-lane vehicle metrics, but is also specifically optimized for the challenges of motorbike-dense environments.

3   Problem Definition and Overview

Urban traffic congestion has long been a pressing issue in developing cities, where rapid urbanization, limited infrastructure, and motorbike-dominated mobility contribute to highly complex traffic dynamics. In cities such as Ho Chi Minh City, conventional approaches relying on car-centric datasets and manually crafted ROI zones fall short in accurately reflecting real-world conditions. A key challenge lies in the lack of scalable and automated systems that can generalize across diverse camera perspectives and dynamically changing traffic flows. We formalize the problem of traffic congestion estimation as a multi-stage learning task that jointly performs road structure segmentation, per-lane vehicle detection and tracking, and congestion level classification. Given an uncalibrated RGB video stream $V = \{I_1, I_2, \ldots, I_T\}$ from a static traffic camera, the objective is to automatically infer the following:

- **Per-lane ROIs:** Derived from instance segmentation of road markings;
- **Vehicle Count and Speed Analysis:** The number and velocity of vehicles passing through each ROI in a predefined period of time;
- **Congestion Level Label** $y \in \{A, B, C, D, E, F\}$: Determined based on the urban traffic LoS standard using real-time traffic density and average speed.

4   Method and Implementations

4.1   Traffic Congestion Analysis

The LoS framework offers a qualitative classification of traffic stream conditions, detailing six tiers of operational performance from LoS A (free-flow conditions) to LoS F (severe congestion). These classifications are derived from a comprehensive set of indicators, including traffic density, average speed, roadway capacity, service volume, and flow rate, as outlined in the Highway Capacity Manual [40]. However, in practical surveillance environments, key metrics such as capacity, service volume, and flow rate are often unavailable or inconsistently reported. Therefore, to enhance applicability, we adopt a simplified assessment strategy based solely on two observable and measurable attributes: vehicle density and speed (km/h). The corresponding LoS thresholds for operational multilane urban roads are presented in Table 1.

Table 1. Six levels of service classification based on speed and density

Level of Service	Maximized Velocity for Vehicles *(km/h)*		Vehicle Density
	50 km/h (2-wheelers)	60 km/h (4-wheelers)	*(units/km/lane)*
A	$\geq$40	$\geq$45	$D \leq 5$
B	35–40	40–45	5–10
C	30–35	35–40	10–14
D	25–30	25–35	14–24
E	20–25	15–20	24–40
F	$\leq$20	$\leq$20	$D \geq 40$

To analyze the traffic congestion levels, we constructed an automatic pipeline developed for analyzing road traffic, encompassing road marking segmentation, vehicle detection and tracking, and LoS determination. As depicted in Fig. 1, this robust pipeline employs DeepLabV3 with enhanced ResNet34 backbone for precise segmentation of road markings and YOLO11s for efficient and accurate vehicle counting and speed estimation. This integrated approach analyses real-time traffic performance, enabling congestion diagnostics at multiple lanes.

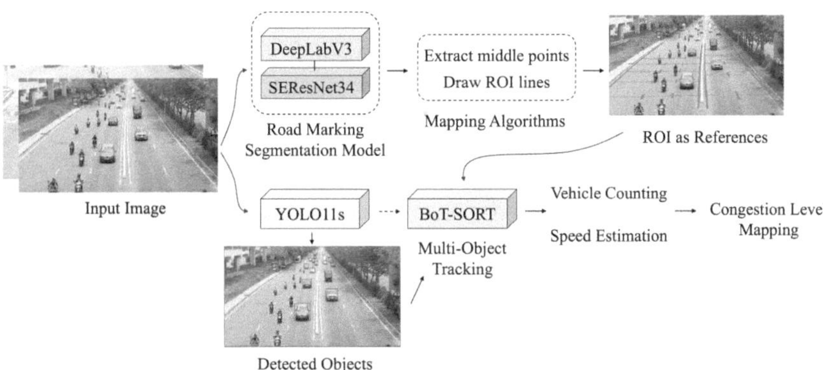

Fig. 1. Overview of our congestion level classification pipeline. Raw images will be segmented to extract ROI lines, eventually used as references for vehicle counting and speed estimation.

4.2 Road Markings Mapping

To generate ROIs, we segment the road lane markings and develop a geometry-based approach to extract reference lines from dash and solid patterns. First, we employ DeepLabV3 [5] as the instance segmentation architecture, integrated with a modified ResNet34 backbone [15]. In contrast to the standard config-uration, we enhance the encoder with two key architectural modifications as shown in Fig. 2. First, we substitute the traditional convolutional layers within the residual blocks with depthwise-separable convolutions to reduce the number of parameters [8,17]. In particular, this technique decomposes the convolution block into two operations: (1) a depthwise convolution that applies spatial filter-ing to each input channel, and (2) a pointwise 1×1 convolution that aggregates over all output channels. Second, we further complement the channel-wise fea-ture with Squeeze-and-Excitation (SE) blocks into each residual unit [19]. These blocks model channel dependencies using a gating mechanism consisting of global average pooling followed by a two-layer fully connected network. This module suppresses irrelevant activations while enhancing informative features.

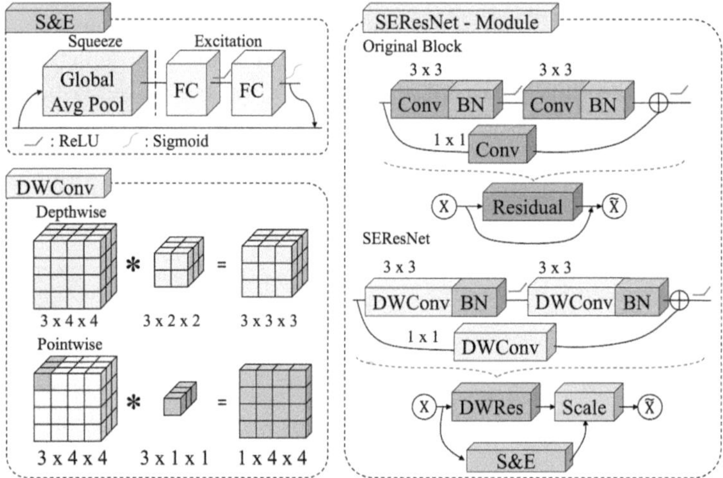

Fig. 2. Modified ResNet module with Squeeze-&-Excitation and Depthwise-separable Convolutional Blocks.

To construct ROI lines, we choose dashed lines as references for both inter-mittent and solid markings. For dashed lines, we extract the midpoint of each dash segment. Then, a horizontal line is drawn between them to complete the ROI at that vertical level. Furthermore, to ensure the consistency of ROI genera-tion, we follow the road marking specifications defined in Appendix G of QCVN 41:2019/BGTVT (Vietnam's National Technical Regulation on Road Signs) [30]. The regulation insists dashed line segments must be between 1 m and 3 m in length, with a mark-to-gap ratio typically set at 1:2 or 1:3. For generalization across varying road scenes and camera perspectives, we use rounded average length segment of 1 m and 2 m as reference values for meter-to-pixel conversion ratio.

4.3   Per-Lane Vehicle Detection and Speed Estimation

The task of determining traffic congestion level comprises two parallel steps: counting the number of vehicles and estimating traffic speed. In our design, we select YOLO11s, a state-of-the-art model for real-time object detection due to its superior balance between accuracy and inference speed [24]. The introduction of C3k2 blocks and C2PSA modules in YOLO11 architecture enhances its ability to extract fine-grained spatial features. Instead of one large convolution within the CSP Bottleneck in YOLOv8, the C3k2 modification customizes the CSP Bottleneck with two convolutions [22]. First, it splits the feature map into two branches. Each branch is independently processed through a shallower bottleneck structure, and their outputs are merged via concatenation or addition. These improvements are designed to address the challenges of detecting smaller and rapidly moving objects.

For speed estimation, the model will localize and track vehicle objects across frames with BoT-SORT [42] and determine if it crosses predefined ROI lines. Each crossing event at the ROI line triggers a vehicle count, allowing for traffic volume estimation per lane. Concurrently, speed estimation utilizes the temporal tracking results from BoT-SORT to compute the pixel displacement over time between consecutive frames. This displacement is then converted to real-world velocity v_i using a calibrated meter-to-pixel ratio (r_{mp}), derived from estimated lane marking lengths, using the following equation:

$$v_i = \frac{d_p}{t_i} \times r_{mp} \tag{1}$$

where d_p is the pixel displacement of tracked vehicles and t_i is the time interval between the first detection of the object and its position at the current frame. The meter-to-pixel ratio is computed by dividing the real-world length of dash line marking (e.g., 1 m) by its corresponding pixel length d_l obtained from segmentation output. Subsequently, we compute the mean velocity of all tracked vehicles within a lane over a defined threshold of 30-second time window and a minimum of 5 objects. This fixed interval serves as a reset cycle for the vehicle counter and congestion label, allowing the system to cope with accumulative speed labels and counts.

4.4   Dataset Collection

Data collection is conducted in Ho Chi Minh City during clear skies and sunny weather. The dataset comprises 53 video sequences, each recorded at 30 fps with a fixed duration of 30 s at a resolution of 1080 × 720. The data was captured from pedestrian overpasses located between Vo Nguyen Giap Street in District 2, and Calmette bridge over Vo Van Kiet Street in District 1. These bridges are approximately 5 m above ground level as indicated at height limit sign on the bridge. All recordings were conducted using a tripod-mounted camera positioned at an effective height of 1.4 m, angled at 40°. To annotate data, we use pretrained YOLOv8 with input of 640 × 640 to pre-annotate vehicle bounding boxes, and then refine manually to address missing or incomplete labels. For speed data, we use Bushnell Velocity Speed Gun 101911 with a measurement accuracy of ± 1 mph to acquire ground truth labels for the traffic flow. The speed gun is angled at $45°$ from the y-axis following the measurement protocol as shown in Fig. 3. We performed three separate measurements on different within each scene and averaged the recorded speeds. This average value is used as the representative ground truth label for the entire traffic flow within the 6-second time window because of the delay in switching between consecutive trials.

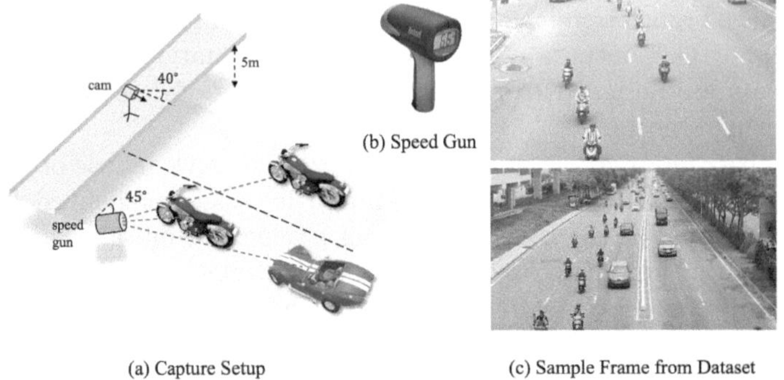

Fig. 3. Overview of data collection setup. (a) Capture setup includes a camera, and a Bushnell speed gun to record speed. (b) Bushnell speed gun with ± 1 mph accuracy. (c) Sample frames from the collected dataset.

5   Experimental Results

To evaluate the effectiveness of proposed traffic congestion prediction, we conduct a series of experiments under different conditions. These evaluations assess the accuracy of road lane markings detection, vehicle object localization, and traffic speed estimation, thereby validating the performance of the system in real-world scenarios.

5.1   Implementation Details

The training and testing tasks are accomplished on a Windows 11 operating system using PyTorch 2.7.1 framework [32], with Intel(R) Xeon(R) Gold 6154 CPU @ 3.00GHz (16 CPUs), 64GB installed memory, a NVIDIA GRID T4-16Q (16GB global memory). Under this setup, the training is performed for 100 epochs with a batch size of 16. For benchmarking purposes, the training hyper-parameters and experimental settings are kept consistent across all evaluated models to ensure a fair competition. RAdam optimization is chosen with an initial learning rate of 0.0003 [26]. A Cosine Annealing learning rate scheduler is employed to progressively reduce the learning rate to a minimum threshold of 1×10^{-6} [28].

Road Lane Markings Detection Performance: We adopt Road Lane Marking Dataset (RLMD) [18] as the primary training corpus for road segmentation model. RLMD offers a diverse collection of 2,137 urban driving images, densely annotated with 25 distinct road marking categories in polygonal mask format. Its inclusion of both directional indicators (e.g., arrows, stop lines) and structural separators (e.g., solid/dashed lines) makes it particularly suitable for lane-level scene understanding. For our application, we extract eight relevant classes detailed in Table 2 to boost model's focus on detecting lane separators (e.g., road lines) and negate irrelevant information of directional indicators.

Table 2. Selected segmentation classes of road lane markings from RLMD dataset

ID	Class Name	Abbr.	Type	R	G	B
4	Solid Single White	SSW	Structural Line	237	28	36
5	Solid Single Yellow	SSY	Structural Line	163	73	164
6	Solid Single Red	SSR	Structural Line	185	122	87
7	Solid Double White	SDW	Structural Line	136	0	21
8	Solid Double Yellow	SDY	Structural Line	112	146	190
9	Dashed Single White	DSW	Structural Line	181	230	29
10	Dashed Single Yellow	DSY	Structural Line	153	217	234
16	Channelizing Line	CL	Auxiliary Line	255	204	153

The RLMD dataset is divided into training, validation, and testing subsets with a ratio of 70:20:10, respectively. Standard augmentation techniques for instance segmentation tasks are employed to enhance the model's generalization, including random horizontal flipping, color jittering, and normalization. Two widely-used metrics to evaluate segmentation performance are adopted: the Mean Intersection over Union (mIoU) and the Dice Score. The mIoU measures the overlap between the predicted and ground truth regions averaged over all classes [27], defined as:

$$\text{mIoU} = \frac{1}{C} \sum_{c=1}^{C} \frac{TP_c}{TP_c + FP_c + FN_c} \tag{2}$$

where TP_c, FP_c, and FN_c are the true positives, false positives, and false negatives for each class c, respectively. The Dice Score captures similarity between prediction and ground truth by emphasizing both precision and recall [37]. The Dice Score between predicted mask P and ground truth G, is given by:

$$\text{Dice}(P, G) = \frac{2|P \cap G|}{|P| + |G|} \tag{3}$$

Table 3 summarizes our results in comparison with classical models (e.g., Mask-RCNN [14], SegNet [3]) and a recent state-of-the-art model (Mask2Former

Table 3. Performance comparison of segmentation models on the RLMD dataset

Model	Inference Time (ms)	Params (M)	Dice Score (%)	mIoU (%)
Mask R-CNN [14]	85	44.3	71.2	65.6
SegNet [3]	39	29.5	68.5	61.0
Mask2Former-ResNet50 [6, 7]	46	84.8	**78.9**	**76.8**
DeepLabV3-MobileNetV2 [5, 33]	14	**24.3**	72.3	68.8
DeepLabV3-SEResNet34	23	27.1	75.4	73.9

with ResNet50 backbone [6,7]). The results demonstrate that our approach significantly outperforms classical baselines in terms of both mIoU (73.9%) and Dice Score (75.4%). The state-of-the-art Mask2Former-ResNet50 reaches the highest Dice Score (78.9%) and mIoU (76.8%), proving significant performance in detecting precise lane markings. However, this model has a higher computational cost (84.8 M parameters) with an inference time of 46 ms, making it relatively slow for real-time applications. On the other hand, DeepLabV3 integrating MobileNetV2 backbone demonstrates the fastest detection ability (14 ms and 24.3 M parameters), but with a cut in accuracy of only 72.3% in Dice Score. Therefore, our model offers a well-balanced trade-off between accuracy and efficiency with an acceptable inference time of 23 ms. Since the implementation of road lane markings detection is simplified, which only triggers at the start or during resets of the tracking module, exact real-time performance is not critical. Thus, the latency of our model is sufficiently low for practical deployment.

Table 4. Evaluation metrics for object localization and speed estimation

Model]	Params (M)	FPS	mAP@50 (%)	MSE	MAE	Error (%)
YOLO11s	9.4	83	81.2	9.77	3.38	8.03
YOLO11s (Finetuned)	9.4	**85**	**83.1**	**8.59**	**3.17**	**7.82**

Traffic Congestion Prediction: For the task of predicting traffic congestion levels, we evaluate the effectiveness of our pipeline using two configurations of YOLO11 object detection model: (1) YOLO11s with pretrained weights, and (2) YOLO11s finetuned on our dataset (see Fig. 4 for tuning results). The metrics used for the two tasks include mean Average Precision at IoU threshold 50 (mAP@50) for object detection, and Mean Absolute Error (MAE), Mean Squared Error (MSE), and Percentage Error (% Error) for speed estimation, and Macro F_1-score for congestion level mapping. The results for the two models are detailed in Table 4.

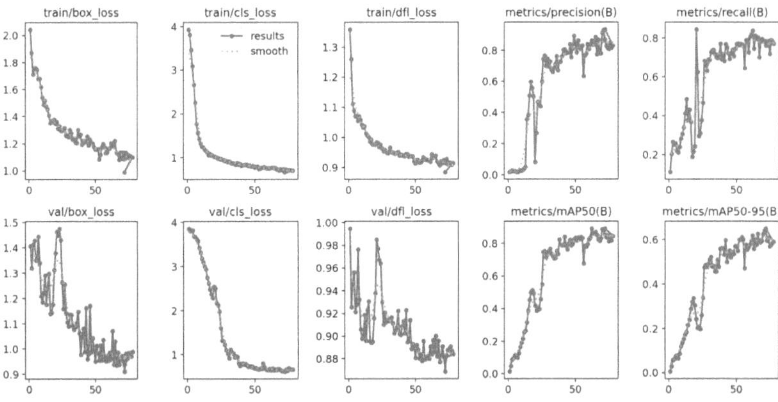

Fig. 4. Training results of YOLO11s on our custom dataset. The last two graphs depict the mAP scores of the model.

Both YOLO11s models are highly capable of detecting vehicle objects effectively. However, the fine-tuned version shows a better adaptability to crowded scenes in our dataset. This yields a slightly higher score in detecting objects (83.1% compared to 81.2%). Frame Per Second (FPS) is also considered to evaluate the result. Even though the two models share the same architecture and an identical number of parameters of 9.4 M, the fine-tuned model interestingly has a better inference speed of 85 FPS compared to 83 FPS for the original one. This is likely due to weight adaptation during finetuning. The finetuned model is then selected for predicting the levels for traffic congestion. As shown in Table 5, the model achieves consistently high F_1-scores across all Levels of Service: the peak performance is at 86.2% for LoS A. However, for increasingly congested scenes more visual complexity in scenes is introduced, posting greater challenges for accurate classification. Figure 5 illustrates the whole traffic congestion level classification, starting from a segmentation mask to drawing ROI lines, and computing vehicle speed and counting.

Table 5. F1 Score for each LoS and the Macro F1 score

LoS	A	B	C	D	E	F	Macro
F1 Score (%)	86.2	85.3	85.6	85.1	84.9	84.7	**85.3**

5.2 Discussion

The effectiveness of the proposed automatic framework for traffic congestion prediction is clearly demonstrated through the performance results. Our method removes the need for manual ROI extraction, making the pipeline more adaptable to different environments. For instance, if a new camera is installed at a

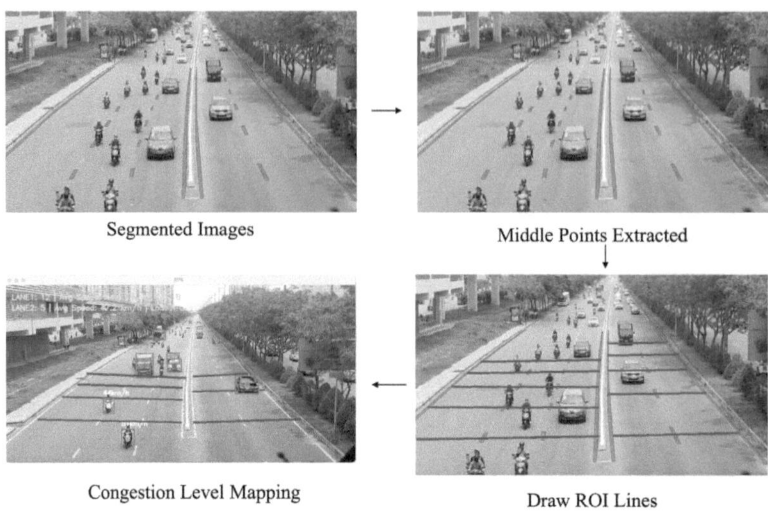

Fig. 5. Congestion Level Mapping from generating masks and drawing ROI lines.

different angle, the system can automatically detect road lane markings without having to redraw ROI manually. However, some limitations remain. ROI lines are only computed once at the beginning, which means that a sudden camera shift can lead to incorrect outputs. Besides, if lane markings are faded, the marking detection fails to map ROI lines correctly, leading to misaligned ROI lines or overlapping each other. Additionally, due to the fixed 30-second time window, short-term traffic fluctuations may not be captured effectively. For instance, the crowded road suddenly becomes clear, while the LoS label persists for the whole 30 s. Besides, since our data collection method is set up during clear skies, this hinders the model's ability to perform in diverse environmental conditions such as rain, fog. Even though we have mimicked the environmental effects by augmentation, it cannot fully capture the complexity of real-world challenges. The error percentage of 7.82% is not sufficiently reliable for high precision monitoring tasks and, therefore, is not yet capable of replacing LiDAR-based systems (e.g. Blickfield LiDAR) or traffic enforcement systems (e.g. Gatso Radar). These issues raise the need for an end-to-end trainable pipeline that integrates and improves all above elements. To tackle the problem of misaligned ROI in sudden camera movements, an error-correction paradigm such as Temporal ROI Align introduced by Gong et al. [13] should be considered to guarantee a consistent reference for meter-to-pixel calibration. Concurrently, an adaptive time window should be implemented to handle abnormal traffic scenarios as described above. We also acknowledge that our dataset is limited to 53 videos and clear-weather conditions, therefore, expanding the dataset and validating on larger public benchmarks will be explored in future work. Real-life applications of traffic congestion mapping must also be taken into thorough consideration, whether for traffic flow monitoring or automated traffic light control system [20]. These improvements are deployed to edge devices and are capable of synchronizing with centralized traffic control systems for an efficient urban traffic management.

6   Conclusion

This work proposed a framework that automatically draws ROI regions and uses them as references for later object counting and vehicle speed estimation for traffic congestion mapping. Road marking segmentation leverages DeepLabV3 model with a modified ResNet34 backbone incorporating depthwise-separable convolutions and squeeze-and-excitation blocks. This architecture aims to preserve the segmentation accuracy while making the model lighter for edge devices. This model scores 76.8% mIoU score, with only 24.3 M parameters on the RLMD dataset. Subsequently, the ROI line is drawn automatically from the middle point of segmented dashed line to the nearest left and right neighbor to form ROI lines. On the other hand, the task of traffic congestion level mapping through vehicle counting and speed estimation is performed by YOLO11s, finetuned on our dataset. The resulting model is evaluated on our dataset, which achieves a mAP@50 of 83.1%, speed estimation error of only 7.82%, and macro F_1-score of 85.3% for traffic congestion mapping. The model proves to be efficient and reliable on resource-constrained environment.

Acknowledgments. We acknowledge Ho Chi Minh University of Technology (HCMUT), VNU-HCM for supporting this study. We also acknowledge Khanh Ngoc Do for her comments on the application manners of the proposed approach.

References

1. Alif, M.A.R.: YOLOv11 for vehicle detection: advancements, performance, and applications in intelligent transportation systems. arXiv:2410.22898 (2024)
2. Azimjonov, J., Özmen, A.: A real-time vehicle detection and a novel vehicle tracking systems for estimating and monitoring traffic flow on highways. Adv. Eng. Inform. **50**, 101393 (2021)
3. Badrinarayanan, V., Kendall, A., Cipolla, R.: SegNet: a deep convolutional encoder-decoder architecture for image segmentation. IEEE Trans. Pattern Anal. Mach. Intell. **39**(12), 2481–2495 (2017)
4. Bisio, I., Garibotto, C., Haleem, H., Lavagetto, F., Sciarrone, A.: Traffic analysis through deep-learning-based image segmentation from UAV streaming. IEEE Internet Things J. **10**(7), 6059–6073 (2022)
5. Chen, L.C., Zhu, Y., Papandreou, G., Schroff, F., Adam, H.: Encoder-decoder with Atrous separable convolution for semantic image segmentation. In: Proceedings of the European conference on computer vision (ECCV), pp. 801–818 (2018)
6. Cheng, B., Misra, I., Schwing, A.G., Kirillov, A., Girdhar, R.: Masked-attention mask transformer for universal image segmentation. In: IEEE/CVF Conference on Computer Vision and Pattern Recognition (CVPR), pp. 1280–1289 (2022)

7. Cheng, B., Schwing, A., Kirillov, A.: Per-pixel classification is not all you need for semantic segmentation. In: Advances in Neural Information Processing Systems, vol. 34, pp. 17864–17875 (2021)

8. Chollet, F.: Xception: deep learning with depthwise separable convolutions. In: Proceedings of the IEEE/CVF Conference on Computer Vision and Pattern Recognition (CVPR), pp. 1251–1258 (2017)

9. Deshmukh, P., Satyanarayana, G., Majhi, S., Sahoo, U.K., Das, S.K.: Swin transformer based vehicle detection in undisciplined traffic environment. Expert Syst. Appl. **213**, 118992 (2023)

10. Dey, M., Prakash, P., Aithal, B.H.: UnetEdge: a transfer learning-based framework for road feature segmentation from high-resolution remote sensing images. Remote Sens. Appl.: Soc. Environ. **34**, 101160 (2024)

11. Gao, Y., Li, J., Xu, Z., Liu, Z., Zhao, X., Chen, J.: A novel image-based convolutional neural network approach for traffic congestion estimation. Expert Syst. Appl. **180**, 115037 (2021)

12. Geiger, A., Lenz, P., Urtasun, R.: Are we ready for autonomous driving? The KITTI vision benchmark suite. In: IEEE Conference on Computer Vision and Pattern Recognition, pp. 3354–3361 (2012)

13. Gong, T., et al.: Temporal ROI align for video object recognition. In: Proceedings of the AAAI Conference on Artificial Intelligence, vol. 35, pp. 1442–1450 (2021)

14. He, K., Gkioxari, G., Dollár, P., Girshick, R.: Mask R-CNN. In: Proceedings of the IEEE International Conference on Computer Vision, pp. 2961–2969 (2017)

15. He, K., Zhang, X., Ren, S., Sun, J.: Deep residual learning for image recognition. In: Proceedings of the IEEE/CVF Conference on Computer Vision and Pattern Recognition (CVPR), pp. 770–778 (2016)

16. He, Y., Jin, L., Wang, H., Huo, Z., Wang, G., Sun, X.: Automatic ROI setting method based on LSC for a traffic congestion area. Sustainability **14**(23) (2022)

17. Howard, A.G., et al.: MobileNets: efficient convolutional neural networks for mobile vision applications. arXiv:1704.04861 (2017)

18. Hsiao, H.C., Cai, Y.C., Lin, H.Y., Chiu, W.C., Chan, C.T.: RLMD: a dataset for road marking segmentation. In: International Conference on Consumer Electronics-Taiwan (ICCE-Taiwan), pp. 427–428 (2023)

19. Hu, J., Shen, L., Sun, G.: Squeeze-and-excitation networks. In: Proceedings of the IEEE/CVF Conference on Computer Vision and Pattern Recognition (CVPR), pp. 7132–7141 (2018)

20. Islam, M.M., Rehena, Z., Mukherjee, N.: An approach for adaptive traffic light control system in its using VANET. In: Second International Conference on Emerging Trends in Information Technology and Engineering (ICETITE), pp. 1–7 (2024)

21. Jayasinghe, O., Hemachandra, S., Anhettigama, D., Kariyawasam, S., Rodrigo, R., Jayasekara, P.: CeyMo: see more on roads-a novel benchmark dataset for road marking detection. In: Proceedings of the IEEE/CVF Winter Conference on Applications of Computer Vision, pp. 3104–3113 (2022)

22. Jegham, N., Koh, C.Y., Abdelatti, M., Hendawi, A.: Evaluating the evolution of yolo (you only look once) models: a comprehensive benchmark study of YOLO11 and its predecessors. arXiv:2411.00201 (2024)

23. Jiaming, L., Kai, X., Boyu, Z., et al.: Estimation of stripping ratio of road lane markings by in-vehicle smartphone camera and deep learning-based segmentation. Intell. Inform. Infrastruct. **5**(2), 106–115 (2024)

24. Jocher, G., Qiu, J., Chaurasia, A.: Ultralytics YOLO (2023). https://github.com/ultralytics/ultralytics

25. Lin, T.-Y., et al.: Microsoft COCO: common objects in context. In: Fleet, D., Pajdla, T., Schiele, B., Tuytelaars, T. (eds.) ECCV 2014. LNCS, vol. 8693, pp. 740–755. Springer, Cham (2014). https://doi.org/10.1007/978-3-319-10602-1_48
26. Liu, L., et al.: On the variance of the adaptive learning rate and beyond. arXiv:1908.03265 (2019)
27. Long, J., Shelhamer, E., Darrell, T.: Fully convolutional networks for semantic segmentation. In: Proceedings of the IEEE/CVF Conference on Computer Vision and Pattern Recognition (CVPR), pp. 3431–3440 (2015)
28. Loshchilov, I., Hutter, F.: SGDR: stochastic gradient descent with warm restarts. arXiv preprint arXiv:1608.03983 (2016)
29. Minh, Q.T., et al.: Traffic flow velocity estimation from single camera data. In: Thai-Nghe, N., Do, T.N., Benferhat, S. (eds.) ISDS 2024. CCIS, vol. 2190, pp. 129–143. Springer, Singapore (2024). https://doi.org/10.1007/978-981-97-9613-7_10
30. Ministry of Transport, Socialist Republic of Vietnam: QCVN41:2019/BGTVT – National Technical Regulation on Traffic Signs and Signals (2019). Original document in Vietnamese
31. Pan, X., Shi, J., Luo, P., Wang, X., Tang, X.: Spatial as deep: spatial CNN for traffic scene understanding. In: Proceedings of the AAAI Conference on Artificial Intelligence, vol. 32 (2018)
32. Paszke, A., et al.: PyTorch: an imperative style, high-performance deep learning library. In: Advances in Neural Information Processing Systems, pp. 8024–8035 (2019)
33. Sandler, M., Howard, A., Zhu, M., Zhmoginov, A., Chen, L.C.: MobileNetv2: inverted residuals and linear bottlenecks. In: Proceedings of the IEEE/CVF Conference on Computer Vision and Pattern Recognition (CVPR), pp. 4510–4520 (2018)
34. Saputri, H.A., Avrillio, M., Christofer, L., Simanjaya, V., Alam, I.N.: Implementation of YOLO v7 algorithm in estimating traffic flow in Malang. Procedia Comput. Sci. **245**, 117–126 (2024)
35. Sasikala, S., Neelaveni, R., Jose, P.S.: OSTM-net: joint scale variation and occlusion handling deep network for real-time vehicle counting and volume estimation. Digit. Signal Process. **149**, 104507 (2024)
36. SP, K., Mohandas, P.: DETR-SPP: a fine-tuned vehicle detection with transformer. Multimed. Tools Appl. **83**(9), 25573–25594 (2024)
37. Sudre, C.H., Li, W., Vercauteren, T., Ourselin, S., Jorge Cardoso, M.: Generalised dice overlap as a deep learning loss function for highly unbalanced segmentations. In: Deep Learning in Medical Image Analysis and Multimodal Learning for Clinical Decision Support, pp. 240–248 (2017)
38. Tahir, N.U.A., Long, Z., Zhang, Z., Asim, M., ELAffendi, M.: PVswin-YOLOv8s: UAV-based pedestrian and vehicle detection for traffic management in smart cities using improved yolov8. Drones **8**(3), 84 (2024)
39. Tran Minh, Q., Pham-Nguyen, H.N., Mai Tan, H., Xuan Long, N.: Traffic congestion estimation based on crowd-sourced data. In: 2019 International Conference on Advanced Computing and Applications (ACOMP), pp. 119–126 (2019)
40. Transportation Research Board: Highway Capacity Manual. National Research Council, Washington, DC (2000)
41. Trinh, T., Nguyen, K.: A Vietnamese benchmark for vehicle detection and real-time empirical evaluation. CTU J. Innov. Sustain. Dev. **14**(3), 45–52 (2022)
42. Yan, S., Fu, Y., Zhang, W., Yang, W., Yu, R., Zhang, F.: Multi-target instance segmentation and tracking using YOLOv8 and bot-sort for video SAR. In: 2023 5th International Conference on Electronic Engineering and Informatics (EEI), pp. 506–510 (2023)

Simultaneous Low-Light Image Enhancement and Deblurring for Surveillance Systems Using Intelligent Cameras

Minh-Thien Duong[1], Van-Ho Tran[2], Vi-Do Tran[1], Hoai-An Trinh[2], Thien M. Tran[3], and My-Ha Le[2(✉)]

[1] Department of Automatic Control, HCMC University of Technology and Education, Ho Chi Minh City, Vietnam
[2] Faculty of Electrical and Electronics, HCMC University of Technology and Education, Ho Chi Minh City, Vietnam
halm@hcmute.edu.vn
[3] Faculty of Mechanical Engineering, HCMC University of Technology and Education, Ho Chi Minh City, Vietnam

Abstract. Most deep learning-based restoration methods tend to concentrate solely on a single type of degradation, aiming either at enhancing low-light images or performing deblurring. Although the performances of these restoration models are impressive, they cannot simultaneously correct these degradations. Therefore, this article introduces a U-shaped network architecture for simultaneous low-light image enhancement and deblurring. The proposed architecture includes three main components: an encoder block that progressively extracts multi-kernel features while reducing spatial resolution, a middle block that processes the most abstract feature representations, and a decoder block that reconstructs the enhanced and deblurred output through upsampling and feature fusion. The experimental results demonstrate the capability of the proposed network can obtain promising results in terms of objective and subjective performances.

Keywords: Low-light image enhancement · Image deblurring · U-shaped network

1 Introduction

Practically, real-world low-light images inevitably encounter blurry issues owing to dynamic scenes, camera shakes, and slow shutter speeds. Previous methods have handled the two tasks separately, i.e., low-light image enhancement [1–4], and image deblurring [5–9]. Each of these approaches independently relies on distinct assumptions tailored to its specific problem. Consequently, a concerted integration of these methods proves insufficient to effectively address the

© The Author(s), under exclusive license to Springer Nature Singapore Pte Ltd. 2026
N. Thai-Nghe et al. (Eds.): ISDS 2025, CCIS 2714, pp. 246–257, 2026.
https://doi.org/10.1007/978-981-95-3358-9_18

combined degradation resulting from both low-light and motion blur. In particular, current low-light enhancement techniques focus on intensity boosting and denoising, while overlooking the spatial degradation caused by motion blurs. Regarding deblurring, current methods are trained on datasets exclusively comprising daytime scenes, rendering them unsuitable for direct application to the more complex task of night image deblurring.

To resolve the issues mentioned earlier, the proposed solution involves training a U-shaped network capable of simultaneously addressing both types of degradation. More specifically, the proposed architecture consists of three main components: (1) an encoder block that progressively extracts multi-kernel features while reducing spatial resolution, (2) a middle block that processes the most abstract feature representations, and (3) a decoder block that reconstructs the enhanced and deblurred output through upsampling and feature fusion. The weights of the proposed network are optimized based on training with low-blur/normal-sharp pairs. Finally, we evaluate the effectiveness of our proposed method by employing various image quality assessment (IQA) metrics. The contributions of this study can be summarized as follows:

- We propose a novel encoder-decoder network with multi-kernel depthwise convolutions and complementary attention mechanisms, especially channel attention for the encoder and spatial attention for the decoder to handle joint low-light enhancement and deblurring effectively.
- We design an innovative middle block with stacked composite shape convolution (CSC) modules using progressive kernel scaling ($k = 3, 5, 7, 9$) to capture complete spatial patterns.
- The experimental results demonstrate that the proposed network outperforms other competitive methods in terms of qualitative and quantitative.
  The rest of this article is organized as follows: Sect. 2 explains the proposed method, Sect. 3 analyzes the experimental results, and Sect. 4 states the conclusion of our study.

2   Methodology

In this section, we present our proposed U-shaped network architecture for simultaneous low-light image enhancement and deblurring. Our approach adopts an encoder-decoder structure with skip connections to simultaneously address both tasks within a single framework. As shown in Fig. 1, the proposed architecture consists of three main components: (1) an encoder block that progressively extracts multi-kernel features while reducing spatial resolution, (2) a middle block that processes the most abstract feature representations, and (3) a decoder block that reconstructs the enhanced and deblurred output through upsampling and feature fusion. The following subsections detail each design of the component.

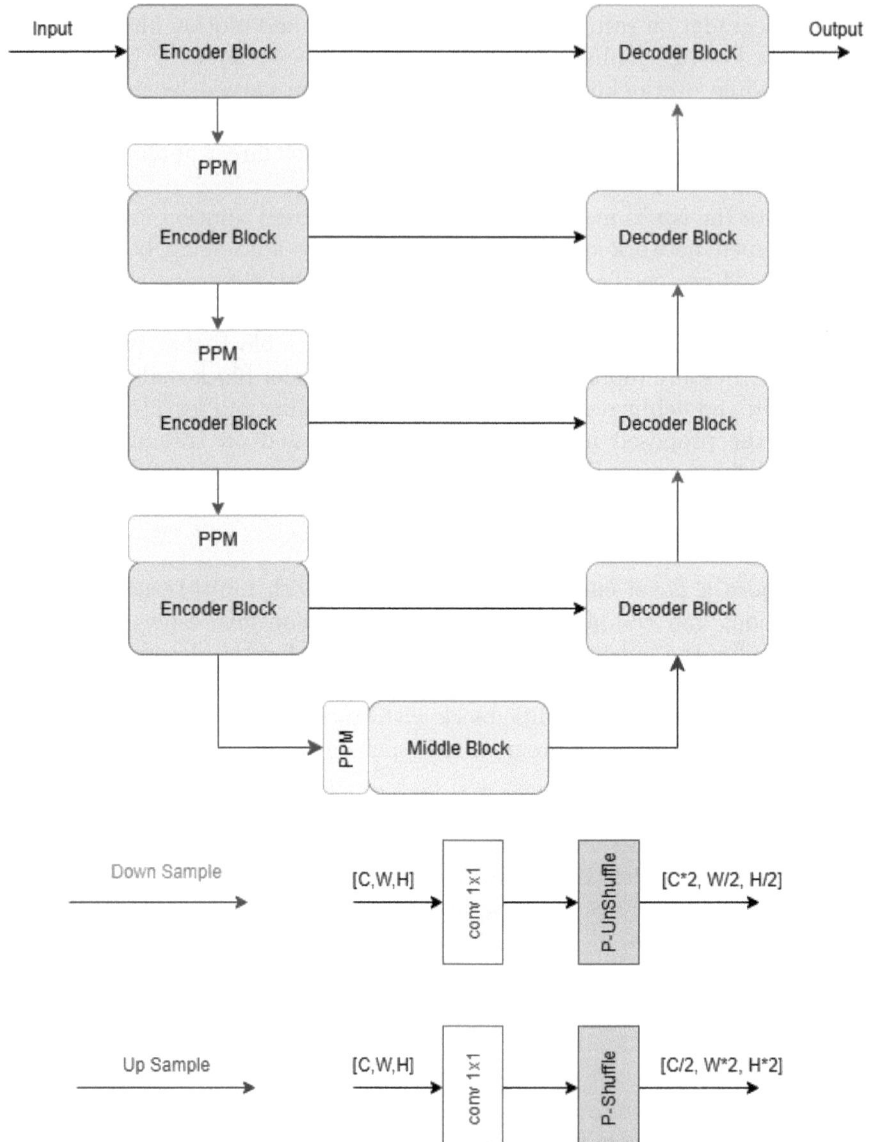

Fig. 1. The proposed U-shaped network architecture for simultaneous low-light image enhancement and deblurring.

2.1   Encoder Block

As shown in Fig. 2a, our encoder block is inspired by the efficient design of NAFNet [6], but we incorporate multi-kernel feature extraction to better extract complex degradation information in low-light and blurry images.

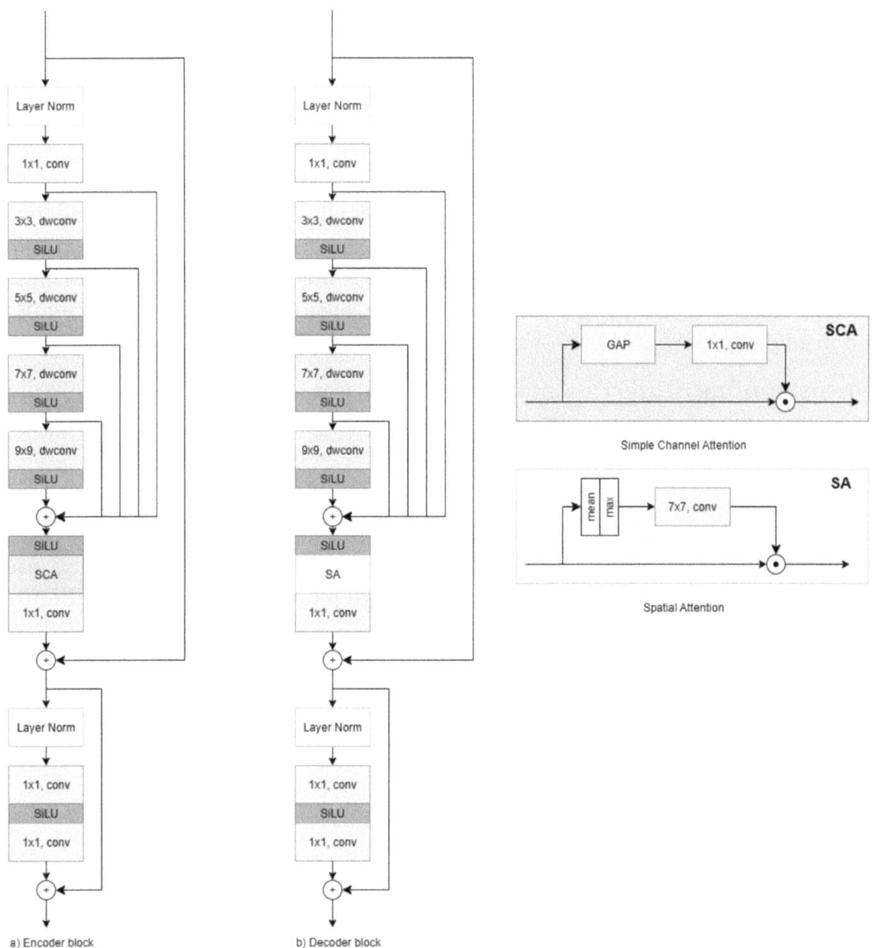

Fig. 2. Architecture of the encoder and decoder blocks.

Feature Extraction: The encoder block begins with Layer Normalization followed by a 1×1 convolution for channel-wise feature transformation. Subsequently, we employ a series of depthwise convolutions with progressively increasing kernel sizes (3×3, 5×5, 7×7, 9×9), each followed by SiLU activation. This multi-kernel approach enables the network to capture both fine-grained local details and broader contextual information simultaneously. Each depthwise convolution operates on the output of the previous layer, creating a sequential feature extraction pipeline.

The output from each depthwise convolution is combined via element-wise addition, which results in a rich multi-kernel feature representation. This feature then passes through the SiLU activation function and the Simple Channel Attention (SCA) [6] module to adaptively recalibrate the features on a per-channel basis, allowing the network to focus on the channels that provide the most information for the overall enhancement and deblurring task.

The refined features then pass through a 1×1 convolution for the final feature transformation, followed by a channel MLP consisting of Layer Normalization, a 1×1 convolution with SiLU activation, and another 1×1 convolution.

The block design incorporates two skip connections: the first skip connection directly adds the input features to the output after SCA module and 1×1 convolution. The second skip connection adds the output from the feature transformation stage (after the first 1×1 conv following SCA) to the final output of the channel MLP. These two skip connections ensure effective information flow and gradient propagation through the refinement stages.

Downsampling: To downsample a feature, we use 1×1 convolution followed by pixel unshuffle operation, transforming features from [C, H, W] to [C$\times$2, H/2, W/2]. This preserves spatial information by reorganizing pixels into channel dimensions.

Multi-scale Context Aggregation: Following downsampling, we integrate a Pyramid Pooling Module (PPM) [10] to extract multi-scale contextual information, enabling the encoder to better capture global scene structure. Figure 3 shows the architecture of the PPM.

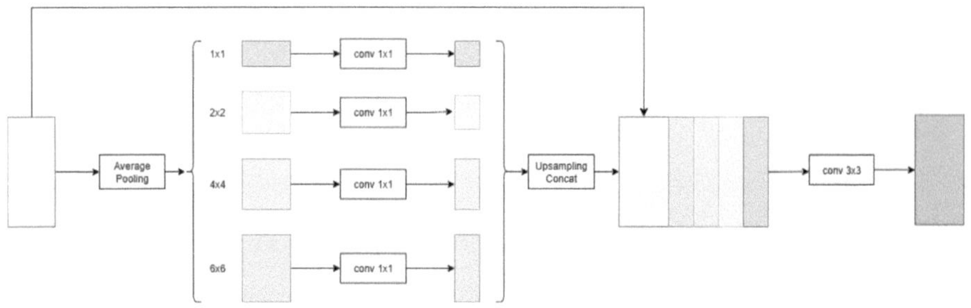

Fig. 3. Architecture of the pyramid pooling module (PPM).

2.2 Decoder Block

As illustrated in Fig. 2b, the decoder block is similar to the encoder architecture but incorporates key modifications specifically designed for spatial reconstruction and detail recovery. While the encoder focuses on information extraction through channel attention, the decoder emphasizes spatial reconstruction through spatial attention mechanisms.

Feature Reconstruction: The decoder block follows the same architectural design as the encoder block. The key distinction lies in replacing Simple Channel Attention (SCA) [6] with Spatial Attention (SA) [11]. This design choice reflects the complementary roles of encoder and decoder: while the encoder focuses on selecting the most informative feature channels for representing degradation patterns, the decoder emphasizes spatial localization to determine where reconstruction should be prioritized. The spatial attention enables the decoder to selectively enhance specific image regions that require restoration, making it particularly effective for joint low-light enhancement and deblurring tasks.

Upsampling: For spatial resolution recovery, we employ a 1×1 convolution followed by a pixel shuffle operation, transforming features from [C, H, W] to [C/2, H $\times$ 2, W $\times$ 2]. This pixel shuffle-based upsampling preserves spatial relationships while effectively redistributing channel information back to spatial dimensions, ensuring fine detail recovery essential for image restoration tasks.

Skip Connection Integration: The decoder integrates skip connections from corresponding encoder stages through element-wise addition, enabling effective fusion of multi-scale features for detail recovery.

2.3   Middle Block

The middle block serves as the bottleneck of our architecture, operating at the lowest spatial resolution to capture the most abstract feature representations. Drawing inspiration from the composite shape convolution (CSC) module presented in [12], we propose a novel middle block designed to integrate diverse kernel shapes, thereby augmenting feature extraction capabilities.

As expressed in Fig. 4, the middle block consists of multiple stacked CSC modules with progressively increasing kernel sizes (k = 3, 5, 7, 9). This approach allows the network to build receptive fields incrementally, where each stage refines and expands upon the previous stage's feature representations. The output from each module is added to the final aggregated output, ensuring that multi-kernel information from all kernel sizes is preserved and preventing information loss during progressive processing.

The Fig. 5 depicts the architecture of the CSC module. First, it begins with a 1×1 convolution followed by SiLU activation for channel adjustment, then processes features through four parallel branches: n $\times$ 1 depthwise convolution, n $\times$ n depthwise convolution, 1 $\times$ n depthwise convolution, and 1 $\times$ 1 depthwise convolution, where n represents the kernel size. These parallel convolutions with different kernel shapes are combined through element-wise addition, followed by SiLU activation for final feature integration.

3   Experiment and Result

3.1   Experimental Settings

Dataset: We conduct experiments on the LOL-Blur dataset [13], which consists of low-light blurry images and normal-light sharp paired images. The dataset contains 10,200 training pairs and 1,800 testing pairs, specifically designed for simultaneous low-light enhancement and deblurring tasks. In addition, we also validated the performance of the proposed method on real-world images. More

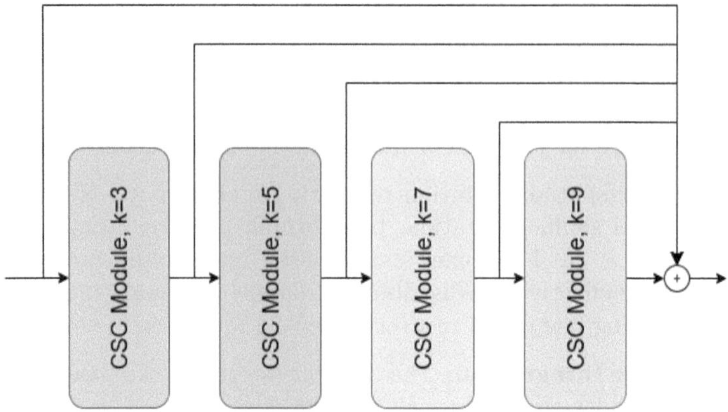

Fig. 4. Architecture of the middle block.

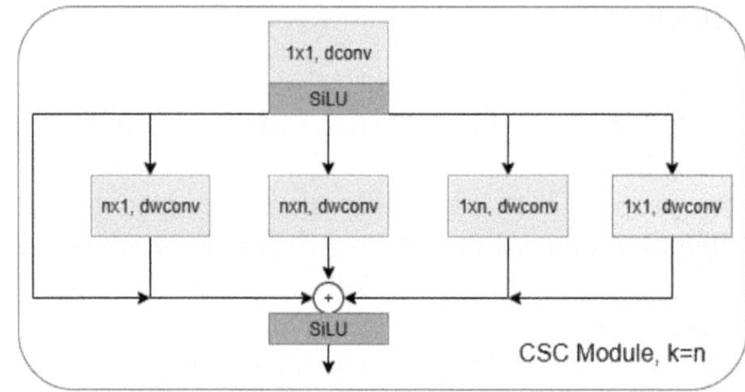

Fig. 5. Architecture of the composite shape convolution module.

specifically, the Real-LOL-Blur dataset [13] integrates 240 wild-captured low-light blurry images with 160 night blurry images from the RealBlur dataset [14].

Training Configuration: Our model is trained end-to-end using the AdamW optimizer with an initial learning rate of $2e^{-4}$ and a final learning rate of $1e^{-7}$. We employ a cosine learning rate scheduler to gradually reduce the learning rate over the training process. The model is trained for 500 epochs with a batch size of 16. To handle memory constraints and improve training efficiency, we use random 256×256 patches cropped from the full-resolution images during training. Data augmentation techniques, including random horizontal flipping and rotation are applied to improve model generalization. A 3.60 GHz Intel® Xeon® CPU E3-1276, 32 GB RAM, and an NVIDIA RTX 3090 GPU are used to implement the proposed method with the PyTorch framework.

Loss Function: We employ a combination of $\mathcal{L}_1$ loss and SSIM loss with an empirically chosen coefficient to train our model. The total loss function is defined as:

$$\mathcal{L}_{total} = \mathcal{L}_1 + 0.4 \times \mathcal{L}_{SSIM}, \tag{1}$$

$$\mathcal{L}_1 = \parallel X - Y \parallel_1, \tag{2}$$

$$\mathcal{L}_{SSIM} = 1 - SSIM(X, Y). \tag{3}$$

where X and Y represent the degradation-free and ground-truth images, respectively. $SSIM(\cdot)$ indicates the SSIM operator [15].

$$SSIM(x, y) = \frac{(2\,\mu_x\mu_y + C_1)(2\sigma_{xy} + C_2)}{(\mu_x^2 + \mu_y^2 + C_1)(\sigma_x^2 + \sigma_y^2 + C_2)}, \tag{4}$$

where μ_x and μ_y are the mean values of two images; σ_x and σ_y are variance values of two images, σ_{xy} is covariance between the two images, and C_1 and C_2 are two constants that prevent the denominator from being zero.

3.2 Experimental Results

This study objectively and subjectively evaluates the proposed network on the LOL-Blur [13] and Real-LOL-Blur datasets [13]. To facilitate comparisons, we constructed three baseline models by thoughtfully combining various existing low-light enhancement and deblurring methods. The baseline methods are specifically categorized as follows:

1. Enhancement → Deblurring: We first apply a low-light enhancement step, and then follow it with a deblurring step. For the enhancement phase, we selected three prominent networks: Retinexformer [2], Zero-DCE [4], and RUAS [3]. Each of these models is designed to improve image visibility in challenging low-light conditions. After the image's brightness and contrast are enhanced, we then feed the output into a cutting-edge deblurring network, MIMO-UNet [7]. This network is specifically chosen for its strong performance in removing blur artifacts, ensuring that the final image is both well-exposed and sharp. This two-stage process aims to tackle both common image degradation issues sequentially.

2. Deblurring → Enhancement: We reverse the order of operations, prioritizing deblurring before proceeding with low-light enhancement. For the crucial deblurring step, we incorporated three competitive networks: NAFNet [6], MIMO-UNet [7], and FFTformer [8]. These models are chosen for their proven effectiveness in restoring sharpness to images degraded by blur. Subsequently, we specifically investigated the impact of Retinexformer to understand its interaction with the initial deblurring process.

3. Single-Pass Enhancement and Deburring: We select the recent state-of-the-art method, LEDNet, which is trained on the LOL-Blur dataset to perform illumination enhancement and blur removal in a single forward pass.

Objective Results: We examined our network on full-reference images with three evaluators: PSNR, SSIM, and LPIPS. Whereas MUSIQ, NRQM, and NIQE are used to objectively evaluate for no-reference images. Table 1 shows the objective results of the LOL-Blur testing dataset. Our proposed method is reported to obtain a promising objective evaluation metric compared to the baseline methods. In addition, it is particularly noteworthy that the proposed method achieves superior performance while maintaining a comparable complexity to other networks. According to Table 2, the proposed network demonstrates superior performance, evidenced by achieving the highest MUSIQ score. This metric indicates that our results are perceptually optimal concerning color contrast and sharpness. Furthermore, the network also yielded the impressive NRQM and NIQE scores, which collectively signify that our generated images possess the highest quality, aligning strongly with human visual perception.

Subjective Results: Several restored images obtained from LOL-Blur testing image by our proposed method are depicted in Fig. 6. The cropped area shows a partially zoomed-in view of representative images. Looking at these images more closely, one can see that the two-stage strategy tended to cause undesirable degradation. In particular, the RUAS → MIMO-Unet baseline was prone to apparent color distortion. While the Retinexformer → MIMO-Unet suffers from the unpleasant artifacts that disrupt visual quality, Zero-DCE →

Table 1. Objective comparison on the LOL-Blur testing dataset regarding the three full-reference metrics. Where baseline Retinexformer → MIMO represents (1), RUAS → MIMO represents (2), Zero-DCE → MIMO represents (3), NAFNet → Retinexformer represents (4), MIMO → Retinexformer represents (5), and FFTformer → Retinexformer represents (6). The symbol "↑" represents higher is better, and "↓" represents lower is better.

Methods	Enhancement → Deblurring			Deblurring → Enhancement			Single forward pass	
	(1)	(2)	(3)	(4)	(5)	(6)	LEDNet	Our
PSNR ↑	18.19	17.81	17.68	14.66	17.02	16.71	25.74	27.13
SSIM ↑	0.68	0.57	0.54	0.50	0.77	0.73	0.85	0.88
LPIPS ↓	0.45	0.52	0.51	0.47	0.27	0.33	0.22	0.21
FLOPS (G)	–	–	–	–	–	–	38.57	13.08
Params (M)	–	–	–	–	–	–	7.4	11.64

Table 2. Objective comparison on the Real-LOL-Blur dataset regarding the three no-reference metrics. Where baseline RUAS → MIMO represents (2), MIMO → Retinexformer represents (5). The symbol "↑" represents higher is better, and "↓" represents lower is better.

Methods	Enhancement → Deblurring	Deblurring → Enhancement	Single forward pass	
	(2)	(5)	LEDNet	Our
MUSIQ ↑	34.39	30.80	39.11	41.20
NRQM ↓	3.32	5.93	5.60	3.27
NIQE ↓	6.81	4.21	4.80	4.16

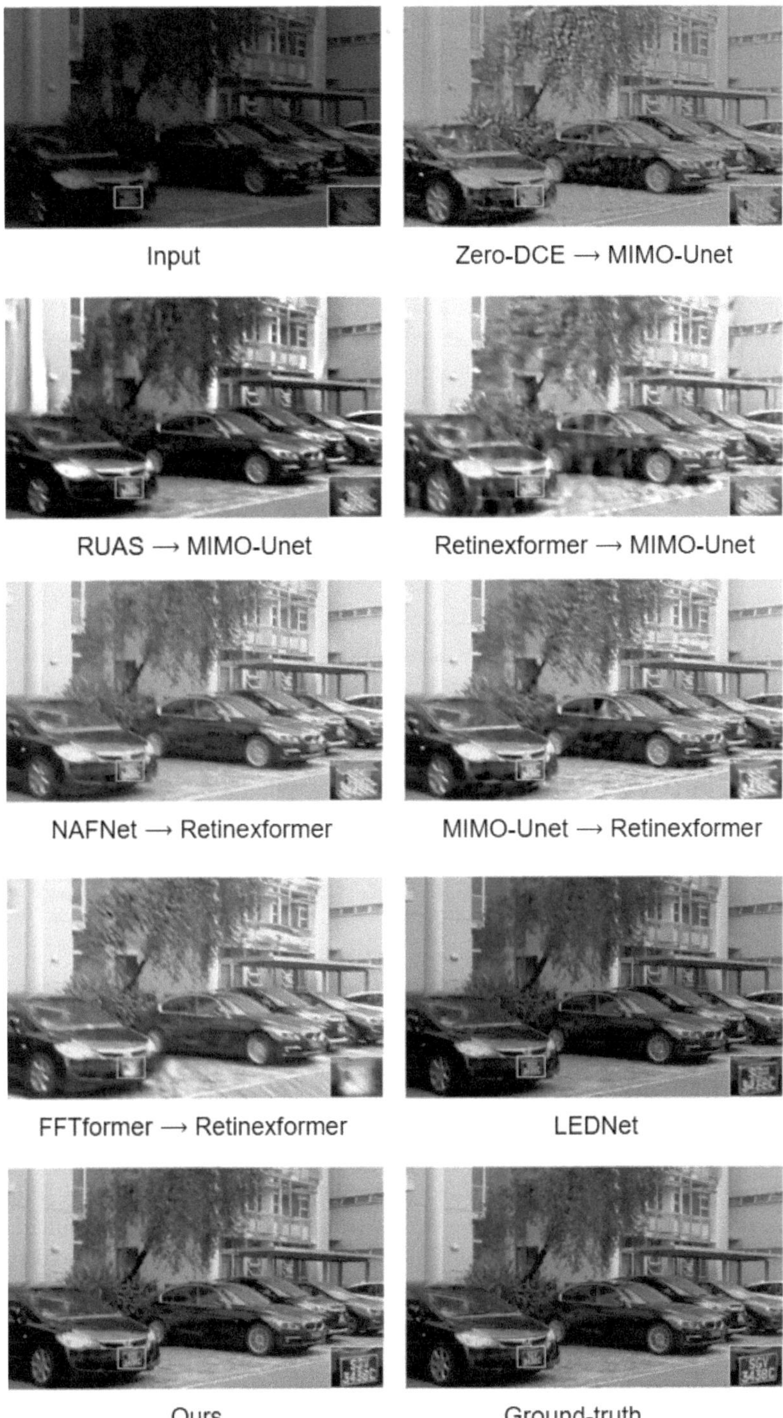

Fig. 6. Subjective comparison with competitive baseline models on a representative image of the LOL-Blur testing dataset.

MIMO-Unet renders under-exposed results along with blurry artifacts. NAFNet → Retinexformer, MIMO-Unet → Retinexformer, FFTformer → Retinexformer can enhance the illumination information, but observable blurry still exists. Although LEDNet produces results with acceptable brightness, it is susceptible to apparent blurry. In general, all comparable approaches left blurry and color distortions in the restored images. On the contrary, our approach is feasible and has outstanding visual quality in low-light and blurry conditions.

The qualitative results of restoring real low-light blurry images using various baseline models are displayed in Figs. 7. The result of the RUAS → MIMO-Unet baseline was prone to apparent color distortion and halo artifacts. MIMO-Unet → Retinexformer renders unwanted distortion in the recovered image. LEDNet failed to generate high-quality restored images. Whereas the proposed network architecture is specifically designed to effectively address the simultaneous degradation of both visibility and texture.

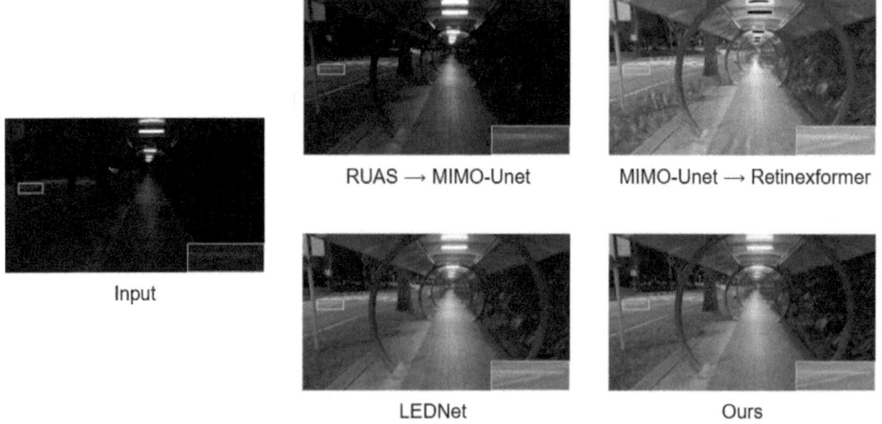

Fig. 7. Subjective comparison with competitive baseline models on a representative image of the Real-LOL-Blur dataset.

4   Conclusions

This article introduces a U-shaped network architecture for simultaneous low-light image enhancement and deblurring. The proposed architecture includes three main components: an encoder block to progressively extract multi-kernel features while reducing spatial resolution, a middle block that processes the most abstract feature representations, and a decoder block to reconstruct the enhanced and deblurred output through upsampling and feature fusion. The experimental results demonstrate that the capability of the proposed network can obtain promising results regarding qualitative and quantitative performance. Additionally, the proposed network achieves moderate complexity, suggesting that it has the potential for surveillance systems using intelligent cameras.

Acknowledgments. This work belongs to the project grant number T2025-179 funded by Ho Chi Minh City University of Technology and Education, Vietnam.

References

1. Duong, M.-T., Lee, S., Hong, M.-C.: Learning to concurrently brighten and mitigate deterioration in low-light images. IEEE Access **12**, 132 891–132 903 (2024)
2. Cai, Y., Bian, H., Lin, J., Wang, H., Timofte, R., Zhang, Y.: Retinexformer: one-stage retinex-based transformer for low-light image enhancement. In: Proceedings of the IEEE/CVF International Conference on Computer Vision (ICCV), pp. 12 504–12 513 (2023)
3. Risheng, L., Long, M., Jiaao, Z., Xin, F., Zhongxuan, L.: Retinexinspired unrolling with cooperative prior architecture search for low-light image enhancement. In: Proceedings of the IEEE Conference on Computer Vision and Pattern Recognition (2021)
4. Guo, C.G., Li, C., Guo, J., et al.: Zero-reference deep curve estimation for low-light image enhancement. In: Proceedings of the IEEE Conference on Computer Vision and Pattern Recognition (CVPR), pp. 1780–1789 (2020)
5. Ho, Q.-T., Duong, M.-T., Lee, S., Hong, M.-C.: EHNet: efficient hybrid network with dual attention for image deblurring. Sensors **24**(20) (2024)
6. Chen, L., Chu, X., Zhang, X., Sun, J.: Simple baselines for image restoration. arXiv preprint arXiv:2204.04676 (2022)
7. Cho, S.-J., Ji, S.-W., Hong, J.-P., Jung, S.-W., Ko, S.-J.: Rethinking coarse-to-fine approach in single image deblurring. In: 2021IEEE/CVF International Conference on Computer Vision (ICCV), pp. 4621–4630 (2021)
8. Kong, L., Dong, J., Ge, J., Li, M., Pan, J.: Efficient frequency domainbased transformers for high-quality image deblurring. In: Proceedings of the IEEE/CVF Conference on Computer Vision and Pattern Recognition (CVPR), pp. 5886–5895 (2023)
9. Ho, Q.-T., Duong, M.-T., Lee, S., Hong, M.-C.: Adaptive image deblurring convolutional neural network with meta-tuning. Sensors **25**(16) (2025)
10. He, K., Zhang, X., Ren, S., Sun, J.: Spatial pyramid pooling in deep convolutional networks for visual recognition. IEEE Trans. Pattern Anal. Mach. Intell. **37**(9), 1904–1916 (2015). https://doi.org/10.1109/TPAMI.2015.2389824
11. Woo, S., Park, J., Lee, J.-Y., Kweon, I.S.: Cbam: convolutional block attention module. In: Proceedings of the European Conference on Computer Vision (ECCV), pp. 3–19 (2018)
12. Guo, X., Dong, Y., Chen, X., et al.: Underwater image restoration via polymorphic large kernel CNNs. In: ICASSP 2025–2025 IEEE International Conference on Acoustics, Speech and Signal Processing (ICASSP), pp. 1–5. IEEE (2025)
13. Zhou, S., Li, C., Loy, C.C.: LedNet: joint low-light enhancement and deblurring in the dark. In: European Conference on Computer Vision (ECCV) (2022)
14. Rim, J., Lee, H., Won, J., Cho, S.: Real-world blur dataset for learning and benchmarking deblurring algorithms. In: Vedaldi, A., Bischof, H., Brox, T., Frahm, J.-M. (eds.) ECCV 2020. LNCS, vol. 12370, pp. 184–201. Springer, Cham (2020). https://doi.org/10.1007/978-3-030-58595-2_12
15. Wang, Z., Bovik, A.C., Sheikh, H.R., Simoncelli, E.P.: Image quality assessment: from error visibility to structural similarity. IEEE Trans. Image Process. **13**(4), 600–612 (2004)

Design of a Solar-Powered AIoT System
for Safety Corridor Violation Detection
in High-Voltage Power Transmission Networks

Huu-Phuoc Nguyen[1], Tan-Phat Tran[2], and Ngoc-Luat Pham[3]($\boxtimes$)

[1] Faculty of Automation Engineering, Can Tho University, Can Tho 900000, Vietnam
[2] The Western Power Transmission N0. 1, Power Transmission Company N0. 4, Can Tho 900000, Vietnam
[3] Faculty of Electronics and Telecommunications, Can Tho University, Can Tho 900000, Vietnam
pnluat@ctu.edu.vn

Abstract. Ensuring the safety corridor of high-voltage transmission lines is critical in construction sites and roadways intersecting with power lines. This study proposes a safety corridor violation warning system based on AIoT technology, integrating the YOLOv8n model deployed on a Raspberry Pi 5, using ONVIF-compatible IP cameras, 4G connectivity, and solar power. The system is compatible with widely available off-the-shelf IP cameras, helping reduce costs and simplify deployment. Experimental results show that the system achieves detection accuracy ranging from 80% to over 90% for heavy machinery such as cranes, pile drivers, and excavators operating near transmission lines. It maintains a processing speed of 4.45 frames per second and operates reliably in outdoor environments. Notably, the system can run continuously 24/7 using solar energy and is designed to remain operational for up to three days without sunlight, ensuring high availability under all weather conditions. Compared to traditional single-point sensor solutions, the proposed system supports wide-area monitoring, real-time alerts, and fully autonomous operation independent of the electrical grid.

Keywords: AIoT System · Power Transmission Lines · Object Detection · Solar-Powered Monitoring

1 Introduction

The safety corridor is a minimum protective distance established to prevent potential harm to humans, property, and technical infrastructure. This concept plays a particularly important role in various fields, such as maintaining safe distances beneath high-voltage power lines, at railway-road intersections, restricted areas, or national borders. In Vietnam, safety corridors have been clearly defined in the legal system. Most recently, the Government issued Decree No. 62/2025/ND-CP, detailing the implementation of the Electricity Law on the protection of power works and safety in the electricity sector.

Violations of safety corridors around power transmission lines are a common and serious issue in many countries. Incidents involving breaches of the protective distance surrounding transmission lines often lead to severe consequences, including injury, death, fire, and widespread power outages. In many cases, the causes stem from individuals arbitrarily constructing buildings, planting trees, or operating lifting equipment near the power lines without timely monitoring or warning measures. This reality shows that traditional monitoring methods such as periodic inspections or visual surveillance are no longer suitable in the context of expanding transmission networks that stretch across complex terrains. Therefore, there is an urgent need to develop automated warning systems capable of operating continuously, accurately, and flexibly under various environmental conditions.

Research efforts have been devoted to developing technical solutions to improve the effectiveness of safety corridor warnings. For example, in the study by Dao Thanh Toan [1], a system using flexible electronic mats was developed to warn of violations of safety lines at metro stations. Meanwhile, Vu Dinh Trung et al. [2] developed a system integrating artificial intelligence (AI) to detect obstacles early at railway crossings, providing timely warnings for train operators. More recently, Huu-Phuoc Nguyen et al. [3] proposed a safety corridor violation warning system using reflective laser sensors combined with LoRa communication and solar power. This system demonstrated high feasibility due to its low cost and independence from the electrical grid; however, it still has limitations in monitoring range (about 300 m) and lacks on-site intelligent processing capabilities.

Many studies have applied IoT, AI, laser sensors, and computer vision to improve safety corridor warning systems. Jinglong Xie et al. [4] proposed an intrusion system using electronic fences and cameras, but it is limited to small areas and requires ultra-broadband. Chunjuan Wei et al. [5] developed a substation protection system with laser sensors and ZigBee, but its SMS-based alerts are unsuitable for real-time use. Neha Bhadwal et al. [6] combined infrared sensors and computer vision for border surveillance, yet the system depends on stable networks. Azfarina Jaafar et al. [7] designed a household-scale intrusion system with laser interfaces and dashboards, but it has not been tested on a larger scale.

In addition, Chen et al. [8] highlighted the role of IoT in enabling real-time monitoring of power lines. Ma et al. [9] proposed detecting external threats by analyzing vanishing points in surveillance images. Shao et al. [10] enhanced YOLOv8 for abnormal object detection near transmission lines, but the model still requires high-end hardware and high power consumption, limiting field deployment. Goh et al. [11] reviewed fault detection methods based on signal and sensor data, providing a foundation for next-generation intelligent warning systems.

From the above analysis, it is clear that although existing solutions have made significant technical contributions, they still face practical limitations in the Vietnamese power sector, such as high investment and maintenance costs, high energy consumption, difficulty in deployment in complex terrain or areas lacking electricity and network infrastructure, and the absence of intelligent edge processing. In this study, we propose an improved safety corridor violation warning system based on the group's previous re-search [3]. The new design overcomes these shortcomings by employing

AI-integrated cameras instead of single-point sensors, incorporating edge computing, supporting flexible two-way wireless communication, and optimizing power usage for fully independent operation. This approach enhances warning effectiveness, broadens monitoring coverage, enables deployment in infrastructure-limited areas, and ensures continuous operation under diverse weather conditions.

2 Materials and Methods

2.1 Proposed System

The overall architecture of the high-voltage power transmission line safety corridor violation warning system is illustrated in Fig. 1. The system comprises five main components: (1) Central Processing Unit, (2) IP Camera, (3) 4G Wi-Fi Router, (4) Warning Speaker, and (5) Solar Power Supply.

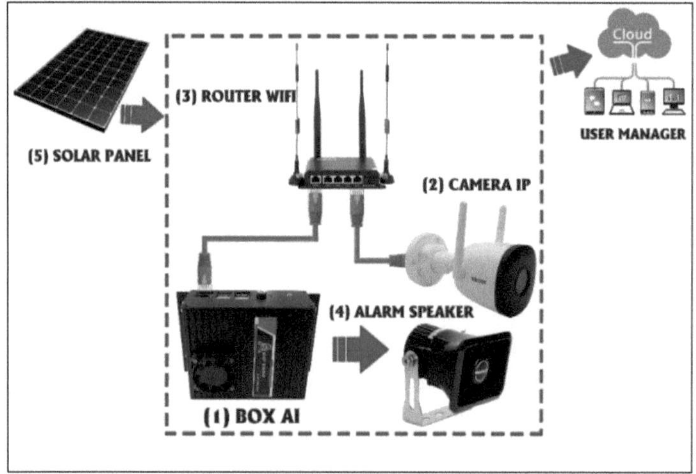

Fig. 1. Overview diagram of the AI-based camera monitoring system.

- **(1) Central Processing Unit (AI Box):** Responsible for receiving image streams from the camera, running the machine learning model, and controlling the automatic warning functions.
- **(2) IP Camera:** A standard security surveillance device that continuously provides image data to the system.
- **(3) 4G Wi-Fi Router:** A device that broadcasts Wi-Fi using a 4G SIM card, enabling data and image transmission to the management unit via mobile Internet.
- **(4) Warning Speaker:** A 30 W loudspeaker used to broadcast pre-recorded voice alerts upon detecting intrusions, helping violators become aware and adjust their behavior.
- **(5) Solar Power Supply:** Generates and stores electrical energy to independently power the entire system.

2.2 Optimal Hardware Selection and Design

2.2.1 Central Processing Unit

The system employs the Raspberry Pi 5 [12] as the central processor running the YOLOv8n model. Compared with platforms like Jetson Nano and LattePanda 3 Delta (Table 1), it provides better machine learning performance, lower power consumption, and more affordable cost.

Table 1. Comparison of Raspberry Pi 5 with other embedded computing platforms

Criterion	Raspberry Pi 5	LattePanda 3 Delta	Jetson Nano
CPU	Quad-core Cortex-A76, 2.4 GHz	Intel Celeron N5105 (Quad-core, 2.0–2.9 GHz)	Quad-core Cortex-A57, 1.43 GHz
GPU	VideoCore VII, 800 MHz	Intel UHD Graphics (24 Execution Units)	128-core Maxwell CUDA
RAM	8 GB LPDDR4x	8 GB LPDDR4 (onboard)	4 GB LPDDR4
Power	5–9 W	10–12 W	10–15 W
Price (USD)	~90 [13]	~250 [14]	~227 [15]

2.2.2 Surveillance Camera

The system employs the IMOU IPC-F22P 2MP outdoor camera [16], which supports the ONVIF protocol and transmits real-time video via RTSP. With Full HD 1080p resolution and IP67 water resistance, it is well-suited for outdoor use. The system is also compatible with any ONVIF-supported IP camera, ensuring adaptability and ease of deployment in real-world conditions.

2.2.3 Wireless Communication Solution

This study employs a Wi-Fi network broadcast from the 4G LTE APTEK L300 router [17], using mobile SIM connectivity. This wireless communication solution is widely adopted for its broadband capability, high transmission speed, scalability, and ease of remote management.

For applications requiring video streaming and real-time monitoring, Wi-Fi via 4G LTE outperforms protocols such as ZigBee, LoRa, RF, or SMS, which suffer from limited bandwidth, high latency, and unsuitability for image transmission. A detailed comparison of these protocols is provided in Table 2.

2.2.4 Power Supply

In this study, the system is powered by solar energy. A 300W mono-type solar panel [18] is connected to a 60A MPPT charge controller [19], which charges a 72Ah LiFePO$_4$

Table 2. Comparison of Wi-Fi over 4G with other wireless communication protocols

Protocol	Range	Bandwidth	Real-Time Response	Image Transmission Capability
Wi-Fi over 4G	Unlimited	High	Good	Good
LoRa	1–2 km	Very Low	Slow	No
Zigbee	Short	Low	Medium	No
SMS	Unlimited	Very Low	Delayed	No
RF	Short	Very Low	Delayed	No

lithium battery pack with a nominal voltage of 12.8 VDC. This battery pack was designed by the research team using 48 LiFePO$_4$ cells of type 32650 (3.2 V–6000 mAh) [20].

When fully charged, the battery pack can supply power for more than three days of continuous operation. Under the assumption of 80% charging efficiency and approximately four hours of sunlight per day, the battery can be fully recharged. These values are calculated based on Eqs. (1) and (2):

$$T = \frac{Q}{P} = \frac{921.6}{11} \approx 3.49 \ (days) \tag{1}$$

$$TC = \frac{Q}{Ws * 0.8} = \frac{921.6}{300 * 0.8} \approx 3.84 \ (hours) \tag{2}$$

whereas,

- T: Duration the system can operate using the battery's energy (in days)
- TC: Time required to fully charge the battery using the solar panel (in hours)
- Q: Battery capacity (Wh = 921.6 Wh)
- P: Power consumption of the entire system (W = 11 W)
- Ws: Power rating of the solar panel (W = 300 W)

2.3   System Software Design

2.3.1   System Operation

The system is programmed primarily using Python version 3.9.7, running on the Raspbian operating system compatible with the Raspberry Pi. The core operation of the system consists of five main steps, as illustrated in Fig. 2.

- **Connect to IP camera:** Automatically discover the IP address of the camera and access the video stream via the ONVIF protocol.
- **Start machine learning model:** Load and run a pre-trained YOLO model to detect specific target objects.
- **Detect the object:** Continuously analyze video frames and assess confidence scores to determine whether to trigger a warning.
- **Warning:** Issue an alert by sending an email with a warning message and image to the system manager, while simultaneously broadcasting a voice message via the on-site speaker.

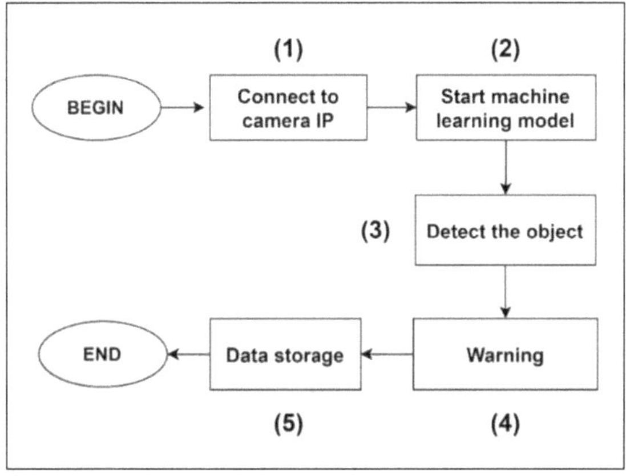

Fig. 2. Flowchart illustrating the main operational steps of the system.

- **Data storage:** Save detected images, timestamps, and locations to the internal storage of the Raspberry Pi.

2.3.2 Remote System Management

For remote monitoring and control, users access the Raspberry Pi via VNC Viewer, which provides a graphical interface for local-like operation. Since monitoring stations use 4G SIM networks without public IPs, traditional methods are impractical. To address this, the system employs Tailscale, a WireGuard-based VPN that enables secure, stable connections without router or NAT configuration. Tailscale is free for personal use, easy to set up, and highly secure.

2.3.3 Machine Learning Model Training

- **Dataset Preparation**

To train the object detection model, the team collected 2,333 real-world images of 10 heavy vehicle classes and one background class, with an average resolution of 1100 × 880 pixels. The dataset was split into 70% training and 30% testing, as shown in Table 3.

- **Model Training**

Model training was conducted on the Kaggle virtual machine platform using YOLOv8n, a lightweight version of the YOLO object detection algorithm developed by Ultralytics. YOLOv8n is well-suited for deployment on resource-constrained devices such as the Raspberry Pi. The model was trained for 300 epochs.

After training, the model was converted to the NCNN format-an open-source inference framework developed by Tencent, specifically optimized for mobile and embedded platforms. NCNN is highly efficient on CPUs, requiring no GPU or specialized hardware, making it ideal for devices like the Raspberry Pi.

Table 3. Image dataset statistics for machine learning model training

No.	Object Class	Train (70%)	Test (30%)
1	Bulldozer	150	64
2	ConcreteMixer	148	64
3	Crane	151	65
4	Excavator	146	63
5	Loader	149	64
6	PileDriving	152	65
7	PumpTruck	147	63
8	Roller	150	64
9	StaticCrane	134	58
10	Truck	148	63
11	Background	158	67
	Total	**1633**	**700**

3　Results and Discussion

3.1　Designed System Model

The system components were installed inside a protective enclosure, with the IP camera, warning speaker, and solar panel mounted externally, as shown in Fig. 3a. In practical deployment, the system is mounted at the height of 10 m above the ground on high-voltage transmission line towers (Fig. 3b) to monitor heavy machinery that poses a risk of safety corridor violations within a range of about 50 m.

3.2　Machine Learning Model Training Results

Figure 4 shows the Precision–Recall (PR) curves of the YOLOv8n model after training, corresponding to each type of vehicle in the test dataset. Each curve represents the detection performance of the model for a specific class, where the x-axis indicates recall (the ratio of correctly detected objects to actual instances), and the y-axis represents precision (the ratio of correctly detected objects to total predictions).

The results show that the classes Excavator, ConcreteMixer, Crane, and Bulldozer achieved high accuracy, with average precision (AP) scores of 0.918, 0.904, 0.869, and 0.838, respectively. Meanwhile, the bold blue line represents the average performance across all classes, with a mean average precision (mAP@0.5) of 0.750, indicating that the model performs relatively well in real-world data scenarios. This provides a strong foundation for deploying the model in safety corridor monitoring systems using object recognition.

Figure 5 presents the normalized confusion matrix, illustrating the prediction results of the YOLOv8n model on a test set of 700 images. Each cell in the matrix represents

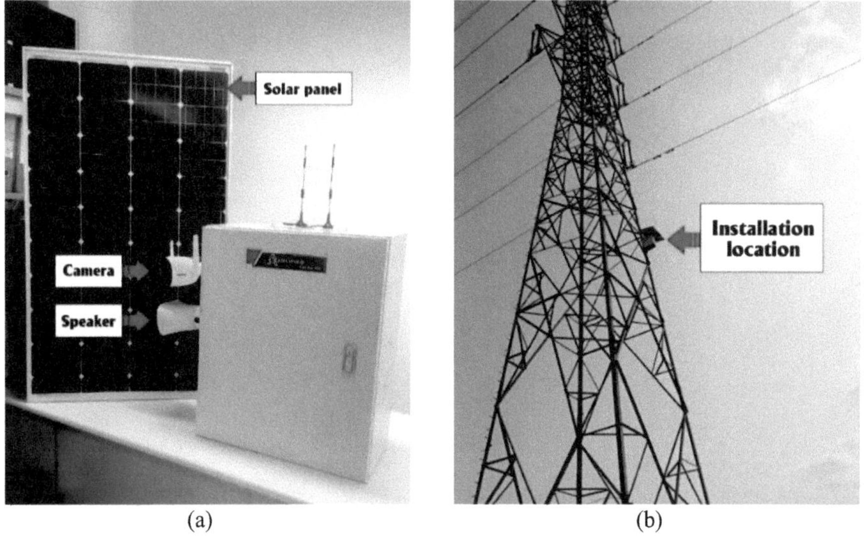

Fig. 3. Actual system (a) and installation location on transmission pole (b)

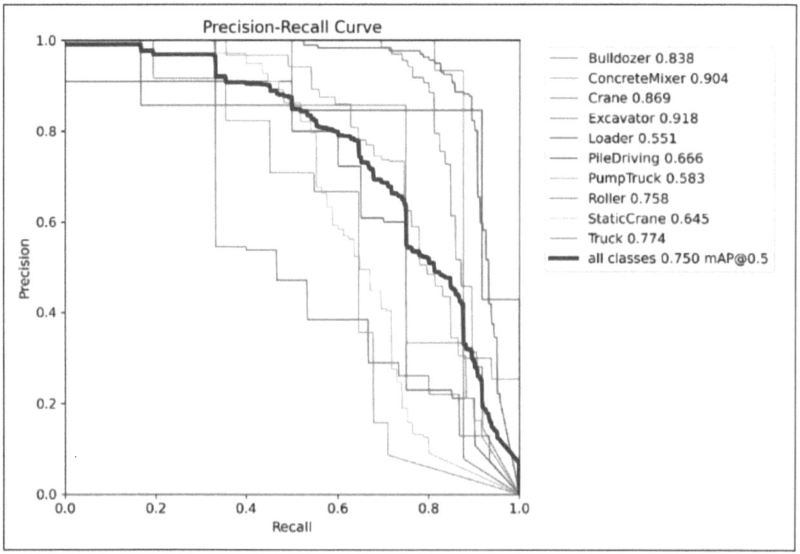

Fig. 4. Precision–Recall curves of object detection model for each vehicle class

the percentage of images belonging to the actual class (x-axis) that were predicted as a different class (y-axis). Darker cells indicate higher prediction accuracy for the corresponding class.

The results reveal that classes such as Excavator (0.89), ConcreteMixer (0.87), Crane (0.81), and Truck (0.75) achieved high prediction accuracy, with over 80% of

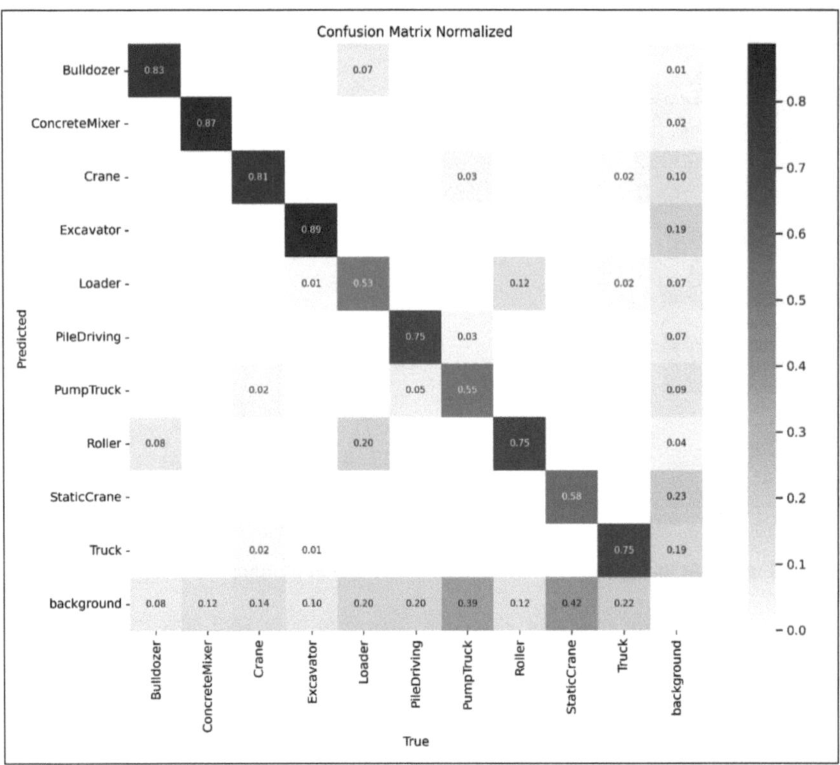

Fig. 5. Confusion matrix of object detection model evaluated on the test dataset

instances correctly classified. In contrast, classes like Loader (0.53), StaticCrane (0.58), and PumpTruck (0.55) showed lower accuracy and were more prone to confusion with visually similar classes or complex backgrounds.

3.3　System Operational Testing

To evaluate the system's object detection capability, the research team deployed it at construction sites to identify heavy vehicles in actual operating conditions. The results demonstrate that the system operates at high speed, with a preprocessing time of 0.0099 s, YOLO inference time of 0.2093 s, and postprocessing time of 0.0056 s, resulting in a total processing time per frame of 0.2248 s, as shown in Table 4. These results confirm that the system can detect potential violations in real-time, meeting practical requirements.

Figure 6a presents sample object detection results from real-world field tests using the YOLOv8n model. Objects such as cranes, trucks, excavators, and static tower cranes were detected with high accuracy, with confidence scores ranging from 0.71 to 0.95.

In this study, the farthest object that the system was able to recognize and that the research team tested was at a distance of 60 m from the camera. As shown in Fig. 6b, a pile-driving machine located approximately 60 m from the camera was detected with

Table 4. Model processing time per image

Preprocess (s)	Inference (s)	Postprocess (s)	Total time (s)
0.0099	0.2093	0.0056	0.2248

Fig. 6. Object detection results on real-world images

an accuracy of 86%, while an excavator at a distance of 40 m was recognized with an accuracy of 92%.

3.4 Discussions

The deployment results highlight several strengths: the lightweight YOLOv8n model achieves high accuracy in detecting heavy machinery within safety corridors; the solar-powered design enables independent operation in remote areas; and secure remote access via VPN (Tailscale) ensures flexible management.

However, the system has limitations. The model sometimes confuses visually similar objects (e.g., StaticCrane vs. Crane, PumpTruck vs. Truck) due to limited data, and its accuracy decreases under poor lighting or weather conditions. Although infrared monitoring is supported, nighttime and rainy scenarios still need more data. In addition, while Raspberry Pi performs on-site inference independently of network quality, 4G transmission may be delayed under weak connections.

Future improvements will focus on expanding datasets, integrating advanced filtering algorithms (distance estimation, motion trajectory, multi-frame verification), and applying active data augmentation. Redundant communication (multi-SIM, dual network), environmental sensors, and multi-source alerts will also be considered to enhance accuracy, reliability, and adaptability.

4　Conclusions

This paper presents an AIoT-based warning system for detecting safety corridor violations along high-voltage transmission lines. The system integrates a YOLOv8n model on a Raspberry Pi 5, using ONVIF-compatible IP cameras, 4G connectivity, and solar power. It achieves high average detection accuracy (mAP@0.5 $= 0.750$), with strong performance on specific classes such as Excavator (0.89), ConcreteMixer (0.87), Crane (0.81), and Truck (0.75). With a fast processing time of 0.2248 s per image, the system meets real-time monitoring requirements. The solar-powered design enables 24/7 operation and sustains functionality for up to 3 days without sunlight. Compared to previous approaches, the system offers advantages in autonomous deployment, cost-effectiveness, and reliability in infrastructure-constrained environments. The proposed system is intended for point-based monitoring at high-risk or frequently violated locations, such as construction sites or intersections with roads.

References

1. Toan, D.T.: Automatic warning of safety line violation at Hanoi Metro station using flexible electronic mats. Transp. Commun. Sci. J. **71**(3), 263–273 (2020). https://doi.org/10.25073/tcsj.71.3.10
2. Trung, V.D., Quang, P.H., Cong, P.H., Chinh, P.Q.: Research and experimentation on the application of artificial intelligence in ensuring safety at railway level crossings. J. Sci. Technol. Univ. Danang **20**(11), 1 (2022)
3. Phuoc, N.H., Khang, N.D., Thanh, T.N.: Optimal design of safety corridor violation warning system based on laser sensor and LoRa communication. J. Sci. Technol. Industr. Univ. Ho Chi Minh City **67**(01), (2024). https://doi.org/10.46242/jstiuh.v67i01.5031
4. Xie, J., Xiao, X., Xu, Y., Jin, B.: An IoT assisted early warning system for smart grid. J. Phys. Conf. Ser. **2218**(1), 012027 (2022). https://doi.org/10.1088/1742-6596/2218/1/012027

5. Wei, C., Yang, J., Zhu, W., Lv, J.: A design of alarm system for substation perimeter based on laser fence and wireless communication. In: Proceedings of the ICCASM, vol. 3, pp. 2–5 (2010). https://doi.org/10.1109/ICCASM.2010.5620690

6. Bhadwal, N., et al.: Smart border surveillance system using wireless sensor network and computer vision. In: International Conference on Automation, Computational and Technology Management, pp. 183–190 (2019). https://doi.org/10.1109/ICACTM.2019.8776749

7. Jaafar, A., et al.: Dynamic home automation security alert system using laser interfaces on webpages and windows mobile. In: IEEE Control and System Graduate Research Colloquium, pp. 153–158 (2016). https://doi.org/10.1109/ICSGRC.2016.7813319

8. Chen, X., Sun, L., Zhu, H., Zhen, Y., Chen, H.: Application of internet of things in power-line monitoring. In: Proceedings of the International Conference on CyberC (2012). https://doi.org/10.1109/CyberC.2012.77

9. Ma, F., et al.: A refined identification method for the hidden dangers of external damage in transmission lines based on the generation of a vanishing point-driven effective region. Processes **12**, 1904 (2024). https://doi.org/10.3390/pr12091904

10. Shao, Y., et al.: TL-YOLO: foreign-object detection on power transmission line based on improved YOLOv8. Electronics **13**, 1543 (2024). https://doi.org/10.3390/electronics13081543

11. Goh, H.H., et al.: Transmission line fault detection: a review. Indon. J. Electr. Eng. Comput. Sci. **8**, 199–205 (2017). https://doi.org/10.11591/ijeecs.v8.i1.pp199-205

12. Raspberry Pi Ltd.: Raspberry Pi 5 Product Brief. https://datasheets.raspberrypi.com/rpi5/raspberry-pi-5-product-brief.pdf. Accessed 21 June 2025

13. Raspberry Pi Ltd.: Raspberry Pi 5. https://lnk.ink/iGF4s. Accessed 21 June 2025

14. LattePanda Team. LattePanda 3 Delta. https://lnk.ink/w8pB5. Accessed 21 June 2025

15. NVIDIA. Jetson Nano Developer Kit. https://lnk.ink/hWVLM. Accessed 21 June 2025

16. Imou. Imou Bullet LC-A32F 4MP Camera. https://www.alibaba.com/product-detail/Imou-Bullet-LC-A32F-4M-D_1601226258942.html. Accessed 21 June 2025

17. Shenzhen Wlink Technology Co., Ltd.: Industrial Grade Wireless Router 4G LTE. https://www.alibaba.com/product-detail/Industrial-Grade-Wireless-Router-4G-LTE_1601199518546.html. Accessed 21 June 2025

18. JCNS Solar. 12/24V 80W–120W Solar Panel Module. https://www.alibaba.com/product-detail/JCNS-12-24v-80w-100w-120w_60693250718.html. Accessed 21 June 2025

19. Renogy. 60A MPPT Solar Charge Controller. https://lnk.ink/QCRxm. Accessed 21 June 2025

20. Shenzhen Power Long Battery Tech Co., Ltd.: High Quality LiFePO4 Cell Battery 32650. https://www.alibaba.com/product-detail/High-Quality-Lifepo4-Cell-Battery-32650_1600902946018.html, last accessed 2025/6/21

A Real-Time Adaptive Traffic Signal Control Framework Using Graph Neural Networks and Multi-Agent Reinforcement Learning with Crowdsourced Event Integration

Do Thanh Thai[1,2] and Quang Tran Minh[1,2(✉)]

[1] Faculty of Computer Science and Engineering, Ho Chi Minh City University of Technology (HCMUT), 268 Ly Thuong Kiet, Dien Hong Ward, Ho Chi Minh City, Vietnam
{dtthai.sdh242,quangtran}@hcmut.edu.vn

[2] Vietnam National University Ho Chi Minh City (VNU-HCM), Linh Xuan Ward, Ho Chi Minh City, Vietnam

Abstract. This paper presents a novel adaptive traffic signal control framework that integrates Graph Neural Networks (GNNs), Multi-Agent Reinforcement Learning (MARL), and real-time crowdsourced event data from Waze to address dynamic urban traffic challenges. The proposed system leverages spatial-temporal graph modeling to predict short-term traffic states, enabling decentralized agents at individual intersections to optimize signal timing policies collaboratively. An event-triggered adaptation mechanism allows rapid policy adjustments in response to incidents such as accidents or roadblocks, enhancing traffic flow resilience. Extensive simulations using real-world traffic data from Ho Chi Minh City demonstrate that the framework significantly reduces vehicle waiting times by over 25%, increases average speeds, and substantially decreases traffic deadlocks compared to fixed-time control strategies. These results underscore the potential of combining graph-based deep learning with reinforcement learning and crowdsourced event integration for effective, scalable urban traffic management.

Keywords: Adaptive Traffic Signal Control · Graph Neural Networks · Multi-Agent Reinforcement Learning · Crowdsourced Traffic Data · Urban Traffic Management

1 Introduction

Modern cities are facing unprecedented traffic challenges as vehicle ownership continues to rise sharply. Congestion not only leads to frustrating delays but also increases fuel consumption and contributes to air pollution. Traditional signal control frameworks like the Sydney Coordinated Adaptive Traffic System

N. Thai-Nghe et al. (Eds.): ISDS 2025, CCIS 2714, pp. 270–285, 2026.
https://doi.org/10.1007/978-981-95-3358-9_20

(SCATS) and the Split Cycle Offset Optimisation Technique (SCOOT), while reliable, often rely on pre-set cycles or local inductive sensors. These systems struggle to adapt swiftly to non-recurrent disruptions such as accidents or sudden traffic surges [1,2].

In recent years, Deep Reinforcement Learning (DRL) has emerged as a powerful paradigm for adaptive traffic signal control. Techniques like the Deep Q-Network (DQN), Double DQN (DDQN), dueling networks, and Soft ActorâĂŞ-Critic (SAC) have demonstrated strong performance in reducing queue lengths and travel delays [3–5]. Further advances leverage MARL, where each intersection operates as an autonomous agent, enabling intelligent coordination across networks, as seen in systems like CoLight [6], PressLight [7], and SEMA2C [8]. Parallelly, the application of GNNs has revolutionized urban traffic modeling. By naturally representing intersections as nodes and roads as edges, GNNs adeptly capture spatial dependencies across the traffic network. Several hybrid models, such as GPlight, Influence Graph Reinforcement Learning (IG-RL), and Graph Attention Network âĂŞ Deep Deterministic Policy Gradient (GAT-DDPG), combine GNN encoding with DRL, delivering significant improvements over classical control methods [9–11].

Despite the notable progress achieved through the integration of DRL and GNNs in traffic signal control, several critical limitations remain. First, most existing approaches largely overlook real-time, crowdsourced event data. Platforms such as Waze [12] provide timely, user-reported information on accidents, road closures, and congestion, which can offer valuable situational awareness that traditional systems, typically reliant on fixed sensors, are unable to capture effectively. Second, current GNN+DRL-based models tend to focus on learning from regular traffic patterns and exhibit limited adaptability when faced with unexpected, non-recurrent events such as collisions or temporary construction work. These systems often struggle to react promptly and optimally under such irregular conditions. Third, scalability remains a persistent challenge. Although some frameworks, such as IG-RL, demonstrate promising results in large-scale simulations involving thousands of intersections, their effectiveness and robustness in real-world deployments, where traffic dynamics are significantly more complex and unpredictable, have yet to be conclusively demonstrated [11].

To address urban traffic management challenges, we propose a real-time adaptive traffic signal control framework that integrates GNN, MARL, and real-time crowdsourced event data from Waze. The key contributions of this study are summarized as follows:

- **Event-aware graph modeling**: We dynamically adjust graph edge weights using live Waze incident reports, enabling the system to detect congestion and blockages promptly.
- **Short-term traffic prediction via GNN**: Spatial-temporal GNN architectures (i.e., GCN) are employed to forecast near-future traffic states, which guide agent decision-making.
- **Distributed MARL controllers**: Autonomous agents, trained with Proximal Policy Optimization (PPO) / Advantage Actor-Critic (A2C) algorithms,

control individual intersections aiming to optimize throughput, reduce delay, and improve resilience.

- **Event-triggered adaptation**: The framework incorporates mechanisms to rapidly adjust control policies in response to real-time incident reports, improving responsiveness and traffic flow recovery.

This combination of crowdsourced event integration, graph-based spatial awareness, and decentralized multi-agent control represents, to the best of our knowledge, a novel and comprehensive approach in adaptive traffic signal control. Through extensive simulation experiments conducted in Simulation of Urban MObility (SUMO) with real-world traffic data from Ho Chi Minh City, our framework demonstrates significant improvements over fixed-time control baselines, particularly in reducing vehicle waiting times by over 25%, increasing average speeds, and substantially minimizing traffic deadlocks (teleport events). These results highlight the potential of the proposed system for effective real-world urban traffic management.

The remainder of this paper is organized as follows. Section 2 reviews the related work on traffic signal control and GNNs. Section 3 presents the proposed adaptive traffic signal control framework using GNNs and MARL. Experimental results and evaluations are discussed in Sect. 4. Finally, Sect. 5 concludes the paper and outlines directions for future research.

2   Related Work

Deep Reinforcement Learning for Traffic Signal Control: Since the introduction of the DQN, DRL has emerged as a core approach for adaptive traffic signal control. Early work demonstrated that DRL can effectively learn control policies from raw traffic states, surpassing traditional fixed-time and actuated methods. A DQN-based framework was shown to reduce vehicle delay by up to 86% under simulation conditions [13]. Subsequent advancements, including dueling DQNs, double DQNs, and Soft ActorâĂŞCritic (SAC), further improved training stability and convergence. A recent large-scale review provided a comprehensive evaluation of DRL-based traffic signal controllers and emphasized the trend toward more expressive and robust policy architectures [14].

Multi-Agent and Graph Neural Network Integration: To address the decentralized and spatially distributed nature of urban intersections, recent approaches have increasingly combined MARL with GNNs. For example, MARL models incorporating Graph Attention Networks (GATs) have achieved over 20% improvements in average travel time compared to the Max-Pressure baseline [7]. Other models integrate short-term traffic forecasting directly into DRL controllers using GNNs [9], while inductive MARL-GNN architectures have been designed for transferability across varying intersection topologies [11]. Spatial awareness has been further enhanced through the use of diffusion convolution modules in decentralized frameworks [15]. Comparative evaluations indicate that decentralized models offer strong scalability and adaptability [16]. Lightweight, region-based models with spatiotemporal coordination have also shown promise for efficient large-scale deployment [17].

Hybrid Architectures and Traffic Adaptivity: Beyond GNNs, hybrid DRL architectures have emerged, combining attention mechanisms, temporal memory, and control-theoretic elements. Hierarchical attention has been applied to capture fine-grained lane dynamics, while models integrating GATs with PPO demonstrate robust performance across varying traffic conditions [16]. Hybrid approaches combining PPO with Model Predictive Control (MPC) have shown improved responsiveness and short-term trajectory planning [18]. Recent innovations include scalable lane-level controllers effective across diverse urban environments [19], and multi-objective DRL frameworks that jointly optimize safety, efficiency, and environmental impact [20]. DRL variants enhanced with channel attention and LSTM modules have also been proposed to reduce queue lengths and carbon emissions [21].

Limitations: Event Awareness and Crowdsourced Input

Despite these advancements, most current systems remain optimized for recurrent traffic patterns and often lack mechanisms for rapid adaptation to non-recurrent events, such as accidents, road closures, or construction activities. Few models incorporate real-time incident streams from crowdsourced sources like Waze, even though such platforms have demonstrated value in enhancing traffic response strategies. Moreover, although GNN-MARL systems offer scalability in simulation, real-world deployments rarely include dynamic, real-time event feedback loops.

In summary, while the integration of DRL, MARL, and GNN has led to significant progress in adaptive traffic signal control, there remains a critical gap in incorporating real-time, crowdsourced event data. To the best of our knowledge, no existing system combines: (1) real-time event input from platforms such as Waze; (2) GNN-based spatial modeling of dynamic traffic networks; and (3) a meta-agent or multi-agent RL architecture capable of fast adaptation to disruptions. This work aims to address these limitations by introducing a unified, event-aware GNN-MARL framework designed for real-world, scalable traffic signal control.

3   Methodology

Problem Definition and Mathematical Modeling: We model the urban traffic network as a directed graph $G = (V, E)$, where V denotes the set of nodes representing traffic intersections and E is the set of edges corresponding to road segments connecting these intersections. The goal is to optimize traffic signal control policies at each intersection to minimize overall vehicle delay, queue lengths, and maximize traffic throughput. At each discrete time step t, the state of node $v \in V$ is described by a feature vector:

$$s_v^t = \left[q_v^t, w_v^t, p_v^t, \ldots \right] \tag{1}$$

where q_v^t denotes the queue length at intersection v, w_v^t is the average waiting time, and p_v^t indicates the current traffic signal phase. Additional features may also include vehicle flow or speed on adjacent edges. The global network state at time t is $S^t = \{s_v^t\}_{v \in V}$.

Each edge $e = (u, v) \in E$ has a dynamic weight w_e^t reflecting traffic conditions and incident impacts, modeled as:

$$[label =] w_e^t = \alpha \cdot f_{\text{flow}}(e, t) + \beta \cdot f_{\text{incident}}(e, t) \tag{2}$$

where $f_{\text{flow}}(e, t)$ represents the traffic flow or congestion level on edge e at time t, and $f_{\text{incident}}(e, t) \in [0, 1]$ quantifies the severity of incidents reported via crowd-sourced platforms such as Waze. The weights α and β, both non-negative, are used to balance the relative influence of flow and incident information.

Graph Neural Network for Spatial-Temporal Traffic State Prediction: To capture spatial dependencies and forecast near-future traffic states, we employ GNNs. For node v at GNN layer l, let the feature vector be $h_v^{(l)} \in \mathbb{R}^d$. A Graph Convolutional Networks (GCNs) layer updates node embeddings as [22]:

$$h_v^{(l+1)} = \sigma \left(\sum_{u \in \mathcal{N}(v) \cup \{v\}} \frac{1}{\sqrt{\hat{d}_v \hat{d}_u}} W^{(l)} h_u^{(l)} \right) \tag{3}$$

where $\mathcal{N}(v)$ denotes the set of neighbors of node v, $\hat{d}_v$ is its degree in the symmetrically normalized adjacency matrix with self-loops, $W^{(l)}$ is a learnable weight matrix at layer l, and $\sigma(\cdot)$ denotes a non-linear activation function, such as ReLU.

Alternatively, GATs compute learnable attention weights $\alpha_{vu}^{(l)}$ [23]:

$$e_{vu}^{(l)} = \text{LeakyReLU} \left(a^{(l)T} \left[W^{(l)} h_v^{(l)} \, \| \, W^{(l)} h_u^{(l)} \right] \right) \tag{4}$$

$$\alpha_{vu}^{(l)} = \frac{\exp(e_{vu}^{(l)})}{\sum_{k \in \mathcal{N}(v)} \exp(e_{vk}^{(l)})} \tag{5}$$

where $a^{(l)}$ is a learnable vector and $\|$ denotes vector concatenation. The final layer's output $h_v^{(L)}$ is used to predict the future state $\hat{s}_v^{t+1}$.

MARL for Traffic Signal Control. Each intersection v is controlled by an independent agent that selects an action $a_v^t \in \mathcal{A}_v$, corresponding to a traffic signal phase. At time t, the agent observes a local state s_v^t, which is augmented by the GNN-predicted future state $\hat{s}_v^{t+1}$. Its policy is parameterized by $\pi_v(a_v^t \mid s_v^t; \theta_v)$, where θ_v denotes the agent's learnable parameters. Agents aim to maximize the expected cumulative discounted reward:

$$J_v(\theta_v) = \mathbb{E}_{\pi_v} \left[\sum_{t=0}^{T} \gamma^t r_v^t \right] \tag{6}$$

with discount factor $\gamma \in [0, 1)$, and reward at time t defined as:

$$r_v^t = -\left(\lambda_1 q_v^t + \lambda_2 w_v^t + \lambda_3 \Delta p_v^t\right) \tag{7}$$

where q_v^t denotes the queue length, w_v^t is the waiting time, and Δp_v^t penalizes frequent traffic signal phase changes. The coefficients $\lambda_i \geq 0$ are used to weight the contribution of each reward component. Policy gradient methods such as PPO or A2C optimize the policy by estimating:

$$\nabla_{\theta_v} J_v(\theta_v) \approx \mathbb{E}_{s_v, a_v \sim \pi_v} \left[\nabla_{\theta_v} \log \pi_v(a_v \mid s_v; \theta_v) A_v(s_v, a_v)\right] \tag{8}$$

where $A_v(s_v, a_v)$ is the advantage function.

Event-Triggered Adaptive Control Using Crowdsourced Data: When a traffic event e is detected from crowdsourced platforms (e.g., Waze) at time t_0, define the indicator function:

$$\mathbb{I}_{\text{event}}(e, t) = \begin{cases} 1, & \text{if event at edge } e \text{ at time } t \\ 0, & \text{otherwise} \end{cases} \tag{9}$$

which modifies edge weights and agent rewards:

$$r_v^t = -\left(\lambda_1 q_v^t + \lambda_2 w_v^t + \lambda_3 \Delta p_v^t\right) - \lambda_4 \cdot \mathbb{I}_{\text{event}}(e, t) \tag{10}$$

with $\lambda_4 \geq 0$ encouraging agents to prioritize intersections affected by incidents. A meta-controller can coordinate multiple agents in incident-affected regions to improve global traffic flow and rapid responsiveness.

Training and Deployment Workflow. The proposed system operates through a three-stage workflow. In the offline training phase, GNN-based traffic predictors and MARL agents are trained within simulation environments such as SUMO, using both historical traffic records and synthetic incident scenarios. During online deployment, real-time traffic and incident data are continuously fed into the GNN to generate short-term state predictions, which are then used by agents to make real-time signal control decisions. To ensure robustness, the system incorporates an adaptive control mechanism in which event-triggered reward adjustments enable agents to rapidly respond to unexpected traffic disruptions.

3.1   System Overview of the Proposed Integrated GNN-MARL Traffic Control System

This subsection outlines the high-level architecture of the proposed intelligent traffic signal control framework, which integrates real-time crowdsourced data acquisition, graph-based spatialâĂŞtemporal modeling, MARL for decision-making, and event-driven adaptability for handling non-recurrent incidents.

As illustrated in Fig. 1, traffic data is collected at high frequency from Waze, then preprocessed and encoded into a graph representation. A Graph Neural Network (GNN) captures the spatial relationships among intersections and generates short-term traffic forecasts, such as queue lengths, delays, and congestion levels. These forecasts, combined with the current traffic state, are provided to a decentralized MARL controller. In this setup, each agent is responsible for managing one intersection and learns a policy $\pi_v(a_v \mid s_v)$ that maps local observations to optimal signal control actions. The agents are trained using actorâŞcritic and policy-gradient methods, typically under centralized training and decentralized execution.

To enhance responsiveness in dynamic traffic environments, the framework includes an event-triggered mechanism. When unexpected incidents, such as accidents or road closures, are reported through Waze, the system updates edge weights in the graph, refines GNN-based predictions, and dynamically adjusts the reward functions of affected agents. This close integration between predictive modeling (GNN) and adaptive decision-making (MARL), guided by real-time event data, enables proactive and scalable traffic signal control. The framework is implemented and evaluated using a realistic traffic simulation platform, SUMO.

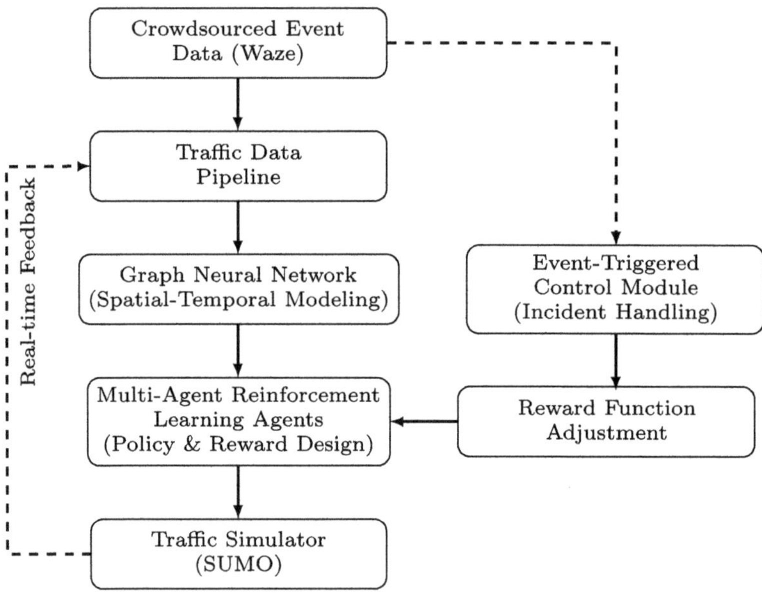

Fig. 1. Block diagram of the proposed ITS system integrating crowdsourced event data, GNN-based traffic modeling, MARL agents with adaptive reward design, and traffic simulation.

3.2 Simulation Environment

The traffic simulation in this study is conducted in a densely congested urban area surrounding Tan Son Nhat International Airport, Ho Chi Minh City, Vietnam. The selected rectangular region spans approximately 5.5km by 2.5km, covering a total area of 13.75km^2. This area serves as both a major access zone to the airport and a key commuter corridor linking multiple urban districts as illustrated in Fig. 2.

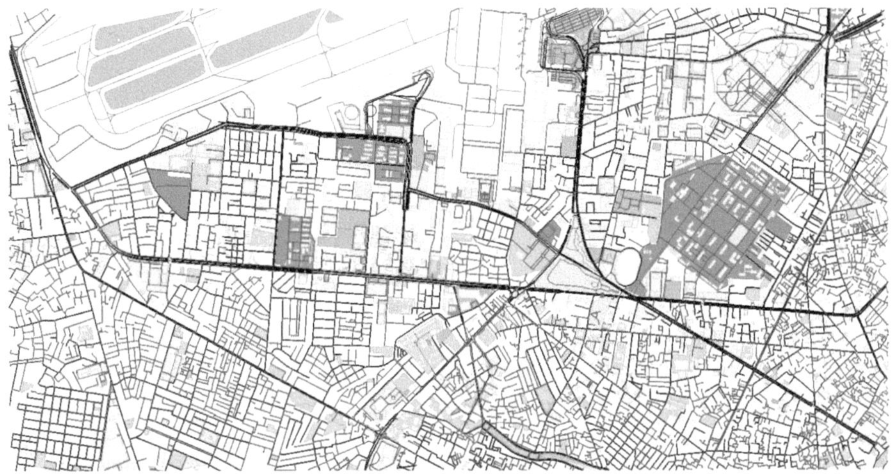

Fig. 2. Study area surrounding Tan Son Nhat International Airport used for simulation, including major arterial roads such as Cong Hoa, Truong Chinh, and the Lang Cha Ca roundabout. The rectangular region (5.5km × 2.5km) captures the airport terminal access zone and high-density commuter flows during peak hours.

Geographically, the area includes several major arterial roads and traffic hubs. The northern boundary features Tan Son Nhat International Airport and the newly operational Terminal 3, which has increased regional travel demand. To the east and southeast lie Cong Hoa and Truong Chinh Streets, two of the city's busiest corridors. The southern and western edges are marked by Hoang Van Thu Park and Lang Cha Ca roundabout, where several high-traffic routes intersect. The area also incorporates the newly constructed Tran Quoc HoanâĂŞCong Hoa connector, designed to alleviate congestion and improve access to Terminal 3. It experiences consistently high traffic density, particularly during peak hours, due to commuter flows between suburban areas (e.g., An Suong roundabout, District 12) and central districts (e.g., District 1 and Phu Nhuan District). Additionally, cross-city traffic from Pham Van Dong Boulevard passes through this zone en route to other key areas. This region was selected due to its strategic importance and representative traffic complexity. Its combination of airport access, daily commuter flows, and new infrastructure makes it a suitable testbed for evaluating adaptive, event-aware traffic control strategies within a SUMO-based simulation framework.

High-Priority Control Zone: Hoang Van Thu Park Area.
Within the defined simulation boundary, the area surrounding Hoang Van Thu Park plays a particularly critical role in shaping overall traffic dynamics. Located

immediately to the south of Tan Son Nhat International Airport, this urban green space is encircled by a dense network of arterial roads that serve both airport access and intra-city commuting, as illustrated in Fig. 3.

Fig. 3. Geographic layout of the Hoang Van Thu Park area and surrounding arterial roads within the simulation boundary.

The park lies at the convergence of several major traffic corridors, including Hoang Van Thu, Phan Dinh Giot, and Nguyen Van Troi Streets, forming a complex junction that transitions into the Lang Cha Ca roundabout. This roundabout is one of the most congested and operationally significant intersections in Ho Chi Minh City, serving as a nexus for flows arriving from the west (Cong Hoa and Truong Chinh streets), the north (airport terminals), and the southeast (District 3 and Phu Nhuan). Its adjacency to both Terminal 1 and Terminal 3 further increases its strategic importance.

Notably, the Hoang Van Thu Park area functions not only as a traffic convergence point, but also as a modal transition zone where various transport services overlap. The area supports a mix of private vehicles, taxis, airport shuttles, ride-hailing services, and buses. This heterogeneity, combined with limited road surface area and frequent pedestrian crossings, introduces highly dynamic and often nonlinear congestion patterns, especially during peak arrival and departure periods at the airport.

From a traffic control perspective, this area presents several unique challenges:

- **Phase coordination complexity:** Multiple signalized intersections within a 200âĂŞ300 m radius require tightly synchronized phase transitions to prevent gridlocks, particularly during surges in inflows from the roundabout.
- **Incident sensitivity:** Minor disruptions such as illegal parking or ride-hailing pick-ups can trigger significant delays due to high density and limited alternative routing.

- **Spillback risk:** Queues near the park often spill back into upstream links, including airport access roads, creating critical interdependencies that demand predictive and preemptive control.

Given these characteristics, the Hoang Van Thu Park area is treated as a high-priority control zone within the simulation. Agents assigned to intersections in this subregion are equipped with enhanced prediction modules, incorporating short-term traffic forecasts from GNNs and real-time disruption signals from crowdsourced data. The objective is to ensure that signal policies adapt not only to routine congestion but also to unplanned incidents and temporal traffic surges.

This focus enables the evaluation of how effectively the proposed multi-agent control framework can maintain flow stability and minimize delays in one of the city's most operationally demanding locations.

To further enhance controllability in this area, the framework also considers the deployment of additional traffic signals at currently unsignalized but high-impact junctions, as illustrated in Fig. 4. These new signal installations are selected based on traffic density, incident frequency, and structural feasibility. Their inclusion allows for finer-grained control and better integration with the wide-area signal network, enabling the proposed system to execute event-aware strategies more effectively.

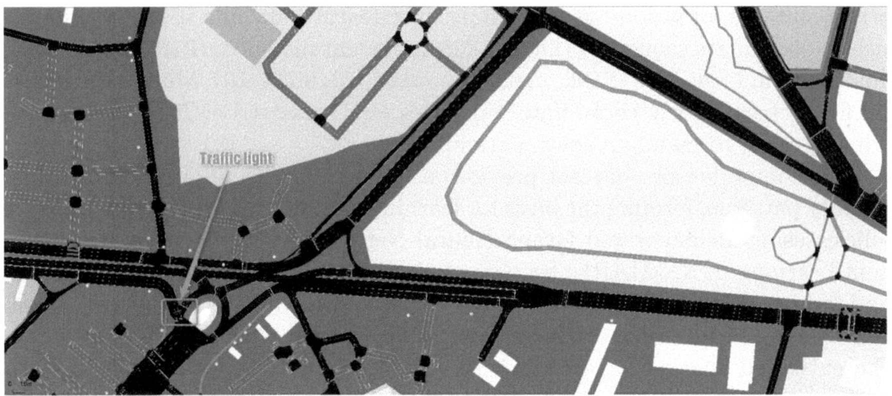

Fig. 4. Traffic convergence zone near Hoang Van Thu Park, showing key intersections, roundabout (Lang Cha Ca), and proposed locations for additional traffic signal installation to support adaptive control strategies.

4   Experiments and Evaluation

4.1   Dataset Description

This study employs a real-world traffic dataset collected from Ho Chi Minh City, Vietnam, sourced via the Waze platform [12], which aggregates user-generated reports to reflect live traffic conditions. The dataset consists of structured JSON objects, each representing a timestamped origin-to-destination (O-D) routing query. These queries include multiple routing alternatives, along with rich metadata describing route characteristics and estimated travel times under various traffic scenarios, including real-time, historical, and free-flow conditions.

To support real-time traffic control applications, data were sampled at a high temporal resolution of five-minute intervals, resulting in twelve observations per hour. Each O-D pair generated approximately 204 routing queries per day, collected across a 17-hour operating window. The data collection campaign spanned 45 d, yielding a corpus of roughly 226,800 trajectory records. Each record includes detailed segment-level information such as geographic paths (latitude-longitude sequences), road segment identifiers, road classification codes, segment lengths (in meters), and street names. To mitigate data sparsity and inaccuracies, we employed preprocessing techniques such as outlier filtering, duplicate removal, and exclusion of routing queries with incomplete or inconsistent metadata. To address incorrect or delayed Waze reports, we applied temporal smoothing and discarded late data outside the real-time decision window.

In addition to spatial details, each record contains a timestamp field from which structured temporal features are derived. These include time-of-day, day-of-week, date index, and week index, allowing the model to capture both short-term variations and long-term seasonal patterns in traffic behavior. The raw JSON data are systematically parsed and converted into a structured tabular format suitable for graph-based learning. Key features include spatial attributes such as coordinate sequences (LATS, LNGS), segment distances (DIST, DIST_gap), and temporal identifiers (TIME, time_gap, time_ID, WEEK_ID). Multi-valued spatial fields, particularly coordinate sequences, are preserved as lists within individual records to maintain spatial granularity.

This comprehensive dataset provides a high-fidelity representation of urban mobility patterns, forming the basis for learning coordinated traffic signal control policies using an integrated Graph Neural NetworkâĂŞMulti-Agent Reinforcement Learning (GNN-MARL) framework. In this setting, traffic intersections are modeled as agents, while the road network is represented as a dynamic graph whose edges encode segment-level connectivity and traffic flow characteristics. The extracted spatio-temporal features enable the GNN components to learn localized structural dependencies, while the MARL agents adaptively optimize signal actions in response to evolving traffic conditions.

4.2 Experimental Setup

We evaluate the proposed Integrated GNN-MARL Traffic Control System using the dataset described in Subsect. 4.1, which reflects real-world traffic conditions in Ho Chi Minh City over a 45-day period, with data sampled at five-minute intervals. Simulations are conducted in the SUMO (Simulation of Urban MObility) environment, configured to mirror a multi-intersection urban road network derived from the original data.

Two traffic signal control strategies are compared: (i) a Fixed-Time Control baseline that operates based on static scheduling, and (ii) our proposed GNN-MARL controller, which leverages graph-based traffic state representation and decentralized reinforcement learning to adapt signal timings in real time. Each simulation scenario spans 3000 s, allowing the system to capture dynamic traffic fluctuations during both peak and off-peak periods. Performance is assessed based on standard metrics such as average vehicle delay, throughput, and queue length.

4.3   Evaluation Metrics

The evaluation employed key performance indicators extracted from simulation logs to quantify system effectiveness. These metrics include Average Vehicle Speed (m/s), representing traffic flow fluidity; Trip Duration (s), the total travel time for vehicles completing their journeys; Waiting Time (s), indicating the duration vehicles remain idle at traffic signals; Time Loss (s), the additional travel time incurred due to congestion or delays; the number of Vehicles Completed, measuring throughput; and Teleports, the count of vehicle repositioning events by the simulator to resolve deadlocks, serving as an indicator of traffic anomalies. Together, these metrics provide a comprehensive assessment of traffic efficiency, congestion levels, and overall system robustness.

4.4   Results and Analysis

Simulation results reveal the clear advantages of the proposed GNN-MARL framework compared to traditional fixed-time control, as summarized in Table 1 and illustrated in Fig. 5 through a percentage improvement chart. The average vehicle speed increased by 1.92%, rising from 13.04 m/s to 13.29 m/s. Although modest in absolute terms, such improvements at the city scale translate into meaningful time savings and reduced emissions. Trip duration was reduced by 3.49%, from 251.83 s to 243.05 s, demonstrating the system's effectiveness in smoothing traffic flow and minimizing stop-and-go patterns. Notably, vehicle waiting time at intersections decreased substantially by 25.20%, from 32.42 s to 24.25 s, reflecting more efficient traffic signal phase allocation and enhanced driver experience. Time loss also declined by 9.68%, indicating a reduction in congestion and improved throughput. The number of vehicles successfully completing their trips increased marginally by 0.03%, confirming that throughput was maintained or slightly improved without compromising safety or inducing gridlocks. Importantly, the frequency of teleport events was dramatically reduced from seven to one, signifying enhanced system stability and smoother traffic conditions.

Table 1. Comparison between Fixed-Time and Proposed GNN-MARL Traffic Control

Metric	Fixed-Time	Proposed Method
Average Speed (m/s)	13.04	**13.29**
Average Trip Duration (s)	251.83	**243.05**
Average Waiting Time (s)	32.42	**24.25**
Time Loss (s)	86.04	**77.71**
Vehicles Inserted	3212	3211
Vehicles Completed	2929	**2930**
Teleports (Total)	7	**1**
Collisions	0	0
Real-Time Factor	19.80	**27.53**
Vehicle Updates per Second	5164.88	**6945.95**

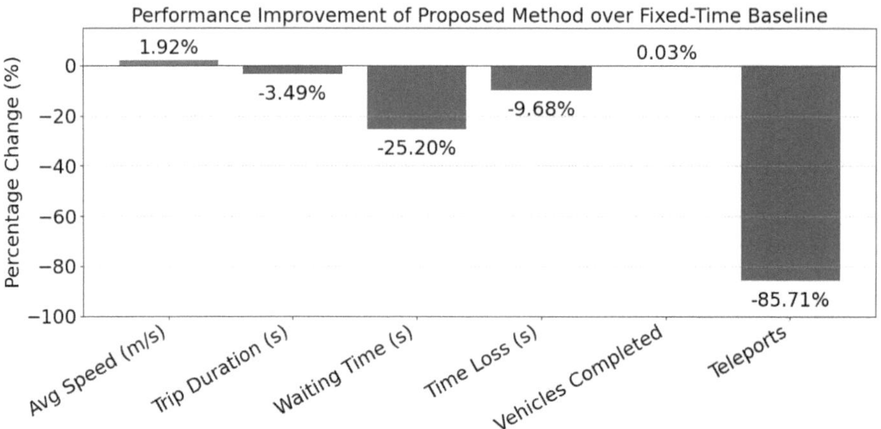

Fig. 5. Percentage change in traffic performance metrics achieved by the proposed GNN-MARL method compared to the fixed-time baseline.

4.5   Discussion

The comparative analysis depicted in Fig. 5 highlights consistent performance improvements achieved by the proposed approach across all evaluated metrics, particularly in reducing waiting time and enhancing traffic stability. The integration of GNNs enables effective modeling of spatial dependencies within the traffic network, while the MARL framework allows dynamic adaptation of control policies in response to evolving traffic conditions. In contrast to static, unresponsive fixed-time control, the hybrid GNN-MARL method exhibits significant advancements in managing urban traffic congestion, minimizing delays, and boosting throughput. The notable decrease in teleport events underscores the system's capacity to mitigate gridlock occurrences, a critical factor for real-world deployment. These results underscore the potential of combining graph-based deep learning with reinforcement learning techniques within Intelligent Transportation Systems, addressing fundamental limitations inherent in traditional traffic control strategies.

5   Conclusion

This paper has presented an adaptive traffic signal control framework that integrates GNNs, MARL, and real-time crowdsourced event data from Waze to improve urban traffic management. By leveraging spatial-temporal traffic prediction and decentralized control, the system effectively reduces congestion, vehicle waiting times, and traffic deadlocks, as demonstrated in extensive SUMO simulations with real-world traffic data from Ho Chi Minh City. The event-triggered adaptation mechanism further enhances the system's responsiveness to dynamic incidents, ensuring robust traffic flow under varying conditions.

Limitations and Future Work
While the proposed framework demonstrates promising results in simulation-based evaluations, several limitations remain. The current implementation has been tested on a limited-scale urban network, and its scalability to larger and more complex traffic systems has yet to be thoroughly evaluated. However, the framework has potential to scale by incorporating detailed analyses of relevant constraints and accurately assessing their impact on large-scale traffic dynamics. These aspects will be addressed in future work to ensure robust scalability and applicability. In addition, the framework primarily targets vehicular traffic, without explicitly modeling multimodal participants such as pedestrians, cyclists, or public transit. Moreover, although real-time crowdsourced event data is incorporated, integration with other real-time traffic sensing modalities (e.g., loop detectors, cameras, GPS traces) remains limited.

Future work will address deployment challenges by scaling the system to larger urban areas, incorporating diverse traffic participants, and integrating richer sensor data for improved robustness and adaptability. The proposed framework is low latency and lightweight, leveraging high-frequency Waze data for real-time conditions and a computationally efficient GNN-MARL architecture. By focusing on representative road segments, it reduces inference workload and fits existing infrastructure like traffic cameras and loop detectors. Additionally, handling non-recurrent events like public gatherings, roadworks, or extreme weather, and validating the system in real-world deployments, will be key to demonstrating its practical feasibility and impact.

Acknowledgments. Do Thanh Thai was funded by the Master, PhD Scholarship Programme of Vingroup Innovation Foundation (VINIF), code VINIF.2024.TS.073.

Disclosure of Interests. No potential conflict of interest was reported by the author(s).

References

1. Sims, A.G.: The Sydney Coordinated Adaptive Traffic System. In: Engineering Foundation Conference on Research Directions in Computer Control of Urban Traffic Systems, Pacific Grove, California, USA (1979)
2. Hunt, P.B., Robertson, D.I., Bretherton, R.D., Winton, R.I.: SCOOT – A traffic responsive method of coordinating signals (No. LR 1014 Monograph). Transport and Road Research Laboratory (1981)
3. Liang, X., Du, X., Wang, G., Han, Z.: Deep Reinforcement Learning for Traffic Light Control in Vehicular Networks. arXiv (2018)
4. Gao, J., Shen, Y., Liu, J., Ito, M., Shiratori, N.: Adaptive Traffic Signal Control: Deep Reinforcement Learning Algorithm with Experience Replay and Target Network. arXiv (2017)
5. Liu, X.Y., Zhu, M., Borst, S., Walid, A.: Deep Reinforcement Learning for Traffic Light Control in Intelligent Transportation Systems. arXiv (2023)
6. Wei, H.: CoLight: Learning network-level cooperation for traffic signal control. In: Proceedings of the 28th ACM International Conference on Information Knowledge Management (CIKM), pp. 1913–1922 (2019)
7. Wei, H., et al.: PressLight: learning max pressure control to coordinate traffic signals in arterial network. In: Proceedings of the 25th ACM SIGKDD International Conference on Knowledge Discovery and Data Mining, pp. 1290–1298 (2019)
8. Wang, Z., Yang, K., Li, L., Lu, Y., Tao, Y.: Traffic signal priority control based on shared experience multi-agent deep reinforcement learning. IET Intel. Transport Syst. **17**(7), 1363–1379 (2023)
9. Hu, X., Zhao, C., Wang, G.: A Traffic Light Dynamic Control Algorithm with Deep Reinforcement Learning Based on GNN Prediction (GPlight). arXiv (2020)
10. Yang, G., Wen, X., Chen, F.: Multi-Agent deep reinforcement learning with graph attention network for traffic signal control in multiple-intersection urban areas. Transp. Res. Rec. (2025)
11. Devailly, F.X., Larocque, D., Charlin, L.: IG-RL: inductive graph reinforcement learning for massive-scale traffic signal control. IEEE Trans. Intell. Transp. Syst. **23**(7), 7496–7507 (2021)
12. Waze. https://waze.com. Accessed June 2025
13. Park, S., Han, E., Park, S., Jeong, H., Yun, I.: Deep Q-network-based traffic signal control models. PLOS ONE **16**(9), e0256405 (2021)
14. Zhao, H., et al.: A survey on deep reinforcement learning approaches for traffic signal control. Eng. Appl. Artif. Intell. **133**, 108100 (2024)
15. Yu, J., Wang, Z., Zhang, R.: DiffusionLight: a multi-agent reinforcement learning approach for traffic signal control based on shortcut-diffusion model. Appl. Intell. **55**(6), 1–25 (2025)
16. Han, M., et al.: Leveraging reinforcement learning for dynamic traffic control: survey and challenges. Commun. Transp. Res. **3**, 100104 (2023)
17. Li, Y., Zhang, Y., Li, X., Sun, C.: Regional multi-agent cooperative reinforcement learning for city-level traffic grid signal control. IEEE/CAA J. Autom. Sinica **11**(9), 1987–1998 (2024)
18. Guo, G., Wang, Y.: An integrated MPC and deep reinforcement learning approach to trams-priority active signal control. Control Eng. Pract. **110**, 104758 (2021)
19. J. Shao, C., Zheng, Y., Chen, Y., Huang, Zhang, Y.: MoveLight: Enhancing traffic signal control through movement-centric deep reinforcement learning. arXiv preprint arXiv:2407.17303 (2024)

20. Mossalam, H., Assael, Y.M., Roijers, D.M., Whiteson, S.: Multi-objective deep reinforcement learning. arXiv preprint arXiv:1610.02707 (2016)
21. Zai, W., Yang, D.: Improved deep reinforcement learning for intelligent traffic signal control using ECA_LSTM network. Sustainability **15**(18), 13668 (2023)
22. Kipf, T.N., Welling, M.: Semi-supervised classification with graph convolutional networks. arXiv preprint arXiv:1609.02907 (2016)
23. Veličković, P.: Graph attention networks. arXiv preprint arXiv:1710.10903 (2017)

Intelligent Systems

Towards Local Learning Algorithm
on Apache Spark for Large Datasets

Quoc-Bao Bui-Vo[1] and Thanh-Nghi Do[1,2(✉)]

[1] College of Information and Communication Technology, Can Tho University,
Can Tho City, Viet Nam
bvqbao@cit.ctu.edu.vn

[2] UMI UMMISCO 209 (IRD/UPMC), Sorbonne University, Pierre and Marie Curie
University - Paris 6, Paris, France
dtnghi@cit.ctu.edu.vn

Abstract. This paper presents a scalable local learning algorithm for large-scale classification tasks, implemented on Apache Spark. Rather than relying on a global non-linear model, our proposed approach classifies each test datapoint using a lightweight model trained on its k-nearest neighbors, assuming that local regions are less complex and more easily separable than the full training dataset. To achieve scalability, both training and test datasets are distributed across a Spark cluster, enabling parallel distance computation and independent local model training for each test datapoint. Experimental results on the ImageNet dataset, which comprises 1,281,167 images across 1,000 classes, show that the algorithm, executed on two PCs, completed the classification task in 14.7 h and achieved an accuracy of 89.58%, demonstrating its effectiveness and efficiency on large-scale data.

Keywords: Classification of large datasets · Local learning algorithm · Apache Spark

1 Introduction

Classification task is one of the most fundamental problems in the field of machine learning and data mining. It refers to assign datapoints into pre-defined classes based on their features. A classification algorithm learns a model from labeled training data to make class predictions. The classification plays an important role in understanding patterns and relationships within data. It serves as the foundation for many theoretical studies in artificial intelligence, statistical learning, and computational intelligence. In practice, data classification is widely applied into various domains such as healthcare, finance, cybersecurity, and marketing (e.g., customer segmentation), making it highly valuable in real-world decision-making processes.

Despite its wide applicability, data classification faces several significant challenges, particularly when dealing with large-scale and complex datasets. Efficiently handling and processing large volumes of data in a reasonable time

N. Thai-Nghe et al. (Eds.): ISDS 2025, CCIS 2714, pp. 289–300, 2026.
https://doi.org/10.1007/978-981-95-3358-9_21

remains a big issue. Recent advances focus on developing scalable and efficient algorithms capable of handling high-dimensional, massive datasets, often under resource and time constraints. Linear classification methods, such as those discussed in [18], and linear support vector machines [9], large-scale logistic regression [12] using Apache Spark [19], emphasize the importance of model simplicity and computational efficiency. These methods leverage distributed frameworks like Apache Spark to scale-up linear classifiers for large datasets. To further enhance scalability, optimization techniques like truncated Newton methods for linear classification have been proposed in [10] to reduce computation in large datasets. Meanwhile, Do and his coleagues [5–7] proposed parallel algorithms of local support vector machines for dealing with classification and regression tasks of large datasets. The studies in [1,13] outline the key challenges and opportunities associated with applying machine learning techniques to big data. Mousaarab et al. [14] highlight the increasing demand for computational resources and robust model generalization in handling complex classification problems within big data environments. As concerns over data privacy and distribution grow, Konecny et al. [11] introduced federated learning as a transformative approach—enabling large-scale model training across decentralized data sources without sharing raw data, thereby opening new avenues for collaborative and privacy-preserving classification.

Our investigation aims to develop the local learning algorithm on Apache Spark for large-scale classification tasks. Instead of training a global model, the algorithm focuses on a small, relevant subset of the training data—specifically, the k-nearest neighbors of a given test datapoint. By assuming approximate local near linear separability, it trains a simple local classifier using only these neighbors, reducing model complexity while still effectively capturing the local decision boundary. To parallelize the local learning algorithm on Apache Spark, the training and test datasets are distributed across the cluster as RDD partitions. For each test datapoint, Spark computes distances to training datapoints in all partitions in parallel to identify its k-nearest neighbors. A lightweight local model is then trained independently for each test datapoint using its neighbors. This parallel design ensures scalability and efficiency when handling large-scale classification tasks. The experimental results on the ImageNet dataset [4], which comprises 1,281,167 images across 1,000 classes, demonstrate that our local learning algorithm, implemented on Apache Spark and executed across two PCs, completed the classification task in 14.7 h, achieving an accuracy of 89.58%.

The remainders of this paper are organized as follows. Section 2 describes the development of the local learning algorithm on Apache Spark to efficiently handle large-scale datasets. Section 3 shows the experimental results before conclusions and future works presented in Sect. 4.

2   Local Learning Algorithm on Apache Spark

Let consider a non-linear binary classification task shown in Fig. 1. The dataset consists of datapoints belonging to two distinct classes: the positive class (represented by squares), and the negative class (depicted as circles). These datapoints are not linearly separable. To address this classification task, a machine learning algorithm tries to learn a nonlinear classification model with a complex decision boundary. One such approach may be a non-linear Support Vector Machine (SVM [17]) with a Radial Basis Function (RBF) kernel. Once the non-linear classifier has been trained, it can be used to predict the class labels of new or unseen test datapoints, such as points denoted by $\mathbf{x}$ or $*$. The model evaluates the position of these test points relative to the learned decision boundary to assign them to either the positive or negative class. This is a common approach for training classification models.

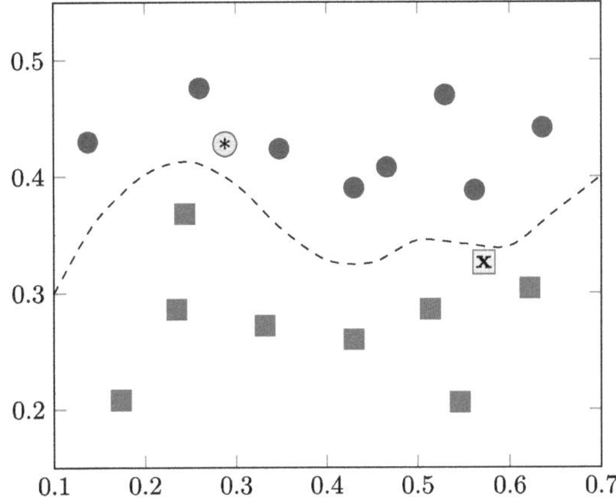

Fig. 1. Non-linear classification task

2.1   Local Learning Algorithm

Our local learning algorithm (illustrated in Fig. 2) addresses the classification task by focusing on a small, relevant subset of the training data. Specifically, for a given test datapoint $\mathbf{x}$, the algorithm first identifies its k-nearest neighbors (e.g., $k = 3$), based on a predefined distance metric, such as Euclidean distance or cosine similarity. This step effectively transforms the global classification problem into a localized one by narrowing the learning scope to only a few datapoints that are most similar to the test datapoint.

By focusing only on the k nearest neighbors, the algorithm assumes that the separation within this local region is simpler than in the entire dataset. It then proceeds to train a local classification model, denoted as H^1, using just these k neighbors. This model is significantly less complex than a global non-linear classifier (such as an SVM with a RBF kernel in Fig. 1), yet it is often sufficient to capture the decision boundary in the immediate vicinity of $\mathbf{x}$.

Once the local model H^1 is trained, it is used to predict the class label of the test datapoint $\mathbf{x}$. Because the model is tailored specifically to the local data distribution around $\mathbf{x}$, it can offer improved generalization for heterogeneous datasets where global patterns may not be consistent across the entire feature space. Similarly, for the testing datapoint $*$, the model H^2 is also trained based on the 3 nearest neighbors of $*$ in the training dataset, in order to classify $*$.

This local learning strategy provides a flexible and computationally efficient alternative to global models, especially in cases where the data exhibits varying structures in different regions of the input space. The local algorithm is summarized in Algorithm 1.

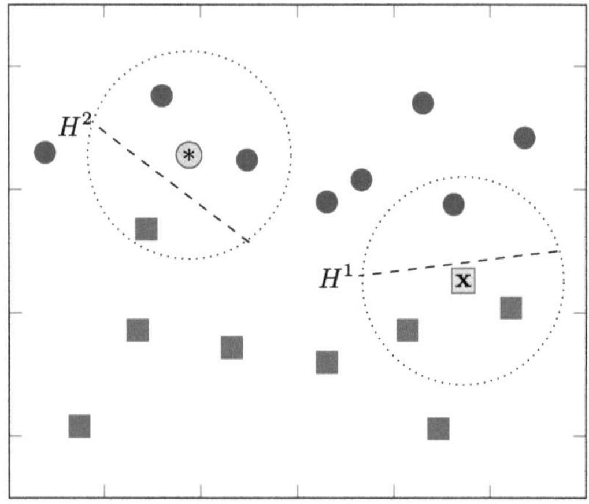

Fig. 2. 3-nearest neighbors local models for classification task.

2.2   Performance Analysis of the Local Learning Algorithm

Let's analyze the performance of local learning algorithms as discussed in the foundational paper by Bottou and Vapnik [3].

The local learning algorithm is a lazy learning strategy where most computation is deferred until prediction time. Let the training dataset consist of m datapoints, in a d-dimensional feature space. During the training phase, there is no actual model training cost. The only computational overhead arises from optional pre-processing steps, such as indexing the training data to efficient find nearest-neighbor at prediction time. For each test datapoint, identifying the k nearest neighbors typically requires $O(\log m)$ time when efficient data structures are used for indexing. Once the local neighborhood is retrieved, a local model such as a linear classifier is trained on these k datapoints. The computational complexity of this step is approximately $O(k \cdot d^2)$. For small values of k, this remains computationally efficient.

Therefore, the total time complexity for making a prediction on a single test datapoint can be expressed as:

$$O(k \cdot d^2 + \log m) \tag{1}$$

Algorithm 1. Local learning algorithm for classifying datapoints

---

input :
        Training dataset D
        Testing dataset T
        Machine learning algorithm ML
        Number of nearest neighbors k
output:
        Predicted classes $\hat{y}$

1 **begin**
2    $\hat{y} = []$
3   **for** $x \in T$ **do**
4   Find k nearest neighbors of $\mathbf{x}$ from D: $<X_{NN}, y_{NN}> = k\mathrm{NN}(\mathbf{x}, D)$
5   Train a local model: $\mathrm{Clf} = \mathrm{ML.fit}(X_{NN}, y_{NN})$
6   Predict the class of $\mathbf{x}$: $p = \mathrm{Clf.predict}(\mathbf{x})$
7       $\hat{y}$.append(p)
8   **end**
9 **end**

The overall algorithmic complexity of the local learning algorithm is linear in the number of test datapoints.

In terms of the generalizability (error rate), the analysis relies on statistical learning theory [17], particularly the VC-dimension concept. The VC-dimension measures the capacity of a class of functions, which serves as a theoretical upper bound on the model's generalization error.

In the case of global models, such as a SVM with a Radial Basis Function (RBF) kernel, the capacity is high because the model tries to fit the entire dataset with a complex boundary. As a result, the associated VC-dimension is also high, which may increase the risk of overfitting, especially in the absence of sufficient regularization.

In contrast, local learning models construct a classifier for each test datapoint using only a small subset of the training data (e.g., its nearest neighbors). When the number of datapoints used to train each local model is small, the effective VC-dimension of the function class applied locally remains low. This reduced complexity in local regions helps improve generalization, as the bound on the generalization error becomes tighter. This leads to the important bound:

$$R(f) \leq R_{emp}(f) + O\left(\sqrt{\frac{h \log(k/h)}{k}}\right) \tag{2}$$

where, $R(f)$ is true risk (expected error), $R_{emp}(f)$ denotes empirical risk on local subset, h refers to the VC-dimension of local model (e.g., for linear classifiers, $h = d+1$), and k corresponds to the number of neighbors (training samples used locally).

Thus, even if the global function is complex, the local models can generalize well, because they are learned from a small neighborhood, using a model class with low VC-dimension. The model adapts to local complexity, avoiding overfitting to the entire dataset. Local linear models maintain a low capacity, yielding better generalization under small k. Nevertheless, local data may be insufficient or noisy, especially near class boundaries. Therefore, if k is too small, the empirical risk may be high due to underfitting. If k is too large, the VC-dimension bound may worsen due to model complexity or local non-linearity. This implies that the number of neighbors k must be carefully selected to minimize the generalization error within local regions, thereby making local learning particularly well-suited for scenarios where training time is limited.

2.3 Local Learning Algorithm on Apache Spark

Given a training dataset D and a testing dataset T, we would like to use the Apache Spark distributed computing model to efficiently search the k nearest neighbors in D and train a local classifier for each test datapoint $\mathbf{x} \in T$. We use the low-level RDD API to implement the kNN algorithm as we have more control over the parallelization. The training dataset is loaded and partitioned into p RDD partitions across the Apache Spark cluster. For each test datapoint, we find its k nearest neighbors on each partition independently and in parallel. We then combine the p lists of k nearest neighbors into a single, global list of neighbors thanks to the $aggregateByKey$ function. We train a local classifier on these neighbors if the latter consists of more than one class. Several optimization strategies are implemented, notably:

- During the kNN searching, the datapoints are not moved or shuffled. We reference a datapoint as a tuple of ($partitionId$, $localId$) where $localId$ is the index of the datapoint in the partition. Once the final list of neighbors is identified, the actual datapoints are collected.
- We process the test datapoints in batches. This helps us accomplish two things: (1) we can train multiple local classifiers in parallel, and (2) we may reduce the number of datapoints collected as there are chances that two datapoints may have some identical neighbors. The Apache Spark implementation of the local learning algorithm is summarized in Algorithm 2.

Algorithm 2. Local learning algorithm on Apache Spark

---

input :
         Training dataset D
         Testing dataset T
         Number of nearest neighbors k
output:
         Predicted classes $\hat{y}$

```
 1  begin
 2  |   ŷ ← []
 3  |   P ← (split D into p RDD partitions)
 4  |   B ← (split T into batches of size β)
 5  |   for batch b ∈ B do
 6  |   |   Broadcast every x ∈ b to all workers
 7  |   |   collecting_points ← {}
 8  |   |   need_classifier ← []
 9  |   |   for datapoint x ∈ b do
10  |   |   |   local_kNN ← P.mapPartitions(p_i => (x, kNN_i))
11  |   |   |   global_kNN ← local_kNN.aggregateByKey()
12  |   |   |   if number of classes in global_kNN ≥ 2 then
13  |   |   |   |   collecting_points.add(global_kNN)
14  |   |   |   |   need_classifier.append((x, global_kNN))
15  |   |   |   else
16  |   |   |   |   p ← The sole class in global_kNN
17  |   |   |   |   ŷ.append(p)
18  |   |   |   end
19  |   |   end
20  |   |   collecting_points.collect()
21  |   |   for (x, global_kNN) ∈ need_classifier do
22  |   |   |   (X_NN, y_NN) ← collecting_points.filter(i => i ∈ global_kNN)
23  |   |   |   classifier ← LogisticRegression.fit(X_NN, y_NN)
24  |   |   |   p ← classifier.predict(x)
25  |   |   |   ŷ.append(p)
26  |   |   end
27  |   end
28  end
```

---

3 Experimental Results

We are interested in the assessment of the local learning algorithm for handling the large-scale classification task. Therefore, it needs to evaluate the performance in terms of training time and classification correctness.

We implemented the local learning algorithm in Python using libraries PySpark, Scikit-learn [16], and joblib. Scikit-learn and joblib are used for parallelizing local models' training. For the local model, Scikit-learn's LinearRegression is used due to its simplicity and the fact that we would like to demonstrate the effectiveness of the local learning algorithm. However, it is not limited to linear models. More complex and non-linear models can be easily integrated into our implementation.

We would like to compare with the best state-of-the-art linear SVM algorithm, LIBLINEAR [9] implemented in C/C++ (the parallel version on multi-core computers with OpenMP [15]). We have developed in C/C++: the k nearest neighbors using the multi-dimensional indexing covertree [2], denoted by covertree-kNN on multi-core computers with OpenMP [15].

To conduct experimental evaluations, we installed Apache Spark in standalone mode on a small-scale PC cluster consisting of two computers. Each machine was configured with Linux Ubuntu 22.04 as the operating system and equipped with an Intel Core i7-14700K processor (20 cores, 28 threads, base clock 3.4GHz, turbo boost up to 5.6GHz, 33MB cache) and 64 GB of main memory. The two nodes were connected via a high-speed Ethernet connection to ensure efficient data exchange and communication during Spark operations. Given the above cluster configuration, the training dataset is split into 56 RDD partitions and the test dataset is processed in batches of size 30.

3.1   Dataset

We carry out the experimental evaluation on the challenging ImageNet ILSVRC2012 dataset [4], which contains 1,281,167 images across 1,000 classes and serves as one of the most widely used benchmarks for image classification.

We use the pre-trained Vision Transformer (ViT [8]), specifically the Base variant with a 16×16 input patch size. The MLP head is removed, and the model is used to extract 768-dimensional feature representations from ImageNet images. The dataset is then split into a training set of 1,024,933 images and a test set of 256,234 images.

We report overall performance using Top-1 accuracy, which measures the percentage of test images correctly classified. Given the 1,000-class setting, the expected accuracy of random guessing is 0.1%.

3.2   Hyper-parameters

We use LIBLINEAR and covertree-kNN as baseline methods for classifying the ImageNet challenge dataset. For LIBLINEAR, we set the regularization parameter to $C = 10^5$, which balances margin maximization and error minimization. The covertree-kNN method is evaluated with $k = 1$. To assess the effectiveness of our proposed kNN-LR algorithm, we experiment with various values of k, analyzing its performance sensitivity with respect to this parameter.

3.3   Classification Results

Although the algorithms are implemented in different programming languages (C/C++ for LIBLINEAR and covertree-kNN, Python for our local learning algorithm), making the speed comparison somewhat unfair, our local learning algorithm still achieves very competitive classification time and performance. Specifically, the results are presented in Table 1 (Fig. 3 and Fig. 4).

Table 1. Classification results

No	Algorithm	Time (min)	Accuracy (%)
1	LIBLINEAR	2,820.25	85.72
2	covertree-kNN	914.01	87.77
3	kNN-LR (k=50)	882.15	89.58

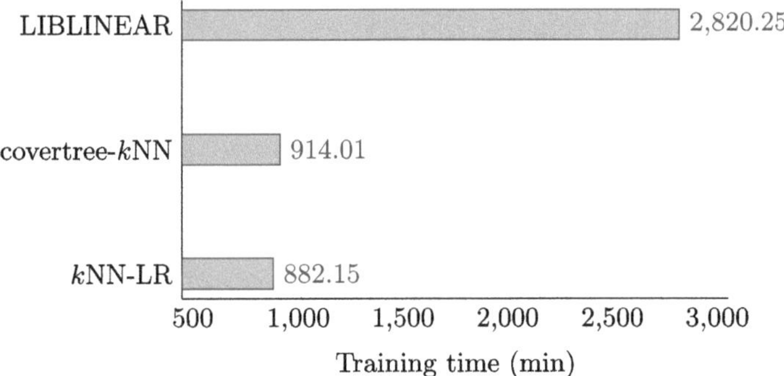

Fig. 3. Training time (min)

We perform experiments for several values of k to evaluate the effects of the number of neighbors k in the algorithm. The results are shown in Fig. 5. The best result in terms of classification time and accuracy is at $k = 50$. As k increases from 50 to 300, the classification time increases dramatically (by approximately 47.94%) and the accuracy drops very slightly (by approximately 0.32%). The accuracy getting worse is expected due to the fact that the local non-linearity slowly appears, as discussed. The significant increase in classification time comes mainly from the collection of the final k nearest neighbors of each test point, where network I/O may be involved (Fig. 5).

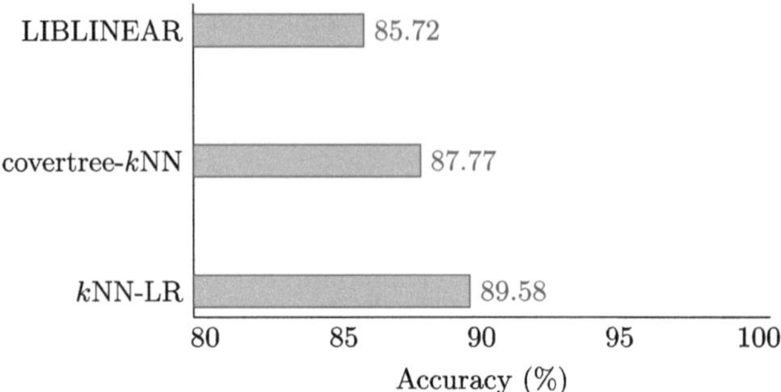

Fig. 4. Overall classification accuracy.

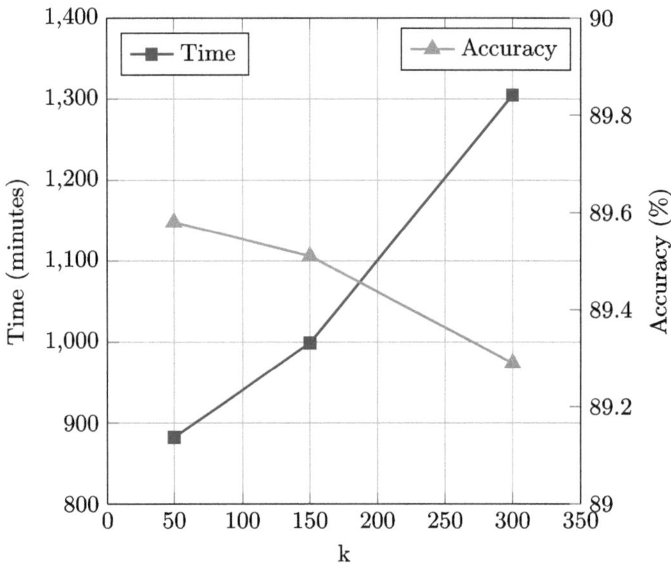

Fig. 5. Effects of k on classification time and accuracy.

4   Conclusion and Future Works

We have introduced a scalable local learning algorithm for large-scale classification tasks, implemented on Apache Spark. Instead of building a global model, our local learning algorithm focuses on training lightweight local classifiers using the k-nearest neighbors of each test datapoint, assuming that local regions exhibit lower complexity and are more easily separable than the entire training dataset. By parallelizing the neighbor search across a Spark cluster and the local model training using joblib, our approach effectively handles large datasets with improved scalability and efficiency. Experimental results on the ImageNet dataset demonstrated the effectiveness of the method, achieving 89.58% accuracy while completing the classification task within 14.7 h across two PCs.

For future work, several promising directions can be explored. First, optimizing the neighbor search process could further reduce computational costs. Second, integrating more sophisticated local models, such as small neural networks or ensemble methods, may improve classification performance. Finally, investigating the use of federated and privacy-preserving techniques in the local learning context could enhance data security while maintaining scalability.

Acknowledgments. This research has received support from the European Union's Horizon research and innovation programme under the MSCA-SE (Marie Skłodowska-Curie Actions Staff Exchange) grant agreement 101086252; Call: HORIZON-MSCA-2021-SE-01; Project title: STARWARS (STormwAteR and WastewAteR networkS heterogeneous data AI-driven management).

References

1. Machine learning on big data: opportunities and challenges. Neurocomputing **237**, 350–361 (2017)
2. Beygelzimer, A., Kakade, S., Langford, J.: Cover trees for nearest neighbor. In: Proceedings of the 23rd international conference on Machine learning, pp. 97–104. ACM (2006)
3. Bottou, L., Vapnik, V.: Local learning algorithms. Neural Comput. **4**(6), 888–900 (1992)
4. Deng, J., Berg, A.C., Li, K., Fei-Fei, L.: What does classifying more than 10,000 image categories tell us? In: Daniilidis, K., Maragos, P., Paragios, N. (eds.) ECCV 2010. LNCS, vol. 6315, pp. 71–84. Springer, Heidelberg (2010). https://doi.org/10.1007/978-3-642-15555-0_6
5. Do, T.-N.: Non-linear classification of massive datasets with a parallel algorithm of local support vector machines. In: Le Thi, H.A., Nguyen, N.T., Do, T.V. (eds.) Advanced Computational Methods for Knowledge Engineering. AISC, vol. 358, pp. 231–241. Springer, Cham (2015). https://doi.org/10.1007/978-3-319-17996-4_21
6. Do, T., Bui, L.: Parallel learning algorithms of local support vector regression for dealing with large datasets. Trans. Large Scale Data Knowl. Centered Syst. **41**, 59–77 (2019)
7. Do, T.N., Poulet, F.: Parallel learning of local SVM algorithms for classifying large datasets. Trans. Large-Scale Data Knowl. Centered Syst. **31**, 67–93 (2016)
8. Dosovitskiy, A., et al.: An image is worth 16×16 words: transformers for image recognition at scale. In: 9th International Conference on Learning Representations, ICLR 2021, Virtual Event, Austria, May 3–7, 2021. OpenReview.net (2021)
9. Fan, R., Chang, K., Hsieh, C., Wang, X., Lin, C.: LIBLINEAR: a library for large linear classification. J. Mach. Learn. Res. **9**, 1871–1874 (2008)
10. Galli, L., Lin, C.: A study on truncated newton methods for linear classification. IEEE Trans. Neural Netw. Learn. Syst. **33**(7), 2828–2841 (2022)
11. Konecný, J., McMahan, H.B., Ramage, D.: Federated optimization: distributed optimization beyond the datacenter. ArXiv **abs/1511.03575** (2015)
12. Lin, C., Tsai, C., Lee, C., Lin, C.: Large-scale logistic regression and linear support vector machines using spark. In: 2014 IEEE International Conference on Big Data (IEEE BigData 2014), Washington, DC, USA, October 27-30, 2014, pp. 519–528. IEEE Computer Society (2014)

13. L'Heureux, A., Grolinger, K., Elyamany, H.F., Capretz, M.A.M.: Machine learning with big data: challenges and approaches. IEEE Access **5**, 7776–7797 (2017)
14. Najafabadi, M.M., et al.: Deep learning applications and challenges in big data analytics. J. Big Data **2**(1), 1–21 (2015). https://doi.org/10.1186/s40537-014-0007-7
15. OpenMP Architecture Review Board: OpenMP application program interface version 3.0 (2008). http://www.openmp.org/mp-documents/spec30.pdf
16. Pedregosa, F.: Scikit-learn: machine learning in Python. J. Mach. Learn. Res. **12**, 2825–2830 (2011)
17. Vapnik, V.: The Nature of Statistical Learning Theory. Springer-Verlag; 2nd edn. (2000)
18. Yuan, G., Ho, C., Lin, C.: Recent advances of large-scale linear classification. Proc. IEEE **100**(9), 2584–2603 (2012)
19. Zaharia, M., Chowdhury, M., Franklin, M.J., Shenker, S., Stoica, I.: Spark: Cluster computing with working sets. In: Proceedings of the 2Nd USENIX Conference on Hot Topics in Cloud Computing, pp. 10–10. HotCloud'10, USENIX Association (2010)

PM2.5 Prediction Model Using CNN with Depthwise Separable Convolution and Bi-LSTM

Thi-Phuong-Trang Nguyen[1]([✉]), Duc-Cuong Nguyen[1], and Tuong Le[2]

[1] Faculty of Information Technology, HCMC University of Foreign Languages - Information Technology (HUFLIT), Ho Chi Minh City, Vietnam
`{trangntp,cuongntp}@huflit.edu.vn`
[2] Faculty of Information Technology, HUTECH University, Ho Chi Minh City, Vietnam
`lc.tuong@hutech.edu.vn`

Abstract. This study presents an improved air quality prediction method using hybrid deep learning approach named PM25-CDSCBL, which is an enhancement of the PM25-CBL model. The PM25-CDSCBL combines CNN (Convolutional Neural Network) with a Depthwise Separable Convolutions layer and Bi-LSTM (Bidirectional Long Short-Term Memory). The model is evaluated and compared with other methods such as LSTM, Bi-LSTM, CNN-LSTM, ARIMA and PM25-CBL on the HCMC Air Quality 2020 dataset. Experimental results show that PM25-CDSCBL achieves superior performance with the lowest MSE (Mean Squared Error), RMSE (Root Mean Squared Error), MAE (Mean Absolute Error), and MAPE (Mean Absolute Percentage Error) among all tested models. Notably, the PM25-CDSCBL model demonstrates high accuracy in predicting PM2.5 concentrations across time steps, with a close match between actual and predicted values. These results confirm the feasibility and effectiveness of PM25-CDSCBL in air quality prediction, especially in the context of increasing urbanization and air pollution in major cities.

Keywords: Deep learning · CNN · Bi-LSTM · Depthwise separable convolutions · PM2.5 concentration prediction

1 Introduction

The success of machine learning (ML), especially deep learning (DL), across various fields has led to a data-driven methodology. DL is a subset of ML algorithms that focuses on how data is represented, making DL an extension of ML. One of the greatest advantages of DL is its ability to learn features from raw data. Optimizing ML with deep learning models enhances the learning process, leading to better results [1, 2]. Each DL model uses different ML techniques, which lead to varying levels of accuracy [3]. With DL's ability to support learning complex feature representations from large datasets, the application of DL models to predictive tasks in fields such as economics [4, 5], education [6], healthcare [7, 8], environmental management [9], energy consumption [10, 11],

N. Thai-Nghe et al. (Eds.): ISDS 2025, CCIS 2714, pp. 301–311, 2026.
https://doi.org/10.1007/978-981-95-3358-9_22

natural language processing [12, 13], and time-series forecasting is increasing [14, 15]. ML and DL-based algorithms are commonly used for time-series prediction problems. Techniques like RNN, CNN, LSTM, and BiLSTM have proven to deliver more accurate results compared to traditional regression models. Recurrent Neural Networks (RNN) are considered effective approaches for time-series forecasting [16]. Some studies utilize variants of RNN, such as Long Short-Term Memory (LSTM) and Bidirectional Long Short-Term Memory (BiLSTM), achieving promising results in various applications. Research by Sahoo et al. [17] indicates that LSTM-RNN is highly effective for continuous time-series processing, with superior results in predicting low-density water volume in the Mahanadi River basin in India. In [18], the authors propose the BOP-BL model, based on BiLSTM, for predicting oil prices, consisting of two blocks, with the first block using three Bi-LSTM layers and the second block featuring a fully connected layer for oil price prediction. In [19], the authors demonstrate that BiLSTM models significantly outperform traditional LSTM models, with BiLSTM reducing error rates by 37.78% in stock market predictions, although BiLSTM models take longer to reach equilibrium. Similarly, in [20], the Bi-LSTM-based DL model (DLBL-WQA) outperforms LSTM in forecasting water quality factors for the Yamuna River in India. Time-series data often contains time-invariant relationships, making CNN-based models suitable for such tasks. CNN is a hierarchical network with convolution layers and fully connected layers, unlike traditional neural networks. Some studies have combined CNN with BiLSTM for time-series forecasting, where CNN extracts and reduces feature dimensions, and BiLSTM is used sequentially to predict specific values. This combination improves performance and reduces prediction time. In [21], the authors propose using CNN and Bi-LSTM to predict electricity consumption, showing improved prediction performance. While CNN and Bi-LSTM models perform well in time-series data detection, they may face high computational complexity and require significant resources during training and prediction. To reduce computational complexity [22, 23], Depthwise Separable Convolution (used in MobileNet) has been widely adopted in CNN models. This technique applies filters on individual points rather than the entire depth of the previous layer, significantly reducing model size and parameter count. In [24], the authors propose a fault diagnosis model for bearings using depthwise separable convolutions in CNN, improving efficiency, accuracy, and computation speed while ensuring low parameter space and noise resilience.

Currently, air pollution is a major concern worldwide, including in Vietnam, due to its severe impact on the environment and human health. The surrounding air contains pollutants such as particulate matter (PM), aerosols, and gaseous pollutants. Among these, PM2.5 is considered the most dangerous silent killer. Several studies have utilized Machine Learning (ML) and Deep Learning (DL) techniques to predict PM2.5 levels [25–30]. Wong et al. [25] developed an ML model to estimate daily PM2.5 concentrations on Taiwan Island from 2000 to 2016, using three algorithms: DNN, RF, and XGBoost. Phong et al. [26] employed a Multilayer Perceptron (MLP) to analyze and predict PM2.5 levels across eight major cities in China, identifying factors influencing PM2.5 fluctuations. Ma et al. [27] used the XGBoost algorithm along with Lasso linear regression (to reduce overfitting) to predict daily PM2.5 concentrations in several cities in China. Pak et al. [28] applied Deep Learning to predict PM2.5 levels in various

Chinese cities, utilizing a CNN-LSTM model, which showed better and more stable prediction performance compared to MLP models. Vo et al. [29] developed the PM25-CBL deep learning model, integrating CNN and Bi-LSTM, to predict PM2.5 levels in the Air Quality HCMC dataset. Experimental results confirmed that the PM25-CBL model outperforms other time-series prediction methods, including LSTM, Bi-LSTM, CNN, CNN-LSTM, CNN-Bi-LSTM, and ARIMA, in terms of MSE, RMSE, MAE, and MAPE. The authors in [30] built a model using CNN and LSTM to predict hourly PM2.5 concentrations for the next 24 h at six locations in HCMC, using real-time data from six air monitoring stations. The study showed that the model's predictions outperformed methods like SGDRegressor and XGBoost. While these models provide high accuracy in PM2.5 prediction, they still face limitations in training time or may not be applicable for predictions at other locations with datasets containing different features.

In this study, we analyze the impact of various factors on PM2.5 concentrations in the Air Quality HCMC dataset from Vietnam. We then develop the PM25-CBL model, integrating CNN and Bi-LSTM, where CNN is combined with Depthwise Separable Convolutions to enhance model performance. The experimental section evaluates the forecasting ability of the proposed method and compares the results with LSTM, Bi-LSTM, CNN-LSTM, ARIMA, and PM25-CBL models. The results show that the PM25-CDSCBL model outperforms other tested methods on the Air Quality HCMC dataset, as measured by MSE, RMSE, MAE, and MAPE. The rest of the paper is organized as follows: Sect. 2 presents the Air Quality HCMC dataset and the PM25-CDSCBL model integrating CNN, Depthwise Separable Convolutions, and Bi-LSTM for PM2.5 prediction. Section 3 describes the experiments and compares the results between models. Finally, Sect. 4 provides the conclusion of this study.

2 Material and Method

2.1 Dataset Construction

The Air Quality in Vietnam dataset for 2020 provides air quality values for several cities across Vietnam [31]. This study focuses on analyzing air quality values in Ho Chi Minh City (HCMC). The dataset includes various air quality indices as well as additional values such as variance, median, and the maximum and minimum values. The Dataset consists of 12249 rows and has seven parameters: temperature, humidity, wind speed, PM2.5 concentration ($\mu g/m^3$), and fog. The statistical summary of the dataset is shown in Table 1. provides a description and basic statistics for the seven air quality parameters in HCMC (Include table with statistical details such as mean, variance, maximum, and minimum values for each of the 7 parameters mentioned).

Table 1. Basic Statistics of Air Quality Parameters in the Air Quality in Vietnam Dataset for Ho Chi Minh City

Variable	Min	Max	Median
temperature	23.0	31.0	27.5
dew	14.5	26.5	24.0
humidity	47.0	100.0	78.0
pm25	5.0	175.0	65.0
pressure	1003.0	1014.0	1009.0
wind gust	2.0	6.6	11.8
wind speed	0.5	5.4	2.3

To improve the model's performance, we apply the Min-Max scaling method, using the formula: $x_{scaled} \frac{x-\min(x)}{\max(x)-\min(x)}$. This normalization technique transforms the numeric values into a common scale, specifically mapping them to a range between 0 and 1. This scaling ensures that the input features have a uniform range, helping the model converge more efficiently and improving its predictive accuracy.

2.2 The PM25-CDSCBL Model

This section presents the overall architecture of a hybrid deep learning approach named the PM25- CDSCBL model for PM2.5 concentration prediction in the Air Quality HCMC dataset. The PM25-CDSCBL model is an extension of the PM25-CBL model [29], incorporating Depthwise Separable Convolutions (DSC) into the architecture alongside Convolutional Neural Networks (CNN) and Bi-directional Long Short-Term Memory (Bi-LSTM) networks. The overview of the PM25-CDSCBL model architecture is shown in Fig. 1.

The PM25-CDSCBL model operates in two primary phases:

- Training Phase: The input variables from the dataset (temperature, humidity, wind speed,....) are passed through a series of Depthwise Separable Convolution (DSC) and Max Pooling layers, followed by the Bi-LSTM module for sequential data processing. The final prediction is made by the Fully Connected Layer.
- Testing Phase: For predictions, only key features (temperature, humidity, wind speed,...) are used in the trained model to predict PM2.5 concentrations.

In CNN with DSC model, 1D SeparableConv layers are used instead of regular 1D Convolution layers. These layers efficiently extract features while reducing the computational burden. In this layer, the input is convolved separately for each channel, followed by a 1x1 convolution to combine the results. This approach helps in reducing the parameter count and making the model more computationally efficient. After the convolutional layers, 1D Max Pooling layers are applied to reduce the dimensionality and computational complexity.

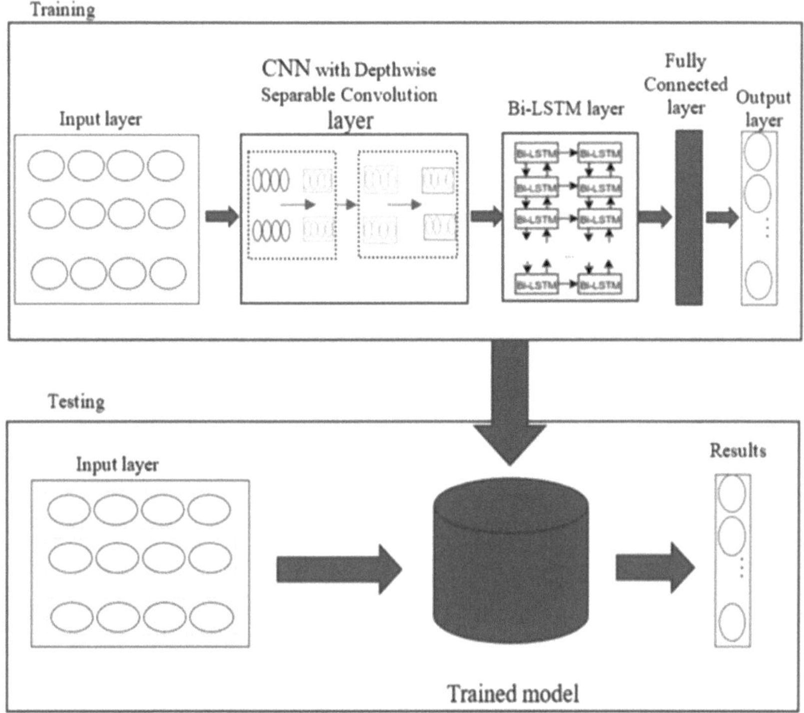

Fig. 1. PM25-CDSCBL model

In Bi-LSTM module, the PM25-CDSCBL model utilizes two Bi-LSTM layers: The first Bi-LSTM layer processes the sequence in the forward direction. The second Bi-LSTM layer processes the sequence in the backward direction, helping the model capture future information as well as past data.

In Fully Connected Layer, the outputs from the Bi-LSTM module are passed into the Fully Connected Layer to make the final prediction of PM2.5 concentrations. This layer reduces the learning parameters and consolidates the information extracted by the previous layers into the final output.

3 Experiment

3.1 Experiment Setup

This section evaluates our approach and several state-of-the-art models in time series prediction for Air Quality HCMC 2020 dataset. The above methods are implemented in the Keras framework and executed in the Ubuntu computer with an Intel Core i7-4790K (4.0 GHz, 8 cores), 32 GB of RAM, and GeForce GTX 1080 Ti. To demonstrate the effectiveness of the proposed model, the Air Quality HCMC 2020 dataset is divided into 5-folds with 20% for testing set and the remaining of this dataset for training set. This study utilizes four following common performance metrics in time series prediction to compare the experimental methods. MSE is the average squared difference between the predicted values and the actual values while RMSE is the square root of the MSE. They are determined by the following equations.

$$MSE = \frac{1}{n} \sum_{1}^{n} (\mathbf{y} - \hat{\mathbf{y}})^2 \tag{1}$$

$$RMSE = \sqrt{\frac{1}{n} \sum_{1}^{n} (\mathbf{y} - \hat{\mathbf{y}})^2} \tag{2}$$

Next, MAE gives the average magnitude of the prediction errors and MAPE provides prediction accuracy in the percentage of a forecasting method. The following equations can obtain them.

$$MAE = \frac{1}{n} \sum_{1}^{n} |\mathbf{y} - \hat{\mathbf{y}}| \tag{3}$$

$$MAPE = \frac{100}{n} \sum_{1}^{n} \left| \frac{\mathbf{y} - \hat{\mathbf{y}}}{\mathbf{y}} \right| \tag{4}$$

3.2 Experiment Results

3.2.1 A Comparison of the Model Scales

Table 2. The configurations of PM25-CBL model.

#No	Layer type	Neurons	Parameters
1	1D Convolution	(None, None, 5, 64)	192
2	1D Max Pooling	(None, None, 5, 64)	0
3	1D Convolution	(None, None, 4, 64)	8256
4	1D Max Pooling	(None, None, 4, 64)	0
5	Flatten	(None, None, 256)	0
6	Bi-LSTM	(None, None, 128)	164,352
7	Dropout	(None, 128)	0
8	Bi-LSTM	(None, 64)	41,216
9	Fully connected layer	(None, 1)	65

Total params: 214,081.

The PM25-CDSCBL model offers superior improvements over the PM25-CBL model. The use of separable convolutions allows PM25-CDSCBL to achieve faster

Table 3. The configurations of PM25-CDSCBL model.

#No	Layer type	Neurons	Parameters
1	1D SeparableConv	(None, None, 5, 64)	66
2	1D Max Pooling	(None, None, 5, 64)	0
3	1D SeparableConv	(None, None, 4, 64)	1,120
4	1D Max Pooling	(None, None, 4, 64)	0
5	Flatten	(None, None, 256)	0
6	Bi-LSTM	(None, None, 128)	41,216
7	Dropout	(None, 128)	0
8	Bi-LSTM	(None, 64)	24,832
9	Fully connected layer	(None, 1)	65

Total params: 67,299.

training and inference times. Which significantly reduces the number of parameters from 214,081 to 67,299 while retaining the model's effectiveness in feature extraction. This reduction in parameters not only makes the model faster but also more resource-efficient, making it an optimal choice for tasks that require computational efficiency. The models setup parameters are shown in Table 2. and Table 3..

A key improvement in the PM25-CDSCBL model (Table 3.) is the replacement of the 1D Convolution layer with 1D Separable Convolution. The Separable Convolution layer significantly reduces the number of parameters while maintaining its effectiveness in feature extraction. Specifically, the 1D SeparableConv layer in PM25-CDSCBL uses only 66 parameters, while the 1D Convolution layer in PM25-CBL (Table 2.) requires 192 parameters. This reduction in parameters helps decrease the model's complexity without compromising the quality of feature extraction, making the model lighter and faster. The total number of parameters in the PM25-CDSCBL model is only 67,299, which is a significant reduction from 214,081 in the PM25-CBL model. Both models use Bi-LSTM layers with a similar number of neurons (128 neurons in Layer 6 and 64 neurons in Layer 8). However, the optimization in PM25-CDSCBL primarily comes from replacing the Convolution and Max Pooling layers with SeparableConv, which reduces the parameters and accelerates the training process without impacting the Bi-LSTM computations.

3.2.2 Results of Experimental Methods

Table 4. presents the performance results of various experimental methods used for predicting air quality in Ho Chi Minh City (HCMC) using the 2020 dataset. The models evaluated include Long Short-Term Memory (LSTM), Bidirectional LSTM (Bi-LSTM), Convolutional Neural Network combined with LSTM (CNN-LSTM), ARIMA, PM25-CBL and PM25-CDSCBL.

ARIMA produced the highest error metrics with a significantly MAPE of 16.19, which suggests that traditional time series models may struggle with the complexity of air quality forecasting. CNN-LSTM showed a slightly higher performance compared to LSTM with an MSE of 1.41, RMSE of 1.17, MAE of 0.97, and MAPE of 3.46. PM25-CBL demonstrated competitive results with an MSE of 1.23, RMSE of 1.09, MAE of 0.89, and MAPE of 3.19, indicating its efficiency in modeling air quality data. PM25-CDSCBL performed the best overall, achieving the lowest MSE of 0.92, RMSE of 0.96, MAE of 0.75, and the lowest MAPE of 2.9, highlighting its superior performance in air quality prediction.

Table 4. Results of experimental methods for Air Quality HCMC 2020 dataset.

#No	Model	MSE	RMSE	MAE	MAPE
1	LSTM	1.48	1.21	1.02	3.63
2	Bi-LSTM	1.31	1.13	0.94	3.36
3	CNN-LSTM	1.41	1.17	0.97	3.46
4	PM25-CBL	1.23	1.09	0.89	3.19
5	ARIMA	1.97	1.40	0.99	16.19
6	PM25-CDSCBL	0.92	0.96	0.75	2.9

Compared to models such as LSTM (MSE 1.48), Bi-LSTM (MSE 1.31), and CNN-LSTM (MSE 1.41), PM25-CBL (MSE 1.23), PM25-CDSCBL shows clear superiority. Deep learning models like LSTM and Bi-LSTM struggle to fully capture the complex nature of air quality data, while PM25-CDSCBL demonstrates superior performance with lower MSE and MAPE scores. This indicates its enhanced capability to model intricate patterns within the data. PM25-CDSCBL excels with the ability to process complex datasets and identify relationships between variables. This allows the model to capture non-linearities in the data that other models like ARIMA or even LSTM might miss. It makes PM25-CDSCBL particularly adept at handling diverse and dynamic data patterns typical in air quality data.

In the Fig. 2, we can observe the convergence speed of a model, with PM25-CBL Model (A), the model's loss decreases rapidly in the initial epochs, showing a fast convergence during the early training phase. After the initial drop, the loss stabilizes and continues to decrease slowly for both training and validation phases, suggesting that the model reaches a plateau or optimal state relatively early in the training process. With PM25-CDSCBL Model (B), the initial drop in loss is similar to the PM25-CBL model but continues to show a noticeable difference between training and validation loss. Both models exhibit relatively fast initial convergence, but PM25-CDSCBL (B) shows more potential for overfitting or slower generalization.

In Fig. 3, the graph illustrates the predicted vs. actual PM2.5 values. The actual values (represented by the blue line) fluctuate between 0.1 and 0.7, with occasional peaks and valleys, capturing the natural variation of PM2.5 levels over time. The predicted values (represented by the red dotted line) closely follow the trend of the actual values, with minor discrepancies. The prediction tends to stay within the same range as the actual values, with slight deviations at certain time steps, but it still closely tracks the

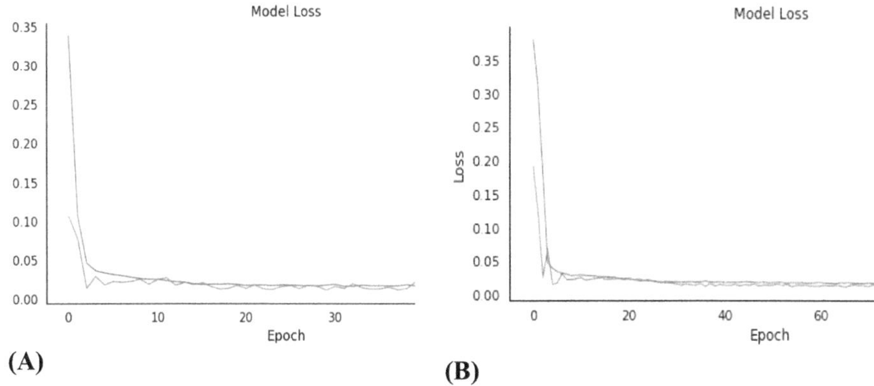

Fig. 2. The loss values of PM25-CBL (A) and PM25-CDSCBL (B) models in training and testing phases.

fluctuations and overall pattern. At time step 10, the actual value is around 0.5, and the prediction is nearly the same (0.48). At time step 30, the actual value peaks at 0.7, with the prediction slightly underestimating it at 0.65. At time step 70, the actual value drops to 0.2, and the prediction follows closely at 0.22, showing good consistency. This suggests that the prediction method is effective in capturing the trends in the data.

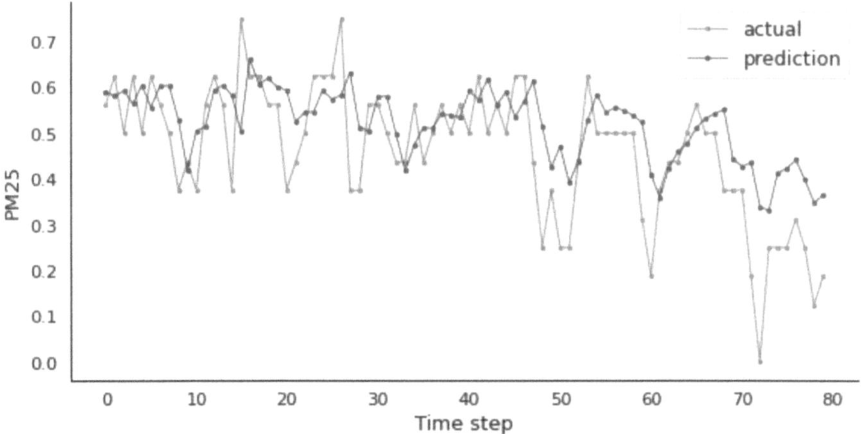

Fig. 3. Illustrating the performance of the proposed prediction method

4 Conclusion

In this study, we propose the PM25-CDSCBL model to improve the results of the PM25-CBL model in the study [29] using the Air Quality HCMC dataset. The PM25-CBL model consists of three modules: CNN, Bi-LSTM, and Fully Connected layers. While the model offers high prediction accuracy, its main drawback is the long training time, which is almost double that of a standard CNN model. The architecture of PM25-CDSCBL combines CNN with Depthwise Separable Convolutions and Bi-LSTM. The integration of Depthwise Separable Convolutions with CNN reduces the number of parameters in the PM25-CBL model from 214,081 to 67,299, significantly improving the training efficiency while maintaining high prediction performance. Additionally, the model applies Bi-LSTM based on the learning process of transition to predict air quality in scenarios with insufficient data. The results show that PM25-CDSCBL achieves the lowest RMSE when compared to LSTM, Bi-LSTM, CNN-LSTM, PM25-CBL, and ARIMA models, demonstrating superior performance in a new testing environment.

Acknowledgments. This research is funded by the Ho Chi Minh City University of Foreign Languages – Information Technology under grant number H2024-08.

References

1. Shlezinger, N., Eldar, Y.C., Boyd, S.P.: Model-based deep learning: on the intersection of deep learning and optimization. IEEE Access **10**, 115384–115398 (2022). https://doi.org/10.1109/ACCESS.2022.3218802
2. Chen, J., Fang, Y., Khisti, A., Özgür, A., Shlezinger, N.: Information compression in the AI era: recent advances and future challenges. IEEE J. Sel. Areas Commun. https://doi.org/10.1109/JSAC.2025.3560359
3. Afzal, S., et al.: Building energy consumption prediction using multilayer perceptron neural network-assisted models; comparison of different optimization algorithms. Energy **282**, 128446 (2023)
4. Le, T.: A comprehensive survey of imbalanced learning methods for bankruptcy prediction. IET Commun. **16**(5), 433–441 (2022)
5. Li, Y., et al.: A dual-optimization wind speed forecasting model based on deep learning and improved dung beetle optimization algorithm. Energy **286**, 129604 (2024)
6. Shou, Z., et al.: Predicting student performance in online learning: a multidimensional time-series data analysis approach. Appl. Sci. **14**(6), 2522 (2024)
7. Nguyen, T.P.T., Nguyen, N.L.-T., Duong, T.H.: Deepo: an ontology-based deep learning system for disease prediction. Int. J. Intell. Inf. Database Syst. **15**(2), 166–182 (2022). https://doi.org/10.1504/IJIIDS.2022.121897
8. El Nahhas, O.S.M., et al.: From whole-slide image to biomarker prediction: end-to-end weakly supervised deep learning in computational pathology. Nat. Protocols **20**(1), 293–316 (2025)
9. Wu, Z., et al.: Prediction of air pollutant concentrations based on the long short-term memory neural network. J. Hazardous Mater. **465**, 133099 (2024)
10. Ladjal, B., et al.: Hybrid deep learning CNN-LSTM model for forecasting direct normal irradiance: a study on solar potential in Ghardaia, Algeria. Sci. Rep. **15**(1), 15404 (2025)
11. Li, C., Shi, J.: A novel CNN-LSTM-based forecasting model for household electricity load by merging mode decomposition, self-attention and autoencoder. Energy 136883 (2025)
12. Vo, M.T., Vo, A.H., Nguyen, T., Sharma, R., Le, T.: Dealing with the class imbalance problem in the detection of fake job descriptions. Comput. Mater. Continua **68**(1), 521–535 (2021)
13. Zhou, H., et al.: Intelligent detection on construction project contract missing clauses based on deep learning and NLP. Eng. Constr. Archit. Manag. **32**(3), 1546–1580 (2025)
14. Sineglazov, V., Horbatiuk, V.: Time series forecasting using recurrent neural networks based on recurrent sigmoid piecewise linear neurons. Appl. Artif. Intell. **39**(1), 2490057 (2025)

15. Huang, Z., Zhang, F., Liu, Y.: A channel-independent network based on wavelet enhancement for long-term time series forecasting. Eng. Appl. Artif. Intell. **154**, 110964 (2025)

16. Hewamalage, H., Bergmeir, C., Bandara, K.: Recurrent neural networks for time series forecasting: current status and future directions. Int. J. Forecast. **37**(1), 388–427 (2021)

17. Sahoo, B.B., Jha, R., Singh, A., et al.: Long short-term memory (LSTM) recurrent neural network for low-flow hydrological time series forecasting. Acta Geophys. **67**, 1471–1481 (2019). https://doi.org/10.1007/s11600-019-00330-1

18. Vo, H.A., Nguyen, T., Le, T.: Brent oil price prediction using Bi-LSTM network. Intell. Autom. Soft Comput. **26**(6), 1307–1317 (2020)

19. Siami-Namini, S., Tavakoli, N., Namin, A.S.: The performance of LSTM and BiLSTM in forecasting time series. In: 2019 IEEE International Conference on Big Data (Big Data), Los Angeles, CA, USA, pp. 3285–3292 (2019). https://doi.org/10.1109/BigData47090.2019.9005997

20. Khullar, S., Singh, N.: Water quality assessment of a river using deep learning Bi-LSTM methodology: forecasting and validation. Environ. Sci. Pollut. Res. **29**, 12875–12889 (2022)

21. Vo, M.T., Vu, D., Nguyen, H., Bui, H., Le, T.: Predicting monthly household water consumption. In: Proceedings of International Conference on Computing and Communication Technologies, Ho Chi Minh City, Vietnam, pp. 720– 724 (2022)

22. Srivastava, H., Sarawadekar, K.: A depthwise separable convolution architecture for CNN accelerator. In: 2020 IEEE Applied Signal Processing Conference (ASPCON), Kolkata, India, pp. 1–5 (2020). https://doi.org/10.1109/ASPCON49795.2020.9276672

23. Liu, B., Zou, D., Feng, L., Feng, S., Fu, P., Li, J.: An FPGA-based CNN accelerator integrating depthwise separable convolution. Electronics **8**, 281 (2019)

24. Ma, S., Liu, W., Cai, W., Shang, Z., Liu, G.: Lightweight deep residual CNN for fault diagnosis of rotating machinery based on depthwise separable convolutions. IEEE Access **7**, 57023–57036 (2019). https://doi.org/10.1109/ACCESS.2019.2912072

25. Wong, P.Y., Lee, H.Y., Chen, Y.C., Zeng, Y.T., Chern, Y.R., et al.: Using a land use regression model with machine learning to estimate ground level PM2.5. Environ. Pollut. **277**, 116846 (2021)

26. Feng, R., Gao, H., Luo, K., Fan, J.R.: Analysis and accurate prediction of ambient PM2. 5 in China using multi-layer perceptron. Atmos. Environ. **232**, 117534 (2020)

27. Ma, J., Yu, Z., Qu, Y., Xu, J., Cao, Y.: Application of the XGBoost machine learning method in PM2. 5 prediction: a case study of Shanghai. Aerosol Air Qual. Res. **20**(1), 128–138 (2020)

28. Pak, U., Ma, J., Ryu, U., Ryom, K., Juhyok, U., et al.: Deep learning-based PM2. 5 prediction considering the spatiotemporal correlations: a case study of Beijing, China. Sci. Total. Environ. **699**, 133561 (2020)

29. Vo, M.T., Vo, H.A., Bui, H., Le, T.: A hybrid deep learning approach for PM2.5 concentration prediction in smart environmental monitoring. Intell. Autom. Soft Comput. **36**, 3029–3041 (2023). http://www.techscience.com/iasc/v36n3/51909

30. Rakholia, R., Le, Q., Vu, K., Ho, B.Q., Carbajo, R.S.: AI-based air quality PM2.5 forecasting models for developing countries: a case study of Ho Chi Minh City, Vietnam. Urban Clim. **46** (2022)

31. Dataset on Air Quality in Vietnam in 2020. https://data.opendevelopmentmekong.net/dataset/timelines-dataset-on-air-quality-in-vietnam. Accessed 20 Feb 2025

Improving Data Quality Post-anonymization in K-Anonymity Models Using a Hybrid Artificial Rabbit and Arctic Puffin Optimization

Le Hoang Minh Nhat, Bach Ngoc Vy, Phung Thi Ly,
and Dinh Nguyen Trong Nghia[(✉)]

Faculty of Information Technology, Ho Chi Minh City University of Industry and Trade,
Ho Chi Minh City, Vietnam
`nghiadnt@huit.edu.vn`

Abstract. In the context of growing demand for privacy-preserving data sharing, k-anonymity has emerged as a widely adopted anonymization technique to protect individual privacy while preserving data utility. However, the generalization processes inherent to k-anonymity often lead to substantial information loss and reduced classification accuracy. To address this trade-off, this paper proposes a novel hybrid optimization model combining Artificial Rabbits Optimization (ARO) and Arctic Puffin Optimization (APO). The proposed approach utilizes a multi-objective fitness function that simultaneously minimizes information loss (IL), preserves classification accuracy (CA), and satisfies the k-anonymity constraint by minimizing violations (M). In the hybrid framework, APO is employed for global exploration to escape local optima, while ARO provides efficient local exploitation of promising solutions. The hybrid model integrates both algorithms in parallel and periodically exchanges their best individuals to enhance convergence and diversity. Experimental evaluations on six real-world dataset – including census, healthcare, and demographic data-demonstrate that the proposed model significantly reduces information loss while maintaining high post-anonymization accuracy. The results show its robustness and adaptability across datasets of varying sizes and complexities, highlighting its potential application in secure and intelligent data-sharing systems.

Keywords: Artificial Rabbits Optimization · Artic Puffin Optimization · K-Anonymity · Data Anonymity · Hybrid Algorithm

1 Introduction

In the era of intelligent systems driven by data, the explosion of personal information and digital transactions in critical domains such as healthcare, finance, and smart cities has significantly increased the demand for secure data sharing. As the digital ecosystem continues to evolve, the challenge of preserving individual privacy while maintaining the utility of data has become a key issue for both researchers and real-world applications. Among the proposed data anonymization techniques, k-anonymity stands out as a widely

N. Thai-Nghe et al. (Eds.): ISDS 2025, CCIS 2714, pp. 312–327, 2026.
https://doi.org/10.1007/978-981-95-3358-9_23

adopted data protection model. This model ensures that each record in a dataset cannot be distinguished from at least $k - 1$ other records based on quasi-identifiers such as age, gender, or geographic region [1]. Generalizing or masking these attributes not only reduces the risk of re – identification but also supports compliance with legal regulations such as GDPR and HIPAA [2].

However, despite its privacy-preserving capabilities, k-anonymity still presents notable limitations, particularly when high levels of generalization lead to severe Information Loss (IL). This, in turn, results in decreased Classification Accuracy (CA), thereby negatively impacting the utility of anonymized data [3]. In applications where high accuracy is crucial – such as medical diagnosis, fraud detection, or behavioral analysis in smart cities – preserving the utility of anonymized data is especially important.

To address the trade-off between privacy and data utility, various approaches have been proposed. In addition to heuristic and metaheuristic methods, greedy algorithms such as DataFly and Mondrian have been employed to generate data partitions that satisfy k-anonymity with reasonable computational cost. In particular, Mondrian applies multidimensional data partitioning and is considered one of the most effective algorithms for handling high-dimensional data [7]. Furthermore, Incognito – a full domain generalization algorithm – has been introduced as an accurate and scalable solution, capable of generating all generalized solutions that satisfy k-anonymity by leveraging the monotonicity and transitivity properties of generalization domains [8].

However, greedy or exact methods like Mondrian and Incognito may struggle with large search spaces or with balancing multiple objective criteria. Consequently, there has been growing interest in modern metaheuristic algorithms, which offer efficient exploration of the solution space. In recent years, nature-inspired algorithms such as Artificial Rabbits Optimization (ARO) [5] and Arctic Puffin Optimization (APO) [6] have emerged as powerful tools for solving constrained optimization problems in large combinatorial spaces.

This paper proposes a hybrid model combining ARO and APO to enhance the quality of anonymized data within the k-anonymity framework. In the proposed model, APO performs global search to avoid local optima, while ARO effectively exploits promising regions through local refinement. The synergy between these two algorithms results in a hybrid model with both strong exploration capabilities and good convergence performance in the generalization search space.

At the core of the model is a multi-objective fitness function designed to simultaneously optimize three aspects: (1) minimizing information loss (IL), (2) maintaining classification accuracy (CA), and (3) ensuring the k-anonymity condition by reducing the number of violations (M). By adjusting the weights among these components, the model can flexibly adapt to varying requirements for balancing data privacy and utility.

The main contributions of this study include: (1) proposing the ARO – APO hybrid model for solving the data anonymization problem based on k-anonymity; (2) constructing a multi-objective fitness function to guide the optimization process, and (3) conducting experiments on real world datasets, demonstrating that the model significantly reduces information loss while maintaining high classification accuracy. These results highlight the potential applicability of the model in secure and intelligent data–sharing systems.

2 Related Works

In the realm of data privacy, k-anonymity stands as one of the most widely adopted techniques for anonymizing data and safeguarding personal privacy. Building upon this foundational model, numerous research directions have emerged, primarily focusing on two key objectives: minimizing information loss and preserving data utility after anonymization.

Several foundational algorithms for k-anonymity were proposed early on, notably DataFly, Incognito, and Mondrian. The DataFly algorithm, developed by Sweeney, employs a dynamic generalization mechanism to ensure that each quasi-identifier combination appears at least k times [9]. However, this method often leads to excessive generalization when k is large, resulting in significant information loss.

To enhance optimality, LeFevre et al. proposed Incognito, which constructs the entire generalization space based on full domain generalization, thereby guaranteeing a globally optimal solution [7]. A drawback of Incognito, however, is its substantial computational cost, especially as the number of attributes and generalization levels increases, making it challenging to apply to large datasets.

To improve performance, LeFevre et al. introduced Mondrian – a greedy partitioning algorithm that applies binary partitioning on quantitative attributes to classify data [8]. This method better preserves the natural distribution of data and performs effectively in practical scenarios. Nevertheless, Mondrian does not guarantee global optimality due to its greedy nature, choosing the median split point might overlook better generalization configurations. Additionally, Mondrian does not handle categorical attributes well and lacks support for advanced constraints. Consequently, this solution can easily lead to an imbalance between privacy and utility, particularly with unevenly distributed data.

Loukides and Shao investigated the trade-off between privacy and data utility during the k-anonymization process, demonstrating that excessive generalization significantly reduces classification accuracy [10]. Aggarwal and Yu also provided an overview of privacy-preserving data mining models and algorithms, emphasizing the necessity of optimization methods in large scale data anonymization [11].

Another branch of research applies Genetic Algorithms (GA) to this problem. Works by Lin et al., for instance, used GA to encode solutions as binary strings representing generalization levels, thereby finding optimal configurations that satisfy privacy conditions and minimize information loss. However, GA can still get stuck in local optimal if not combined with stronger global search techniques [4].

Recently, many studies have focused on improving the quality of anonymized data by integrating optimization algorithms and machine learning models. Su et al. proposed a multi-dimensional k-anonymity algorithm capable of defending against distribution skew and similarity attacks, significantly enhancing security while maintaining data utility [1].

Karagiannis et al. developed a medical data sharing model based on k-anonymity, demonstrating high effectiveness when combined with a role-based access control system architecture [2]. Meanwhile, Yu et al. applied the k-anonymity model to trajectory data, using partitioning and density-based techniques to ensure anonymization in spatial data [3]. Furthermore, Dwork introduced Differential Privacy as an alternative to k-anonymity, ensuring privacy by adding controlled noise to statistical queries [12]. However, this technique entails a more significant trade-off in accuracy.

The overview of these works reveals a growing trend in combining modern optimization algorithms with k-anonymity to address the triple objective problem: privacy, information loss, and classification accuracy. However, no prior research has utilized a combination of Artificial Rabbit Optimization (ARO) and Arctic Puffin Optimization (APO) for the k-anonymity anonymization problem. Therefore, this study proposes a hybrid ARO – APO optimization model, aiming to enhance the quality of anonymized data while ensuring privacy requirements and practical applicability.

3   Problem Description

3.1   Basic Definitions

In this paper, the term "data" refers to information related to individuals, organized in a table format with rows (or records) and columns (or fields). Each row is called a tuple, representing the relationship between a set of values pertinent to an individual. Tuples within a table are not necessarily unique. Each column is referred to as an attribute, denoting a field or semantic category of information, which is a set of possible values. Therefore, an attribute also functions as a domain. Attributes in a table are unique. Consequently, when observing a table, each row is an ordered n-tuple $(d_1, d_2, d_3, \dots)$ where v_i is a value from the domain of the i – th column, with $1 \leq i \leq n$ and n being the number of columns [9].

Definition 1. (k-anonymity Problem): Given a dataset D with n records, where each record comprises a set of attributes $A = \{QI \cup S\}$, QI represents the set of quasi-identifying attributes, and S is the set of sensitive attributes. The objective is to transform the data in D to produce a new dataset D' such that every record shares the same values across the QI attributes with at least $k - 1$ other records, thereby ensuring anonymity [9].

Definition 2. (Failure to achieve k-anonymity – FTK): FTK is the number of records in dataset D' where the combination of values on the QI attributes appears fewer than k times. This indicates that some groups of records still fail to meet the anonymity criteria. FTK reflects the extent to which the k-anonymity condition is not met, meaning certain data groups do not have at least k identical records, which increases the risk of individual information disclosure [9].

Definition 3. (Classification Accuracy – CA): CA assesses how well the classification performance of the data is maintained after applying k-anonymity. It's a crucial metric for evaluating the effectiveness of a classification model post-anonymization. CA is defined as the percentage of records correctly predicted by the classification model relative to the total number of records, calculated using the formula:

$$CA = \frac{Number\ of\ correctly\ predicted\ records}{Total\ number\ of\ records} \times 100\% \tag{1}$$

For example, if a dataset of 1000 anonymized records have 850 correctly classified records, then $CA = 850/1000 \times 100\% = 85\%$. A high CA indicates that the data retains its analytical value, a vital factor in balancing data privacy and utility [9].

Definition 4. (Information Loss – IL): IL quantifies the degree of data alteration after applying k-anonymity, measured by the level of generalization applied to the quasi-identifying attributes compared to the original data. IL is determined using the Normalized Certainty Penalty (NCP) index for each attribute, where for any quasi-identifying attribute A, it's calculated by:

$$NCP(A) = \frac{Number\ of\ levels\ of\ generalization\ applied\ to\ attribute\ A}{Maximum\ number\ of\ generalization\ levels\ of\ the\ problem\ attribute\ set} \tag{2}$$

Subsequently, the IL index of a solution X (representing a generalization configuration of the QI attributes) is typically calculated by aggregating (e.g., by averaging) the NCP values of all attributes. This provides a single measure of loss of information after the anonymization process, calculated using the formula:

$$IL(X) = \frac{1}{n}\sum_{i=1}^{n} NCP(A_i) \tag{3}$$

For instance, if the "Age" attribute is generalized to level 3, and the maximum generalization level for the problem's attribute set is 6, then $IL(Age) = 3/6 = 1/2$. A lower IL value indicates less information loss, meaning the data better preserves its utility a core factor in optimizing the trade-off between privacy and utility [9].

Definition 5. (Security Parameter – M): M is an index that measures the level of security of a dataset after k-anonymity anonymization. If every record in D' satisfies the condition that each combination of values on the QI attributes appears with at least k other records (i.e., k-anonymity is achieved), then $M = 0$. Conversely, each violation of this condition will increase M. A lower value of M signifies a higher level of data security [2].

$$M = \sum_{record\,r\in D'} if\,(count(r_{QI}) < K) \tag{4}$$

3.2 Problem Inputs and Constraints

The input for this problem is an original dataset organized in a tabular format, where each row represents an individual and each column is an attribute. This data includes both numerical and categorical attributes. For k-anonymity, the input dataset consists of I records, with each record containing k attributes. The parameter k is an integer used to define the minimum number of records required in each equivalence group to ensure anonymity. Additionally, parameters for the hybrid APO – ARO algorithm are used to optimize the data grouping and anonymization process.

Hard Constraints (HC): These are mandatory constraints that must be strictly adhered to. If violated, the solution is discarded as infeasible. Hard constraints include:

- Each group in the anonymized dataset must contain at least k records to satisfy k-anonymity.
- Sensitive Attributes (SA) must remain unchanged.

Soft Constraints (SC): These constraints are not mandatory, but they should be optimized to improve the quality of anonymization. They can be violated if no other feasible options exist, but such violations will incur a penalty. Soft constraints include:

- Minimizing the degree of information loss (IL) to preserve data utility.
- Maintaining the highest possible classification accuracy (CA).

3.3 Problem Modeling

Mathematically, the dataset comprises records $T = \{t_1, t_2, ..., t_n\}$, where each record t_i consists of a set of quasi-identifying (QI) attributes $Q = \{q_1, q_2, ..., q_m\}$ and one sensitive attribute (SA) S. The objective is to partition T into equivalence classes $E_1, E_2, ..., E_n$ such that $|E_j| \geq k \forall j$, satisfying the k-anonymity condition.

Each solution is represented by a generalization vector:

$$X = (g_1, g_2, ..., g_m) \tag{5}$$

Here, g_i denotes the generalization level for the $i - $ th QI attribute, chosen from a predefined set of values. However, sensitive attributes remain unchanged and are not part of the solution vector. The ultimate objective is to achieve k-anonymity with minimal information loss while ensuring the utility of the transformed data.

3.4 Building the Fitness Function

Creating a weighted sum objective function is an effective way to balance three key criteria in data anonymization: Information Loss (IL), security (M), and Classification Accuracy (CA). The objective function is defined as follows:

$$F(X) = \alpha IL(X) + \lambda M(X) + \gamma(1 - CA(X)) \tag{6}$$

Here's what each part represents:

- α, λ, γ are the corresponding weight coefficients. They allow you to adjust the priority of each component during the optimization process.
- IL(X) represents the information loss, measured as the average of NCP values calculated using formula (3).
- M(X) is the security parameter, defined as a binary variable in formula (4).
- CA(X) is the classification accuracy, reflecting the ability to maintain analytical effectiveness after data anonymization. The term $(1 - CA(X))$ is included in penalizing solutions with low accuracy.

Choosing a large value for the lambda coefficient acts as a strong penalty mechanism for solutions that don't satisfy the k-anonymity requirement, regardless of how low IL(X) might be. Similarly, the gamma coefficient ensures that solutions significantly reducing classification accuracy are also eliminated from the search space. Meanwhile, alpha helps prioritize minimizing information loss to a reasonable extent.

The objective function directs the APO–ARO hybrid algorithm to first ensure security and classification accuracy, then optimize information loss. Improper coefficient settings (e.g., too small λ or γ) may yield low IL but violate k-anonymity or reduce CA, undermining safety and utility. This approach balances the three key criteria: utility (IL), accuracy (CA), and security (M).

4 APO – ARO Hybrid Algorithm

The random initialization of a population of individuals X_i^t for both algorithms is described by the following equation. Here, X_i^t is the i th individual at time t, rand generates a random number between 0 and 1, ub and lb represent the upper and lower bounds in the solution space respectively, and N is the number of individuals in the population:

$$\overrightarrow{X_i^t} = rand * (ub - lb) + lb, i = 1,2, 3...N \tag{7}$$

4.1 Artic Puffin Optimization (APO)

The APO algorithm is an optimization algorithm that simulates the survival process of Arctic puffins. It consists of three phases: population initialization, exploration, and exploitation behavior, controlled by parameter C and coefficient B which determine behavior switching. Here, T and t are the total number of iterations and the current iteration, respectively [6]:

$$B = 2 * log(1/rand) * (1 - t/T) \tag{8}$$

Exploration Behavior
If $B > C$, the population enters this phase to explore a vast space through two strategies:

$$Aerial\ search\ strategy: \begin{cases} \overrightarrow{Y_i^{t+1}} = \overrightarrow{X_i^t} + \left(\overrightarrow{X_i^t} - \overrightarrow{X_r^t}\right) * L(D) + R \\ R = round(0.5 * (0.05 + rand)) * \alpha \end{cases} \tag{9}$$

$$Diving\ and\ hunting\ strategy: \begin{cases} \overrightarrow{Z_i^{t+1}} = \overrightarrow{Y_i^{t+1}} * S \\ S = tan((rand - 0.5) * \pi) \end{cases} \tag{10}$$

Here, $\overrightarrow{Y_i^{t+1}}$ is the new position of puffin individual i in the search space, $L(D)$ is generated according to a Levy distribution with D being the number of dimensions, rand is a random number [0,1], and R and S are the additional coefficient and velocity coefficient, respectively [6].

Underwater Exploitation Behavior
Conversely, if $B \leq C$, the population executes this phase, comprising three main strategies to enhance hunting efficiency:

$$\textit{Foraging collection:} \begin{cases} \overrightarrow{W_i^{t+1}} = \{\overrightarrow{X_{r1}^t} + F * L(D) * \left(\overrightarrow{X_{r2}^t} - \overrightarrow{X_{r3}^t}\right) rand \geq 0.5 \\ \overrightarrow{X_{r1}^t} + F * \left(\overrightarrow{X_{r2}^t} - \overrightarrow{X_{r3}^t}\right) rand < 0.5 \end{cases} \tag{11}$$

$$\textit{Enhanced search:} \begin{cases} \overrightarrow{Y_i^{t+1}} = \overrightarrow{W_i^{t+1}} * (1 + f) \\ f = 0.1 * (rand - 1) * \frac{T-t}{T} \end{cases} \tag{12}$$

$$\textit{Predator avoidance:} \begin{cases} \overrightarrow{Z_i^{t+1}} = \{\overrightarrow{X_i^t} + F * L(D) * \left(\overrightarrow{X_{r1}^t} - \overrightarrow{X_{r2}^t}\right) rand \geq 0.5 \\ \overrightarrow{X_i^t} + \beta * \left(\overrightarrow{X_{r1}^t} - \overrightarrow{X_{r2}^t}\right) rand < 0.5 \end{cases} \tag{13}$$

Here, $\overrightarrow{X_{r1}^t}, \overrightarrow{X_{r2}^t}, \overrightarrow{X_{r3}^t}$ with variables r_1, r_2, r_3 are random integers from 1 to $N - 1$ (excluding i), using a cooperative coefficient $F = 0.5$ for exploring the surroundings, f is an adaptive coefficient, and β is a uniformly distributed random number between $(0,1)$ [6].

4.2 Artificial Rabbits Optimization (ARO)

The Artificial Rabbits Optimization (ARO) algorithm is an optimization technique inspired by the survival behaviors of rabbits in nature. It features two core strategies: encircling foraging and random hiding, which are switched based on an energy factor. In this context, t is the current iteration, T is the maximum number of iterations, and r_1 is a random number within the range $[0, 1]$ [5]:

$$A(t) = 4 \times \left(1 - \frac{t}{T}\right) \times ln\frac{1}{r_1} \tag{14}$$

Foraging Behavior
Condition and Purpose: If $A(t) > 1$, the algorithm executes this behavior. It explores new regions within the search space to discover potential solution areas. The update formula is:

$$\overrightarrow{v_i}(t + 1) = \overrightarrow{X_j}(t) + R \times (\overrightarrow{X_i}(t) - \overrightarrow{X_j}(t)) + round(0.5 \times (0.05 + r_4)) \times n_1 \tag{15}$$

where:

$$\begin{cases} \textit{Run operator: } R = L \times c \\ \textit{Random vector: } c(k) = \begin{cases} 1 \; if \; k = g(l) \\ 0 \qquad else \end{cases} \\ \textit{Run length: } L = (e - e^{-(\frac{t}{T})^2}) \times sin(2\pi r_2) \end{cases} \tag{16}$$

Here, $X_j(t)$ is the position of a randomly selected rabbit j (where $j \neq i$), r_4 is a random number in the range $[0,1]$, n_1 is a random number from a standard Gaussian distribution $N(0,1)$, , e is the base of the natural logarithm (approximately 2.718), r_2 is a random number in $[0, 1]$, $k = 1, 2, \ldots, d$ is the dimension index of the problem, $g = randperm(d)$ is a random permutation of integers from 1 to d, $l = \lceil r3 \cdot d \rceil$ is the number of dimensions chosen for transformation, $r_3 \in (0,1)$ is a random number, and $\lceil . \rceil$ is the ceiling function [5].

Hiding Behavior

Condition and purpose: If $A(t) \leq 1$, this behavior is activated. It simulates rabbits creating multiple burrows around their nest and randomly selecting one for hiding, aiming for local optimization around the current position.

$$\text{Creating burrows:} \quad \begin{cases} \overrightarrow{b_{i,k}}(t) = X_i(t) + H \times \left(g_k \times \overrightarrow{X_i}(t) \right) \\ \quad H = \frac{T-t+1}{T} \times r_5 \end{cases} \tag{17}$$

Here, $\overrightarrow{b_{i,k}}(t)$ represents the burrows around the current position for each dimension $k = 1, 2, \ldots, d$. Each rabbit i creates d burrows (candidates) around its current position $\overrightarrow{X_i}(t)$, and H is the hiding coefficient [5].

$$\text{Hiding:} \quad \overrightarrow{v_i}(t+1) = \overrightarrow{X_i}(t) + R \times \left(r_6 \times \overrightarrow{b_{i,r}}(t) - \overrightarrow{X_j}(t) \right) \tag{18}$$

Here, $\overrightarrow{v_i}(t+1)$ is the selected burrow candidate. g_k is a randomly chosen unit vector for transformation along dimension k, the index r ranges from 1 to d, and r_6 is a random number in the range $[0, 1]$ [5].

4.3 APO - ARO Hybrid Algorithm

This hybrid approach runs ARO and APO in parallel with two separate populations. Periodically (every k iterations), they exchange their best individuals to maintain diversity and accelerate convergence. ARO explores search space through a dynamic balance mechanism, while APO uses a predator-avoidance hunting strategy for both exploration and solution refinement. After every k iterations, the best individual from ARO is inserted into the APO population, and vice-versa: Individual exchange: $X_{APO} \leftarrow Best_{ARO}$ and $X_{ARO} \leftarrow Best_{APO}$.

This sharing mechanism effectively combines the strengths of both algorithms, improving solution quality and convergence speed, especially for complex optimization problems. The method doesn't build a direct hybrid equation but rather integrates them in parallel, facilitating the exchange of solution information between the two.

Pseudocode

```
Initialize N individuals for APO and ARO
Evaluate X̄ᵢᵗ_APO and X̄ᵢᵗ_ARO, find the individual with the best fitness value in X̄ᵗ current_iteration ←
0

while (current_iteration < T)
  for It ← 1:k
    if current_iteration + It > T
      break
    Update using APO
  end for
  // Exploration and Exploitation Phase:
  for It ← 1:k
    if current_iteration + It > T
      break
    Update using ARO
  end for
  // Exchange best individuals between the two populations
  best_APO ← Find the best individual in X̄ᵗ_APO
  best_ARO ← Find the best individual in X̄ᵗ_ARO
  rand_index_APO ← Random index in X̄ᵗ_APO
  rand_index_ARO ← Random index in X̄ᵗ_ARO
  X̄ᵗ_APO[rand_index_APO] ← best_ARO   // Corrected: APO gets ARO's best
  X̄ᵗ_ARO[rand_index_ARO] ← best_APO   // Corrected: ARO gets APO's best

  // Update global best fitness value
  best_fitness_APO ← Best fitness value in X̄ᵗ_APO
  best_fitness_ARO ← Best fitness value in X̄ᵗ_ARO
  X̄ᵗ ← Individual with min(best_fitness_APO, best_fitness_ARO)

  current_iteration ← current_iteration + k
end while
return X̄ᵗ
```

To evaluate the performance of the proposed hybrid strategy, experiments were conducted on six benchmark functions presented in Table 1, including both unimodal and multimodal functions, to examine its exploration and exploitation capabilities in complex search spaces. The experiments were implemented in Python on a computer with an Intel Core i5-10700 CPU (2.9 GHz), 16 GB of RAM, running Windows 11. NumPy and Pandas libraries were used for computation and analysis. The main parameters were set as follows: population size $N = 50$, maximum number of iterations $T = 1000$, and dimensionality of 70 (Fig. 1). All algorithms were initialized with the same population to ensure fairness. The objective functions were defined with their respective search bounds [lb,ub].

Table 1. Mathematical formulas of benchmark functions for hybrid algorithm experiments

Function	Formula	Range	Dimension	Objective
(Sphere)	$f(x) = \sum_{i=1}^{n} x_i^2$	$[-5, 5]$	70	0
(Schwefel's Problem 2.22)	$f(x) = \sum_{i=1}^{n} |x_i| + \prod_{i=1}^{n} |x_i|$	$[-10, 10]$	70	0
(Max Absolute)	$f(x) = max_{1 \leq i \leq n} |x_i|$	$[-100, 100]$	70	0
(Generalized Power)	$f(x) = \sum_{i=1}^{n} |x_i|^i$	$[-2, 2]$	70	0
(Weighted Sphere)	$f(x) = \sum_{i=1}^{n} i.x_i^2$	$[-100, 100]$	70	0
(Composite Quadratic)	$f(x) = \sum_{i=1}^{n} (x_i^2 + \frac{D}{2} \times x_i^2 + \frac{D}{2} \times x_i^4$	$[-100, 100]$	70	0

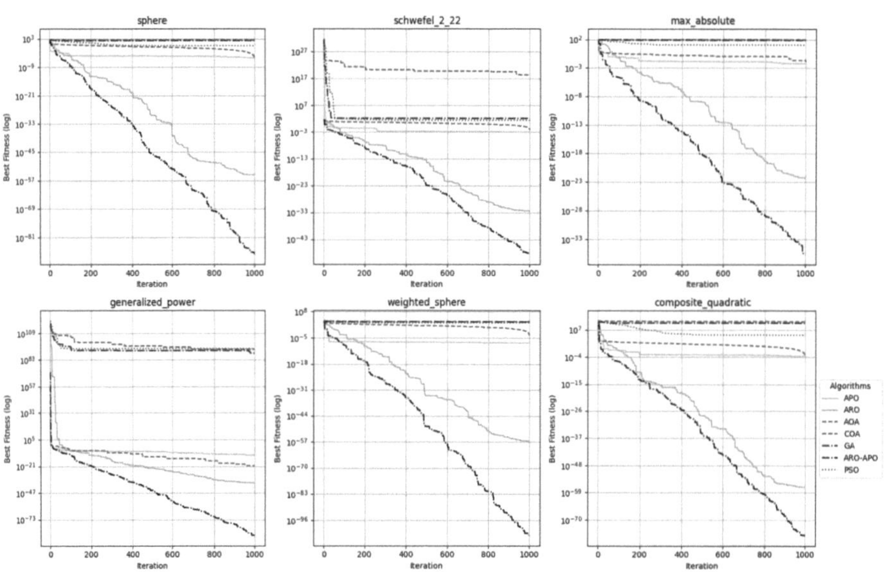

Fig. 1. Experimental results with a dimensionality of 70

5   Experimental Model

Loukides and Shao's research focuses on the trade-off between data utility and privacy in k-anonymity. They didn't explicitly state k as a fixed value, but they discussed selecting k values ranging from 2 to 10. They consider $k = 5$ a popular choice in healthcare applications, as it strikes a balance between privacy and data utility. Specifically, they mention that $k = 5$ is often used in medical datasets to ensure a sufficiently low re-identification risk, aligning with standards like HIPAA [10]. Therefore, this study's experiments use a k value of 5. Simultaneously, the objective function's coefficients are set as follows: $= 1$, $\lambda = 1000$, and $\gamma = 100$. This aims to heavily penalize solutions that violate k-anonymity, ensuring hard constraint satisfaction. For soft constraints, the priority order is accuracy, followed by information loss.

All experiments in this study were performed on a computer with the configuration mentioned in Sect. 4.3. The experiments were implemented in Python 3.12.6, utilizing data science libraries like NumPy and SciPy for data computation and processing, and Scikit-learn to implement the Random Forest machine learning model, which assists in pre-processing and evaluating the model before and after anonymization. The datasets used for the experiments include six sets, each representing a different data characteristic:

The Adult Dataset (ADULT), also known as the Census Income Dataset, comprises 30,162 records extracted from the 1994 U.S. census data. Its attributes include gender, age, race, marital status, education level, nationality, social class, occupation, and salary. All of these attributes are treated as quasi-identifiers (QIDs).

The California Housing Prices Dataset (CAHOUSING) contains 20,640 records. This dataset uses 11 attributes, with features like latitude, longitude, average house age, average income, and average house value designated as QIDs.

The Contraceptive Method Choice Dataset (CMC) consists of 1,473 data rows from a 1987 Indonesian contraceptive methods survey, recording demographic and economic information of women who had not previously been pregnant regarding the contraceptive methods they used. This set includes 10 attributes, where attributes such as age, education level, and number of children are considered QIDs.

The Mammographic Mass Dataset (MGM) has 830 records containing data derived from mammography analysis. It includes 6 attributes, with information on age, shape, assessment score, margins, and density serving as QIDs.

The INFORMS Data Mining Dataset (INFORMS) provides 102,580 rows of patient information from an anonymous hospital, released as part of a data mining competition. This dataset comprises 16 attributes, with information on birth month, birth year, years of education, income, and race as QIDs. The data fields have been digitized to prevent identity leakage. This dataset is used to test anonymization methods and evaluate their privacy effectiveness.

The Italia Healthcare Dataset (ITALIA) is a small dataset with 100 samples of patients with rare diseases in Italy, consisting of 4 attributes, where age, place of birth, and zip code are QIDs. This dataset is used to evaluate the effectiveness of k-anonymity algorithms that are highly complex and have unfeasible running times on larger datasets.

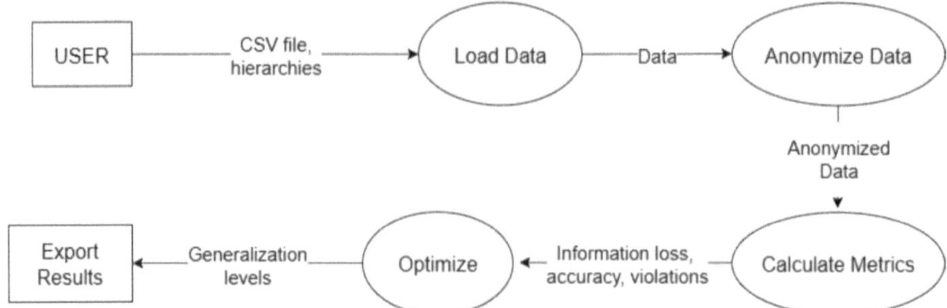

Fig. 2. A comprehensive architectural diagram of the system can be generated and exported

The data anonymization system described in the diagram operates as shown in Fig. 2 through a sequential process combined with optimization loop. Initially, the user uploads a dataset in CSV format along with the hierarchy structures required for anonymization. The system then processes and loads this data in preparation for subsequent steps. Once the data is ready, anonymization techniques are applied to generate an anonymized dataset. However, the quality of this dataset must be evaluated using key metrics such as information loss, accuracy, and compliance with k-anonymity. The evaluation results serve as the basis for the optimization step, where the system adjusts the level of generalization applied to the data. This optimization process may be repeated multiple times until a satisfactory balance between privacy protection and data utility is achieved. Finally, the optimized anonymized dataset, together with a detailed evaluation report, is exported in CSV format and supporting documentation, allowing users to securely employ the data for research or analytical purposes without compromising sensitive information.

6   Results and Discussion

(See Table 2).

Table 2. The experimental results from six datasets (k = 5)

	Fitness	IL	CA before anonymous	CA after anonymous
ADULT	0.3889	38.89%	82.98%	100.00%
CAHOUSING	0.895	72.24%	96.87%	82.74%
CMC	0.5258	44.44%	95.48%	91.86%
MGM	0.5789	35.00%	84.74%	77.11%
INFORMS	0.9207	31.43%	54.71%	39.36%
ITALIA	1.3333	50.00%	10.00%	16.67%

The experimental findings, conducted on datasets varying in characteristics and scale, underscore the adaptability and robustness of our anonymization method. It effectively balances the dual objectives of personal information privacy and analytical utility, demonstrating consistently promising results across various data types.

For instance, the ADULT dataset showcased remarkable improvement in post-anonymization accuracy, reaching a perfect 100%, a significant leap from its original 82.98%. This unexpected gain suggests that the generalization process may, in some cases, enhance the signal-to-noise ratio in the data, thereby improving machine learning model performance.

However, there are still some cases called noise. These cases can be seen that when the k value increases, the performance of the model also increases gradually compared to the previous k value and sometimes exceeds the threshold of the baseline model. This is an interesting finding that in some rare cases, the anonymized data gives better results than the original data when classified by the model, some other cases have the phenomenon of k value increasing with classification performance. We analyze and try to find an explanation for these cases, this may happen because when the hierarchy tree of the data sets acts as a metadata that helps the anonymization algorithm synchronize with the class structure of the data set, thus helping the classification model to better separate the classes of the data set. However, if we look at it in a general and theoretical way, the larger the k value, the more obvious the classification performance will be.

While the CAHOUSING dataset incurred a notable information loss of 72.24%, it still retained a strong post-anonymization accuracy of 82.74%. This illustrates the model's ability to preserve analytical value despite aggressive generalization, especially in complex, high-dimensional data.

With smaller datasets like CMC and MGM, the model achieved a notable equilibrium between accuracy and privacy. The CMC dataset experienced a negligible accuracy drop of less than 4% after anonymization. Similarly, MGM maintained a high accuracy of 77.11% with a relatively low information loss of 35%. These results confirm the method's applicability in medical and socio-demographic contexts.

The INFORMS dataset yielded the favorable fitness score (0.9207) and the lowest information loss (31.43%). This performance reflects the model's strong optimization capabilities, even when handling complex data structures. Despite a decrease in post-anonymization accuracy, the outcomes still indicate substantial potential for protecting patient data while largely preserving its analytical value.

Finally, although the ITALIA dataset showed a low post-anonymization accuracy of 16.67%, it remains valuable as a stress-testing benchmark, particularly for evaluating the model's behavior under stringent anonymization constraints and in challenging generalization scenarios.

7 Conclusion and Future Work

Protecting individual privacy while ensuring effective data utility is a critical need in today's era of data explosion. Sectors such as healthcare, finance, and e-government, in particular, require data anonymization solutions that not only comply with legal regulations but also preserve analytical accuracy for effective decision-making. In this study, we proposed an optimization model combining two bio-inspired algorithms: Artificial Rabbits Optimization (ARO) and Arctic Puffin Optimization (APO). Our goal was to improve the quality of data after applying k-anonymity, a widely used anonymization technique.

The proposed model uses a multi-objective fitness function to simultaneously optimize three crucial factors: (1) minimizing information loss (IL), (2) maintaining classification accuracy (CA), and (3) satisfying the k-anonymity constraint by reducing anonymity violations (M). The synergy between APO's global exploration capabilities and ARO's efficient local exploitation creates a robust search mechanism that effectively balances exploration and exploitation. Additionally, the exchange of best individuals between the two algorithmic populations at each iteration contributes to faster convergence and improved solution quality.

We conducted experiments on six real world datasets, diverse in size, format, and application domain. The results show outstanding effectiveness from our model: some datasets like ADULT and CMC maintained or even increased accuracy after anonymization; complex structured datasets like INFORMS still exhibited low information loss; and smaller datasets like ITALIA helped test the model's generalization limits. Overall, the experimental results demonstrate that the ARO – APO model not only protects personal information but also maintains acceptable data utility for analysis.

However, our model does have certain limitations. First, adjusting the weights of the objective function requires experience and specialized knowledge to achieve optimal performance in different real-world scenarios. Second, the current model applies only to k-anonymity and hasn't yet been extended to advanced privacy models like l-diversity or t-closeness, which demand more complex criteria.

In future research, we plan to expand the model to support a wider range of privacy-preserving standards. We also intend to integrate it with machine learning and deep learning techniques for automated parameter tuning, enhancing its adaptability and efficiency in large-scale, real-time data environments. Furthermore, we intend to develop a user-friendly software implementation with an intuitive interface to facilitate broader adoption of the model in smart data management systems.

References

1. Su, B., Huang, J., Miao, K., Wang, Z., Zhang, X., Chen, Y.: K-anonymity privacy protection algorithm for multi-dimensional data against skewness and similarity attacks. Sensors **23**(3), 1554 (2023)
2. Karagiannis, S., Ntantogian, C., Magkos, E., Tsohou, A., Ribeiro, L.: Mastering data privacy: leveraging k-anonymity for robust health data sharing. Int. J. Inf. Secur. **23**, 2189–2201 (2024)
3. Yu, W., Shi, H., Xu, H.: A trajectory k-anonymity model based on point density and partition. arXiv preprint arXiv:2307.16849 (2023)

4. Lin, J., Wang, X., Zhang, Y.: A genetic algorithm-based approach for data anonymization. In: Li, J., Wang, X. (eds.) PRICAI 2014. LNCS, vol. 8862, pp. 273–285. Springer, Heidelberg (2014)
5. Wang, L., Cao, Q.: Artificial rabbits optimization: a new bio-inspired meta-heuristic algorithm for solving engineering optimization problems. Eng. Appl. Artif. Intell. **114**, 105082 (2022)
6. Wang, W., Tian, W., Xu, D., Zang, H.: Arctic puffin optimization: a bio-inspired metaheuristic algorithm for solving engineering design optimization. Adv. Eng. Softw. **195**, 1–15 (2024)
7. LeFevre, K., DeWitt, D.J., Ramakrishnan, R.: Incognito: efficient full-domain k-anonymity. In: Proceedings of the 2005 ACM SIGMOD International Conference on Management of Data, pp. 49–60. ACM, New York (2005)
8. LeFevre, K., DeWitt, D.J., Ramakrishnan, R.: Mondrian multidimensional k-anonymity. In: Proceedings of the 22nd International Conference on Data Engineering (ICDE), pp. 25–35. IEEE, Los Alamitos (2006)
9. Sweeney, L.: Protecting privacy when disclosing information: k-anonymity and its enforcement through generalization and suppression. Technical report, MIT Laboratory for Computer Science (1997)
10. Loukides, G., Shao, J.: Data utility and privacy protection trade-off in k-anonymization. In: Proceedings of the 2008 International Workshop on Privacy and Anonymity in Information Society (PAIS 2008), pp. 36–45. ACM, New York (2008)
11. Aggarwal, C.C., Yu, P.S.: A general survey of privacy-preserving data mining models and algorithms. In: Aggarwal, C.C., Yu, P.S. (eds.) Privacy-Preserving Data Mining, pp. 11–52. Springer, Boston (2008)

FMN-OE: Evaluation And Elimination Of Hyperbox Overlap

Dinh-Minh Vu[1] , Thanh-Son Nguyen[2(✉)] , Trinh-Hoang Vu[3] ,
Thi-Duong Vu[1], Duc-Luu Nguyen[1], Quang-Huy Hoang[1], and Ba-Dung Le[4]

[1] Hanoi University of Industry, No. 298 Cau Dien Street, Tay Tuu Ward, Hanoi,
Vietnam
{minhvd,duongvt,luund,huyhq}@haui.edu.vn
[2] Academy of Finance, No. 58, Le Van Hien Street, Dong Ngac Ward, Hanoi,
Vietnam
sonnt@hvtc.edu.vn
[3] University of Engineering and Technology Vietnam National University, No. 144,
Xuan Thuy St., Cau Giay Ward, Hanoi, Vietnam
hoangvt06012003@gmail.com
[4] Vietnam Academy of Science and Technology, No. 18 Hoang Quoc Viet St., Nghia
Do Ward, Hanoi, Vietnam
l_bdung@yahoo.com

Abstract. This study aims to introduce a novel semi-supervised learning algorithm in fuzzy min-max neural networks, known as FMN-OE, focusing on calculating and eliminating the potential overlap between hyperboxes in the network. Overlap between hyperboxes can lead to ambiguity and uncertainty in recognition and classification, potentially reducing the performance of FMNN. To address this issue, the proposed methods involve adjusting and shrinking overlapping hyperboxes to eliminate the overlap. However, this adjustment may result in the loss of valuable information when removing data points that provide additional context from the previously evaluated hyperbox. To mitigate this drawback, FMN-OE evaluates whether the expansion adjustments will create overlapping regions between hyperboxes. If overlap is detected, FMN-OE adjusts the hyperboxes to expand them while striving to avoid creating overlapping regions. Finally, a new hyperbox is generated from the training data that causes overlap. FMN-OE determines supplementary information for this new hyperbox based on fuzziness and/or additional information provided by experts during the training process. The performance and superiority of the proposed method on synthetic and Benchmark datasets are demonstrated by comparing it with existing unsupervised and semi-supervised fuzzy clustering methods operating under similar conditions.

Keywords: Fuzzy neural network · Semi-Supervised learning ·
Evaluation hyperbox · Overlap elimination · Fuzzy clustering

1 Introduction

Data clustering is a crucial technique in data mining and machine learning, used to group objects within a data set into clusters such that objects in the same cluster share similar characteristics, while objects in different clusters exhibit significant differences [4]. Traditional data clustering consists of two major models: supervised clustering and unsupervised clustering. Semi-supervised clustering combines elements of both supervised and unsupervised clustering, incorporating some additional information during the clustering process. Clustering methods can also be categorized into crisp clustering and fuzzy clustering [10], probabilistic clustering, among others. In crisp clustering (soft hard clustering), each data point belongs to exactly one cluster. In fuzzy clustering, data points can belong to multiple clusters, with associated membership values. Recently, semi-supervised fuzzy clustering, an extension of fuzzy clustering, has gained research interest by utilizing semi-supervised learning techniques. Semi-supervised fuzzy clustering leverages prior knowledge to guide the clustering process, thereby enhancing cluster quality. Probabilistic clustering is a data analysis method in which objects are grouped according to the probability that they belong to different clusters [9].

Auxiliary information plays a crucial role in guiding, supervising, and controlling the clustering process. This supplementary information can be established either before or during the training process [20]. Among these, one form of auxiliary information is the labels associated with a small subset of samples in the training dataset [24]. Notable semi-supervised fuzzy clustering algorithms include eSFCM [24], SSSFC (Semi-Supervised Standard Fuzzy Clustering) [23], the GSOM (Growing Self-Organizing Map) [3], GFMM (General FMM) [8], and SCFMN [19].

Pedrycz, W. highlighted that, in real-world applications, the effectiveness of clustering can be significantly enhanced by leveraging prior information. Even a small percentage of labeled samples can substantially improve clustering outcomes [14].

Data clustering has been applied across various fields such as commerce, biology, spatial data analysis, urban planning, web mining, and biomedical informatics. Model selection is a critical step in addressing real-world clustering problems. The choice of a clustering model heavily depends on the clustering objective. Additionally, factors such as the priority given to cluster quality versus computational time, the characteristics of the data, and other considerations also influence the selection of the clustering model.

In healthcare, many scientists and doctors are interested in applying data clustering techniques for prediction and, more specifically, for disease diagnosis [18]. Diagnosing diseases based on medical test results can be framed as a pattern recognition problem, which has garnered significant attention from researchers. Intelligent techniques such as data mining, fuzzy logic, artificial neural networks, genetic algorithms, and others are commonly used. The use of fuzzy min-max neural networks (FMN) is considered an effective approach [16]. FMN leverages the advantages of integrating fuzzy logic, artificial neural networks (NN), and fuzzy min-max theory (FMM) to effectively handle uncertain information. Another reason for the widespread adoption of FMN is its ability to generate simple fuzzy rules that are quantitatively defined based on data.

The FMN model was first introduced by Simpson in 1992 [17] with a supervised learning approach (FMCnN - FMM Classification NN). In 1993, Simpson presented a variant with an unsupervised learning approach (FMCgN - FMM Clustering NN) [15]. Later, a semi-supervised variant called GFMM (General FMN) was developed [8]. GFMM is a hybrid algorithm capable of handling both classification and clustering tasks. The semi-supervised learning technique was subsequently enhanced to improve the quality and applicability of FMN [21].

For an overview of FMN and its evolution, the studies by Alhroob [2] and Sayaydeh [1] provide a comprehensive review of advancements based on the original FMN since its introduction. These studies highlight the strengths, weaknesses, and practical applications of FMNN.

One of the limitations of many models evolved from FMN is their difficulty in identifying and adjusting overlaps or managing overlapping regions between hyperboxes (HyBs) during network training. To address this issue, we propose the FMN-OE (semi-supervised learning in Fuzzy Min-max Neural networks with Overlap Elimination) model, which aims to minimize overlaps between HyBs during training and enhance the learning process with auxiliary information provided by experts. Additionally, FMN-OE has the capability to autonomously identify supplementary information during network training. The main contributions of this paper are (i) the improvement of the learning method with semi-supervised learning to optimize FMN; and (ii) the evaluation of the proposed method using datasets from the UCI and CS machine learning repositories.

The remainder of the paper is organized as follows: Sect. 2 discusses the FMN model. Section 3 introduces the new FMN-OE algorithm. Section 4 describes the experiments conducted on standard datasets and compares the results with SCFMN [19], MSCFMN [13], FMM-GA [22], and several other methods. Section 5 discusses the advantages, limitations of FMN, and potential future research directions.

2   Related Work

2.1   Fuzzy Min-Max Neural Network

FMN is developed based on FART and LCA [17]. FMN leverages the strengths of combining fuzzy logic, artificial neural networks, and FMM theory to address clustering and classification problems. FMN is an incremental neural network model that partitions data into fuzzy HyBs, enabling it to handle large-scale datasets [12]. Incremental learning is an effective technique in knowledge discovery, as it allows for the reuse and enhancement of information with a single pass through the dataset without the need to retrain from scratch on the entire dataset, especially when dealing with large volumes of data [11]. Furthermore, FMN provides soft decisions in fuzzy data classification and clustering.

FMN is built upon fuzzy HyBs, where each HyB represents a subpartition of data in the data space, bounded by min and max points. Each fuzzy HyB contains data points with different membership values, allowing the model to handle data that does not entirely belong to a single cluster, but can instead belong to multiple clusters with varying degrees of membership. During classification, FMN calculates the membership value of a data point with respect to each HyB. A data point can belong to multiple HyBs with different degrees of membership, rather than being strictly assigned to a single cluster. The membership value is determined based on the data point's position relative to the HyB's boundary points. The winning HyB is selected as the one with the highest membership value.

Fuzzy Hyperbox Membership Function. In the FMN model, each fuzzy HyB is bounded in the data space by the min point V and the max point W (denoted as $B_{(V,W)}$), which are determined by equations (1) and (2). The size of the HyB B, denoted as θ_B, is calculated using equation (3). The membership value of a data point A with respect to the HyB $B_{(V,W)}$ (denoted as $b_{(A,B)}$), is computed using equation (4). The data point A is considered to be contained within the HyB B if $b_{(A,B)} = 1$) (in which case, A is said to have full membership in B). Conversely, if $b_{(A,B)} < 1$), A is not contained within B. Figure 1 illustrates the representation of HyBs B_1, and B_2 with samples A_1, A_2, and A_3 in a 2-D: $b_{(A_1,B_1)} = 1$, $b_{(A_2,B_2)} = 1$, $b_{(A_1,B_2)} < 1$, $b_{(A_2,B_1)} < 1$, $b_{(A_3,B_1)} < 1$, $b_{(A_3,B_2)} < 1$.

$$V = (v_i \mid i = 1, \ldots, n), \quad V \in \mathbb{R}^n \tag{1}$$

$$W = (w_i \mid i = 1, \ldots, n), \quad W \in \mathbb{R}^n \tag{2}$$

$$\theta_B = \frac{1}{n} \sum_{i=1}^{n} (w_i - v_i) \tag{3}$$

$$b_{(A,B_{(V,W)})} = \frac{1}{n} \sum_{i=1}^{n} [1 - f(a_i - w_i, \gamma) - f(v_i - a_i, \gamma)] \tag{4}$$

where: $A \in R^n, A = (a_i \mid i = 1, ..., n)$; γ is a parameter that adjusts the rate at which b decreases during the contraction of the HyB; $f(x,y)$ is a two-parameter function used to modulate the attenuation of the membership value μ_j, based on the correlation between the data sample A_h and the boundary of the FH B_j, its mathematical formulation is given in Equation (5):

$$f(x,y) = \begin{cases} 0, & \text{if } xy < 0 \\ xy, & \text{if } 0 \leq xy \leq 1 \\ 1, & \text{if } xy > 1 \end{cases} \tag{5}$$

The architecture of FMNs. The FMCnN neural network utilizes a feed forward structure composed of three layers: F_A, F_B, and F_C (Fig. 2(a)) [17]. The F_A layer contains n nodes, with each node corresponding to a feature of the data. The F_C layer consists of k nodes, each representing an output class/cluster.

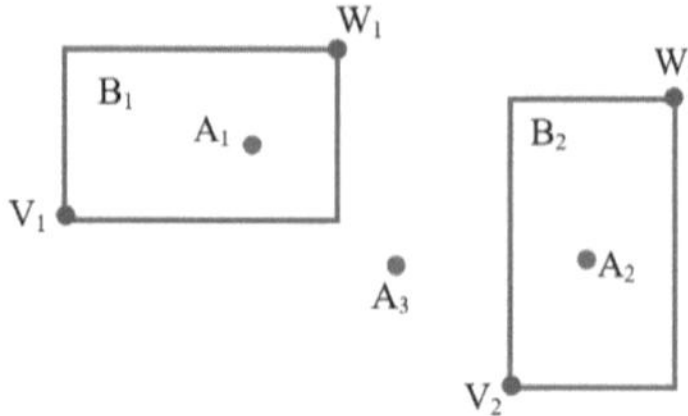

Fig. 1. An example of HyB representation with input samples in 2-D space.

The FMCgN structure consists of two layers: F_A and F_B (Fig. 2(b)) [15]. The input layer, F_A, contains n nodes, with each node corresponding to a feature of the data. The output layer, F_B, consists of k nodes, with each node corresponding to a HyB. Each output HyB represents a cluster.

Figure 2(c) illustrates the structure of a neuron in the F_B layer, where each neuron represents a HyB.

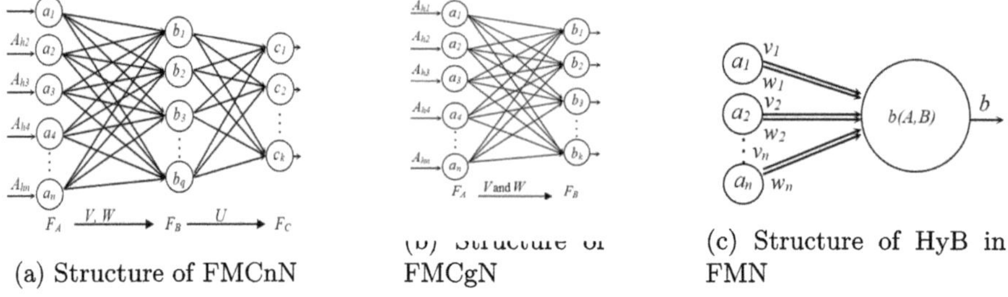

(a) Structure of FMCnN

(b) Structure of FMCgN

(c) Structure of HyB in FMN

Fig. 2. Provides an overview of the FMN neural network model.

Fuzzy Min-Max Clustering Algorithm. The learning algorithm of the FMN involves expanding and contracting HyBs to adjust their min-max values within the sample space. The learning process begins by sequentially selecting each data sample from the training dataset to form and refine the HyBs. The learning algorithm in FMNs consists of three main steps [15]: *(1) create/expand HyB; (2) HyB overlap checking; and (3) HyB overlap elimination.* Steps 1 through 3 are repeated for each input data point.

2.2 Review of FMN-Related Models

SCFMN Model. SCFMN [19] enhances the learning method of FMN by training the network to automatically compute supplementary information, which is not provided by experts. SCFMN uses a fuzzy parameter Beta to supervise the data partitioning process in the feature space. SCFMN employs a comprehensive search method with multiple passes through the training dataset (exhaustive search). SCFMN significantly improves the performance of FMN and GFMM and reduces the number of HyBs compared to similar models. However, SCFMN has some drawbacks: (i) the β parameter is determined based on empirical experience through multiple training iterations to find the optimal value, which directly affects the algorithm's performance; and (ii) the input order affects the training time, although it has been shown not to impact clustering quality.

MSCFMN Model. MSCFMN [13] inherits the advantages and operational mechanisms similar to SCFMN. The advantage of MSCFMN over SCFMN is that it requires only a single pass through the training dataset to form HyBs and compute supplementary information for them. To minimize the need for multiple passes through the training data in SCFMN, MSCFMN determines additional information for unlabeled HyBs during training based on their membership degree to previously labeled HyBs, using the fuzzy parameter Beta. This approach may lead to incorrect labeling if the data distribution within the HyBs is uneven or if the HyBs have low utilization rates. This represents a limitation of MSCFMN.

3   FMN-OE Model

This section presents the concept of FMN-OE. FMN-OE is designed to generate new HyBs instead of creating overlapping regions and then adjusting them, as seen in other variants and the original FMN. This approach gives FMN-OE an advantage in handling overlapping regions, which are often the boundaries of the data (noisy data regions).

The Idea of FMN-OE. The learning algorithm of FMN-OE is based on evaluations of FMN variants regarding the management of overlapping regions [1]. During the training process of FMN, the expansion of HyBs to incorporate additional training data points can create overlapping regions between HyBs. There are two approaches to handling these overlapping regions: (i) accepting the overlap between HyBs and creating new HyBs to manage the overlapping region, which increases the network's structural complexity; (ii) shrinking the HyBs to eliminate the overlap, which may result in the loss of important information and degrade network performance [5].

FMN-OE differs from these models in its approach to handling HyB overlap during training. Instead of managing overlapping areas, FMN-OE avoids creating overlaps altogether. When a potential overlap is detected, FMN-OE expands the HyB to the maximum possible extent without creating an overlap. The remaining unclassified region is then formed into a new HyB, which is assigned a label and reviewed by experts. Figure 3 illustrates the idea of FMN-OE.

Assume that A_h is a training data point in the FMN neural network (with two initial HyBs in the network, $B_{1(V_1,W_1)}$ and $B_{2(V_2,W_2)}$). The learning algorithm computes $b_{(A_h,B_1)}, b_{(A_h,B_2)}$ and checks the expansion condition (Fig. 3(a)).

Assuming B_1 is the winning HyB, FMN adjusts B_1 based on the characteristics of A_h to expand B_1 and include A_h within it (Fig. 3(b)). After expanding B_1, an overlap occurs between B_1 and B_2 (Fig. 3(b)). FMN then shrinks both B_1 and B_2 to eliminate the overlap (Fig. 3(c)). This overlap adjustment can lead to the loss of previously trained and informative data points (Fig. 3(d)). Unlike FMN, FMN-OE does not expand B_1 based solely on A_h. Instead, FMN-OE expands B_1 while considering the position of B_2 and creates a new HyB (Fig. 3(e)).

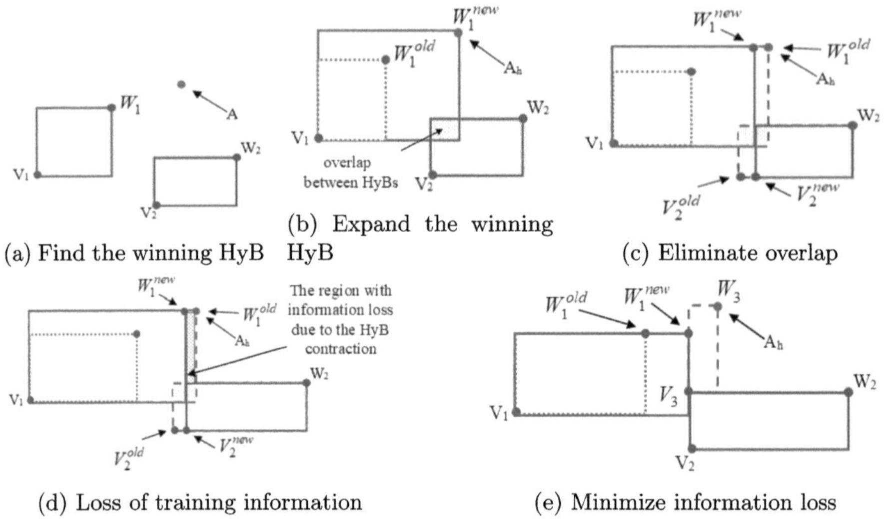

(a) Find the winning HyB

(b) Expand the winning HyB

(c) Eliminate overlap

(d) Loss of training information

(e) Minimize information loss

Fig. 3. An illustrates the idea of FMN-OE.

Main Steps of FMN-OE. The FMN-OE neural network training algorithm operates similarly to MSCFMN [13] and consists of two phases:

Phase 1. Calculating additional information and identifying the data points that receive this additional information: This phase operates similarly to the additional information determination phase of MSCFMN. The learning algorithm performs: (i) the calculation of additional information, in some cases, additional information can be provided by experts; (ii) the identification of data points that will receive this additional information.

Phase 2. Perform network training to partition the data, prioritizing the processing of data points that were supplemented with additional information in the previous phase. The new supplementary information in this phase is provided by experts. The steps of the FMN-OE neural network training algorithm are detailed in Table 1 .

$$b_{(A_h, B_j)} = \max \left\{ b_{(A_h, B_j)} \mid j = 1, 2, \ldots, |B| \right\} \tag{6}$$

Table 1. The steps of the FMN-OE neural network training algorithm.

Input c, θ_{max}, $D = \{A_h | h = 1, ..., |D|\}$

---

1 **for** $Ah \in D$

2   finds an B_j satisfying (6 and (7):

3   if there exists a B_j that satisfies conditions (6) and (7), check whether B_j is overlaps with B_k ($B_k \in B$, $k \neq j$) after expansion.

4   - if there is no overlap, perform the expansion of B_j;
    - if there is overlap:

4.1   + Create B_{new} ($B_{new} = A_h$); consult experts regarding the labeling of B_{new};

4.2   + Expand HyB B_j so that it does not create an overlapping region with B_k;

5   else Create B_{new} ($B_{new} = A_h$); consult experts regarding the labeling of B_{new}.

6 **end for**

Return B;

$$\theta_{(A_h, B_j)} \leq \theta_{\max} \tag{7}$$

$$\theta_{(A, B_j)} = \frac{1}{n} \sum_{i=1}^{n} \left(\max(w_{ji}, a_i) - \min(v_{ji}, a_i) \right) \tag{8}$$

The complexity of FMN-OE algorithm. Let $m = |D|$ be the number of training samples, n the data dimensionality, and $h_t = |B|$ the number of HyBs at iteration t.

Checking or updating one HyB costs $O(n)$. Thus, the cost for one sample is

$$T(A_h) = O(h_t n). \tag{9}$$

Summing over m samples yields

$$T(m) = \sum_{t=1}^{m} O(h_t n) = O(m \bar{h} n), \tag{10}$$

where $\bar{h} = \frac{1}{m} \sum_{t=1}^{m} h_t$.

In the best case $h_t = O(1)$,

$$T(m) = O(mn), \tag{11}$$

while in the worst case $h_t = O(t)$,

$$T(m) = O(m^2 n). \tag{12}$$

4   Experimental Results

Data Description. The experimental data used in our study includes the Benchmark dataset from the CS [7] and UCI Machine Learning Repository [6]. Details of each dataset are provided in Table 2. We selected these datasets to compare performance with several previously published algorithms.

Table 2. Information on training datasets.

#	Datasets	Number of objects	Number of attributes	Number of clusters
1	Thyroid	7200	21	3
2	Iris	150	4	3
3	PID	768	8	2
4	Sonar	208	60	2
5	Wine	178	13	2
6	Flame	240	2	2
7	Jain	373	2	2
8	R15	600	15	2
9	Spiral	312	2	3
10	Pathbased	317	2	3
11	Aggregation	788	2	7

Experimental Method. The objective of the experiment is to evaluate the performance, quantity, and distribution of fuzzy HyBs generated by the newly proposed FMN-OE algorithm. We compare the performance of FMN-OE with several previously published methods that operate based on the original FMN model. We conduct practical experiments and evaluations to demonstrate the capabilities of FMN-OE in real-world applications, particularly in environments characterized by scarce or uncertain information, guided by suggestions from experts in the field.

The experiments utilize the *"k-fold cross-validation"* method, with k=10 for evaluation. The experimental results are the average of 10 runs.

To evaluate and provide a basis for comparing FMN-OE with other methods, we use the Accuracy metric (*Acy*) [22], calculated using equation (13).

$$Acy = \frac{1}{m} \sum_{i=1}^{m} H \left(A_i^l = \overline{A_i^l} \right) \tag{13}$$

where: represents the true label of A_i, while represents the cluster labels assigned to A_i based on the clustering results; m is the total number of samples in the test set; H(y)=1 if $A_i^l = \overline{A_i^l}$, H(y)=0 if $A_i^l \neq \overline{A_i^l}$.

Experimental scenarios:

- Scenario 1: Evaluate the performance of the FMN-OE model by altering the order of the training dataset across 10 experimental runs (10-*fold* cross-validation) using the accuracy metric and the mean value of accuracy.

- Scenario 2: Compare the performance of the FMN-OE model with several other models based on the accuracy metric and network training time.

Implement and Evaluation Results. Table 3 and Fig. 4 present the Accuracy metric values of the FMN-OE model across 10 experimental runs on the benchmark datasets. The results indicate that the performance of the FMN-OE model remains consistent throughout the experiments, with the Accuracy metric being minimally affected by the input data order in the training datasets.

Table 3. Acc (%) values over 10 experimental runs.

Datasets	1	2	3	4	5	6	7	8	9	10
PID	71.65	71.85	70.96	71.82	71.73	71.99	70.92	71.68	70.95	71.65
Sonar	75.01	73.48	74.13	74.30	74.23	75.63	74.03	74.12	73.24	74.13
Thyroid	92.58	93.45	92.55	93.87	94.07	92.14	93.21	93.12	92.87	92.60
Wine	94.28	94.15	94.23	93.97	94.05	94.25	93.88	94.42	94.37	94.03
Iris	95.23	96.45	96.33	95.36	95.35	95.56	94.25	95.68	96.76	94.93
Flame	98.75	98.75	97.65	97.87	97.98	98.75	98.45	98.76	97.92	97.92
Jain	99.42	99.21	98.65	99.73	99.47	99.47	99.73	100.00	100.00	99.73
R15	99.50	99.72	99.50	99.63	99.39	99.50	99.50	99.50	99.50	99.17
Spiral	97.77	97.93	97.92	97.86	98.13	97.91	97.93	98.01	99.05	99.36
Pathbased	97.81	98.08	98.41	98.61	98.43	97.84	98.02	99.44	98.41	99.38
Aggregation	97.72	97.85	97.97	98.22	97.92	98.23	98.73	97.72	98.22	97.56

Fig. 5 illustrates the variation of the Accuracy metric with respect to the hyperbox size parameter θ. The results indicate that as θ increases, the Accuracy tends to decrease across all datasets. Conversely, reducing θ generally improves Accuracy. However, when θ becomes too small, the model suffers from "overfitting", leading to a decline in Accuracy and less reliable results.

From Table 4 to Table 6 present the performance comparison results between the models based on the Accuracy metric and training time. From Fig. 6 to Fig. 8 provide visual charts of the comparison results.

The results in Table 4 and the visual representation in Fig. 6 show that the FMN-OE model outperforms SCMFN and MSCFMN. FMN-OE's superiority can be attributed to two main reasons: (i) FMN-OE receives additional expert input during the training process, and (ii) it avoids information loss (especially for data points that have received expert input) by not performing HyB contraction to eliminate overlap. Notably, FMN-OE can perform clustering even with datasets that have non-spherical distributions or clusters with significantly different sizes (such as Spiral, Pathbased, Aggregation), whereas SCFMN and

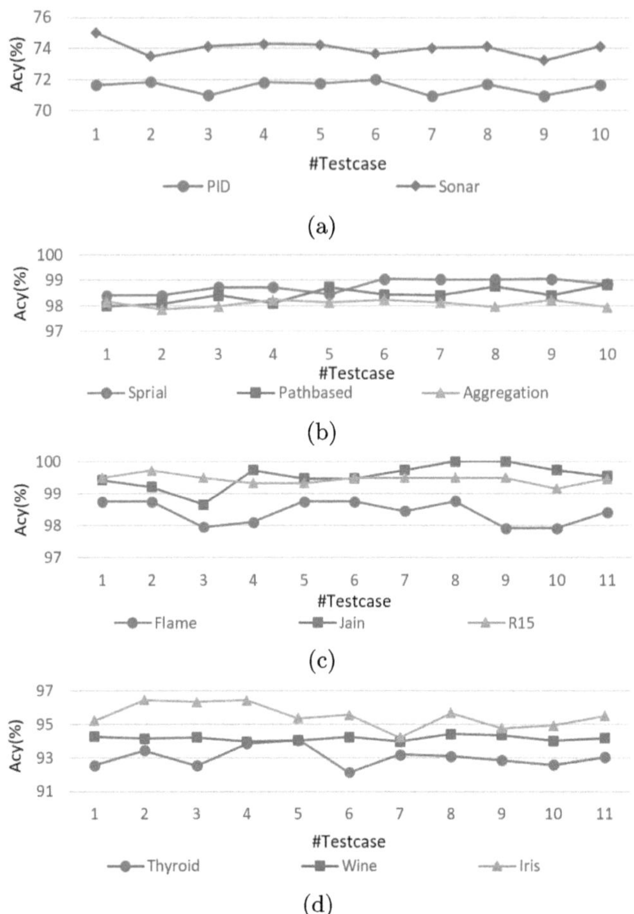

Fig. 4. Variation in the Accuracy metric across experimental runs with $k=10$.

MSCFMN are unable to do so. This is because FMN-OE can receive supplementary information from experts to guide the training process, while SCFMN and MSCFMN rely entirely on self-determined supplementary information.

The results in Table 5 and the visual representation in Fig. 7 show that the FMN-OE model has a similar Accuracy metric value to FMN-GA on the PID, Sonar, Thyroid, Wine, and Iris datasets, and slightly higher on the Flame, Jain, and R15 datasets. This indicates that the FMN-OE training process effectively leverages the supplementary information associated with the samples. In contrast, FMN-GA loses some training information due to the adjustments made to eliminate overlap between HyBs.

Table 6 presents the performance comparison results of FMN-OE with K-means, SC, CSPA, HPGA, NMFC, WC, EGWCA, and eGWCAFMM based on

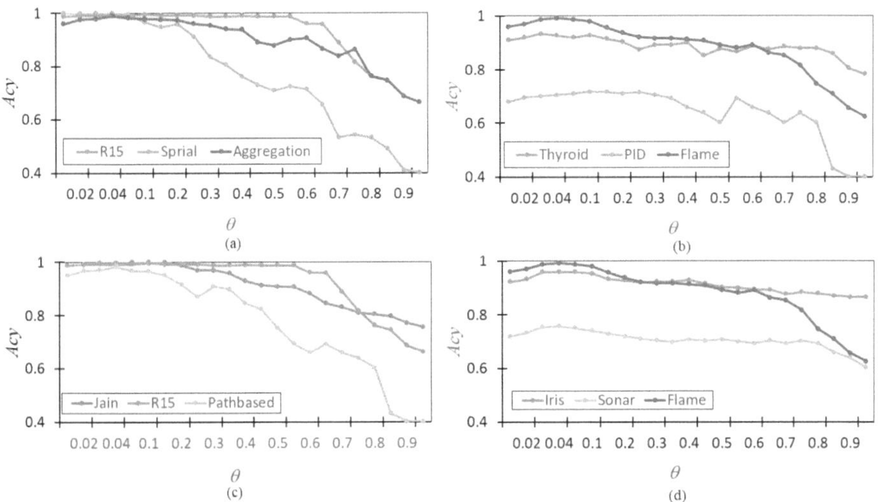

Fig. 5. The results in the figures demonstrate how accuracy varies with different values of theta.

Table 4. Comparison of *Acy* values (%) among the models.

#	Datasets	SCFMN	MSCFMN	FMN-OE
1	PID	70.57	70.12	71.52
2	Sonar	73.91	73.54	74.03
3	Thyroid	92.65	92.63	93.05
4	Wine	93.33	93.86	94.15
5	Iris	94.12	94.67	95.50
6	Flame	97.93	97.95	98.35
7	Jain	98.72	98.54	99.50
8	R15	99.50	99.64	99.70
9	Spiral	–	–	98.72
10	Pathbased	–	–	98.41
11	Aggregation	–	–	97.97

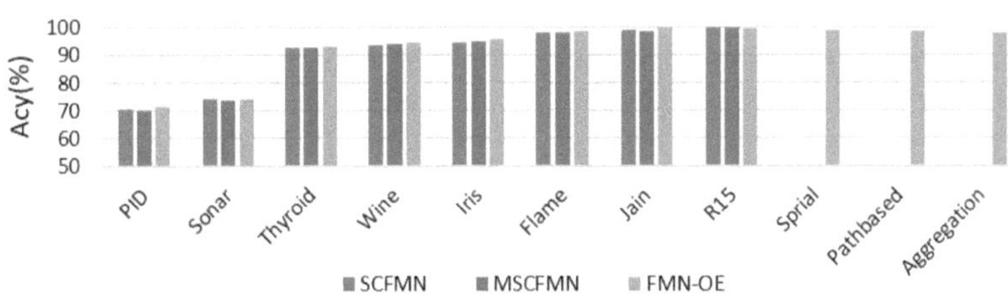

Fig. 6. A visual chart comparing the performance based on the Acy metric.

Table 5. *Acy* values (%) on the benchmark datasets.

Datasets	FMM-GA	FMN-OE
PID	70.44	71.52
Sonar	73.43	74.03
Thyroid	92.63	93.05
Wine	93.33	94.17
Iris	95.42	95.50
Flame	96.32	98.36
Jain	96.54	99.54
R15	97.83	99.46

the Accuracy metric. The results indicate that FMN-OE outperforms the other methods.

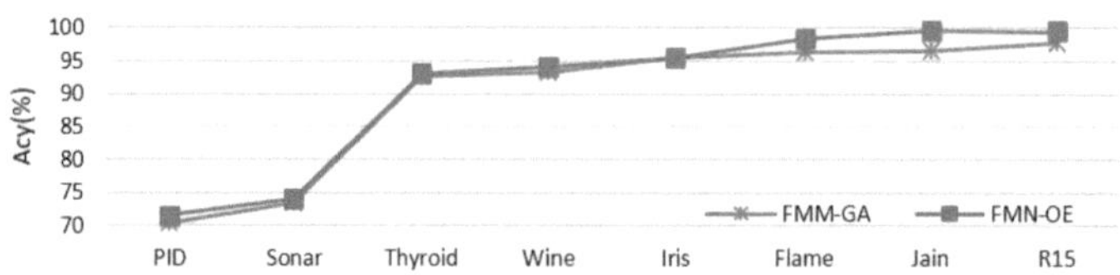

Fig. 7. Graph comparing the Accuracy metric between FMM-GA and FMN-OE.

Table 6. Comparison of *Acy* values across the methods in [22]

Methods	*Acy*(%)	Methods	*Acy* (%)	Methods	*Acy* (%)
K-means	83	WC	89	HPGA	69
SC	91	EGWCA	92	FMN-OE	**95**
CSPA	86	eGWCAFMM	90	NMFC	89

Figure 8 provides a visual comparison of the training time (Time) between FMN-OE with SCFMN, and MSCFMN . The chart shows that FMN-OE has a shorter training time. This shorter training time for FMN-OE can be attributed to the fact that it only requires a single pass through the training dataset. In contrast, SCFMN performs multiple passes through the training data to exhaustively cover all cases, while MSCFMN requires multiple passes over the HyBs to determine supplementary information for the data points.

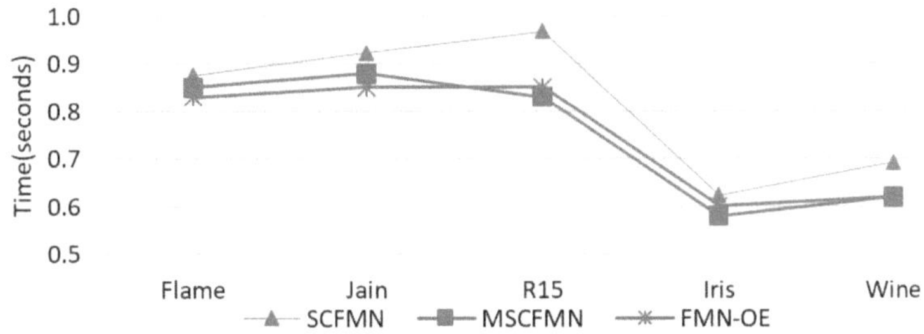

Fig. 8. The graphic compares the training time.

5 Conclusion and Future Works

This study introduces a new model called FMN-OE. FMN-OE improves upon the learning algorithm of Fuzzy Min-Max Neural Networks (FMN) and its previous variants by addressing HyB overlap and utilizing y information provided by experts during training. Experimental results demonstrate that FMN-OE: (i) stabilizes HyBs by preventing overlap within the network, (ii) is not affected by the order of data sample in the training dataset, and (iii) performs better than several other methods. However, similar to other FMN variants, the adaptive learning process of FMN-OE requires time and experience through"trial and error" to determine input parameters. This is also a limitation of FMN and its variants, and it represents a potential direction for future research.

References

1. Al Sayaydeh, O.N., Mohammed, M.H., Lim, C.P.: Survey of fuzzy min–max neural network for pattern classification variants and applications. IEEE trans. fuzzy syst. 27(4):635–645, 2018
2. Alhroob, E., Mohammed, M.F., Lim, C.P., Tao, H.: A critical review on selected fuzzy min-max neural networks and their significance and challenges in pattern classification. IEEE Access, 7:56129–56146, 2019
3. Allahyar, A.: Hadi Sadoghi Yazdi, and Ahad Harati. Constrained semi-supervised growing self-organizing map. Neurocomputing **147**, 456–471 (2015)
4. Bezdek, J.C.: Pattern recognition with fuzzy objective function algorithms. Springer Science and Business Media, 2013
5. Davtalab, R., Dezfoulian, M.H., Mansoorizadeh, M.: Multi-level fuzzy min-max neural network classifier. IEEE Trans. Neural Netw. Learn. Syst. 25(3):470–482, 2013
6. Dua, D., Graff, C.: UCI machine learning repository, (2024). Accessed 14 Oct 2024
7. Fränti, P., Sieranoja, S.: Clustering datasets, (2024). Accessed 14 Oct 2024
8. Gabrys, B., Bargiela, A.: General fuzzy min-max neural network for clustering and classification. IEEE Trans. Neural Netw. **11**(3), 769–783 (2000)
9. Gao, J., Wang, Z., Lei, Z., Rong-Long Wang., Wu, Z., Gao, S.: Feature selection with clustering probabilistic particle swarm optimization. Int. J. Mach. Learn. Cybern. pp. 1–19, 2024

10. Li, F., Yue, P., Su, L.: Research on the Convergence of Fuzzy Genetic Algorithm Based on Rough Classification. In: Jiao, L., Wang, L., Gao, X., Liu, J., Wu, F. (eds.) ICNC 2006. LNCS, vol. 4221, pp. 792–795. Springer, Heidelberg (2006). https://doi.org/10.1007/11881070_105
11. Luo, C., Li, T., Chen, H., Liu, D.: Incremental approaches for updating approximations in set-valued ordered information systems. Knowl.-Based Syst. **50**, 218–233 (2013)
12. Martınez-Rego, D., Fontenla-Romero, O., Alonso-Betanzos, A.: Nonlinear single layer neural network training algorithm for incremental, nonstationary and distributed learning scenarios. Pattern Recogn. **45**(12), 4536–4546 (2012)
13. Minh, V.D., Ngan, T.T., Tuan, T.M., Duong, V.T., Cuong, N.T.: An improvement in integrating clustering method and neural network to extract rules and application in diagnosis support. Iranian J. Fuzzy Syst. **19**(5), 147–165 (2022)
14. Pedrycz, W., Waletzky, J.: Fuzzy clustering with partial supervision. IEEE Trans. Syst. Man, and Cybern. Part B (Cybernetics), 27(5):787–795, 1997
15. SIMPSON PK.: Fuzzy min-max neural networks-part ii: clustering. IEEE Trans. Fuzzy Syst. 1:32–45, 1993
16. Quteishat, A., Lim, C.P.: Application of the fuzzy min-max neural networks to medical diagnosis. In: International Conference on Knowledge-Based and Intelligent Information and Engineering Systems, pp. 548–555. Springer, 2008
17. Simpson, P.K.: Fuzzy min-max neural networks. i. classification. IEEE trans. neural netw. 3(5):776–786, 1992
18. Singh, A., Pandey, B.: Intelligent techniques and applications in liver disorders: a survey. Int. J. Biomed. Eng. Technol. **16**(1), 27–70 (2014)
19. Tran, T.N., Vu, D.M., Tran, M.T., Le, B.D.: The combination of fuzzy min–max neural network and semi-supervised learning in solving liver disease diagnosis support problem. Arab. J. Sci. Eng. 44:2933–2944, 2019
20. Vu, D., Nguyen, T., Phung, T., Vu, T.: Choosing data points to label for semi-supervised learning based on neighborhood. In: International Conference on Advances in Information and Communication Technology, pp. 134–141. Springer, 2023
21. Vu, D., Nguyen, V., Le, B.D.: Semi-supervised clustering in fuzzy min-max neural network. In: Advances in Information and Communication Technology: Proceedings of the International Conference, ICTA 2016, pp. 541–550. Springer, 2017
22. Wang, J., et al.: Patient admission prediction using a pruned fuzzy min–max neural network with rule extraction. Neural Comput. Appl. 26:277–289, 2015
23. Yasunori, E., Yukihiro, H., Makito, Y., Sadaaki, M.: On semi-supervised fuzzy c-means clustering. In: 2009 IEEE International Conference on Fuzzy Systems, pp. 1119–1124. IEEE, 2009
24. Zhang, H., Jing, L.: Semi-supervised fuzzy clustering: a kernel-based approach. Knowl.-Based Syst. **22**(6), 477–481 (2009)

Enhancing Cryptocurrency Forecasting Using Temporal Features On High-Frequency Data

Hoang-Sang Le[1,2], Dinh-Chuong Vu[1,2], Hoang-Hai Pham[1,2],
and Duy-Hoang Tran[1,2(✉)]

[1] Faculty of Information Technology, University of Science, Ho Chi Minh City,
Vietnam
tdhoang@fit.hcmus.edu.v
[2] Vietnam National University, Ho Chi Minh City, Vietnam

Abstract. Forecasting cryptocurrency price and trend using high frequency data remains a challenging task due to the inherent noise and complexity of market microstructures. While deep learning models have achieved notable success, those based primarily on standard open, high, low, close, volume (OHLCV) or limit order book (LOB) data often suffer from overfitting and limited generalization to out-of-sample scenarios. This paper explores whether incorporating cyclical temporal features—such as minute-of-hour, hour-of-day, and day-of-week—can improve prediction performance. We apply this approach to high-frequency Ethereum data and evaluate its effectiveness using three deep learning architectures: LSTM, CNN, and a hybrid model. Across both minute-ahead price prediction and short-term trend prediction tasks, our results show that integrating these periodic time features leads to consistent and significant improvements in forecasting accuracy. The source code is available and maintained in the GitHub repository (https://github.com/LHSang6403/HFT-Time-Cyclic-Improvement).

Keywords: Quantitative Trading · Deep Learning · Feature Engineering · Time Series Forecasting · Temporal Feature

1 Introduction

High-Frequency Trading (HFT) has fundamentally reshaped modern financial markets by leveraging algorithmic speed and sophisticated data analysis to exploit fleeting price inefficiencies [1]. The emergence of cryptocurrencies has introduced a new, formidable frontier for these strategies [2]. Unlike traditional equity markets, cryptocurrency markets operate continuously, 24 h a day, 7 d a week, and are characterized by pronounced volatility, decentralized governance,

H. S. Le and D.-C. Vu—Equal contribution.

and unique market microstructure dynamics [3,4]. This non-stop trading environment provides a unique and ideal laboratory for investigating the predictive power of temporal cycles, as any observed periodicities are not artifacts of institutional open/close schedules but rather reflect uninterrupted, fundamental market behaviors.

Predictive modeling in this domain is a formidable challenge. High-frequency data, particularly from the limit order book (LOB), is inherently noisy, non-stationary, and high-dimensional, with complex spatio-temporal dependencies between price levels and order volumes [5–7]. These characteristics make forecasting intrinsically difficult and render models susceptible to common pitfalls such as overfitting and poor generalization on out-of-sample data [1,8]. In response, deep learning (DL) has emerged as a state-of-the-art tool, with architectures like Convolutional Neural Networks (CNNs) and Long Short-Term Memory (LSTM) networks demonstrating a powerful capacity to learn hierarchical, non-linear patterns directly from raw market data [9,10].

Much of the existing research on DL-based forecasting has advanced along two primary avenues: proposing novel neural architectures of increasing complexity, or developing handcrafted features from market microstructure data to capture supply-demand imbalances [6,9,11]. However, a significant research gap persists. While the importance of seasonality and cyclical effects in stock investment markets and foreign exchange data are well-documented in lower-frequency financial analysis [12], the systematic incorporation and rigorous validation of recurring temporal periodicities—such as intraday and weekly cycles—as potent predictive features remains largely under-explored within the specific context of high-frequency cryptocurrency forecasting [3,4,13].

This paper addresses a key gap in high-frequency cryptocurrency forecasting by systematically investigating the impact of incorporating cyclical temporal features into deep learning models. Our methodology centers on a feature engineering approach that appends sine and cosine transformations of minute-of-hour, hour-of-day, and day-of-week to traditional inputs derived from OHLCV bars and LOB statistics. This aligns with emerging research suggesting that thoughtful, context-aware feature design can yield greater performance gains than merely increasing architectural complexity [1]. To evaluate the effectiveness of this approach, we conduct a comprehensive benchmarking study on high-frequency Ethereum data [14], testing three distinct deep learning architectures—LSTM, CNN, and a hybrid CNN-LSTM model [15,16]. Performance is assessed on two predictive tasks: minute-ahead price prediction and short-term trend prediction, ensuring the generality of our findings across both precise numerical forecasting and signal-driven directional prediction [8]. The primary contributions of this work are threefold:

- A systematic evaluation of diverse feature sets in high-frequency trading.
- A novel cyclic time-encoding technique to enhance deep model performance.
- A comprehensive benchmarking of modern deep learning architectures on both price and trend prediction tasks.

Our findings demonstrate that the inclusion of cyclical temporal features yields statistically significant and substantial improvements in predictive accuracy across all tested models and tasks. This result underscores that the systematic modeling of temporal periodicities is a crucial, yet previously under-appreciated, pathway to enhancing forecasting performance in high-frequency cryptocurrency markets.

2 Related Work

Despite advances in high-frequency forecasting, challenges with robustness and generalization remain. A key area of progress lies in refining input features, as the quality and structure of inputs often play a more critical role than model complexity.

The choice of input features is a critical design decision that defines the universe of information from which a model can learn. The most foundational approach relies on OHLCV bars, which aggregate trading activity into fixed time intervals [15]. Bao et al. [17] has focused on feeding this raw OHLCV data directly into deep learning models like LSTM, relying on the network's hierarchical feature extraction capabilities. Piravechsakul et al. demonstrated parallel and historically dominant approach augments OHLCV data with a wide array of handcrafted technical indicators (e.g., RSI, MACD) to explicitly quantify market concepts like momentum and volatility [18]. However, at high frequencies, the predictive power of these indicators can diminish, often adding more noise and increasing the risk of overfitting.

To gain a more timely predictive edge, the research frontier has decisively moved towards the LOB, the most fundamental data structure in modern electronic markets [5]. The LOB provides a real-time, high-resolution snapshot of latent supply and demand, containing a wealth of microstructural information critical for short-term forecasting [6,7]. Two paradigms have emerged for exploiting this data. The first treats the LOB as an image-like structure, applying CNN to learn spatial patterns across price levels directly from raw LOB snapshots [9,10]. The second paradigm focuses on engineering features from the LOB based on market microstructure theory, such as order flow imbalance, which is known to be a primary driver of short-term price changes [6,11]. While powerful, both approaches have led to models that often struggle to generalize beyond the specific data on which they were trained.

Temporal periodicity is a well-documented yet underutilized source of predictive information in high-frequency deep learning. Classical finance research has long established the presence of calendar anomalies—such as day-of-the-week and month-of-the-year effects—across various markets, challenging the strictest forms of the efficient market hypothesis [12]. Notably, similar patterns have been observed in cryptocurrency markets, where studies report weekday-dependent variations in efficiency and returns [3,13]. For example, Caporale and Plastun [4] found that Bitcoin yields significantly higher returns on Mondays, with a strategy exploiting this effect producing statistically significant profits. This reveals a gap in current research: while deep learning in high-frequency trading emphasizes complex models and noisy microstructure data, it largely overlooks robust temporal signals identified in empirical finance. This paper bridges that divide by integrating cyclical temporal features into deep learning models and systematically evaluating their contribution to forecasting performance.

3   Methodology

3.1   Problem Definition

This paper investigates how different features influence two key prediction problems in HFT: price prediction and trend prediction. As Fig. 1 illustrates, we developed two distinct predictive pipelines: one for price forecasting and another for trend forecasting. Both pipelines use sliding windows of historical data as input, with each data point consisting of a d-dimensional feature vector. Sliding windows are defined as follows:

$$X_{t-T+1:t} = \left[x_{t-T+1}, x_{t-T+2}, \ldots, x_t\right] \in \mathbb{R}^{T \times d} \tag{1}$$

where T represents the length of the sliding window (e.g., $T = 20\,\text{min}$).

Price prediction: Price prediction aims to forecast the exact closing price of the subsequent data point. This task demands a precise prediction of a continuous value, specifically, of what the market closing price will be at the very next time step. The success of such a prediction hinges on the model's ability to discern subtle patterns and relationships within the historical data, translating them into an accurate future price estimate. The target variable is defined as follows:

$$y_{t+1}^{\text{price}} = \text{close}_{t+1} \tag{2}$$

Trend prediction: Trend prediction aims to forecast the overarching direction of price movement over a specified future horizon, denoted as H time steps. Instead of pinpointing an exact future price, this task focuses on classifying whether the price will generally move "up", "down", or remain "flat" within that H-time step window (e.g., $H = 15\,\text{min}$). This provides a broader perspective on market sentiment and potential directional changes, which can be crucial for strategic trading decisions. The label targets are defined as follows:

$$y_{t+H}^{\text{trend}} = \begin{cases} up, & \text{if } \text{close}_{t+H} > \text{close}_t + \varepsilon \\ down, & \text{if } \text{close}_{t+H} < \text{close}_t - \varepsilon \\ flat, & \text{if } |\text{close}_{t+H} - \text{close}_t| \leq \varepsilon \end{cases} \tag{3}$$

where ε is a small threshold (e.g., $10^{-4} \times \text{close}_t$).

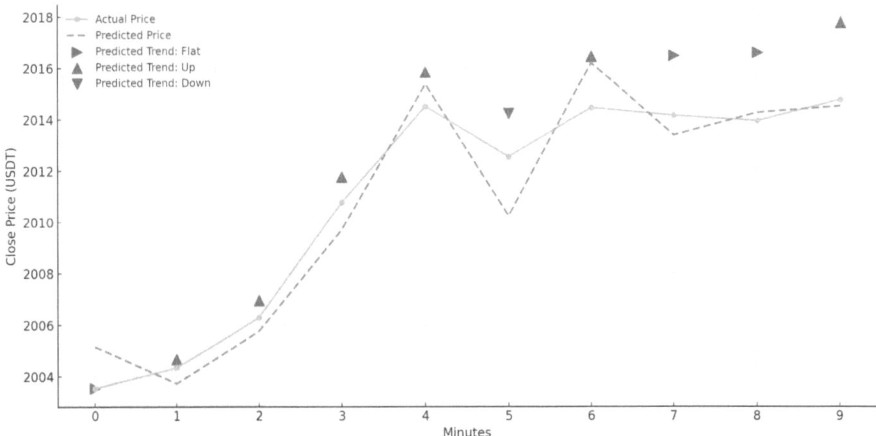

Fig. 1. High-Frequency Trading Data Prediction.

3.2   Feature Engineering

This section details the various features incorporated into our analysis. We specifically leverage historical market data, encompassing both OHLCV and LOB information, in conjunction with cyclical time-based features.

OHLCV. A common approach to process high-frequency trade data is to aggregate individual trades into fixed time intervals, from which standard OHLCV features are computed. For each given interval, let $\{(p_{i,j}, v_{i,j})\}_{j=1}^{n_i}$ represent the sequence of trade prices and corresponding volumes that occurred within that specific interval. The derived bar features are defined as follows:

$$open_i = p_{i,1} \qquad (4) \qquad\qquad high_i = \max_{1 \leq j \leq n_i} p_{i,j} \qquad (5)$$

$$low_i = \min_{1 \leq j \leq n_i} p_{i,j} \qquad (6) \qquad\qquad close_i = p_{i,n_i} \qquad (7)$$

$$volume_i = \sum_{j=1}^{n_i} v_{i,j} \qquad (8)$$

where $open_i$ denotes the first trade price, $high_i$ and low_i represent the highest and lowest prices, $close_i$ is the final trade price, and $volume_i$ is the total traded volume within the i-th interval. These OHLCV bars provide a compact summary of price range and liquidity over fixed time intervals. Aggregating data into such bars helps smooth out sub-second noise and yields fixed-frequency inputs suitable for modeling. OHLCV bars are widely used in both practice and research to capture intraday price dynamics [7]. For instance, interval-based returns or technical indicators derived from these values can serve as informative features for predictive models.

Limit Order Book (LOB). LOB offers detailed, time-stamped snapshots of market supply and demand. LOB features capture the instantaneous state of market microstructure—such as liquidity depth and order flow pressure—which are critical for forecasting short-term price movements. Empirical evidence shows that imbalances between buy and sell orders are key drivers of price changes over short horizons [6]. Recent deep learning models (e.g., Zhang et al.) process raw LOB data directly, learning complex patterns without manual feature engineering [9]. Nonetheless, handcrafted LOB-derived features remain widely used in high-frequency trading models to summarize essential dynamics of supply, demand, and price pressure. At each time t, the LOB records the top L bid and ask price levels:

- $p_{t,i}^b$, $q_{t,i}^b$: price and volume at the i-th best bid (with $p_{t,1}^b$ the highest bid and $q_{t,1}^b$ the corresponding volume).
- $p_{t,i}^a$, $q_{t,i}^a$: price and volume at the i-th best ask (with $p_{t,1}^a$ the lowest ask and $q_{t,1}^a$ the corresponding volume).
- $V_t^b = \sum_{i=1}^L q_{t,i}^b$, $V_t^a = \sum_{i=1}^L q_{t,i}^a$: total visible bid and ask volume over L levels.

Bid prices are non-increasing $(p_{t,1}^b \geq p_{t,2}^b \geq ... \geq p_{t,L}^b)$ and ask prices are non-decreasing $(p_{t,1}^a \leq p_{t,2}^a \leq ... \leq p_{t,L}^a)$. This unified notation is used throughout to define all engineered LOB features, including:

- Weighted Average Price (WAP):

$$y_{\mathrm{wap}_1}(t) = \frac{q_t^{a_1} p_t^{b_1} + q_t^{b_1} p_t^{a_1}}{q_t^{a_1} + q_t^{b_1}} \quad (9) \qquad y_{\mathrm{wap}_2}(t) = \frac{q_t^{a_2} p_t^{b_2} + q_t^{b_2} p_t^{a_2}}{q_t^{a_2} + q_t^{b_2}} \quad (10)$$

- Volume-Weighted Average Price (VWAP):

$$s_t^{b_i} = \frac{q_t^{b_i}}{V_t^b + V_t^a} \quad (11) \qquad s_t^{a_i} = \frac{q_t^{a_i}}{V_t^b + V_t^a} \quad (12)$$

$$y_{\mathrm{buy\_vwap}}(t) = \sum_{i=1}^L s_t^{b_i} p_t^{b_i} \quad (13) \qquad y_{\mathrm{sell\_vwap}}(t) = \sum_{i=1}^L s_t^{a_i} p_t^{a_i} \quad (14)$$

- Depth Spreads:

$$y_{\mathrm{buy\_spread}}(t) = p_t^{b_1} - p_t^{b_L} \quad (15) \qquad y_{\mathrm{sell\_spread}}(t) = p_t^{a_L} - p_t^{a_1} \quad (16)$$

- Normalized Price Spread:

$$y_{\mathrm{price\_spread}}(t) = \frac{2\left(p_t^{a_1} - p_t^{b_1}\right)}{p_t^{a_1} + p_t^{b_1}} \quad (17)$$

Cyclical Time-based Features. Encoding temporal components—such as minute of the hour, hour of the day, and day of the week—using sine and cosine functions is a common technique in time-series analysis. This approach effectively captures cyclical patterns inherent in financial data. For example, Kristjanpoller et al. [3] observed that market efficiency metrics displayed distinct weekday-dependent patterns across various cryptocurrencies, indicating that efficiency varies by day. Similarly, Aharon and Qadan [13] investigated calendar anomalies in Bitcoin and found that Mondays are often associated with higher returns and volatility, with this day-of-week effect evolving over time. Despite such findings, few studies have explored these cyclical effects in the context of HFT. In this study, we explicitly incorporate cyclical time features into our forecasting model. For each timestamp, we compute sine and cosine encodings of the minute (m), hour (h), and weekday (d) components. Following standard practice, we define:

$$x_{\min,\sin} = \sin\!\left(\tfrac{2\pi m}{60}\right) \quad (18) \qquad x_{\min,\cos} = \cos\!\left(\tfrac{2\pi m}{60}\right) \quad (19)$$

$$x_{\text{hour},\sin} = \sin\!\left(\tfrac{2\pi h}{24}\right) \quad (20) \qquad x_{\text{hour},\cos} = \cos\!\left(\tfrac{2\pi h}{24}\right) \quad (21)$$

$$x_{\text{dow},\sin} = \sin\!\left(\tfrac{2\pi d}{7}\right) \quad (22) \qquad x_{\text{dow},\cos} = \cos\!\left(\tfrac{2\pi d}{7}\right) \quad (23)$$

where $m = 0,\ldots,59$ (minute of the hour), $h = 0,\ldots,23$ (hour of the day), and $d = 0,\ldots,6$ (day of the week). Illustrative examples of these encodings are presented in Fig. 2.

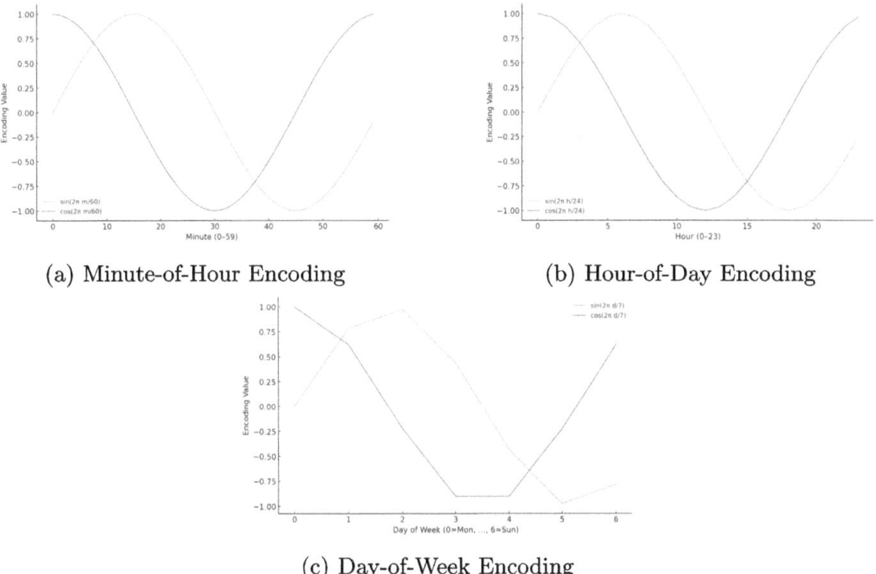

(a) Minute-of-Hour Encoding   (b) Hour-of-Day Encoding

(c) Day-of-Week Encoding

Fig. 2. Time cyclic formulas: (a) Minute-of-Hour Encoding, (b) Hour-of-Day Encoding, (c) Day-of-Week Encoding.

These six cyclical time features are appended to the standard OHLCV bars and LOB-derived variables in our model. We then evaluate the model's short-term forecasting performance—both in price prediction and trend prediction tasks—with and without the inclusion of time encodings. This allows us to assess whether incorporating minute-level cyclical temporal information enhances predictive accuracy.

3.3 Models

Given the complexity and high volatility of financial time-series data, we explored multiple deep learning architectures to assess their suitability for high-frequency forecasting. A comprehensive evaluation was conducted using three prominent model types: Long Short-Term Memory (LSTM) networks [18], Convolutional Neural Networks (CNN) [2], and a hybrid CNN-LSTM architecture [9]. Each architecture was tested under a range of hyperparameter configurations and instantiated from ten distinct random initializations to evaluate performance robustness.

LSTM were chosen for their strength in capturing long-term temporal dependencies in time-series data. The LSTM model includes two stacked LSTM layers with 64 and 32 hidden units, respectively, separated by a dropout layer with a rate of 0.2 to mitigate overfitting. CNN were selected for their ability to efficiently detect local patterns and short-term dependencies within sequential input windows. The CNN model architecture consists of two 1D convolutional layers with 64 filters and a kernel size of 3, each followed by ReLU activation and max-pooling layers. These are connected to a fully connected dense layer with 64 units and a sigmoid output layer for binary trend advertising tasks. The hybrid CNN-LSTM model combines the strengths of both architectures: convolutional layers are used for local feature extraction, followed by LSTM layers to capture sequential dependencies.

All models were implemented using the PyTorch framework and trained with GPU acceleration. The training process used the Adam optimizer with a learning rate of 5×10^{-4}, a batch size of 64, and a maximum of 30 epochs. Additionally, class imbalance in the trend forecasting task was addressed by applying inverse class frequency weighting to the loss function.

4 Experiments

4.1 Dataset

All experiments were conducted using a high-frequency Ethereum (ETH/USDT) price dataset spanning from February 2022 to October 2023 [14]. The raw tick-by-tick trade data was resampled into uniform one-minute intervals to generate both OHLCV and limit order book (LOB) features. Prior to modeling, standard data preprocessing steps were applied, including outlier removal and forward-filling to address missing values at the minute level. This process yielded approximately 770,000 one-minute data points for analysis.

To prevent look-ahead bias, the dataset was split chronologically, with the first 70% used for training and the remaining 30% reserved for testing, so that all test samples occurred strictly after the training period. This ensures the model never sees future values during training, preserving realistic forecasting conditions. Moreover, our sliding window procedure for sequence input was designed so that no window crosses from the training horizon into the test horizon, and no forward-looking features were included. All numerical features were normalized to the [0,1] range using a Min-Max scaler fitted exclusively on the training set and applied unchanged to the test set.

4.2   Experimental Settings

Each predictive model configuration was trained and evaluated under identical conditions to ensure a fair and consistent comparison. We employed the deep learning architectures described in Sect. 3.3 for both forecasting tasks, modifying only the output layer and loss function to suit each pipeline: a single continuous output for price advertising and a three-class output for class advertising. For the price prediction task, we used LSTM, CNN, and CNN-LSTM models to capture temporal dependencies within 20-minute input sequences. These models were optimized using the Mean Squared Error (MSE) loss function, with an adaptive learning rate scheduler applied to reduce the learning rate upon stagnation in validation performance. To prevent overfitting, early stopping was employed when no improvement was observed over successive epochs. For the trend prediction task, the same neural network architectures were adapted to predict one of three trend classes (up, down, and flat). To account for variability due to stochastic initialization and mini-batch sampling, each experiment was repeated twenty times with different random seeds. This 20-run evaluation protocol was consistently applied across all model and feature configurations, allowing us to report average performance metrics and ensure the robustness and statistical reliability of our findings.

All experiments were conducted on a high-performance computing cluster at VNU-HCM University of Science, utilizing the Slurm job scheduling system. Each training job was executed in a standardized environment equipped with an NVIDIA A100 GPU and 32 GB of system RAM. This consistent hardware configuration across all runs ensured that observed performance differences stemmed from variations in model architecture or input features, rather than inconsistencies in computational resources.

4.3   Baseline

To evaluate the impact of our proposed cyclical time-encoding features, we benchmark performance against several well-established baselines from prior work in financial time-series modeling. Our analysis focuses on two core forecasting tasks in high-frequency trading: price prediction and trend prediction. These tasks are assessed across various traditional feature sets and deep learning models.

Feature sets. We compare our approach—enhancing models with cyclical time features—against traditional feature sets as follows:

- **Set 1: OHLCV.** Using only OHLCV features as the baseline. [18].
- **Set 2: LOB.** Using only LOB-based features as the baseline. [9,14].
- **Set 3: OHLCV+LOB.** Combining both OHLCV and LOB features as the baseline.

Models. We employ three prominent deep learning models commonly used in financial forecasting:

- **LSTM Model.** Long Short-Term Memory networks capture sequential dependencies and are widely used in stock and crypto forecasting [15,17,18].
- **CNN Model.** Convolutional Neural Networks extract local patterns and have been applied to LOB time-series data [9,10].
- **Hybrid CNN-LSTM Model.** Hybrid model that stacks convolutional layers for spatial feature extraction with LSTM layers for temporal modeling [2].

4.4   Evaluation Metrics

To comprehensively evaluate model performance, we employed two distinct sets of metrics corresponding to the two core forecasting tasks: price prediction and trend prediction. For the price prediction task, which involves forecasting the future close price as a continuous variable, we used four widely recognized regression metrics: Mean Squared Error (MSE), Root Mean Squared Error (RMSE), Mean Absolute Error (MAE), and the Coefficient of Determination (R^2). These metrics provide complementary perspectives on prediction accuracy, capturing both the magnitude of errors (MSE, RMSE, MAE) and the proportion of variance explained by the model (R^2).

For the trend prediction task, which classifies future price movement into three categories—upward, downward, and flat—we adopted a standard set of classification metrics to assess directional accuracy. These include Accuracy, which measures the overall correctness of predictions; Precision, which reflects the proportion of true positives among all predicted positives for each class; and Recall, which captures the proportion of true positives identified among all actual instances of each class. Together, these metrics provide a robust assessment of the model's prediction capabilities, particularly its ability to distinguish between nuanced market behaviors such as minor fluctuations versus significant directional shifts.

5   Experimental Results

To evaluate the effectiveness of temporal features, we compared models trained on baseline feature sets with those enhanced by cyclic temporal encoding. Experiments were conducted across three architectures—LSTM, CNN, and a hybrid CNN-LSTM—and two forecasting tasks: price prediction and trend prediction. Each model was trained and evaluated over twenty independent runs per configuration to account for the stochastic nature of deep learning. Detailed results for each evaluation metric are presented below.

5.1   Price Prediction Results

Across all model architectures, incorporating sinusoidal time-cyclic encoding consistently enhances price advertising performance. As shown in Table 1, the LSTM model sees notable improvements when time features are added: with OHLCV inputs, MSE decreases by 7.7%, RMSE by nearly 4%, MAE by 4.8%, and R^2 rises by 1.1%; with LOB inputs, error metrics drop more sharply (MSE by 45.1%, RMSE by 25.9%, MAE by 33.4%), and R^2 increases by 0.04%. The combined input set yields the largest gains, with MSE reduced by 61.7%, RMSE by 38.2%, MAE by 53.9%, and R^2 improved by 0.12%. Similar trends are observed in the CNN model, though with smaller gains: up to 28.3% MSE reduction and modest R^2 increases. The hybrid CNN-LSTM architecture also benefits from time-cyclic encoding, showing MSE drops of up to 35.5% and consistent improvements across all error metrics and input sets. These results confirm that integrating cyclical temporal features substantially improves predictive accuracy across deep learning models and data types.

Table 1. Price prediction performance

(a) LSTM model

Feature Set	MSE	RMSE	MAE	R^2Score
Set 1: OHLCV	6.2163	2.4933	1.7569	0.9889
*Set 1**: OHLCV + Time Cyclic Encoding	**5.7369**	**2.3952**	**1.6723**	**0.9994**
Set 2: LOB	13.6915	3.7002	3.2828	0.9991
*Set 2**: LOB + Time Cyclic Encoding	**7.5191**	**2.7421**	**2.1863**	**0.9995**
Set 3: OHLCV + LOB	13.9891	3.7402	3.8909	0.9984
*Set 3**: OHLCV + LOB + Time Cyclic Encoding	**5.3470**	**2.3124**	**1.790**	**0.9996**

(b) CNN model

Feature Set	MSE	RMSE	MAE	R^2Score
Set 1: OHLCV	6.9329	2.6330	1.6563	0.9995
*Set 1**: OHLCV + Time Cyclic Encoding	**6.8501**	**2.6173**	**1.6483**	**0.9996**
Set 2: LOB	4.4038	2.0985	1.6291	0.9997
*Set 2**: LOB + Time Cyclic Encoding	**3.1565**	**1.7767**	**1.2963**	**0.9998**
Set 3: OHLCV + LOB	3.8441	1.9606	1.4515	0.9997
*Set 3**: OHLCV + LOB + Time Cyclic Encoding	**3.1413**	**1.7724**	**1.2605**	**0.9998**

(c) Hybrid CNN-LSTM model

Feature Set	MSE	RMSE	MAE	R^2Score
Set 1: OHLCV	6.8753	2.6221	1.9114	0.9995
*Set 1**: OHLCV + Time Cyclic Encoding	**5.5039**	**2.3460**	**1.7742**	**0.9996**
Set 2: LOB	9.4204	3.0693	2.4392	0.9994
*Set 2**: LOB + Time Cyclic Encoding	**6.5368**	**2.5567**	**1.9463**	**0.9996**
Set 3: OHLCV + LOB	8.0409	2.8356	2.2808	0.9995
*Set 3**: OHLCV + LOB + Time Cyclic Encoding	**5.1833**	**2.2767**	**1.7778**	**0.9997**

5.2 Trend Prediction Results

In the trend forecasting, incorporating time-cyclic encoding consistently improves performance across all models, As shown in Table 2. For the LSTM, directional accuracy increases by 6.1% on OHLCV inputs, with precision and recall gains of up to 5.5%. LOB inputs yield smaller improvements (1% accuracy gain, 0.9% in precision/recall), while combined inputs produce moderate boosts (0.76% in accuracy, 1.6-5.4% in precision/recall). Similarly, the CNN classifier benefits from time encoding, with a 5.5% accuracy gain on OHLCV inputs and 1.4-5.6% improvements in class-wise precision and recall. On LOB data, the CNN shows a 0.87% accuracy increase and 0.65-0.91% gains in other metrics; combined inputs yield more modest improvements. The hybrid CNN-LSTM model also achieves the gains, especially on OHLCV inputs where accuracy rises by 1.9% and precision/recall improve by up to 4.1%. Across all models and input types, these results confirm that time-cyclic encoding enhances forecasting accuracy and helps better capture temporal dynamics in price movement trends.

Table 2. Trend prediction performance

(a) LSTM model

Feature Set	Accuracy	Precision	Recall
Set 1: OHLCV	0.4644	0.5052	0.4473
*Set 1**: OHLCV + Time Cyclic Encoding	**0.4928**	**0.5162**	**0.4720**
Set 2: LOB	0.4909	0.5075	0.5053
*Set 2**: LOB + Time Cyclic Encoding	**0.4959**	**0.5121**	**0.5099**
Set 3: OHLCV + LOB	0.4981	0.5062	0.2875
*Set 3**: OHLCV + LOB + Time Cyclic Encoding	**0.5019**	**0.5144**	**0.3029**

(b) CNN model

Feature Set	Accuracy	Precision	Recall
Set 1: OHLCV	0.4473	0.4981	0.4473
*Set 1**: OHLCV + Time Cyclic Encoding	**0.4719**	**0.5052**	**0.4722**
Set 2: LOB	0.5059	0.5075	0.5055
*Set 2**: LOB + Time Cyclic Encoding	**0.5103**	**0.5121**	**0.5088**
Set 3: OHLCV + LOB	0.5034	0.5062	0.5034
*Set 3**: OHLCV + LOB + Time Cyclic Encoding	**0.5093**	**0.5089**	**0.5037**

(c) Hybrid CNN-LSTM model

Feature Set	Accuracy	Precision	Recall
Set 1: OHLCV	0.4788	0.4942	0.4689
*Set 1**: OHLCV + Time Cyclic Encoding	**0.4881**	**0.4975**	**0.4881**
Set 2: LOB	0.5124	0.5050	0.5121
*Set 2**: LOB + Time Cyclic Encoding	**0.5191**	**0.5157**	**0.5124**
Set 3: OHLCV + LOB	0.5073	0.5105	0.5073
*Set 3**: OHLCV + LOB + Time Cyclic Encoding	**0.5120**	**0.5183**	**0.5167**

6   Conclusion

This research systematically examines the effectiveness of incorporating cyclic temporal encoding into high-frequency trading (HFT) forecasting models across both price and trend prediction tasks. We evaluate a range of feature combinations, comparing traditional OHLCV and limit order book (LOB) data with enriched sets that include sine and cosine transformations of minute, hour, and day-of-week. Across LSTM, CNN, and hybrid CNN-LSTM architectures, the addition of these time-based features consistently enhances predictive performance. In price forecasting, models using cyclic time encoding achieve significantly lower prediction errors than those relying solely on traditional inputs, underscoring the value of capturing intraday and weekly periodicity. In trend advertising, time-enhanced models deliver higher directional accuracy and more balanced precision-recall metrics, particularly improving the detection of downward trends—crucial for trading and risk management. These results highlight the utility of explicitly modeling temporal cycles in high-frequency data, offering a promising direction for improving deep learning-based forecasting in HFT applications and motivating further exploration into temporal feature engineering.

While this study demonstrates the effectiveness of cyclic temporal encoding in improving deep learning models for high-frequency cryptocurrency forecasting, several directions remain for future exploration. First, further research could investigate the integration of additional temporal hierarchies, such as month-of-year, holiday effects, or market session-based features, to capture broader seasonal and structural patterns. Second, future studies may explore the application of adaptive or learned time embeddings, which could provide more flexible representations of temporal information compared to fixed sinusoidal encodings. Third, extending the analysis to other asset classes (e.g., equities, forex) and across multiple cryptocurrencies would help assess the generalizability of these findings. Finally, future experiments could also investigate Transformer-based models, which have recently shown strong performance in financial time-series forecasting. Incorporating cyclical temporal encodings into such attention-based architectures may further enhance forecasting accuracy and provide a meaningful comparison against established CNN and LSTM baselines.

Acknowledgment. This research is supported by research funding from the Faculty of Information Technology, University of Science, Vietnam National University - Ho Chi Minh City. This research used the GPUs provided by the Intelligent Systems Lab at the Faculty of Information Technology, University of Science, VNU-HCM.

References

1. Prata, M., et al.: and Novella Bartolini. A benchmark study, Lob-based deep learning models for stock price trend prediction (2023)

2. Nejad, A., Kalhor, A., Hosseini, R., Araabi, B.N.: Deep high-frequency cryptocurrency trend detection: an approach for data stationarity. SSRN Electron. J. 2024. Available at SSRN: https://ssrn.com/abstract=4796336
3. Werner Kristjanpoller and Benjamin Miranda Tabak: Day of the week effect on the cryptomarket: a high-frequency asymmetric multifractal analysis. XXPhys. A **658**, 130306 (2025)
4. Maria, G., Caporale., Plastun, A.: The day of the week effect in the cryptocurrency market. Finance Res. Lett. 31, 2019
5. Briola, A., Bartolucci, S., Aste, T.: Deep limit order book forecasting, 2024
6. Cont, R., Kukanov, A., Stoikov, S.:The price impact of order book events. J. Financ. Econometrics, 12(1):47–88, 06 2013
7. Zhang, L., Hua, L.: Major issues in high-frequency financial data analysis: a survey of solutions. Math. 13(3), 2025
8. Li, W., et al.: A benchmark study for limit order book (lob) models and time series forecasting models on lob data. In: Proceedings of the International Conference on Learning Representations (ICLR), 2025. Available at OpenReview: https://openreview.net/forum?id=MhD9rLeU31
9. Zhang, Z., Zohren, S., Roberts, S.: Deeplob: deep convolutional neural networks for limit order books. IEEE Trans. Signal Process. **67**(11), 3001–3012 (2019)
10. Tsantekidis, A., Passalis, N., Tefas, A., Kanniainen, J., Gabbouj, M., Iosifidis, A.: Forecasting stock prices from the limit order book using convolutional neural networks. In: 2017 IEEE 19th Conference on Business Informatics (CBI), volume 01, pp. 7–12, 2017
11. Sirignano, J., Cont ,R.: Universal features of price formation in financial markets: perspectives from deep learning, 2018
12. Plastun, A., Bouri, E., Havrylina, A., Ji, Q.: Calendar anomalies in passion investments: Price patterns and profit opportunities. Res. Int. Bus. Financ. **61**, 101678 (2022)
13. Aharon, D.Y., Qadan, M.: Bitcoin and the day-of-the-week effect. Finance Res. Lett. 31, 2019
14. Zong, C., Wang, C., Qin, M., Feng, L.: Xinrun Wang, and Bo An. Memory augmented context-aware reinforcement learning on high frequency trading, Macrohft (2024)
15. He, Z., Zhang, H., Por, L.Y.: A Comparative Study on Deep Learning Models for Stock Price Prediction. 07 2024
16. Shah, J., Vaidya, D., Shah, M.: A comprehensive review on multiple hybrid deep learning approaches for stock prediction. Intell. Syst. Appl. **16**, 200111 (2022)
17. Bao, W., Yue, J., Rao, Y.: A deep learning framework for financial time series using stacked autoencoders and long-short term memory. PLOS ONE, 12(7):1–24, 07 2017
18. Piravechsakul, P., Kasetkasem, T., Marukatat, S., Kumazawa, I.: Combining technical indicators and deep learning by using LSTM stock price predictor. In: 2021 18th International Conference on Electrical Engineering/Electronics, Computer, Telecommunications and Information Technology (ECTI-CON), pp. 1155–1158, 2021

Edge-Optimized CNN Accelerator: A Flexible and Lightweight Architecture

Tuan-Kiet Tran[1,2]($\boxtimes$), Minh-Tuyen Huynh[1,2], Duc-Hung Pham[1,2], Cong-Kha Pham[3], and Huu-Thuan Huynh[1,2]($\boxtimes$)

[1] University of Science, Ho Chi Minh City, Vietnam
{trtkiet,hhthuan}@hcmus.edu.vn
[2] Vietnam National University, Ho Chi Minh City, Vietnam
[3] University of Electro-Communications (UEC), Tokyo, Japan

Abstract. Deep learning is transforming real-time biomedical applications, but deploying these powerful models on edge devices faces significant hurdles. These devices have strict limits on computational resources, memory, and energy. We've developed a lightweight, hardware-optimized 1D Convolutional Neural Network (CNN) architecture specifically for classifying electrocardiogram (ECG) signals. By strategically using knowledge distillation and quantization, we've dramatically compressed the model. It now occupies just 0.08 MB with roughly 66,000 parameters, yet it maintains an exceptional 99.6% classification accuracy. Our CNN accelerator design also demonstrates impressive efficiency. It achieves 594.49 GOP/s/MeLUT in logic efficiency and 32.94 GOP/s/W in power efficiency. These results show that our architecture makes efficient, low-power biomedical inference on edge devices entirely feasible, especially benefiting deployments where resources are scarce.

Keywords: ECG classification · CNN accelerator · Edge devices

1 Introduction

Modern technological advancements have made edge computing a crucial paradigm, especially for real-time biomedical applications. Edge devices with on-device intelligence can directly analyze physiological signals like ECGs, EMGs, and respiratory patterns right at the point of care [1]. This local processing enables early detection and warnings for health issues, while also reducing system latency, saving communication bandwidth, and enhancing data privacy by minimizing the transmission of raw medical data to centralized servers [2].

However, deploying deep learning models on these edge platforms presents significant challenges due to tight constraints on computational power, memory, and energy [3]. These limitations necessitate the design of lightweight yet efficient AI network models that can operate reliably under hardware and power constraints while maintaining high inference accuracy for real-time biomedical applications.

N. Thai-Nghe et al. (Eds.): ISDS 2025, CCIS 2714, pp. 357–371, 2026.
https://doi.org/10.1007/978-981-95-3358-9_26

Traditionally, ECG classification has explored methods like Support Vector Machines (SVM) [4], k-Nearest Neighbors (KNN) [5], and Decision Trees [6]. While effective, these techniques require manual feature extraction and specific domain expertise, limiting their scalability and adaptability.

More recently, Convolutional Neural Networks (CNNs) have emerged as a powerful alternative. They can directly learn discriminative features from raw input data in an end-to-end manner, automatically capturing complex temporal and morphological patterns in biomedical signals like ECGs without manual feature engineering.

While 2D-CNNs [7] are prevalent in image tasks, applying them to 1D signals like ECGs requires converting the waveform into a 2D format (e.g., spectrograms). This additional preprocessing increases complexity and computational demands, which is problematic for resource-constrained or real-time systems. In contrast, 1D-CNNs are naturally suited for time-series signals. By performing convolution along the time axis, 1D-CNNs efficiently capture local sequential patterns with fewer parameters and minimal memory usage, making them ideal for efficient, low-latency edge device deployment.

Previous hardware accelerators for CNN models often rely on floating-point arithmetic, demanding substantial memory and computational resources [8]. Furthermore, while High-Level Synthesis (HLS) tools accelerate FPGA development, they can lack the architectural control needed for optimal speed and resource efficiency [9]. Many existing CNN accelerators also heavily depend on DSP blocks for arithmetic operations, which increases area and power consumption and limits portability to Application-Specific Integrated Circuit (ASIC) platforms [10,11].

To overcome these challenges, we propose a lightweight, hardware-efficient 1D-CNN model specifically for ECG classification. Our model is optimized using compression techniques and quantization, including knowledge distillation and quantization-aware training, to minimize resource usage while maintaining high classification accuracy.

Complementing the model, we introduce a flexible and efficient hardware accelerator designed explicitly for low-resource environments. This accelerator utilizes a systolic array structure, enabling quantized convolution and activation operations entirely through general-purpose logic without requiring DSP blocks. This DSP-free design not only reduces power and area consumption but also significantly enhances portability across both FPGA and ASIC platforms.

Our main contributions are:

- We designed a lightweight 1D-CNN model specifically for ECG signal classification. It uses knowledge distillation and quantization to create a compact and computationally efficient structure perfect for edge deployment.
- We developed a flexible CNN accelerator based on a systolic array using only custom logic elements, without DSPs. The design supports high hardware reusability and efficiently handles various kernel sizes and input/output channel configurations, making it adaptable to different CNN models.

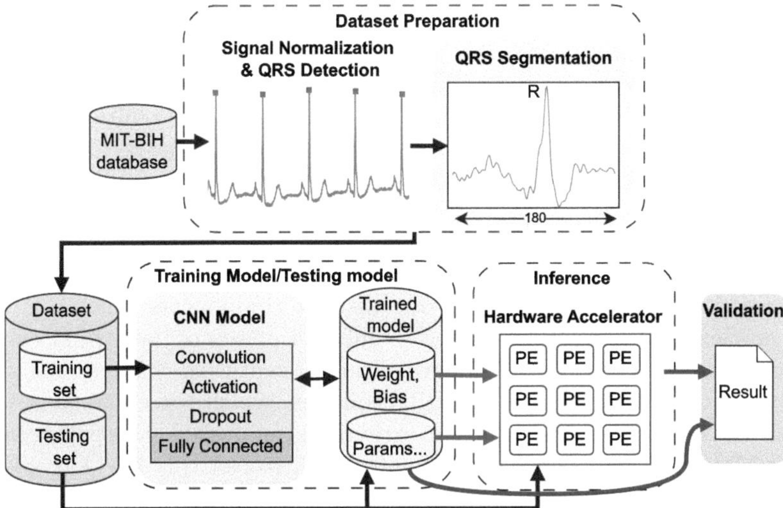

Fig. 1. The overall of CNN hardware accelerator design process.

- We implemented and validated our accelerator on an FPGA platform. This demonstrates its effectiveness for real-time biomedical signal processing in resource-constrained edge environments.

This paper is structured into five sections: Sect. 1 provides an introduction and a literature review. Section 2 details the proposed CNN model architecture, including dataset preparation, preprocessing, and model design. Section 3 presents the hardware accelerator's architecture, covering fixed-point quantization, processing element structure, and the general CNN acceleration architecture. Section 4 showcases our experimental results, including model evaluation and hardware performance analysis. Finally, Sect. 5 concludes the paper.

2   Proposed Methodology

The general design flow of the proposed CNN-based hardware accelerator can be divided into two main stages, as illustrated in Fig. 1. In the first stage, raw ECG signals are preprocessed through noise filtering, normalization, and segmentation to generate clean and reliable input. Then a 1D-CNN model is developed and trained for ECG classification. Once satisfactory accuracy is obtained, the model is further optimized and quantized to INT8 precision to match the resource constraints of the edge platforms. In the second stage, the optimized model is implemented on a custom hardware accelerator designed in HDL, which employs a systolic array architecture to efficiently execute convolution operations. The accelerator is synthesized and deployed on the FPGA platform, enabling high-throughput and low-latency inference while maintaining efficient resource utilization. The systolic array architecture not only maximizes parallelism but also ensures predictable dataflow, making the design well-suited for real-time biomedical signal processing at the edge. Finally, the hardware accelerator is evaluated by comparing inference results against the software baseline, together with detailed measurements of resource usage and power consumption.

2.1 Dataset and Preprocessing

ECG is a time series signal that captures the heart's electrical activity through a characteristic waveform, in which the distinct components such as P wave, QRS complex, and T wave correspond to different phases of the cardiac cycle. It is widely used to detect cardiac abnormalities by capturing characteristic waveform components, particularly the QRS complex, which represents ventricular depolarization and is a key indicator in arrhythmia analysis. In this paper, we utilize the MIT-BIH Arrhythmia Database from PhysioNet [12], a widely adopted benchmark for ECG classification. The dataset consists of 48 annotated two-lead ambulatory ECG recordings from 47 patients, each approximately 30 min in duration and sampled at 360 Hz.

To prepare the dataset for training and validation, we applied z-score normalization to reduce inter-patient variation of the ECG signals. A QRS complex detection algorithm is then deployed to identify the R-peaks, which serve as temporal anchors for heartbeat segmentation. Each heartbeat is extracted as a fixed-length window of 180 samples centered on the R-peak, ensuring consistent alignment and capturing the full morphological structure of the QRS complex along with adjacent P and T waves. These segmented heartbeats are used to train a 1D-CNN for classifying five clinically significant beat types: normal beat (N), right bundle branch block (R), left bundle branch block (L), premature ventricular contraction (V), and atrial premature beat (P). Each class has unique timing characteristics around the QRS region, which is why focusing on this area for segmentation works so well for model training. By honing in on the most important part of the signal, we can boost both classification accuracy and efficiency.

2.2 The Architecture of 1D-CNN Model

In this section, we proposed a CNN model, which consists of two main variants: a large CNN model (Fig. 2) and a small CNN model (Fig. 3). The large CNN model is designed to achieve high classification accuracy by utilizing a deeper architecture with more convolutional layers and parameters. The primary structure of the large CNN model is composed of repeated calculation blocks (CB), denoted as Block A, Block B, and Block C. Each CB consists of a sequence of convolutional layers, batch normalization, and ReLU activation functions. The model begins calculation with an input layer that receives QRS segments; hierarchical features from the input data are extracted progressively, followed by these blocks. The convolutional layers within each block are responsible for capturing local patterns and temporal dependencies in the ECG signals, which are essential for distinguishing between different heartbeat types. The batch normalization layers are applied after each convolutional layer to stabilize and accelerate the training process, while the ReLU activation function introduces non-linearity, enabling the model to learn complex representations. After passing through the sequence of blocks, the extracted features are aggregated and fed into a fully connected (FC) layer that produces the final classification output. This deep and modular architecture allows the large CNN model to achieve superior classification performance, particularly in scenarios where high accuracy is required. The detailed configuration of the CBs, including filter counts, kernel sizes, and other key parameters, is summarized in Table 1.

The small CNN model, illustrated in Fig. 3, is a cut-down and optimized version of the large model. The small CNN model focuses on reducing complexity and efficiency while maintaining classification accuracy. Although it follows the core structure of the large model, it has fewer convolutional layers and parameters, resulting in a reduced number of model parameters, which significantly reduces storage requirements and computational effort. To preserve classification accuracy, Knowledge Distillation [13] is employed to transfer learned representations from the large (teacher) model to the small (student) model. Quantization-Aware Training (QAT) [14] is also utilized, allowing weights and activations to be quantized into fixed-point formats during training, which further reduces computational and memory requirements for hardware deployment. The small model maintains essential components such as convolutional layers, batch normalization, and ReLU activations, but with each convolutional layer using **32 filters** of size **1x10** and the first fully connected layer (FC_1) reduced to **96 nodes**. Other layers retain their original settings. Additionally, when quantized to an 8-bit fixed-point, the convolution, ReLU, and BatchNorm operations are fused into a single ConvReLU layer, with normalization parameters precomputed during training. This architecture design ensures that small CNN models achieve competitive accuracy while minimizing resource usage and supporting efficient inference on edge hardware.

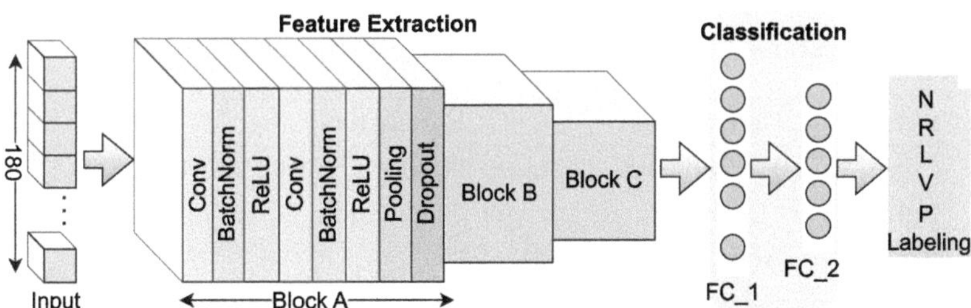

Fig. 2. The architecture of the large CNN model.

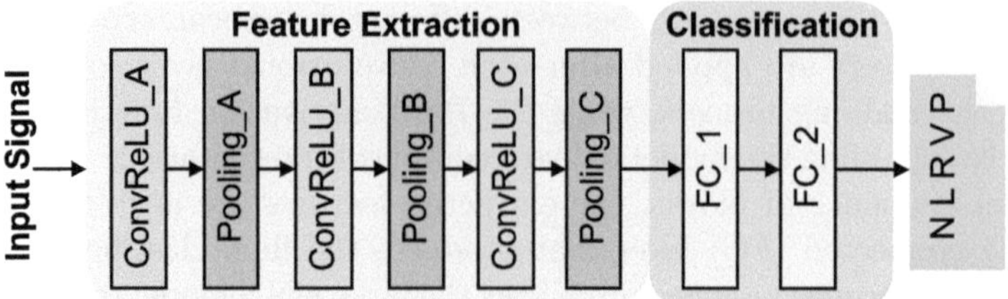

Fig. 3. The architecture of the light weight CNN model.

3 Proposed Accelerator Achitecture

3.1 Fixed-Point Calculation of Quantized CNN Model

The proposed quantized 1D-CNN model performs inference primarily through quantized convolution and activation operations, which are inherently suited for hardware implementation. The quantization process reduces the model's reliance on floating-point arithmetic by mapping real-valued inputs into low-bit integer representations, thus improving computational efficiency and reducing memory bandwidth. This process can be mathematically described as shown in Eq. 1:

$$x_q = \mathrm{round}\left(\frac{x_r}{s_x} + z_x\right); \quad y_r = \sum x_r w_r + b \tag{1}$$

Here, x_q and x_r are the quantized and real-valued inputs, respectively; s_x and z_x represent the input scale factor and zero-point. The convolution result y_r is computed from the real-valued input and weights. In the quantized domain, the convolution can be rewritten as Eq. 2:

$$s_y(y_q - z_y) = \sum s_x(x_q - z_x)s_w(w_q - z_w) + s_b(b_q - z_b) \tag{2}$$

Table 1. Parameters of the CBs and FCs layers in the large CNN model.

Layer	Description	Parameters
Conv	1D Convolutional layer	96 filters, kernel size 10, stride 1
BatchNorm	1D Batch normalization	96 channels
ReLU	Activation function	–
Pooling	Pooling layer	Max pooling, Kernel size 2
Dropout	Regularization	Rate 0.2
FC_1	Fully connected layer	288 units
FC_2	Output layer	5 units (classes)

This equation can be rearranged to isolate y_q as:

$$y_q = z_y + \frac{s_b}{s_y}(b_q - z_b) + \frac{s_x s_w}{s_y}T \tag{3}$$

where $T = Psum - z_w S_x - z_x S_w + K z_x z_w$, and $Psum = \sum x_q w_q$ is the accumulated sum over quantized inputs and weights. In many practical cases and in work [14], zero-points for weights and bias are set to zero ($z_b = z_w = 0$), and the scale for bias is defined as $s_b = s_x s_w$, simplifying the equation to:

$$y_q = z_y + \alpha(T + b_q), \quad \text{where } \alpha = \frac{s_x s_w}{s_y} \tag{4}$$

The convolution result is then passed through a quantized ReLU activation, as described in Eq. 5:

$$y_r' = s_y'(y_q' - z_y') = \begin{cases} 0 & \text{if } y_r < 0 \\ y_r & \text{if } y_r \geq 0 \end{cases} \tag{5}$$

By applying Eq. 4 to the activation, the final output after quantized ReLU is:

$$y_q' = \begin{cases} z_y' & \text{if } Psum + b_q < N \\ z_y' + \alpha'(T + b_q) & \text{otherwise} \end{cases} \tag{6}$$

Here, α' is a rescaled factor, and $N = z_x S_w$ represents a threshold derived from input zero-points and sum of weight. Importantly, all quantization parameters (α', N, z_y') can be precomputed offline, significantly reducing the runtime overhead and simplifying hardware control logic.

To efficiently implement these computations in hardware, we propose an optimized algorithm that performs quantized convolution followed by ReLU activation, as summarized in Algorithm 1. The algorithm includes channel unrolling with a fixed parallelism factor ($unroll\_size = 8$), allowing multiple output channels to be computed in parallel. This approach maximizes data reuse and improves throughput, making it highly suitable for edge deployment.

3.2 CNN Accelerator Architecture

The overall architecture of the proposed CNN accelerator is shown in Fig. 4, which is designed strictly following the Algorithm 1. It is designed to efficiently execute quantized convolution and ReLU operations in real-time while maximizing parallelism and throughput. The accelerator follows a top-down modular structure that begins with a global buffer for managing input data and ends with an output FIFO for transferring final results. The hardware is managed by a central controller.

Algorithm 1: The CONV_ReLU operation with channel unrolling

Input: x_q, w_q, b_q, z_y, α, N

Data:

- x_q: Input data of shape $(number\_channel, input\_length)$
- w_q: Kernel weights of shape $(number\_channel, number\_filter, kernel\_length)$
- b_q: Bias values of shape $(number\_channel)$
- z_y: Output zero-point of shape $(number\_channel)$
- α: Output scaling factor of shape $(number\_channel)$
- y_q: Out data of shape $(number\_channel, output\_length)$

Output: y_q

Result: Y: Output after Convolution-ReLU

for $c \leftarrow 0$ **to** $number\_channel - 1$ **do**

    for $i \leftarrow 0$ **to** $output\_length - 1$ **do**

        for $k \leftarrow 0$ **to** $number\_filter - 1$ **do**

            for $l \leftarrow 0$ **to** $uncroll\_size - 1$ **do**

                // Unroll loop of convolution

                $Psum[l] \leftarrow b_q[c + l]$;

                for $j \leftarrow 0$ **to** $kernel\_length - 1$ **do**

                    $Psum[l] \leftarrow Psum[l] + x_q[c + l][i + j] \times w_q[c + l][k][j]$;

        for $l \leftarrow 0$ **to** $uncroll\_size - 1$ **do**

            // Unroll loop of ReLu and output scaling

            if $Psum[l] - N[c] < 0$ **then**

                $y_q[c + l][i] \leftarrow z_y[c]$;

            else

                $y_q[c + l][i] \leftarrow z_y[c + l] + \alpha[c] \times (Psum[c + l] - N[c])$;

    $c \leftarrow c + unroll\_size$;

return y_q;

At the system level, input feature maps are first delivered to a global buffer via a shared input bus. Within this buffer, data coordination is managed by an arbiter, which a centralized controller governs to determine the appropriate destination of each incoming data stream. The arbiter dynamically routes data to the corresponding functional modules, namely the input PISO (Parallel-In Serial-Out), weight FIFO, and partial sum FIFO (Psum FIFO), ensuring conflict-free scheduling and synchronized data distribution across the accelerator pipeline. The Psum FIFO is responsible for providing initial accumulation values to the convolution engine and is initialized with 32-bit bias values corresponding to each output channel for the first computation. These 32-bit biases are preloaded and serve as the starting point for the partial sum accumulation in each Processing Element (PE) row, effectively eliminating the need for a separate bias addition stage. At the heart of the accelerator is a systolic array composed of 8×5 PEs, designed to perform quantized multiply-accumulate (MAC) operations in parallel. The architecture follows a weight-stationary dataflow, where weights remain resident within PEs once loaded, enabling repeated reuse across multiple convolution steps without additional memory access. Since each weight

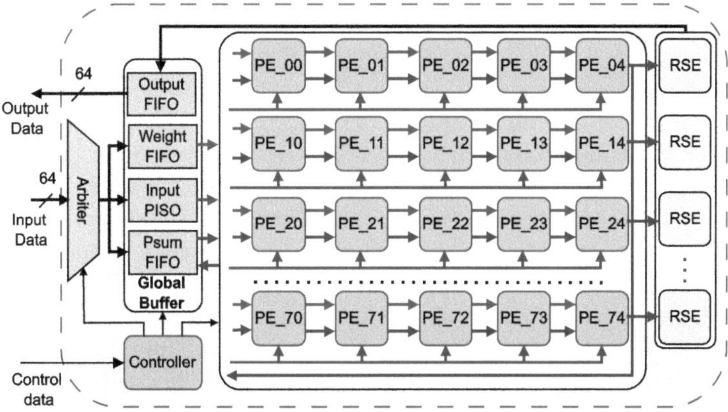

Fig. 4. The architecture of the CNN hardware accelerator: PE array, RSE array, and data movement paths.

value is represented with 8-bit precision, loading a full set of weights for an entire PE column (8 weights per clock cycle) requires a data width of $8 \times 8 = 64$ bits. Consequently, the design provisions a 64-bit bus for both the input and output data paths to support efficient, single-cycle transfer of weight vectors and to match the bandwidth requirements of the PE array. In this setup, weights are serially shifted into the first column of the PE array along the vertical axis (blue arrow), completing the loading process within five cycles for a five-column structure. Simultaneously, input feature values are broadcast across each PE row (red arrow), and bias-initialized partial sums are propagated through the system using the green path. As the convolution progresses, these partial sums flow from PE to PE within each row, with the final accumulated result at the last PE being forwarded back to the Psum FIFO. This mechanism not only supports single-stage convolution for small kernels but also enables multi-stage accumulation for larger kernel sizes or deeper layers, where intermediate results can be stored and reused. The use of this architecture also supports minimizing latency and bandwidth usage through carefully coordinated data movement and bus width matching.

Following the convolution stage, results are forwarded to a corresponding array of ReLU-Scale Elements (RSEs) configured as an 8×1 vector. Each RSE performs a quantized ReLU activation and output scaling based on precomputed quantization parameters. The processed outputs are written to an output FIFO and subsequently passed to downstream modules or stored externally.

Processing Element (PE) Architecture. Each PE, as illustrated in Fig. 5a, is optimized for executing quantized MAC operations in a pipelined manner. The PE receives 8-bit quantized input and weight values, which are first latched into internal registers to ensure proper timing alignment. These values are then multiplied using a dedicated multiplier to produce a 16-bit product, which is temporarily stored before proceeding to the next pipeline stage. In the accumulation stage, the product is added to a 32-bit partial sum using an adder. This partial sum may originate from the previous PE or be initialized with a bias value at the beginning of the computation. The accumulated result is then stored in a register to stabilize the output and prepare it for further processing. To support flexible kernel sizes and enable computation reuse, each PE includes a multiplexer that selects between the accumulated result and a previous partial sum based on a control signal. The selected value is then propagated as the output partial sum to the next stage or to the activation unit. This two-stage pipelined architecture, which comprises multiplication and accumulation, allows one MAC result to be produced every clock cycle after pipeline initialization. The use of internal registers and configurable control logic enhances computational efficiency and scalability, allowing the PE array in the CNN IP to be reused for multiple computations while maintaining speed.

ReLU-Scale Element (RSE) Architecture. As illustrated in Fig. 5b, the RSE receives the final accumulated partial sums from the PE array and performs a quantized ReLU activation followed by output scaling. To determine whether the result should be suppressed by ReLU, the RSE does not perform a direct comparison between $Psum_i$ and the precomputed threshold N. Instead, it computes the subtraction $Psum_i - N$ and examines the sign bit of the result. If the sign bit is set (i.e., the result is negative), the value is considered less than the threshold, and the output is clamped to the quantized zero-point. Otherwise, the result is scaled by a predetermined scale factor α ($Scale\_i$) to produce the final quantized output. This approach facilitates efficient hardware implementation by replacing explicit comparison logic with a simple arithmetic subtraction followed by sign-bit evaluation. Both the subtraction and scaling operations are fully pipelined, enabling the RSE to generate one output per clock cycle and thus maintain high computational throughput.

Pipeline Execution and Parallelism. To further enhance throughput, the architecture is deeply pipelined at both intra-PE and inter-PE levels. Figure 6 depicts the execution timeline of a single PE row. Intra-PE pipelining enables

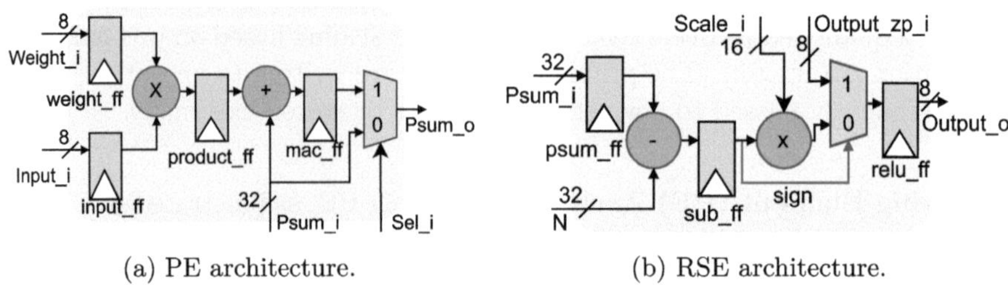

(a) PE architecture.   (b) RSE architecture.

Fig. 5. Modular structure of computational units in the accelerator.

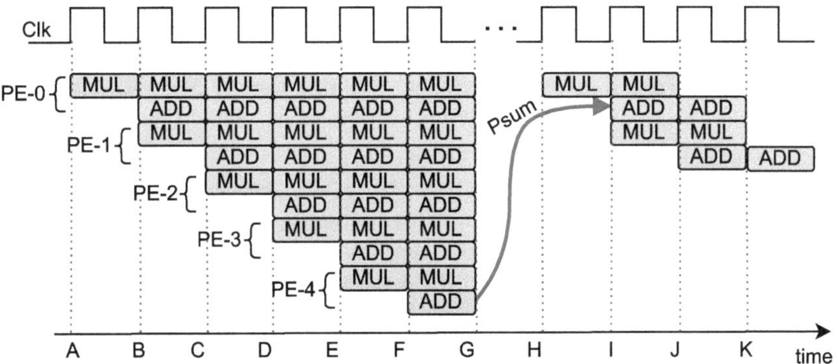

Fig. 6. Timeline of pipelined PE array processing for one row.

overlapping of multiplication and addition stages, while inter-PE pipelining ensures concurrent processing across all five PEs. Once the pipeline is filled (after five cycles), the row outputs one result per clock cycle. This model is replicated across all eight rows, providing full spatial parallelism and maximizing resource utilization.

Kernel Scaling and Flexibility. The proposed architecture is designed with flexibility to support kernel sizes larger than the physical width of the PE array. For example, a convolution with a kernel size of 1×7 is executed in two phases: during the first phase, the initial 1×5 segment is processed by the full set of five PEs in the row, and the resulting partial sum is stored in Psum FIFO locally. In the second phase, the remaining 1×2 portion of the kernel is computed by reloading the stored partial sum into only two PEs (e.g., PE-0 and PE-1), which perform the remaining computations. To efficiently handle such scenarios without incurring unnecessary overhead, each PE incorporates a multiplexer that allows the partial sum to be bypassed directly to the last active PEs in the row. This bypass mechanism is activated when fewer than five PEs are required in the second phase, enabling the architecture to dynamically reconfigure the data path and avoid idle computation stages. By using a modular design and deep pipeline mechanism, this multi-phase accumulation process is seamlessly integrated into the data path, allowing the accelerator to handle arbitrary kernel sizes while maintaining high computational efficiency. These architectural features make the design highly adaptable to a broad spectrum of CNN workloads, particularly in edge-AI scenarios where both performance and resource utilization are critical.

4 Experimental Results

4.1 Model Training and Evaluation

Table 2 summarizes the heartbeat class distribution in the MIT-BIH Arrhythmia Database, along with the evaluation results of the proposed CNN model.

The dataset includes five main classes: Normal (N), Left bundle branch block (L), Right bundle branch block (R), Premature ventricular contraction (V), and Paced beat (P). To balance the dataset, the number of Normal samples was reduced to match the other classes (7,000–8,000 each). The "Used" column indicates the selected sample count per class.

The proposed CNN model was trained using cross-entropy loss with the Adam optimizer and data augmentation (random cropping and jittering) to improve generalization. To enable edge deployment, the model was first reduced from its original size of **2.43 MB** to **0.26 MB** using knowledge distillation and then further quantized to INT8, yielding a final model size of only **0.08 MB**. This represents a total size reduction of approximately **96.7%**.

Despite the significant compression, the quantized model achieves a high classification accuracy of **99.6%** on the test set. Evaluation metrics across all five heartbeat classes remain consistently strong, with an average sensitivity of 99.65%, accuracy of 99.86%, and F1-score of 99.65%, confirming the model's robustness and its suitability for real-time ECG classification on resource-constrained edge devices.

4.2   Hardware Performance Analysis

To evaluate the proposed CNN accelerator, we have transfered the calculation process of the inference of the 1D CNN model, introduced earlier, on the hardware. The model processes 180 input samples, which is the most extended data length in the model's layers, using a 1×10 convolution kernel, producing 171 outputs. With eight output channels and one input channel, the total number of MAC operations per inference is $171 \times 10 \times 8 = 13,680$, equivalent to **27,360** arithmetic operations. To efficiently execute this workload on the 8×5 PE array, the 1×10 kernel is decomposed into two sequential 1×5 convolutions, each computed in parallel across eight channels. This split strategy improves pipeline utilization and reduces hardware complexity. The first 1×5 convolution block requires 177 cycles, including 1 cycle for input loading, 5 cycles for PE array warm-up, and 171 cycles for computation. The second block takes 173 cycles, including 171 compute cycles and 2 cycles to fill the pipeline for the entire RSE array. Thus, each inference completes in a total of **350 cycles**. At

Table 2. Label Distribution and Evaluation Metrics for the MIT-BIH Dataset.

Symbol	Count	Used	Sensitivity	Accuracy	Precision	F1-Score
N	76324	8215	0.9964	0.9984	0.9964	0.9964
L	8170	8170	0.9970	0.9972	0.9904	0.9936
R	7342	7342	0.9973	0.9991	0.9980	0.9977
V	7278	7278	0.9881	0.9970	0.9958	0.9919
P	7076	7076	0.9993	0.9996	0.9986	0.9989
Average			**0.9956**	**0.9983**	**0.9958**	**0.9957**

Table 3. Comparison of CNN accelerators on FPGA platforms.

Reference	[10]	[15]	[11]	[16]	[17]	This work
Platform	ZCU706	ZC706	XC7A15T	ZYNQ-7100	ZCU102	**VC707**
Method	1D-CNN	1D-CNN	2D-CNN	LMA	1D-CNN	**1D-CNN**
Precision	INT16	INT16	INT8	INT16	INT16	**INT8**
Frequency (MHz)	200	200	150	200	250	**150**
LUT	125350	3200	6480	41000	24685	**10444**
DSP	449	48	21	352	40	**0**
BRAM	530	24.5	20	270	4.5	**14.5**
eLUT[†]	590270	32320	25160	312360	38765	**19724**
Power (W)	3.8	0.58	0.93	6.6	0.827	**0.356**
GOP/s	115.2	16.28	7.85	116.7	9.85	**11.73**
GOP/s/MeLUT	195.16	503.71	312.00	373.61	254.10	**594.49**
GOP/s/W	30.32	28.07	8.44	17.68	11.91	**32.94**

[†] $eLUT = \#LUT + \#DSP \times 280 + \#BRAM \times 640$.

an operating frequency of 150 MHz, the accelerator achieves a throughput of approximately **11.73 GOP/s**, demonstrating the effectiveness of the pipelined split-convolution design in maximizing hardware utilization and computational efficiency.

The proposed 1D-CNN model was synthesized and implemented on the Xilinx VC707 FPGA platform, targeting edge-level biomedical signal processing applications. After synthesis in Vivado, the design achieves a maximum operating frequency of 150 MHz and applies 8-bit integer (INT8) quantization to both weights and activations. A key feature of the hardware architecture is that it completely eliminates the use of dedicated DSP blocks by adopting an efficient logic-based MAC design. As a result, the implementation requires only **10,444** LUTs and **14.5** BRAMs, with all arithmetic operations executed using general-purpose logic. This DSP-free approach not only simplifies resource allocation on the FPGA but also enhances portability, making the architecture highly suitable for ASIC deployment where custom MAC units can be efficiently realized using multiplication and addition circuits. In addition, Vivado's power estimation reports a total power consumption of just **0.356** W, confirming the design's efficiency and suitability for energy-constrained edge devices.

To enable fair comparison across FPGA-based CNN accelerators, we adopt the effective logic utilization metric eLUT[†], which converts heterogeneous hardware resources into LUT equivalents. Based on the most popular estimation that one DSP equals 100 slices and one BRAM equals 200 slices, and assuming each slice contains 4 LUTs, the practical logic contribution is scaled by utilization factors of 0.7 for DSPs and 0.8 for BRAMs [18]. Accordingly, the total equivalent LUTs are computed as $eLUT = \#LUT + \#DSP \times 280 + \#BRAM \times 640$, providing a unified and architecture-independent estimate of hardware cost.

Table 3 provides a comprehensive comparison between the proposed design and several state-of-the-art FPGA-based CNN accelerators for ECG classification. Most existing works rely on 16-bit fixed-point arithmetic and make extensive use of dedicated DSP blocks. For instance, [10] uses 449 DSPs and 125k LUTs, while [15] and [16] also require 48 and 352 DSPs, respectively. In contrast, the proposed design adopts an INT8 quantization scheme and avoids the use of DSPs entirely. All computations are implemented using general-purpose logic and BRAMs, enhancing portability and making the architecture highly amenable to ASIC implementation. Despite significantly reduced resource usage, consuming only 10,444 LUTs, 14.5 BRAMs, and 0 DSPs, our design achieves competitive performance and demonstrates superior hardware efficiency. It delivers a logic efficiency of **594.49 GOP/s/MeLUT**, outperforming all other designs in the table, including [10] (195.16), [15] (503.71), [11] (312.00), [16] (373.61), and [17] (254.10). Moreover, in terms of energy efficiency, the proposed design reaches a power efficiency of **32.94 GOP/s/W**, which is also the highest among all compared implementations. The next best power efficiencies [10] at 30.32 and [15] at 28.07 remain notably lower. These results confirm the effectiveness of the proposed accelerator design in delivering high-performance inference while maintaining minimal resource and power consumption. The combination of low-bandwidth quantization, full-DSP-free logic implementation, and optimized pipelining makes it particularly well-suited for deployment in edge environments where area and energy constraints are critical.

5   Conclusion

This paper presents a lightweight and flexible CNN accelerator architecture tailored for efficient edge deployment. We've compressed our 1D-CNN model using knowledge distillation and quantized it to 8-bit precision. This dramatically shrinks the model size and reduces computational complexity, all while maintaining high classification performance.

On the hardware front, our flexible and lightweight CNN accelerator is built entirely with general-purpose logic and on-chip memory, completely avoiding dedicated DSP blocks. This DSP-free design not only enhances portability and simplifies ASIC integration, but the architecture's modularity also provides significant flexibility. Its reusable and reconfigurable computation units efficiently support diverse CNN configurations and kernel sizes. We've also optimized the design for high data reuse and efficient memory access, leading to lower bandwidth requirements and improved power efficiency. Collectively, these features make our proposed architecture an excellent fit for scalable, low-power inference in edge AI systems.

Acknowledgments. This research is funded by University of Science, VNU-HCM under grant number: DTVT 2024-08.

References

1. Rocha, A., et al.: Edge ai for internet of medical things: a literature review. Comput. Electr. Eng. **116**, 109202 (2024)
2. Awad, A.I., Fouda, M.M., Khashaba, M.M., Mohamed, E.R., Hosny, K.M.: Utilization of mobile edge computing on the internet of medical things: a survey. ICT Express **9**(3), 473–485 (2023)
3. Syu, J.-H., Lin, J.C.-W., Srivastava, G., Yu, K.: A comprehensive survey on artificial intelligence empowered edge computing on consumer electronics. IEEE Trans. Consum. Electron. **69**(4), 1023–1034 (2023)
4. Geweid, G.G., Chen, J.D.: Automatic classification of atrial fibrillation from short single-lead ecg recordings using a hybrid approach of dual support vector machine. Expert Syst. Appl. **198**, 116848 (2022)
5. Li, Z., Zhang, H.: Fusing deep metric learning with knn for 12-lead multi-labelled ecg classification. Biomed. Signal Process. Control **85**, 104849 (2023)
6. Hael, M.A.: Unbiased recursive decision tree for supervised functional data classification with applying on electrocardiogram signals. Int. J. Data Sci. Anal. **16**(4), 441–454 (2023)
7. Mewada, H.: 2D-wavelet encoded deep cnn for image-based ecg classification. Multimed. Tools Appl. **82**(13), 20 553–20 569 (2023)
8. Nevarez, Y., et al.: Cnn sensor analytics with hybrid-float6 quantization on low-power embedded fpgas. IEEE Access **11**, 4852–4868 (2023)
9. Mangaraj, S., Oraon, S. Ari, P., Swain, A.K., Mahapatra, K.: Fpga accelerated convolutional neural network for detection of cardiac arrhythmia. In: 2024 IEEE 4th International Conference on VLSI Systems, Architecture, Technology and Applications (VLSI SATA), pp. 1–6. IEEE, 2024
10. Zhang, C., Wang, X., Yong, S., Zhang, Y., Li, Q., Wang, C.: An energy-efficient convolutional neural network processor architecture based on a systolic array. Appl. Sci. **12**(24), 12633 (2022)
11. Peng, P., et al.: Design of an efficient cnn-based cough detection system on lightweight fpga. IEEE Trans. Biomed. Circuits Syst. **17**(1), 116–128 (2023)
12. Goldberger, A.L., et al.: Physiobank, physiotoolkit, and physionet: components of a new research resource for complex physiologic signals. Circulation **101**(23), e215–e220 (2000)
13. Hinton, G.: Distilling the knowledge in a neural network, arXiv preprint arXiv:1503.02531, 2015
14. Jacob, B., et al.: Quantization and training of neural networks for efficient integer-arithmetic-only inference. In: Proceedings of the IEEE Conference on Computer Vision and Pattern Recognition, pp. 2704–2713, 2018
15. Zhang, C., Li, J., Guo, P., Li, Q., Zhang, X., Wang, X.: A configurable hardware-efficient ecg classification inference engine based on cnn for mobile healthcare applications. Microelectron. J. **141**, 105969 (2023)
16. Lu, T., Zhao, B., Xie, M., Ma, Z.: Fpga design and implementation of ecg classification neural network. In: 2023 IEEE 3rd International Conference on Computer Communication and Artificial Intelligence (CCAI). IEEE, pp. 85–91, 2023
17. Pham, H.L., Tran, T.D., Le, V.T.D., Nakashima, Y.: Mina: a hardware-efficient and flexible mini-inceptionnet accelerator for ecg classification in wearable devices. IEEE Trans. Circuits Syst. I: Regul. Pap. (2025)
18. Liu, W., Fan, S., Khalid, A., Rafferty, C., O'Neill, M.: Optimized schoolbook polynomial multiplication for compact lattice-based cryptography on fpga. IEEE Trans. Very Large Scale Integr. (VLSI) Syst. **27**(10), 2459–2463 (2019)

DWACE: Enhancing Graph Clustering via Autoencoder and Contrastive Learning Refinement of DeepWalk Embeddings

Nguyen Xuan Thao Mai[1], Cong Phap Huynh[1], Supeeti Kunchan[2],
and Dai Tho Dang[1](✉)

[1] Vietnam - Korea University of Information and Communications Technology, The University of Da Nang, Da Nang, Vietnam
{thaomnx.21ad,hcphap,ddtho}@vku.udn.vn

[2] Faculty of Industrial Technology and Management, King Mongkut's University of Technology North, Bangkok, Thailand
supeeti.k@itm.kmutnb.ac.th

Abstract. Graph-structured data is essential in numerous domains. The quality of node representations plays a crucial role in graph mining tasks. Random-walk-based techniques, such as DeepWalk (DW) [1], are effective for this task. However, the embeddings they generate often exhibit noise and redundancy. As a result, clustering efforts may face potential obstacles. To tackle this issue, we introduce the DWACE (DeepWalk Autoencoder Contrastive Enhancement) framework. This proposal enhances these embeddings, thus increasing clustering efficacy. DWACE employs a nonlinear autoencoder to denoise and compress the raw embeddings while preserving their structural integrity. It additionally integrates a contrastive learning objective to preserve the consistency of embeddings and enhance intra-community coherence. Our investigations on five benchmark datasets demonstrate that DWACE surpasses baseline approaches, attaining state-of-the-art performance in clustering.

Keywords: DWACE · Graph clustering · Graph-structured · DeepWalk

1 Introduction

Graph-structured data is widely used to represent problems in many domains, such as biological networks, social networks, and knowledge graphs. This structure models entities as nodes and relationships as edges. Learning efficient node representations is crucial for downstream tasks, including community detection, node classification, and link prediction. The quality of node embeddings affects the performance of downstream tasks. High-quality node embeddings are designed to preserve both local structure and global properties. Local structure is neighborhood connections and community information. Global properties are long-range dependencies and the overall graph structure. According to recent studies, combining local and global graph structures often yields better results than focusing on either aspect [1–3].

N. Thai-Nghe et al. (Eds.): ISDS 2025, CCIS 2714, pp. 372–383, 2026.
https://doi.org/10.1007/978-981-95-3358-9_27

Random walk-based embedding techniques have attracted considerable attention. Researchers widely use these methods [4], like DW, Node2Vec, and LINE, because of their ability to capture both local and global graph structures. These methods utilize random walks to generate sequences of nodes and employ skip-gram models for learning embeddings. Thereby, they yield high performance in various downstream tasks, such as node classification and community detection. However, they may contain noise or redundant information [4, 10]. To solve this issue, some studies have used nonlinear autoencoders to denoise and compress embeddings while preserving structural information [5, 6]. Recently, contrastive learning has attracted much attention due to its discriminative ability of graph representations [7–9].

This paper introduces DWACE, which integrates nonlinear autoencoders and contrastive learning in an end-to-end manner to refine random-walk-based embeddings for clustering. First, DWACE applies a multilayer autoencoder to the DW embeddings, minimizing reconstruction loss to denoise and compress the representations. Next, a contrastive objective based on Deep Graph Infomax and GRACE maximizes the consistency between noisy views of the same node to improve the embedding quality further.

The main contributions of this work are as follows:

– We propose DWACE, which integrates autoencoder-based denoising with contrastive learning to refine node representations.
– We evaluate the effectiveness of DWACE via experiments on benchmark datasets. The experimental results indicate that DWACE is more effective than the state-of-the-art, including DW and Node2vec.

The remainder of this paper is organized as follows: Sect. 2 presents related work. Section 3 describes the proposed DWACE. Section 4 presents experiments and evaluation. Section 5 presents the conclusion and future directions.

2   Related Work

Nowadays, in the research community, many studies pay attention to low-dimensional node embeddings that preserve graph topology. Shallow embedding techniques learn node representations by modeling weighted edges or random-walk co-occurrence statistics. These approaches are highly efficient and scalable, thus, it makes them widely applicable. However, the limitation of the approaches is that noise and redundancy limit their effectiveness for clustering tasks.

Currently, the research community focuses on many studies that utilize low-dimensional node embeddings to preserve graph topology. Shallow embedding techniques learn node representations by modeling weighted edges or random-walk co-occurrence statistics. These approaches are highly efficient and scalable, which makes them widely applicable. However, the limitation of the approaches is that noise and redundancy limit their effectiveness for clustering tasks. Recently, some approaches have leveraged graph convolutional architectures or explicitly encoded structural information during training to address these limitations. [7, 11–14]. Despite these advances, random-walk-based embeddings remain popular due to their simplicity and scalability, motivating research into advanced post-processing techniques.

Contrastive learning has emerged as a practical approach for unsupervised learning of graph representations. Velickovic et al. [7] proposed Deep Graph Infomax (DGI). This method aims to maximize mutual information between patch-level and graph-level representations. GRACE [8] extended this idea by leveraging node-level contrastive signals under random perturbations. GraphCL [9] employs contrastive learning at the graph level by utilizing augmented views. GCA [15] improves contrastive signals by introducing adaptive augmentation based on graph structure. While these methods have demonstrated strong performance, most contrastive learning approaches focus on end-to-end GCN-based models and have not been widely applied to refine random-walk-based embeddings. For example, the embeddings produced by DW or Node2Vec can be used for clustering. This gap motivates our proposed approach.

3 Proposed Approach

To address the limitations of noisy and redundant node embeddings generated by random-walk-based methods, we propose the DWACE approach. This proposed framework integrates nonlinear autoencoding and contrastive learning to refine graph embeddings for clustering tasks. The main steps of graph clustering are as follows: Data Collection - Data Preprocessing - DWACE - Model GAE – Clustering (Fig. 1).

3.1 Data Collection

The procedure begins with the Data Collection phase, wherein a graph-structured dataset is selected. These datasets typically originate from domains such as social networks or citation networks, from which they form the basis for representation learning and community detection tasks.

3.2 Data Preprocessing

It creates the adjacency matrix, which is used for the autoencoder loss, and assigns ground truth or creates a default label array.

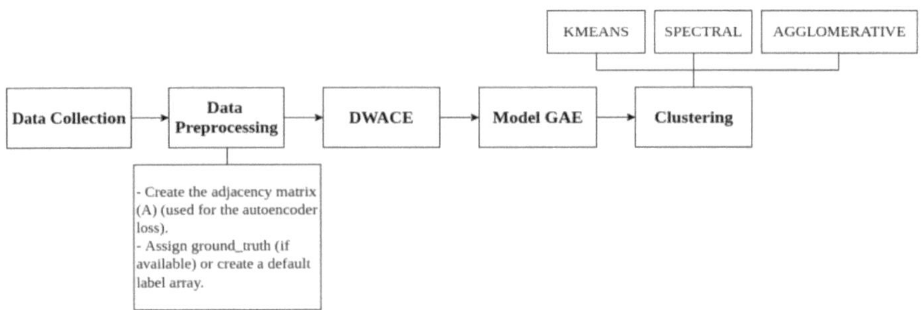

Fig. 1. Schema of DWACE

DWACE is created by comprising three sequential works: DW embedding, autoencoder-based refinement, and contrastive projection learning (Fig. 2).

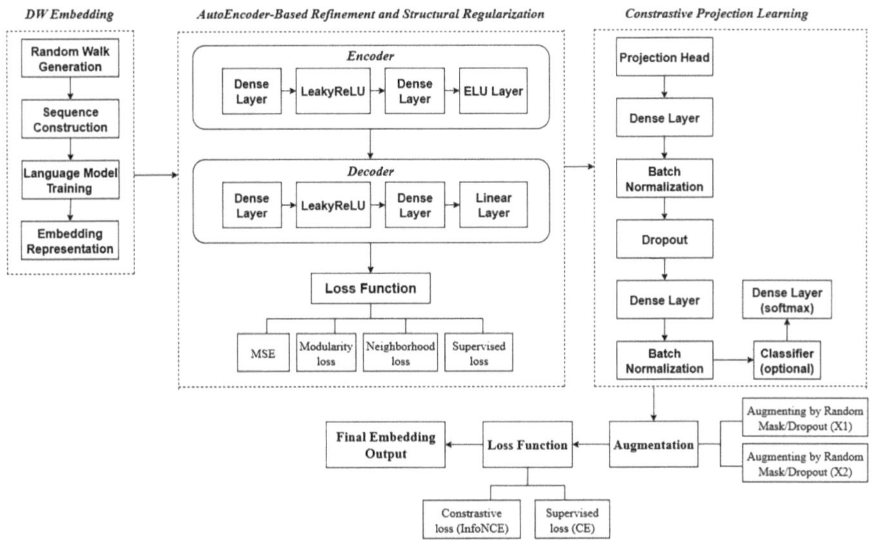

Fig. 2. Embedding Generating in DWACE

- DW Embedding

The process begins by generating initial node embeddings using DW [1]. DW takes multiple short random walks from each node to generate sequences that capture local and global structural information. These are then used to learn embeddings using a neural language model. Although considered good, DW produces multidimensional embeddings that often contain noise and redundancy.

- Autoencoder-Based Refinement and Structural Regularization

To reduce dimensionality and enhance structural fidelity, DW embeddings are passed through a multi-layer autoencoder [5, 6, 14]. The encoder consists of two dense layers with nonlinear activations: Dense $\rightarrow$ LeakyReLU $\rightarrow$ Dense $\rightarrow$ ELU, followed by Batch Normalization and Dropout, enabling non-linear compression of the input space and improving generalization. The use of LeakyReLU addresses the "dying ReLU" issue by allowing small gradients when neurons are inactive [19].

The decoder mirrors the encoder's structure and aims to reconstruct the original DW embeddings. The reconstruction is guided by minimizing Mean Squared Error (MSE) between input and output embeddings [20]. To encourage community-aware and structurally meaningful representations, we incorporate additional loss components:

- Modularity loss, preserving the density of intra-community connections [14];
- Neighborhood-preserving loss, maintaining local topological proximity [5];
- Supervised cross-entropy loss, applied when ground-truth labels are available, facilitating label-aware representation learning [3].

When labels exist, a classifier head is appended to the encoder's output. It supports supervised training in parallel with self-supervised learning.

- Contrastive Projection Learning

To enhance discriminability and robustness, we propose a contrastive learning module that utilizes available real-world labels. For each node, two random views are generated via random omission or masking, simulating dynamic noise. These views are encoded independently and passed through a projection head consisting of two fully connected layers with Batch Normalization, projecting the embeddings into a contrastive space [7–9].

The contrastive module is trained using the InfoNCE loss, which pulls positive pairs (i.e., different views of the same node) closer while pushing negative pairs (i.e., different nodes) apart in the latent space [8, 9, 15]. Simultaneously, if labels are available, the output of the classifier head is supervised using cross-entropy loss. We use a fomula to compute t overall loss function for this stage as following:

$$\mathcal{L} = \lambda_1 L_{constrastive} + \lambda_2 L_{supervised}$$

where $\lambda 1$ and $\lambda 2$ are tunable hyperparameters balancing the two objectives.

After training, the output from the projection head is retained as the final node embedding, which is optimized for compactness, structural consistency, and robustness.

3.3 Model GAE

The refined embeddings serve as the input to a Graph Autoencoder (GAE) architecture. We use GCN [11] as encoders. This model can learn from the topological structure and feature information via neighborhood aggregation [13].

3.4 Clustering

This study applies clustering algorithms, such as K-Means [16], Spectral Clustering [17], or Agglomerative Clustering [18], to the final node embeddings. These algorithms are used because of their effectiveness [10, 14]. The quality of clustering is evaluated using standard metrics for community detection.

4 Experiments and Results

We evaluate the effectiveness of DWACE by comparing it with DW and Node2vec on five widely used datasets for community detection: Football, Karate Club, Dolphins, Email-Eu-Core, and Facebook. They are chosen for their diverse structural properties, varying sizes, and established ground-truth community labels [10, 14]. The Karate Club network captures social interactions within a university karate club, studied initially by Zachary; the Dolphins network represents associations among a dolphin community; the Football dataset encodes games between U.S. college football teams with conferences as communities; the Email-Eu-Core dataset represents email exchanges within a European research institution; and the Facebook dataset reflects friendship links in a Facebook ego-network. Agglomerative, K-means, and Spectral are used (Table 1).

Table 1. Datasets

Dataset	Number of nodes	Number of edges	Number of communities
Karate Club	34	78	2
Dolphins	62	159	–
Football	115	613	12
Email-EU-Core	1005	16706	–
Facebook	4039	88234	–

Initial node embeddings are produced using DeepWalk with an embedding dimension of 128 and context and stride parameters set to [1]. The embeddings are refined using a nonlinear autoencoder [5, 6], trained for 50 epochs with a batch size of 64. This method uses a pooling loss that mixes mean squared error (MSE), modulus, and neighborhood-preserving loss, along with supervised cross-entropy when labels are present. The autoencoder output is refined using two types of graph autoencoders: GAE-GCN and GAE-SAGE, employing the GCN and GraphSAGE encoders, respectively. The embeddings are examined utilizing three algorithms: K-Means, Agglomerative Clustering, and Spectral Clustering.

The autoencoder's output was input into a contrastive projection head, which underwent training for 100 epochs utilizing the InfoNCE loss [8, 9]. A random mask [15] was used to create two augmented views for each sample, utilizing a temperature parameter of 0.005. Each experiment was conducted 20 times. The mean, standard deviation, minimum, and maximum values are utilized for evaluation.

Table 2 shows that DWACE consistently attains superior or equivalent outcomes compared to DeepWalk and Node2vec across five datasets utilizing K-means clustering. DWACE excels in critical measures, including Modularity, ARI, and NMI, particularly in the contexts of Karate, Football, and Facebook. Besides, it also maintains low Conductance and high Coverage. In Foot-ball, DWACE attains the top results across all parameters with negligible volatility. It excels with Dolphins and Facebook, and it maintains competitiveness on Email-EU-Core. DWACE demonstrates efficacy in community detection when integrated with K-means clustering.

Table 2. Experimental results on DWACE, DW and Node2vec with KMeans

Dataset	Method	Modularity	ARI	NMI	Conductance	Coverage
Karate	DW	0.39 ± 0.09 (min=0.10, max=0.43)	0.49 ± 0.39 (min=0.00, max=1.00)	0.49 ± 0.38 (min=0.00, max=1.00)	0.15 ± 0.18 (min=0.00, max=0.82)	0.52 ± 0.41 (min=0.00, max=0.87)
	Node2vec	0.23 ± 0.10 (min=0.08, max=0.45)	0.33 ± 0.10 (min=0.19, max=0.47)	0.51 ± 0.05 (min=0.44, max=0.59)	0.63 ± 0.13 (min=0.29, max=0.79)	0.40 ± 0.15 (min=0.19, max=0.73)
	DWACE	0.39 ± 0.09 (min=0.17, max=0.44)	0.59 ± 0.20 (min=0.00, max=0.79)	0.62 ± 0.17 (min=0.00, max=0.82)	0.29 ± 0.21 (min=0.00, max=0.71)	0.71 ± 0.24 (min=0.00, max=0.87)
Dolphins	DW	0.49 ± 0.03 (min=0.44, max=0.52)	-	-	0.26 ± 0.06 (min=0.17, max=0.45)	0.77 ± 0.06 (min=0.65, max=0.89)
	Node2vec	0.47 ± 0.02 (min=0.42, max=0.51)	-	-	0.43 ± 0.05 (min=0.32, max=0.52)	0.64 ± 0.04 (min=0.59, max=0.73)
	DWACE	0.50 ± 0.02 (min=0.43, max=0.52)	-	-	0.25 ± 0.03 (min=0.17, max=0.30)	0.77 ± 0.04 (min=0.72, max=0.89)
Football	DW	0.56 ± 0.05 (min=0.37, max=0.59)	0.56 ± 0.20 (min=0.00, max=0.72)	0.78 ± 0.22 (min=0.00, max=0.88)	0.26 ± 0.07 (min=0.00, max=0.30)	0.69 ± 0.17 (min=0.00, max=0.87)
	Node2vec	0.56 ± 0.02 (min=0.52, max=0.59)	0.58 ± 0.10 (min=0.39, max=0.72)	0.84 ± 0.04 (min=0.75, max=0.89)	0.28 ± 0.02 (min=0.25, max=0.30)	0.73 ± 0.02 (min=0.70, max=0.79)
	DWACE	0.58 ± 0.01 (min=0.57, max=0.60)	0.65 ± 0.03 (min=0.58, max=0.70)	0.86 ± 0.01 (min=0.83, max=0.87)	0.29 ± 0.01 (min=0.28, max=0.30)	0.72 ± 0.01 (min=0.71, max=0.73)
Email-EU-Core	DW	0.18 ± 0.11 (min=0.01, max=0.35)	-	-	0.29 ± 0.12 (min=0.00, max=0.43)	0.73 ± 0.25 (min=0.00, max=0.98)
	Node2vec	0.29 ± 0.05 (min=0.20, max=0.36)	-	-	0.38 ± 0.05 (min=0.26, max=0.43)	0.62 ± 0.09 (min=0.55, max=0.83)
	DWACE	0.22 ± 0.11 (min=0.02, max=0.33)	-	-	0.36 ± 0.10 (min=0.14, max=0.58)	0.69 ± 0.16 (min=0.568, max=0.98)

(continued)

Table 2. (*continued*)

Facebook	DW	0.27 ± 0.07 (min=0.04, max=0.30)	-	-	0.05 ± 0.09 (min=0.00, max=0.44)	0.93 ± 0.22 (min=0.00, max=0.99)
	Node2vec	0.28 ± 0.06 (min=0.09, max=0.32)	-	-	0.04 ± 0.02 (min=0.03, max=0.11)	0.98 ± 0.01 (min=0.96, max=0.99)
	DWACE	0.31 ± 0.10 (min=0.09, max=0.57)	-	-	0.04 ± 0.02 (min=0.02, max=0.10)	0.98 ± 0.01 (min=0.93, max=0.99)

Table 3 displays the experimental outcomes for DWACE, DW, and Node2vec with Agglomerative clustering. With Agglomerative clustering, DWACE consistently performs as well as or better than DW and Node2vec. For example, on the Karate dataset, DWACE achieves superior ARI and NMI scores. It shows the best overall performance on the Football dataset, with the highest Modularity, ARI, and NMI. DWACE also demonstrates good performance on the Dolphins, Facebook, and Email-EU-Core datasets.

The outcomes presented in Table 4 illustrate communities generated by Spectral clustering. DWACE performs better on the Karate and Football datasets. It achieves higher ARI and NMI scores on the Karate dataset. On the Football dataset, DWACE outperforms the other methods on all metrics. On the Dolphins dataset, DWACE generates the most well-defined communities. It achieves the highest Modularity and lowest Conductance scores. On the Email-EU-Core dataset, Node2vec also has a slightly better Modularity score, DWACE has the same Conductance and better Coverage.

On the Facebook dataset DWACE maintained a good performance in Conductance and achieved the highest average Coverage.

Table 3. Experimental results on DWACE, DW and Node2vec with Agglomerative

Dataset	Method	Modularity	ARI	NMI	Conductance	Coverage
Karate	DW	0.25 ± 0.22 (min=-0.02, max=0.44)	0.46 ± 0.37 (min=0.00, max=0.88)	0.48 ± 0.34 (min=0.05, max=0.84)	0.34 ± 0.22 (min=0.13, max=0.82)	0.79 ± 0.14 (min=0.22, max=0.87)
	Node2vec	0.23 ± 0.10 (min=0.09, max=0.45)	0.32 ± 0.10 (min=0.19, max=0.47)	0.51 ± 0.05 (min=0.45, max=0.59)	0.62 ± 0.14 (min=0.29, max=0.79)	0.40 ± 0.15 (min=0.21, max=0.73)
	DWACE	0.36 ± 0.13 (min=-0.02, max=0.44)	0.58 ± 0.20 (min=0.00, max=0.77)	0.61 ± 0.16 (min=0.05, max=0.73)	0.31 ± 0.20 (min=0.13, max=0.71)	0.74 ± 0.18 (min=0.30, max=0.87)
Dolphins	DW	0.49 ± 0.03 (min=0.40, max=0.51)	-	-	0.25 ± 0.06 (min=0.16, max=0.44)	0.78 ± 0.06 (min=0.62, max=0.89)
	Node2vec	0.48 ± 0.02 (min=0.43, max=0.50)	-	-	0.41 ± 0.04 (min=0.32, max=0.48)	0.66 ± 0.04 (min=0.57, max=0.74)
	DWACE	0.49 ± 0.03 (min=0.39, max=0.51)	-	-	0.26 ± 0.04 (min=0.17, max=0.34)	0.77 ± 0.05 (min=0.69, max=0.89)
Football	DW	0.52 ± 0.13 (min=0.00, max=0.59)	0.55 ± 0.20 (min=0.00, max=0.71)	0.79 ± 0.22 (min=0.02, max=0.89)	0.29 ± 0.06 (min=0.13, max=0.51)	0.74 ± 0.07 (min=0.71, max=0.98)
	Node2vec	0.55 ± 0.02 (min=0.51, max=0.58)	0.56 ± 0.09 (min=0.38, max=0.71)	0.85 ± 0.05 (min=0.73, max=0.91)	0.29 ± 0.02 (min=0.25, max=0.30)	0.73 ± 0.02 (min=0.72, max=0.78)
	DWACE	0.56 ± 0.01 (min=0.55, max=0.58)	0.60 ± 0.03 (min=0.58, max=0.64)	0.87 ± 0.01 (min=0.87, max=0.88)	0.29 ± 0.00 (min=0.29, max=0.29)	0.71 ± 0.01 (min=0.70, max=0.72)
Email-EU-Core	DW	0.17 ± 0.12 (min=0.00, max=0.37)	-	-	0.33 ± 0.11 (min=0.14, max=0.50)	0.77 ± 0.19 (min=0.54, max=0.99)
	Node2vec	0.25 ± 0.05 (min=0.18, max=0.33)	-	-	0.39 ± 0.06 (min=0.27, max=0.45)	0.62 ± 0.10 (min=0.50, max=0.84)

(continued)

Table 3. (*continued*)

Dataset	Method	Modularity	ARI	NMI	Conductance	Coverage
	DWACE	0.21 ± 0.10 (min=0.01, max=0.33)	-	-	0.33 ± 0.07 (min=0.20, max=0.43)	0.65 ± 0.15 (min=0.53, max=0.99)
	DW	0.26 ± 0.09 (min=0.00, max=0.32)	-	-	0.08 ± 0.14 (min=0.02, max=0.50)	0.98 ± 0.02 (min=0.92, max=0.99)
Facebook	Node2vec	0.28 ± 0.06 (min=0.10, max=0.33)	-	-	0.03 ± 0.02 (min=0.02, max=0.12)	0.98 ± 0.01 (min=0.96, max=0.99)
	DWACE	0.30 ± 0.09 (min=0.09, max=0.53)	-	-	0.04 ± 0.03 (min=0.02, max=0.12)	0.98 ± 0.02 (min=0.91, max=0.99)

Table 4. Experimental results on DWACE, DW and Node2vec with Spectral

Dataset	Method	Modularity	ARI	NMI	Conductance	Coverage
	DW	0.26 ± 0.20 (min=0.00, max=0.43)	0.48 ± 0.40 (min=0.00, max=1.00)	0.51 ± 0.37 (min=0.05, max=1.00)	0.33 ± 0.20 (min=0.13, max=0.81)	0.85 ± 0.16 (min=0.23, max=0.96)
Karate	Node2vec	0.25 ± 0.11 (min=0.08, max=0.44)	0.28 ± 0.08 (min=0.18, max=0.45)	0.47 ± 0.06 (min=0.35, max=0.58)	0.60 ± 0.16 (min=0.30, max=0.79)	0.42 ± 0.16 (min=0.21, max=0.72)
	DWACE	0.33 ± 0.13 (min=0.00, max=0.44)	0.51 ± 0.19 (min=0.00, max=0.79)	0.56 ± 0.15 (min=0.05, max=0.82)	0.36 ± 0.20 (min=0.13, max=0.74)	0.70 ± 0.19 (min=0.30, max=0.96)
	DW	0.47 ± 0.03 (min=0.39, max=0.52)	-	-	0.28 ± 0.06 (min=0.17, max=0.41)	0.73 ± 0.06 (min=0.60, max=0.84)
Dolphins	Node2vec	0.46 ± 0.01 (min=0.43, max=0.48)	-	-	0.43 ± 0.03 (min=0.36, max=0.49)	0.61 ± 0.02 (min=0.57, max=0.65)
	DWACE	0.49 ± 0.01 (min=0.46, max=0.51)	-	-	0.27 ± 0.05 (min=0.16, max=0.34)	0.73 ± 0.05 (min=0.68, max=0.86)

(*continued*)

Table 4. (*continued*)

	DW	0.53 ± 0.14 (min=0.00, max=0.60)	0.58 ± 0.21 (min=0.00, max=0.77)	0.78 ± 0.21 (min=0.05, max=0.89)	0.29 ± 0.07 (min=0.11, max=0.52)	0.74 ± 0.05 (min=0.69, max=0.89)
Football	Node2vec	0.56 ± 0.03 (min=0.49, max=0.59)	0.63 ± 0.11 (min=0.39, max=0.78)	0.84 ± 0.06 (min=0.72, max=0.89)	0.29 ± 0.02 (min=0.25, max=0.36)	0.71 ± 0.03 (min=0.64, max=0.78)
	DWACE	0.58 ± 0.01 (min=0.56, max=0.59)	0.71 ± 0.04 (min=0.62, max=0.78)	0.87 ± 0.01 (min=0.84, max=0.89)	0.29 ± 0.01 (min=0.29, max=0.34)	0.71 ± 0.00 (min=0.70, max=0.72)
Email-EU-Core	DW	0.20 ± 0.10 (min=0.00, max=0.37)	-	-	0.34 ± 0.10 (min=0.14, max=0.49)	0.70 ± 0.13 (min=0.56, max=0.96)
	Node2vec	0.30 ± 0.06 (min=0.22, max=0.39)	-	-	0.37 ± 0.05 (min=0.24, max=0.43)	0.64 ± 0.09 (min=0.55, max=0.86)
	DWACE	0.26 ± 0.09 (min=0.03, max=0.40)	-	-	0.37 ± 0.07 (min=0.20, max=0.55)	0.68 ± 0.10 (min=0.53, max=0.98)
	DW	0.13 ± 0.15 (min=0.00, max=0.54)	-	-	0.10 ± 0.13 (min=0.00, max=0.50)	0.97 ± 0.05 (min=0.81, max=0.99)
Facebook	Node2vec	0.22 ± 0.16 (min=0.00, max=0.45)	-	-	0.07 ± 0.08 (min=0.00, max=0.27)	0.97 ± 0.05 (min=0.80, max=0.99)
	DWACE	0.12 ± 0.16 (min=0.00, max=0.45)	-	-	0.10 ± 0.12 (min=0.01, max=0.42)	0.98 ± 0.04 (min=0.84, max=1.00)

5 Conclusions

In this paper, we introduced DWACE, a framework that combines a contrastive learning objective to promote intra-community consistency with a nonlinear autoencoder for denoising and dimensionality reduction. This framework is designed to refine random-walk-based node embeddings. Experimental results indicate that DWACE obtains higher modularity, ARI, and NMI than DW and Node2vec. DWACE also achieves reduced conductance on almost all datasets, indicating that cluster boundaries are more defined and sharper.

In the future, we will improve DWACE to accommodate dynamic and heterogeneous graphs and investigate advanced contrastive augmentation and supervision strategies.

References

1. Perozzi, B., Al-Rfou, R., Skiena, S.: DeepWalk: online learning of social representations. In: Proceedings of the 20th ACM SIGKDD International Conference on Knowledge Discovery and Data Mining, pp. 701–710 (2014)
2. Grover, A., Leskovec, J.: node2vec: scalable feature learning for networks. In: Proceedings of the 22nd ACM SIGKDD International Conference on Knowledge Discovery and Data Mining, pp. 855–864 (2016)
3. Tang, J., et al.: LINE: Large-scale information network embedding. In: Proceedings of the 24th International Conference on World Wide Web, pp. 1067–1077 (2015)
4. Zhu, W., et al.: A survey on network representation learning: from shallow to deep models. IEEE Access **7**, 177634–177650 (2019)
5. Chen, W., et al.: Autoencoder enhanced network embedding. In: Proceedings of the IEEE International Conference on Data Mining Workshops (ICDMW), pp. 20–27 (2018)
6. Yang, L., et al.: NEAR: network embedding via autoencoder and random walk. In: Proceedings of the AAAI Workshop on Network Representation Learning (2019)
7. Veličković, P., et al.: Deep graph infomax. In: Proceedings of the International Conference on Learning Representations (ICLR) (2019)
8. Zhu, Y., et al.: Graph contrastive learning with augmentations. In: Advances in Neural Information Processing Systems (NeurIPS) (2020)
9. You, Y., et al.: GraphCL: contrastive self-supervised learning of graph representations. In: Proceedings of the 37th International Conference on Machine Learning (ICML) (2020)
10. Cai, M., Zheng, X.: A survey on community detection algorithms for social networks. Complexity 2020, Article ID 1481532 (2020)
11. Kipf, T., Welling, M.: Semi-supervised classification with graph convolutional networks. In: Proceedings of the International Conference on Learning Representations (ICLR) (2017)
12. Hamilton, W., Ying, Z., Leskovec, J.: Inductive representation learning on large graphs. In: Advances in Neural Information Processing Systems (NeurIPS) (2017)
13. Zhou, J., et al.: Graph neural networks: a review of methods and applications. AI Open **1**, 57–81 (2020)
14. Li, X., et al.: Community-preserving network embedding. In: Proceedings of the AAAI Conference on Artificial Intelligence (2018)
15. Zhu, Z., et al.: Adaptive augmentation for contrastive learning on graphs. In: Proceedings of the 38th International Conference on Machine Learning (ICML) (2021)
16. MacQueen, J.B.: Some methods for classification and analysis of multivariate observations. In: Proceedings of the 5th Berkeley Symposium on Mathematical Statistics and Probability, pp. 281–297 (1967)
17. von Luxburg, U.: A tutorial on spectral clustering. Stat. Comput. **17**(4), 395–416 (2007)
18. Gan, G., Ma, C., Wu, J.: Data Clustering: Theory, Algorithms, and Applications. SIAM, Philadelphia (2007)
19. Maas, A.L., Hannun, A.Y., Ng, A.Y.: Rectifier nonlinearities improve neural network acoustic models. In: Proceedings of the 30th International Conference on Machine Learning (ICML) (2013)
20. Hyndman, R.J., Koehler, A.B.: Another look at measures of forecast accuracy. Int. J. Forecast. **22**(4), 679–688 (2006)

Defective and Non-defective Printed Lot Information Classification Based on Optical Character Recognition

Viet-Phuong Le[1]([envelope]) [iD], Thi-Kim-Ngoc Tran[1] [iD], Tan-Viet-Khoa Nguyen[1] [iD], and Van-Toi Nguyen[2]([envelope]) [iD]

[1] An Giang Vocational College, Binh Duc, An Giang, Vietnam
`levietphuong@agvc.edu.vn`
[2] Faculty of Electrical and Electronic Engineering, PHENIKAA University, Hanoi 12116, Vietnam
`toi.nguyenvan@phenikaa-uni.edu.vn`

Abstract. Lot information plays an essential role in tracking product information. Errors in printing lot information (such as lot numbers, manufacturing dates, and expiration dates) on product packaging often result from inaccuracies in the printing line, printer quality, or human error. These issues can negatively affect a company's brand if such defective products are released to the market. In this paper, an integrated automatic classification system is proposed to classify defective and non-defective printed lot numbers on the existing lot printing line in a factory. This study focuses on the use of optical character recognition (OCR) models, combined with image preprocessing, model training, and fine-tuning, to improve the accuracy of recognizing printed information. The test dataset was collected from lot information printed on real product packaging in the factory. Experimental results demonstrate that the proposed method is effective in recognizing lot information on product packaging. Furthermore, by applying classification and comparing the OCR results with the original reference lot information, the system can efficiently detect defective samples with a noticeable false negative rate on the test set. It is expected that the proposed approach will provide a low-cost and practical inspection solution that can be easily deployed in real-world manufacturing environments.

Keywords: Optical Character Recognition · Expired Date Recognition · Sorting Mechanism

1 Introduction

Food, beverages, and medicine are essential parts of daily life. All necessary information about the product is usually provided on the packaging, such as the expiration date, ingredients, and so on. Among these, details like the manufacturing date, expiration date, and lot number are crucial as it relates to the product's shelf life. Therefore, during production, it is important that companies must ensure product information is printed completely and clearly on the packaging, allowing consumers can read and understand it easily. However, this is not a simple task in practice due to various factors, such as errors in the printing line, poor printer quality, and human mistakes.

Currently at Agimexpharm – a pharmaceutical joint stock company in An Giang, Vietnam – the elimination of products with incorrect or unclear information printed on the packaging is a concern addressed right from the production stage. The company uses a semi-automatic system to print and check the printed information, as presented in Fig. 1. With the above semi-automatic process, errors can occur not only due to the printing equipment or sensors, but also due to human factors when placing products on the conveyor belt. Currently, after printing, a noticeable number of workers are employed to manually inspect the printed information on each product to ensure the printing quality meets the standard before releasing them to the market. This problem leads to several limitations, such as human error, slow production speed, and increased labor costs.

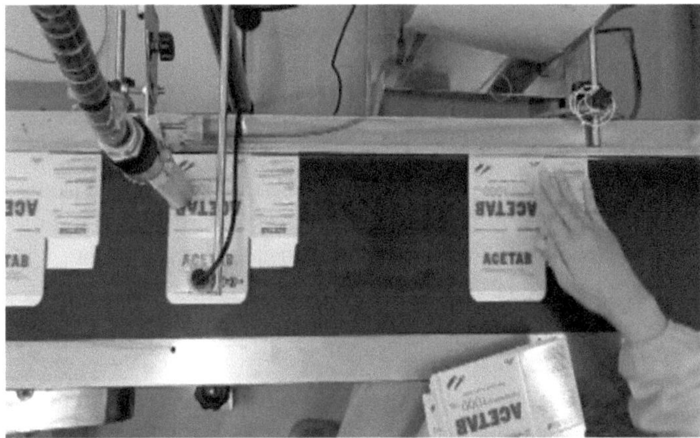

Fig. 1. Semi-automatic system for printing lot information on product packaging.

Present studies have addressed certain challenges in recognizing expiration dates; however, these efforts have mostly remained at the experimental stage in laboratories and have not been tested in industrial environments or implemented as real automated systems. In addition, a major reason lies in the decisions made by individual manufacturers regarding the characteristics of the expiration date, such as date format, font style, etc. As a result, expiration dates are printed in many variations, making them difficult for users to read and interpret. Printers may also malfunction during the production process, leading to printing errors. Furthermore, the diversity of packaging designs and complex backgrounds on product boxes can sometimes lead to recognition errors. Indeed, lighting conditions can further contribute to the problem by creating reflections or shadows that obscure the expiration date, making it unreadable or invisible.

Taking into consideration the practical needs discussed above, in this paper, we propose an automated classification system integrated into the existing lot information printing line at the factory as shown in Fig. 2. Specifically:

(1) Printing blanks: Packaging boxes that require expiration date printing are placed onto the conveyor in a fixed cycle continuously to prepare for the printing stage.
(2) Printing module: The blanks on the conveyor move toward the printing module. When a sensor detects a blank, the print head prints the required information at a fixed position. During this process, printing errors may occur, such as uneven ink, smudged or blurred information, or misaligned printing positions that do not meet the required standards.
(3) Decision module: This module identifies whether the printed lot information is defective or meets quality standards, and sends a signal to the Sorting module.
(4) Sorting module: This is an actuator mechanism controlled by a PLC. After receiving a signal from the Decision Module indicating whether the product is defective or acceptable, it activates the actuator to sort the product accordingly.

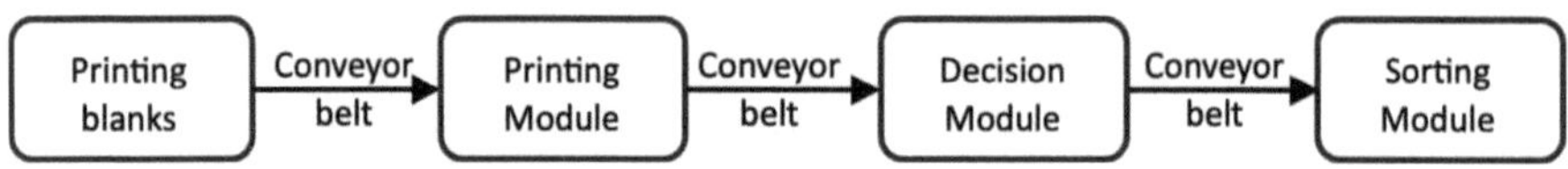

Fig. 2. Block diagram of the proposed system

This paper focuses on the use of a character recognition model based on Tesseract OCR, combined with image pre-processing, training, and model fine-tuning to improve the accuracy of recognizing printed expiration date characters.

2 Related Works

Several studies have focused on the recognition of expiration dates on products. In [1], traditional computer vision techniques was introduced to identify expiration dates which is based on SURF and RANSAC techniques. Then, OCR was applied at different scales and orientations of the input images. Tanaka et al. [2] developed a method for recognizing expiration dates in which Support Vector Machines (SVM) and OCR were applied to extract the expiration date region. Zaafouri et al. [3] proposed a method based on Gabor features and Principal Component Analysis (PCA) for dimensionality reduction. Finally, Gabor Collaborative Representation-based Classification (GCRC) was used to recognize the expiration date characters. Hosozawa et al. [4] employed an open-source OCR engine, using pre-processing techniques such as dilation and binarization to improve OCR performance. To enhance OCR accuracy, Scazzoli et al. [5] proposed a method using Hough Transform to correct the rotation angle of the expiration date, followed by a sliding window technique to detect the expiration date.

Recent studies have shown that neural networks offer promising performance in expiration date recognition. Gong et al. [6] proposed an automated expiration date recognition system in which the expiration date region is detected using an FCN (Fully Convolutional Network) on the input image. Then, MSER features and Canny edge detection are used to separate characters from the background before recognition. Finally, an open-source OCR engine is applied to recognize the expiration date characters, with component analysis and edge-based preprocessing used to improve OCR performance. Muresan et al.

[7] developed an expiration date recognition process for water bottles in a controlled environment. First, images are segmented and the bottle region is extracted using Mask R-CNN. Then, image processing operations are applied to extract each character from the bottle region, and a CNN is implemented to recognize these characters. There are many recent research works using CNN for implementation such as [8–10, 22, 23]. Khan [8] proposed a network for recognizing expiration date digits using CNN layers without the need for image preprocessing. The digit-containing region is used as the network input. Florea and Rebedea [9] proposed a deep learning solution for expiration date recognition. First, the authors use a text detection model to identify candidate text regions, then apply a CRNN (Convolutional Recurrent Neural Network) to recognize the characters. The recognized text is then filtered to select the expiration date using logical date criteria. Gong et al. [10] focused on detecting the expiration date region and then deployed a CRNN to recognize the expiration date. Ashino and Takeuchi [11] used a Deep Neural Network (DNN) to recognize expiration date digits, where numbers and characters are detected in the image before character recognition.

A.C. Seker and S.C. Ahn [12] used an approach for expiration date detection and recognition, addressing the limitations of previous studies and distinguishing between 13 different date formats. Unlike earlier methods, the authors applied a neural network model to decompose the expiration date by identifying the day, month, and year separately. Experimental results showed that the proposed method achieved a recognition accuracy of 97.74% for expiration dates in various formats and challenging cases.

In Vietnam, to the best knowledge of the authors, there has been no prior research and implementation related to the detection and recognition of expiration dates on product packaging. In addition, the application of automation in pharmaceutical manufacturing is still a relatively new field. One of the strict requirements in a pharmaceutical production model is maintaining a sterile environment, so minimizing human involvement in the production line is one of the key challenges that need to be addressed. This study focuses on a character recognition model based on OCR, combined with image preprocessing, model training, and fine-tuning to improve the recognition accuracy of printed expiration date characters. The test dataset was collected from real printed products in the factory. The proposed method has proven effective in recognizing lot information printed on product packaging. Furthermore, by applying detection rules and comparing with reference data, the system is able to classify with a noteworthy false negative rate (FNR) on the test set. This solution helps businesses save inspection costs and is readily deployable in real-world industrial environments.

3 Methodology

3.1 Overview of the Proposed Method of the Decision Module

The proposed Decision Module shown in Fig. 3 including the following steps:

(1) Image Capture: After the product passes through the printing module on the conveyor belt, a camera is used to capture the image region containing the lot information.
(2) Expdate Region Extraction: A method based on matching local feature points is used to detect the region containing the lot information.

(3) Character Recognition: The extracted image region is then converted into text using a fine-tuned OCR model.

(4) Defective/non-defective Decision: The recognized lot information is compared with the reference lot information to be printed, along with factors such as the confidence score of each character and the skew angle of the text line.

(5) Sending Decision Signal to Sorting Module: A digital signal indicating whether the product is defective or non-defective is sent to the sorting module.

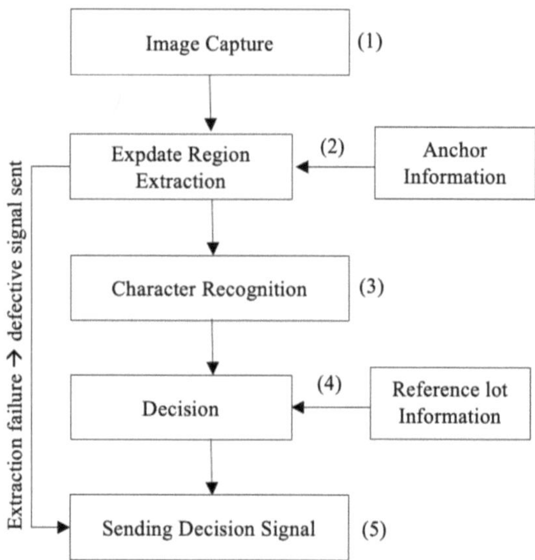

Fig. 3. Proposed method of the Decision Module.

The following sections will explain the proposed classification module in more detail.

3.2 Expdate Region Extraction

Local feature key point matching approach had been widely used for object localization in previous studies [13–16]. Motavating from that, we propose a method to detect and locate the printed lot information regions using a feature point matching method.

First, the system identifies an anchor region on the product packaging. This anchor is a pre-printed, unique, and non-overlapping region on the packaging. Similarly to the work by [14], the keypoints from both the anchor image and the captured image are extracted and described by a local feature; then, keypoints are matched in the local feature space using the nearest neighbor rule, with ambiguity rejection based on the two nearest neighbours. Second, the matched keypoints are filtered based on homography using RANSAC. Finally, the anchor region is localized thanks to the transformation between the pairs of matched keypoints. After localizing the anchor, the lot information region is then located based on its relative position to the anchor region as shown in Fig. 4.

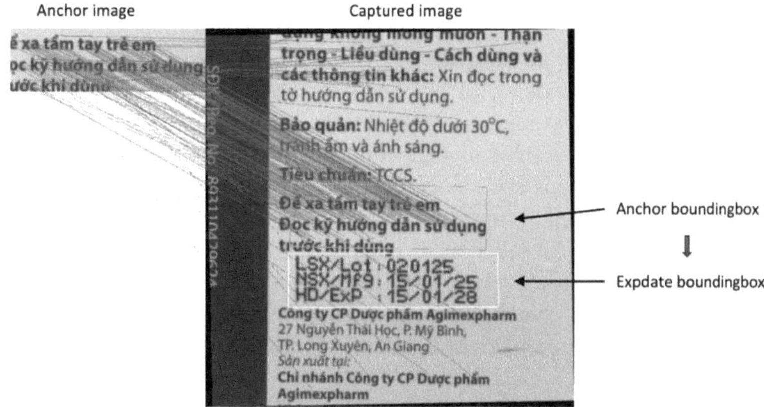

Fig. 4. Extracting the lot information region thanks to the localization of the anchor region

3.3 Character Recognition

Optical Character Recognition (OCR) is a technique that allows the conversion of images containing text into editable or searchable text. With the rapid development of deep learning models, many modern OCR methods have been proposed, of which Tesseract, EasyOCR, TrOCR and CRNN are prominent.

Tesseract OCR is an open-source OCR tool developed by Google that supports over 100 languages and allows for custom model training [17]. The modern Tesseract system uses LSTM (Long Short-Term Memory) to process character sequences, enabling the recognition of continuous text. Parameters such as psm (Page Segmentation Mode) and oem (OCR Engine Mode) can be adjusted to improve recognition results.

EasyOCR is an OCR library built on PyTorch, integrating CRAFT model for text detection and CRNN for character recognition. The library supports more than 80 languages and works effectively on real images [18].

TrOCR is an advanced model from Microsoft Research that applies Transformer architecture in both feature extraction (Vision Transformer) and decoding (Transformer Decoder), allowing input image processing to be completely end-to-end, without character separation or intermediate processing [19].

CRNN is a hybrid model of CNN and RNN with CTC loss function, allowing to recognize character sequences in images without having to align each specific character. This model is especially effective with natural or handwritten texts [20].

For OCR to work effectively, it is required that image quality should be sufficient- Various preprocessing techniques are used to improve accuracy, including Thresholding (Binarization): Converts a grayscale image into a black-and-white image to highlight the text. Common methods include Otsu Threshold and Adaptive Threshold. Denoising: Uses Gaussian or Median filters to remove small noise particles, making the text clearer and easier to separate from the background. Morphological operations: Dilation helps connect broken character strokes (commonly seen in dot-matrix fonts) and Erosion to removes small noisy regions.

Recognizing inkjet-printed text on product packaging is a specific and challenging task due to the following reasons: Dot-matrix fonts consist of discrete pixels, resulting in broken and unconnected strokes, complex image backgrounds, often in green, orange, or yellow colors. The text may be blurred, misaligned, or distorted due to curved surfaces (such as bottles, cans, or pouches). In previous studies, Dot-matrix text requires specially trained models, as Tesseract's default settings perform poorly with this font type. Some research groups used CRNN models trained on real packaging images to improve accuracy. Preprocessing steps such as contour smoothing, contrast enhancement, and skew correction are essential to improve OCR performance.

To address the OCR problem for packaaging inkjet-printed text, this study integrates advanced image processing techniques and employs a deep learning model equipped with a fine-tuned Tesseract-based approach. A custom training dataset is developed to reflect the real-world characteristics of dot-matrix fonts, accounting for a wide range of variations in appearance.

3.4 Defective/Non-defective Decision

After the recognition step, the lot information is compared with the initialized reference lot information. The comparison is performed using the Gestalt Pattern Matching method, designed by Ratcliff & Metzener [21]. The system then eliminates defective products based on the OCR confidence score and the tilt of the text line. Finally, the defective or non-defective result is sent to the Sorting Module as a digital signal to control the mechanical system for removing defective products.

4 Experimental Setup

4.1 Dataset

The data used for training and experimentation plays a crucial role in determining the model's performance. In this study, images were captured from real product packaging, featuring a variety of fonts and backgrounds, including all types of errors commonly found in the factory. A total of 1,569 images were collected, of which 1,261 were used for training and 308 for testing. Figure 5 illustrates some examples of the dataset.

Fig. 5. Some samples of the dataset

4.2 Evaluation Metrics

Character Recognition

To evaluate character recognition, we used two metrics – Character Error Rate (CER) and Word Error Rate (WER) as presented in Eq. 1 and Eq. 2, respectively.

CER is calculated based on the Levenshtein distance, by counting the minimum number of character-level operations required to transform the input reference text into the OCR output.

$$CER = \frac{S + D + I}{N} \tag{1}$$

where:

S: Number of substitutions wrong character used.

D: Number of deletions (character missing).

I: Number of insertions (extra character added).

N: Total number of characters in the reference text.

The denominator N can be calculated using the formula: $N = S + D + C$, where C is the number of correct characters.

The result of this formula represents the percentage of characters in the OCR output that are incorrect compared to the input reference text. The lower the CER value, the better the performance of the OCR model (a perfect model has $CER = 0$).

WER measures the word error rate in the entire recognized text and is often closely related to CER (as long as the error rate is not excessively high), although the WER value is always higher than CER.

$$WER = \frac{S_w + D_w + I_w}{N_w} \tag{2}$$

where

S_w: Number of Substitutions (wrong words used).

D_w: Number of Deletions (words missing).
I_w: Number of Insertions (extra words added).
N_w: Total number of words in the reference.

The denominator N_w can be calculated using the formula: $N_w = S_w + D_w + C_w$ where C_w is the number of correct words.

Classification System

The classification system consists of two classes: packaging with defective printed lot information and packaging non-defective printed lot information. The defective class (the class of interest) is considered Positive, while the non-defective class is considered Negative. Accordingly:

- True Positive (TP): Defective samples that are correctly predicted by the system as defective samples.
- True Negative (TN): Non-defective samples that are correctly predicted by the system as non-defective samples.
- False Positive (FP): Non-defective samples that are incorrectly predicted by the system as defective samples.
- False Negative (FN): Defective samples that are incorrectly predicted by the system as non-defective samples.

The Accuracy, Precision, Recall, F1-score, and FNR (False Negative Rate) metrics are calculated using the following formulas:

$$\text{Accuracy} = \frac{\text{TP} + \text{TN}}{\text{TP} + \text{TN} + \text{FP} + \text{FN}} \tag{3}$$

$$\text{Precision} = \frac{\text{TP}}{\text{TP} + \text{FP}} \tag{4}$$

$$\text{Recall} = \frac{\text{TP}}{\text{TP} + \text{FN}} \tag{5}$$

$$\text{F1} - \text{Score} = 2 \times \frac{\text{Precision} \times \text{Recall}}{\text{Precision} \times \text{Recall}} \tag{6}$$

$$\text{FNR} = \frac{\text{FN}}{\text{FN} + \text{TP}} \tag{7}$$

4.3 Experiment on Character Recognition

For training, we extracted each line of text, as shown in Fig. 6. The data was then augmented by varying brightness, blurriness, etc., to ensure richness and diversity for accurate and objective model training and evaluation. A total of 7,852 real-world samples were extracted from the 1,261 training images mentioned in Subsect. 4.1, each representing a line of lot information. We split the dataset into 7,066 samples for training and 786 samples for testing. All data samples were labeled with the corresponding text content. The training was conducted only with character classes present in the collected dataset.

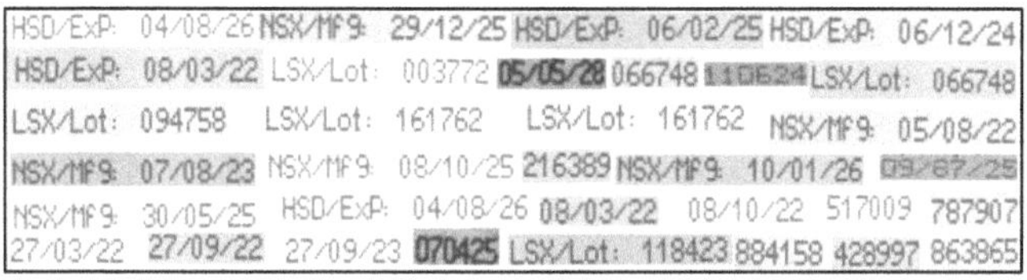

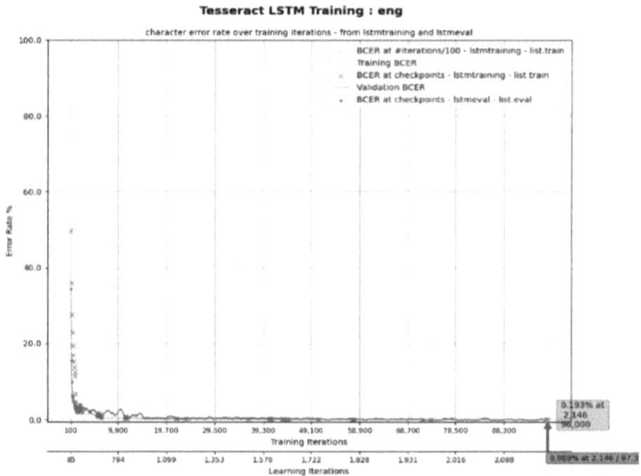

Fig. 6. Some samples used for training the OCR model.

First, we evaluated the character and word recognition error rates, using CER and WER as the evaluation metrics.

We performed OCR using the fine-tuned model. The parameters tested included:

- --oem 1: using the LSTM-based OCR Engine.
- --psm: tested with mode 6 (Assume a single uniform block of text).

Fig. 7. The result with Tesseract OCR

As presented in Fig. 7, the results demonstrate that the fine-tuned Tesseract model achieves and the lowest character error rate when compared to other common OCR models such as EasyOCR (finetune), Tesseract, EasyOCR, TrOCR, and CRNN, as in Fig. 8. This result indicates that the trained model can be effectively used for ExpDate character recognition. However, due to character errors, spacing issues, and incorrect word merging, we enhance the accuracy by matching the patterns "NSX/Mfg:", "LSX/Lot:", and "HD/Exp:" with the extracted results and adjust the content accordingly. After extracting the text, the system compares the OCR results with the initialized reference information to calculate the similarity (%) between the two text strings using Rapid Fuzzy [21].

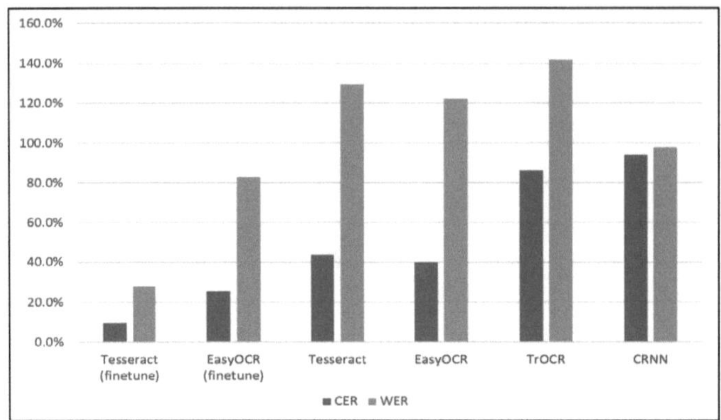

Fig. 8. Performance comparison with common OCR models

4.4 Experiment on Classification System

After training and fine-tuning the OCR model, the next experiment was to evaluate the classification performance of the entire system. We used 308 test images to evaluate the classification system, including 148 defective samples and 160 non-defective samples. Figure 9 illustrates some lot information region samples from the test set used to evaluate the classification system.

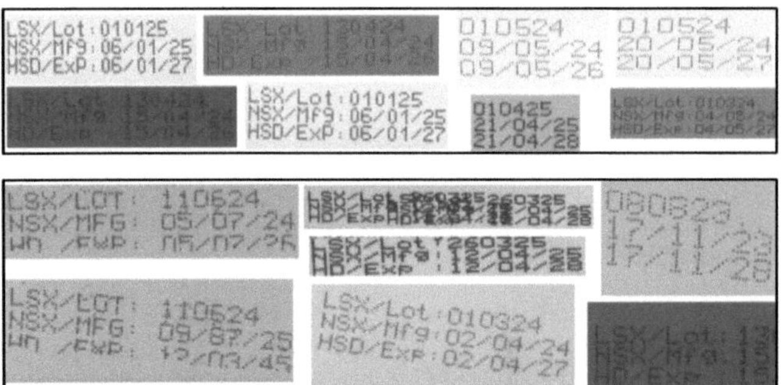

Fig. 9. Some lot information region samples from the testing dataset of the classification system: non-defective samples (above) defective samples (below)

In this experiment, all steps in the proposed Decision Module discussed in Sect. 3 were executed. Test images with OCR results having a similarity score of 95% or higher (compared to the ground truth) were further evaluated based on recognition confidence and text line skew (less than $\pm 3°$) to determine whether the sample was defective or not.

We evaluated the accuracy using a dataset of real-world images containing complete information on lot number, manufacturing date, and expiration date. The results are as follows:

Table 1. Confusion matrix

	Predicted: Defective	Predicted: non-Defective
Defective	147 (TP)	1 (FN)
non-Defective	1 (FP)	159 (TN)

Table 2. Performance of the classification system

Measurement	Our method	Seker's method [12]
Recall	**99.32%**	96.48%
Precision	99.32%	**100.00%**
F1-Score	**99.32%**	98.21%
False Negative Rate (FNR)	**0.68%**	3.52%
Accuracy	**99.35%**	96.75%

The evaluation results on the experimental dataset demonstrate that the error classification system performs with high effectiveness and accuracy (Table 1). The model achieved a Recall of 99.32%, indicating its strong ability to detect nearly all errors without omission. The F1-score, also at 99.32%, reflects a well-balanced performance between Recall and Precision. Importantly, the false negative rate was only 0.68%, a critical metric in applications where failing to detect errors is unacceptable. The overall classification accuracy reached 99.35%. The experimental results presented in Table 2 shows that our method achieves better performance than Seker's method [12] on most evaluation metrics. These findings suggest that our proposed method is highly suitable for deployment in real-world environments requiring accurate and reliable error detection and classification.

The results show that the fine-tuned Tesseract model on dot-matrix datasets achieves outstanding performance with a character error rate of 9.8% and a word error rate of 27.8%. This is the only model capable of accurately recognizing the distinctive characteristics of dot-matrix printed text. Fine-tuning enables the model to effectively adapt to the sparse and noisy nature of this print style, which is not commonly represented in standard training datasets.

To further improve recognition performance, rule-based post-processing can also be incorporated to reduce the word error rate to an acceptable level.

5   Conclusion

This study proposes a system for classifying printing defects in lot information printed on product package. Experiments conducted on real-world product package demonstrated high effectiveness, achieving a low false negative rate of 0.68%, indicating strong potential for practical deployment in manufacturing environments with minimal cost. The classification system uses fine-tuned Tesseract OCR for recognizing dot-matrix characters, which achieves the best performance comperison on common OCR models on the testing dataset such as CRNN, EasyOCR, and TrOCR. Future studies could expand the collection of real-world error samples to improve accuracy across different printers and machines and to support the deployment of the system in real-time production environments.

References

1. Peng, E., Peursum, P., Li, L.: Product barcode and expiry date detection for the visually impaired using a smartphone. In: Proceedings of the 2012 International Conference on Digital Image Computing Techniques and Applications (DICTA), Fremantle, Australia, 3–5 December 2012, pp. 1–7. IEEE Press (2012). https://doi.org/10.1109/DICTA.2012.6411673
2. Tanaka, N., Doi, Y., Matsumoto, T., Takeuchi, Y., Kudo, H., Ohnishi, N.: A system helping the blind to get merchandise information. In: Miesenberger, K., Karshmer, A., Penaz, P., Zagler, W. (eds.) Computers Helping People with Special Needs. ICCHP 2012. LNCS, vol. 7383, pp. 565–572. Springer, Berlin, Heidelberg (2012). https://doi.org/10.1007/978-3-642-31534-3_87
3. Zaafouri, A., Sayadi, M., Fnaiech, F., al Jarrah, O., Wei, W.: A new method for expiration code detection and recognition using Gabor features based collaborative representation. Adv. Eng. Inform. $29(4)$, 1072–1082 (2015). https://doi.org/10.1016/j.aei.2015.07.002
4. Hosozawa, K., et al.: Recognition of expiration dates written on food packages with open source OCR. Int. J. Comput. Theory Eng. $10(5)$, 170–174 (2018). https://doi.org/10.18178/ijcte.2018.10.5.1217
5. Scazzoli, D., Bartezzaghi, G., Uysal, D., Magarini, M., Melacini, M., Marcon, M.: Usage of Hough transform for expiry date extraction via optical character recognition. In: Advances in Science and Engineering Technology International Conferences (ASET), pp. 1–6. IEEE (2019). https://doi.org/10.1109/ICASET.2019.8714437
6. Gong, L., et al.: A novel unified deep neural networks methodology for use-by date recognition in retail food package image. Signal Image Video Process. $15(3)$, 449–457 (2021). https://doi.org/10.1007/s11760-020-01777-6
7. Muresan, M.P., Szabo, P.A., Nedevschi, S.: Dot matrix OCR for bottle validity inspection. In: 2019 IEEE 15th International Conference on Intelligent Computer Communication and Processing (ICCP), pp. 395–401. IEEE, Cluj-Napoca, Romania (2019). https://doi.org/10.1109/ICCP48234.2019.8959674
8. Khan, T.: Expiry date digit recognition using convolutional neural network. Eur. J. Electr. Eng. Comput. Sci. $5(1)$, 85–88 (2021). https://doi.org/10.24018/ejece.2021.5.1.266
9. Florea, V., Rebedea, T.: Expiry date recognition using deep neural networks. Int. J. User-Syst. Interact. $13(1)$, 1–17 (2020). https://doi.org/10.37789/ijusi.2020.13.1.1
10. Gong, L., Yu, M., Duan, W., Ye, X., Gudmundsson, K., Swainson, M.: A novel camera based approach for automatic expiry date detection and recognition on food packages. In: Maglogiannis, I., Iliadis, L., Pimenidis, E. (eds.) Artificial Intelligence Applications and Innovations. AIAI 2018. LNCS, vol. 11197, pp. 133–142. Springer, Cham (2018). https://doi.org/10.1007/978-3-319-92007-8_12
11. Ashino, M., Takeuchi, Y.: Expiry-date recognition system using combination of deep neural networks for visually impaired. In: Miesenberger, K., Kouroupetroglou, G. (eds.) Computers Helping People with Special Needs. ICCHP 2020. LNCS, vol. 12377, pp. 510–516. Springer, Cham (2020). https://doi.org/10.1007/978-3-030-58796-3_60

12. Seker, A.C., Ahn, S.C.: A generalized framework for recognition of expiration dates on product packages using fully convolutional networks. Expert Syst. Appl. **203**, 117310 (2022). https://doi.org/10.1016/j.eswa.2022.117310
13. Le, V.P., Visani, M., De Tran, C., Ogier, J.: Logo spotting for document categorization. In: Proceedings of the 21st International Conference on Pattern Recognition (ICPR), Tsukuba, Japan, 11–15 November 2012, pp. 3484–3487. IEEE Press (2012)
14. Le, V.P., Visani, M., De Tran, C., Ogier, J.-M.: Improving logo spotting and matching for document categorization by a post-filter based on homography. In: Proceedings of the 12th International Conference on Document Analysis and Recognition (ICDAR), Washington, DC, USA, August 2013, pp. 270–274. IEEE Press (2013)
15. Le, V.P.: Logo detection, recognition and spotting in context by matching local visual features. PhD Thesis, Université de La Rochelle (2015)
16. Le, V.P., De Tran, C.: Key-point matching with post-filter using SIFT and BRIEF in logo spotting. In: Proceedings of the IEEE RIVF International Conference on Computing & Communication Technologies – Research, Innovation, and Vision for the Future (RIVF), pp. 89–93. IEEE Press (2015)
17. Smith, R.: An overview of the Tesseract OCR engine. In: Proceedings of the 9th International Conference on Document Analysis and Recognition (ICDAR), pp. 629–633. IEEE (2007)
18. Baek, Y., et al.: Character region awareness for text detection. In: Proceedings of the IEEE Conference on Computer Vision and Pattern Recognition (CVPR), pp. 9365–9374 (2019)
19. Li, M., et al.: TrOCR: transformer-based Optical Character Recognition with Pre-trained Models. arXiv preprint arXiv:2109.10282 (2021)
20. Shi, B., Bai, X., Yao, C.: An end-to-end trainable neural network for image-based sequence recognition and its application to scene text recognition. IEEE Trans. Pattern Anal. Mach. Intell. **39**(11), 2298–2304 (2016)
21. Ratcliff, J.W., Metzener, D.E.: Pattern Matching: The Gestalt Approach. Dr. Dobb's Journal, July 1988
22. Lotfy, M., Soliman, G.: CNN-optimized text recognition with binary embeddings for Arabic expiry date recognition. J. Electr. Syst. Inf. Technol. **11**(1), 11 (2024). Springer, Cham
23. Zaki, H., Soliman, G.: Automated dotted Arabic expiration date extraction using optimized convolutional autoencoder and custom CRNN. J. Electr. Syst. Inf. Technol. **12**(1), 24 (2025). Springer, Cham

A Method for Enhancing Images Quality Based on Machine Learning

Sinh Van Nguyen[1,2]([✉]) [iD], Vinh Xuan Nguyen[1,2], and Hanh Le Thi Ngoc[1,2] [iD]

[1] International University HCM–VNU, Ho Chi Minh City, Vietnam
[2] Vietnam National University of Ho Chi Minh City, Ho Chi Minh City, Vietnam
nvsinh@hcmiu.edu.vn

Abstract. Image quality enhancement remains a critical challenge in image processing techniques, especially for images captured under suboptimal lighting conditions that can generate low contrast, color distortion, and noise. These degradations not only impact visual aesthetics but also hinder the performance of downstream image processing tasks. The primary objective of image quality enhancement is to improve the visual quality of such images to benefit subsequent processing. Despite extensive research, achieving high-quality enhanced images remains challenging. Traditional image quality enhancement techniques often address only overexposure or underexposure, potentially failing when both issues are present. Deep learning has recently been increasingly adopted in image processing, demonstrating significant potential for enhancing image quality with underexposure, overexposure, or a combination of both. In this paper, we first review key traditional and machine learning-based image quality enhancement techniques developed in recent years. Next, we contribute by creating a new dataset to facilitate learning. We then propose an improved denoising step for input images and integrate two Residual Blocks into the "Color Shift Estimation and Correction" network architecture to enhance feature extraction. Furthermore, we introduce a novel loss function, MSE_{LOSS}, aimed at ensuring both pixel-level accuracy and perceptual realism in the enhanced images, which leads to improved visual quality. Finally, we employ a Color Shift Estimation and Correction method to train our model using both public and our newly constructed dataset. Experimental results on our new dataset demonstrate the effectiveness of our proposed approach in generating high-quality enhanced images with improved color fidelity and well-preserved details.

Keywords: Image Processing · Quality Enhancement · Illumination & Resolution · Deep Learning

1 Introduction

Image quality enhancement is a fascinating field that not only improves the quality and usability of visual data but also supports restoring objects in images. This

N. Thai-Nghe et al. (Eds.): ISDS 2025, CCIS 2714, pp. 398–416, 2026.
https://doi.org/10.1007/978-981-95-3358-9_29

is a fundamental process in digital image processing that aims to improve the quality of an image. The improved image is then applied in different fields such as medical image processing and photography. Whether it is bringing out details in a low-light or over-exposed photo, or clarifying the edges of objects in a scanned document. Besides, enhancing images plays a crucial role in extracting valuable information and improving the overall visual experience.

Improving Image quality involves applying various traditional techniques to adjust brightness, contrast, and sharpness [1,2], reduce noise, correcting distortions [1,3], and apply histogram equalization [4,5]. During the improvement steps, noise removal and enhancing the effects of illuminating light from the image using Retinex methods [6], as well as adjusting the color balance [1,3,6]. Besides traditional methods, machine learning has made significant strides in the field of image processing in recent years. These methods have proven effective in improving the quality of images, whether they are underexposed, overexposed, or both [7–9].

Addressing the ongoing need for improved image quality enhancement, this paper investigates a deep learning-based approach for image quality enhancement. Our research focuses on the specific challenges of color distortion and detail loss in poorly illuminated images. First, we introduce an additional dataset specifically designed to facilitate more effective training of deep learning models for challenging scenarios involving poor illumination. Second, we propose an enhanced and refined image quality enhancement pipeline. Specifically, to improve the feature extraction capabilities within the DeepWBNet architecture, we strategically incorporate two Residual Blocks. This architectural modification aids in the crucial refinement of features essential for the accurate generation of weight maps and visually consistent pseudo-normal images, thereby substantially contributing to the model's ability to perform effective and accurate color shift estimation and correction. Furthermore, we integrate a new loss function that encourages the model to achieve not only pixel-level accuracy but also to learn perceptually relevant features. Finally, we employ the Color Shift Estimation and Correction (CSEC) method as the core training strategy for our proposed model, leveraging the benefits of our new dataset and architectural innovations.

The main contributions of this paper include a new image quality enhancement dataset, an improved DeepWBNet architecture with integrated Residual Blocks, the incorporation of LPIPS loss for perceptual quality, and a comprehensive evaluation demonstrating state-of-the-art performance. The remainder of this paper is organized as follows: Sect. 2 provides a review of related work in image processing techniques to enhance image quality. Section 3 details the construction of our new dataset and describes our proposed image quality enhancement method. Section 4 presents the implementation, experimental results and discussion. Section 5 provides concludes the paper and future work.

2   Related Works

This section presents the techniques related to image quality enhancement area in both traditional methods and machine learning-based methods.

Brightness Adjustment: Brightness adjustment is a fundamental technique in image processing used to alter the overall lightness or darkness of an image. This is typically achieved by adding or subtracting a constant value to the pixel values of the image [1]. Mathematically, brightness adjustment operations can be expressed as follows: $(x, y)' = (x, y) + b$, where $f(x,y)$ is the original pixel value at position (x, y), $f(x,y)'$ is a new pixel value, and b is a brightness adjustment factor. If b is a positive value then the image will be brighter and if b is a negative value then the image brightness level will decrease.

Contrast Stretching: Contrast stretching, also known as normalization, is an image quality enhancement technique used to improve the contrast of an image by expanding the range of intensity values. This method aims to make the dark areas darker and the bright areas brighter, thereby enhancing the overall visual quality of the image [1]. Contrast can be divided into three types: low contrast, good contrast and high contrast. Images with low contrast are mostly bright or dark, with pixel values concentrated in one region of the histogram. If values are on the left, the image is darker; if on the right, the image is lighter; if in the middle, the image is neither too bright nor too dark.

Gamma (Power-Law): Transformations. The gamma correction method will produce a brighter and more natural image. Unlike brightness adjustment, which is linear, gamma correction is nonlinear. When depicted in graphical form, the function of the gamma is curve-shaped. The darkest and brightest areas of the gamma graph will not have much effect on different gamma arrangements. However, the middle area of the graph will have an effect by following the arrangement. The equations of gamma can be defined as follows: $f(x, y)' = f(x, y)^{1}/\gamma$, where $f(x, y)'$ is the image after the gamma correction process and $f(x, y)$ is the image before the gamma correction process. The symbol γ is the gamma correction factor with a value range of $(0 < \gamma < 1)$.

Linear Gray Level Transformations: The gray transformation method is a spatial-domain image quality enhancement algorithm. The principle of this method is to transform the gray values of single pixels into other gray values by means of a mathematical function, which is usually called a mapping-based approach. Such a method enhances an image by modifying the distribution and dynamic range of the gray values of the pixels [3].

A linear transformation of gray values, also known as a linear stretching, is a linear function of the gray values of the input image, and the formula is as follows [3]:

$$g(x, y) = C.f(x, y) + R \tag{1}$$

where $f(x, y)$ and $g(x, y)$ represent the input and output images, respectively, and C and R are the coefficients of the linear transformation.

Logarithmic Transformations: A logarithmic transformation means that each pixel's value in the output image is related to the logarithm of the corresponding pixel's value in the input image. This transformation is ideal for very dark images because it can enhance the lower gray values while reducing the range of the higher gray values. The typical form of logarithmic transformations is as formula:

$$g(x, y) = log(1 + c.f(x, y)) \tag{2}$$

where c is a control parameter.

Histogram Equalization (HE): Histogram equalization is a widely used technique in image quality enhancement that aims to improve the contrast of an image by redistributing its pixel intensity values. Histogram equalization utilizes the image's histogram to enhance its quality. Given a greyscale image, the goal is to compute a transformation that, when applied to the gray values of the original image, produces a uniform distribution of the intensity values. This method makes hidden details in dark areas visible again, significantly enhancing the visual quality of the input image [3,10].

Adaptive Histogram Equalization (AHE): The basic idea of AHE is to separate an image into several sub-blocks, and each sub-block is processed by histogram equalization, respectively. The AHE algorithm can be described as follows: (1) Set the size of a window, and select a sub-block of the input image according to the window. (2) Apply HE algorithm to the sub-block, and record the output. (3) Move the window horizontally or vertically and repeat #1 and #2 until all pixels in the input image are modified. (4) Organize all the enhanced sub-blocks into one image as output.

Contrast Limited Adaptive Histogram Equalization (CLAHE): The Contrast Limited Adaptive Histogram Equalization (CLAHE) method operates on the same principle as traditional histogram equalization. In CLAHE, the im-age is divided into several sub-images of size (nxn). Histogram equalization is then applied to each sub-image individually, based on the divisions made earlier. CLAHE is a kind of adaptive histogram equalization in which the contrast amplification can be limited to reduce the problem of noise amplification [11].

Mean Filter (Average Filter, Blur Filter, Box Filter Kernels): Mean filter is used to smoothing the image by calculating the average value of pixels in the image. The process involves considering the surrounding pixels. The pixels to be processed are included in an N × N matrix. It is based around a kernel, which represents the shape and size of the neighborhood to be sampled when calculating the mean. Often a 3 × 3 square kernel is used, or a 5 × 5 squares. Mathematically, the mean filtering has the same weight as the neighboring pixel defined as follows:

$$f(x, y) = \frac{1}{mn} \sum_{k=1}^{m} \sum_{l=1}^{n} U(x + k - 1, y + l - 1) \tag{3}$$

where $f(x, y)$ represents the image of the result. while $U(x, y)$ represents the input image used. The upper bound value of m and n represent the size of the row and column of the mean filtering.

Gaussian Filter: The Gaussian filter is a linear filter that uses the Gaussian function to set pixel weights, and is widely used for smoothing, blurring, and eliminating noise. In the Gaussian filter, the linear process is calculated by multiplying each neighboring pixel by a corresponding weight and summing the results to obtain the value for a specific coordinate point, denoted as (x, y). The mechanism of the linear spatial filter is to move the center of a filter mask from one point to another. In each pixel (x, y), the result of the filter at that point is the sum of the multiplication of the filter coefficients and the corresponding neighbor pixels in the filter mask range. Gaussian filters have two types of filters: one-dimensional and two-dimensional with the forms as below.

$$G(x) = \frac{1}{\sqrt{2\pi}\sigma}e^{\left(-\frac{x^2}{2\sigma^2}\right)} \tag{4}$$

where σ is the standard deviation of the distribution.

$$G(x) = \frac{1}{2\pi\sigma^2}e^{\left(-\frac{x^2+y^2}{2\sigma^2}\right)} \tag{5}$$

where σ is the standard deviation of the same distribution as the one-dimensional gaussian function. For x and y are expressed as coordinate points rows and columns in image pixels.

Non-Linear Filter - Median Filter: The median filter is widely used as it is very effective at removing noise while preserving edges. The median filter works by moving through the image pixel by pixel, replacing each value with the median value of neighbouring pixels. The median is calculated by first sorting all the pixel values from the matrix neighbor into numerical order, and then replacing the pixel being considered with the middle (median) pixel value. Generally the point to be processed along with the points around it is inserted into a matrix of size N x N. This matrix is called matrix neighbor (neighboring matrix), which slides, pixel by pixel, over the entire image.

Non-Linear Filter - Conservative Filter: Conservative filter is one technique to reduce existing noise on the image. In the median filter, the filter process uses the middle value of the processed neighboring pixels and pixels. In the conservative filter, the values used are the minimum and maximum values but excluding the processed middle pixels. The calculation process is performed as follows: If the middle pixel's value is within the range of the surrounding pixels' minimum and maximum values, it remains unchanged. If it exceeds the maximum value, it is replaced with the maximum value. If it is less than the minimum value, it is replaced with the minimum value.

Retinex Methods: The Retinex theory is based on the perception of color by the human eye and the modeling of color invariance [3]. The essence of this theory is to determine the reflective nature of an object by removing the effects of the illuminating light from the image. Based on Retinex theory, the human visual system processes information in a specific way during the transmission of visual information, thus removing a series of uncertain factors such as the intensity of the light source and unevenness of light. Consequently, only information that reflects essential characteristics of the object, such as the reflection coefficient, is retained. Based on the illumination-reflection model, an image can be expressed as the product of a reflection component and an illumination component: $I(x, y) = R(x, y)L(x, y)$. If L(x, y) can be estimated from I(x; y), then the reflection component can be separated from the total amount of light, and the influence of the illumination component on the image can be reduced, thus enhance the image [3].

Single-Scale Retinex (SSR): The SSR algorithm obtains a reflection image by estimating the ambient brightness. The formula is as follows: $\log R_i(x, y) = \log I_i(x, y) - \log[G(x, y) * I_i(x, y)]$, where $I(x, y)$ represents the input image, $R(x, y)$ represents the reflection image, i represents the various color channels, (x, y) represents the position of a pixel in the image, $G(x; y)$ represents the Gaussian surround function, and $*$ represents the convolution operator.

Multi-Scale Retinex (MSR): Maintain a balance between dynamic range compression and color constancy. The MSR algorithm:

$$MSR = \log R_i(x, y)$$
$$= \sum_{k=1}^{N} \omega_k \left\{ \log I_i(x, y) - \log \left[G_k(x, y) * I_i(x, y) \right] \right\} \tag{6}$$

where i represents the three color channels; k represents the Gaussian surround scales; N is the number of scales, generally 3; and the ω parameters are the scale weights. Unlike the SSR algorithm, the MSR algorithm leverages the advantages offered by multiple scales, leading to enhanced image details, improved contrast, better color consistency, and an overall enhanced visual effect.

Multiscale Retinex With Color Restoration (MSRCR): During image quality enhancement, the SSR or MSR algorithm is applied separately to the R, G, and B color channels. This can alter the relative proportions of these channels compared to the original image, leading to color distortion. To address this issue, the MSRCR algorithm has been developed. It includes a color recovery factor (C) for each channel, calculated based on the proportional relationships among the three color channels in the input image. This factor is then used to correct the output image's colors, eliminating color distortion. The algorithm takes advantage of the convolution operation with Gaussian functions. Dynamic range compression and color constancy are achieved for features at large, medium and small scales, thus yielding a relatively ideal visual effect. Nowadays, machine learning-based approaches are increasingly being applied and have proven to be effective solutions in image quality enhancement, and this paper will introduce some of these methods.

URetinex-Net: URetinex-Net was introduced in 2022, a Retinex-based deep unfolding network, which unfolds an optimization problem into a learnable network to decompose a low-light image into reflectance and illumination layers [12]. URetinex-Net includes three modules, i.e. initialization module, unfolding optimization module, and illumination adjustment module. The Retinex-based approach allows for precise manipulation of reflectance and illumination layers for better image quality; however, its complex network architecture requires significant computational resources and expertise. Besides, the U-Net architecture, initially designed for biomedical image segmentation by [13], which has been widely used as feature extractor across various domains. Its encoder-decoder structure with skip connections enables effective multi-scale feature learning. Over time, U-Net has been adapted for diverse applications, such as the backbone to extract brightened and darkened features for color correction tasks [7], content feature analyzer in image/video resizing systems [14,15]. These adaptations often modify the network's depth, width, or connectivity, demonstrating U-Net's flexibility and effectiveness in both low-level and high-level learning tasks.

Local Color Distributions Prior: When the illumination of the input image contains both over- and underexpo-sure problems, these existing methods may not work well because they are typically designed to address either the over- or under-exposure problem in the input image. In 2022, Wang Haoyuan, Ke Xu, and Rynson WH Lau. Introduced the "Local color distributions prior" for image quality enhancement. They observe from the image statistics that the local color distributions (LCDs) of an image depending on the local illuminations, suffering from both problems tend to vary across different regions of the image. Base on this observation, they proposed LCDs as a prior for locating and enhancing the two types of regions (over-/underexposed regions) [16]. First, they utilize the LCDs to depict these regions and introduce a novel local color distribution embedded (LCDE) module that formulates LCDs in multiple scales to model correlations across different areas. Second, they propose a dual-illumination learning mechanism to enhance both types of regions. Third, they create a new dataset to support the learning process, following the camera image signal processing (ISP) pipeline to produce standard RGB images with both under- and over-exposures from raw data [16].

Color Shift Estimation-and-Correction (CSEC): Color Shift Estimation and Correction (CSEC) method [7] was introduced by Yiyu Li et al. to enhance images with both over- and under-exposures by learning to estimate and correct colors of input image. This method first derive the color feature maps of the brightened and darkened versions of the input image via a UNetbased network, then use a pseudo-normal feature generator to produce pseudo-normal color feature maps. There is a COlor Shift Estimation (COSE) module to estimate the color shifts between the derived brightened (or darkened) color feature maps and the pseudo-normal color feature maps. The COSE module used for correcting the estimated color shifts of the over- and under-exposed regions separately.

In addition, this method propose a novel COlor MOdulation (COMO) module to modulate the separately corrected colors in the over- and under-exposed regions to produce the enhanced image.

A research based on a combination of GNN and GAN is presented in [17]. GNN is first used to learn the underlying structure and features of the 3D data; then, GAN is used to generate high-quality, refined versions of the data. This approach proves to be effective in addressing the challenges associated with this type of data. Nguyen et al. [18] proposed a deep learning method for face recognition that augments data in the training process. After pre-processing data based on image processing techniques, Retnet-v1 is used to improve recognition accuracy.

3   Our Proposed Method

3.1   System Overview

Our proposed framework is outlined in Fig. 1, which covers the major phases of training data preparation, pre-processing data, and training. In this system, our contributions include a custom dataset, improved pre-processing, enhancements to model architecture, and loss function. To support our model's learning under non-uniform illumination, we construct a new dataset of 350 unique scenes, each with under-exposed and over-exposed versions, totaling 700 images. To reduce noise and improve input quality, we apply a denoising filter during the pre-processing phase. In the model's architecture, we further enhance the baseline model CSEC [7] by adding two Residual layers, which strengthen feature extraction and improve correction performance. Finally, in terms of optimization, we introduce an additional Mean Square Error (MSE) loss term to complement the original loss functions.

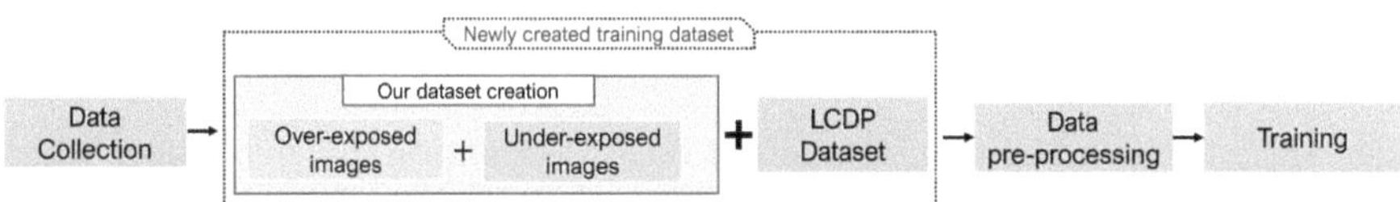

Fig. 1. Overview of our process.

3.2   Training Data Preparation

The LCDP dataset [16] has recently been publicly used in the research of image quality enhancement. This dataset provides 1,733 image pairs of under-exposed and over-exposed scenes, partitioned into 1,415 for training, 100 for validation, and 218 for testing (available at github.com/onpix/LCDPNet). However, the LCDP dataset has drawbacks in capturing the diversity of real-world lighting conditions, lacks sufficient variation in non-uniform or naturally complex illumination, which are commonly encountered in practical applications.

To overcome the above limitations of LCDP and enhance the model's generalizability, we newly construct a dataset and combine it with the public LCDP set. More specifically, we generate an additional set comprising 700 image pairs, 350 under-exposed and 350 over-exposed, derived from 350 distinct ground truth scenes. These scenes were captured using two devices, Canon EOS Rebel SL1 camera and a Samsung M20 smartphone. The images were subsequently transformed using the OpenCV library to simulate realistic lighting imbalances. More specifically, we apply following linear transformation formulas respectively:

$$g(x, y) = alpha * f(x, y) + beta, \text{with alpha} = 1.0, \text{beta} = -80, \qquad (7)$$

and

$$g(x, y) = alpha * f(x, y) + beta, \text{with alpha} = 1.0, \text{beta} = 80, \qquad (8)$$

where $f(x, y)$ represents original input. In the context of images, $f(x, y)$ is color value of a pixel at coordinates (x, y). The dataset is then split into 80% for training, 10% for validation, and 10% for testing. By introducing more diverse images and challenging lighting variations, our dataset enriches LCDP and enables our model to better learn diverse pattern in real-world conditions.

Prior to training, a median filter with a 3×3 kernel is applied to all images in the dataset as a preprocessing step. The median filter is a widely adopted technique in image processing for noise reduction, particularly effective in eliminating salt-and-pepper noise while preserving important image features such as edges. Unlike linear filters that may blur fine details, the median filter operates by sliding a window across the image and replacing each pixel's intensity with the median value of the neighboring pixels within the window. To be specific, applying a median filter to a single image is formulated as follows:

$$medianBlur(InputArray\ src, OutputArray\ dst, int\ ksize), \qquad (9)$$

where src is the source (input) image; dst is the destination (output) image where the blurred image will be returned; $ksize$ specifies the size of the kernel (or window). And, $medianBlur$ is openCV library used to smoothen input images. This non-linear approach ensures that edge sharpness is maintained while suppressing outlier noise, resulting in cleaner input data that improves the stability and performance of the training process [1]. The algorithm for applying a median filter (Algorithm 1) to a folder of images using the OpenCV library is as follows:

Algorithm 1. applyMedianFilterFoder

1: **Input:** Directory name (`directoryName`), File filter (`fileFilter`)
2: **Output:** 0 for success, -1 for error
3: **Create:** `directoryOutput` ← `directoryName` + "output/"
4: **Find:** `filePaths` ← List of all files in `directoryName` matching `fileFilter`
5: `count` ← Number of elements in `filePaths`
6: **for** i = 0 to `count` - 1 **do**
7:     `fullFilePath` ← `filePaths`[i]
8:     `filename` ← Extract base filename from `fullFilePath`
9:     **Read:** `image` ← Image from `directoryName` + `filename`
10:     **if** `image` is empty **then**
11:         **return** −1
12:     **end if**
13:     **Create:** `image_dst` ← Empty image object
14:     **Apply:** Median blur filter (kernel size 3) to `image` and store in `image_dst`
15:     **Save:** `image_dst` to `directoryOutput` + `filename`
16: **end for**
17: **return** 0

3.3   Proposed Training Model

As aforementioned, we adopt the Color Shift Estimation and Correction (CSEC) framework (outlined in Fig. 2), proposed by [7], as the foundation for our enhancement pipeline and introduce improvements to both the network architecture and loss functions to further enhance image quality. In the CSED model, it begins by employing a U-Net-based feature extractor to generate two intermediate feature representations: a darkened feature map $\mathcal{F}_D$ and a brighted feature map $\mathcal{F}_B$. These feature maps are used along with the input image I_x to derive a pseudo-normalized feature map $\mathcal{F}_N$, which serves as a reference for estimating lighting variations. The tweaks of our design are elaborated as follows.

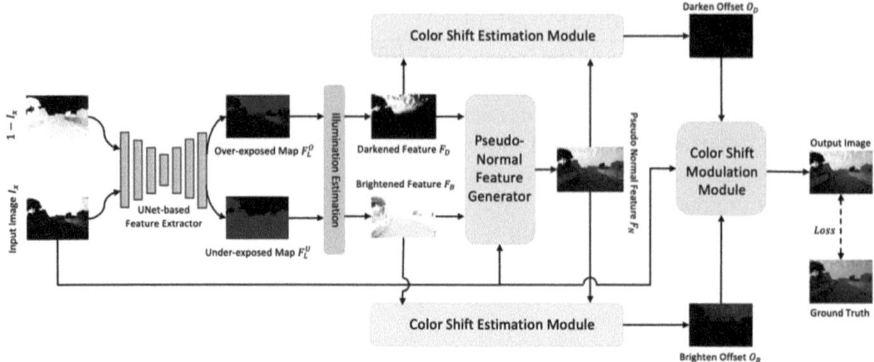

Fig. 2. Overview of Color Shift Estimation and Correction (CSEC) model [7]. First, use the UNet-based feature extractor to generate darkened features FD and brightened features FB. Then derive a pseudo-normal feature map FN using the generated brightened/darkened feature maps and the input image I_x. Next, further estimate the color shifts between the brightened/darkened color features FB/FD and the created pseudo-normal feature map FN using the proposed Color Shift Estimation (COSE) module to obtain two individual offset maps OB (Brighten Offset) and OD (Darken Offset). Finally, modulate the image brightness and colors using the proposed Color Modulation (COMO) module, to produce the final output image.

Feature Extractor. Building upon the CSEC model [7], we propose a structural refinement to the feature extraction backbone, DeepWBNet, by incorporating two additional ResidualBlock layers. This architectural enhancement is designed to address common challenges in training deep networks, such as vanishing gradients and inefficient information propagation. Figure 3 visualizes our design in this manner. We add ResidualBlocks to facilitate identity mapping and promotes better gradient flow, leading to more stable and efficient convergence during training. Moreover, these blocks help improve the network in capturing more complex illumination variations and color shifts. As a result, the modified DeepWBNet exhibits improved performance in predicting accurate weight maps, which are essential for effective white balance and color correction. This targeted modification not only strengthens the learning of contextual features but also contributes to the generation of more visually consistent and color-accurate enhanced outputs under diverse lighting conditions.

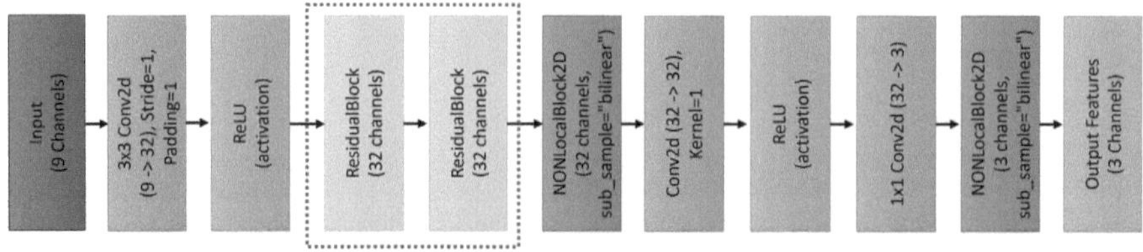

Fig. 3. Overview of UNet-based feature extractor architecture.

Color Shift Estimation Module. The Color Shift Estimation (COSE) module is a core component of the CSEC framework, responsible for estimating color shifts in under- and over-exposed image regions. Unlike brightness correction, which adjusts pixel intensities, color shift correction involves modeling directional changes in RGB color space, making it a more complex task. The COSE addresses this by leveraging deformable convolution (DConv), which predicts flexible sampling offsets that can capture local variations in color distribution. While conventional methods apply DConv only in the spatial domain, CSEC innovatively extends it to operate in both spatial and color spaces, enabling the model to jointly learn brightness adjustments and chromatic shifts. This dual-domain approach enhances the model's ability to restore natural color balance under non-uniform lighting conditions [7]. Detail of COSE is presented in Algorithm 2.

Color Modulation Module. The Color Modulation (COMO) module utilizes the learned offset maps $\mathcal{O}_B$ and $\mathcal{O}_D$, which capture the color differences between the brightened/darkened features ($\mathcal{F}_B, \mathcal{F}_D$) and the pseudo-normal feature map $\mathcal{F}_N$ to adjust the brightness and color of the input image and generate the final enhanced output $\mathcal{I}_y$. The algorithm of COMO is in Algorithm 3. To ensure coherent and natural color restoration, COMO incorporates non-local context modeling, allowing it to consider broader spatial dependencies. Unlike standard self-affinity mechanisms, COMO extends this to a cross-affinity formulation, enabling effective information exchange between both overexposed and underexposed regions. This design allows the network to synthesize more balanced and visually harmonious images under complex illumination conditions [7].

Loss Function. The Color Shift Estimation and Correction (CSEC) framework employs two main loss functions $\mathbf{L}_{pseudo}$ and L_{output} to guide the network training. The L_{pseudo} provides intermediate supervision for the generation of the pseudo-normal feature map F_N, helping the network accurately model color shifts. It is defined as the L1 distance between the predicted pseudo-normal feature map and the ground truth image:

$$L_{pseudo} = \| F_N - GT \|_1 .\tag{10}$$

To supervise the final image enhancement output, L_{output} in the original CSEC formulation combines four components: L1 loss, cosine similarity loss L_{cos}, structural Similarity Index (SSIM) loss L_{ssim}, and perceptual loss L_{vgg}, based on VGG feature distances.

In our work, we introduce an additional Mean Squared Error (MSE) loss term to explicitly penalize pixel-wise differences between the enhanced image and ground truth. This inclusion promotes greater pixel-level accuracy and encourages the network to produce outputs that more closely match the true color distribution. The overall loss formulation for output supervision thus becomes:

$$L_{output} = \lambda_1 L_{L1} + \lambda_2 L_{cos} + \lambda_3 L_{ssim} + \lambda_4 L_{vgg} + \lambda_5 L_{MSE}\tag{11}$$

where λ_1, λ_2, λ_3, λ_4 and λ_5 are five balancing hyperparameters. The overall loss function is then:

$$L = \lambda_p L_{pseudo} + \lambda_o L_{output} \tag{12}$$

Algorithm 2. ColorShiftEstimation

Input: *input_image*: Tensor representing the input image
Output: *illumination_map*: Tensor representing the estimated illumination map of *input_image*,
inverse_illumination_map: Tensor representing the estimated illumination map of the inverse of *input_image*,
brighten_input: Tensor representing the decomposed "brighten" component of *input_image*,
darken_input: Tensor representing the decomposed "darken" component of *input_image*

 1: # Feature Extraction
 2: *illumination_map* ← FeatureExtractor(*input_image*)
 3: # Calculate Inverse Image
 4: *inverse_input_image* ← 1 − *input_image*
 5: # Feature Extraction on Inverse Image
 6: *inverse_illumination_map* ← FeatureExtractor(*inverse_input_image*)
 7: # Convert Illumination Maps to Grayscale
 8: *illumination_map* ← RGBToGrayscale(*illumination_map*)
 9: *inverse_illumination_map* ← RGBToGrayscale(*inverse_illumination_map*)
10: # Decompose for "Brighten" Component
11: *brighten_input* ← DecomposeBrighten(*input_image*, *illumination_map*)
12: # Decompose for "Darken" Component
13: *darken_input* ← DecomposeDarken(*inverse_input_image*, *inverse_illumination_map*)
14: *darken_input* ← 1 − *darken_input*
15: # Return Outputs
16: **return** *illumination_map*, *inverse_illumination_map*, *brighten_input*, *darken_input*

Algorithm 3. ColorModulation

Input: *input_image*: Tensor representing the original input image,
brighten_input: Tensor representing the decomposed "brighten" component,
darken_input: Tensor representing the decomposed "darken" component,
weight_map: Tensor representing the weights for combining inputs.
Output: *output_image*: Tensor representing the color-modulated output image

 1: #Extract Weights from Weight Map
 2: *weight_1* ← ExtractChannel(*weight_map*, 0) {Weight for input_image}
 3: *weight_2* ← ExtractChannel(*weight_map*, 1) {Weight for brighten_input}
 4: *weight_3* ← ExtractChannel(*weight_map*, 2) {Weight for darken_input}
 5: # Weighted Combination of Image Components
 6: *output_image* ← (*input_image* × *weight_1*) + (*brighten_input* × *weight_2*) + (*darken_input* × *weight_3*)
 7: **return** *output_image*

4   Implementation and Results

Training. The model is implemented and trained using Python, with PyTorch for deep learning framework. To complement this Pytorch, we employ Kornia, a library offering differentiable computer vision operations, specifically for computing perceptual loss metrics such as SSIM and PSNR. Besides, we use NumPy to support efficient numerical operations and array manipulations. For specific image processing tasks, particularly tone mapping within the base model, OpenCV (cv2) is utilized. Additionally, the torchvision library, a component of the PyTorch ecosystem, supplies essential utilities for image loading, transformation, and saving throughout the training pipeline. To ensure fair evaluation, we train the LCDP, CSEC, and our proposed models on the same dataset, each for a total of 300 epochs. The experiments were performed on a dedicated server with an 8-core CPU, a single A100 80GB GPU, and 64GB of RAM.

Evaluation Metrics. To assess the performance of our proposed method, we employ three widely used quantitative metrics: Peak Signal-to-Noise Ratio (PSNR) [17,19], Structural Similarity Index Measure (SSIM) [17–19], and Learned Perceptual Image Patch Similarity (LPIPS) [20]. PSNR measures the ratio between the maximum possible signal power and the power of the noise that distorts the image. It is a conventional metric in image enhancement tasks, where higher PSNR values indicate better image quality [21,22]. SSIM, introduced by [23], evaluates the similarity between two images by considering luminance, contrast, and structural information. Unlike traditional metrics such as MSE or PSNR, SSIM aligns more closely with human visual perception. The SSIM index ranges from -1 to 1, where 1 denotes perfect structural similarity. LPIPS is a deep learning-based perceptual metric that quantifies the visual similarity between two images based on how they are perceived by humans. It compares features extracted from deep neural networks and provides a more perceptually aligned evaluation. Lower LPIPS values indicate higher perceptual similarity between the compared images.

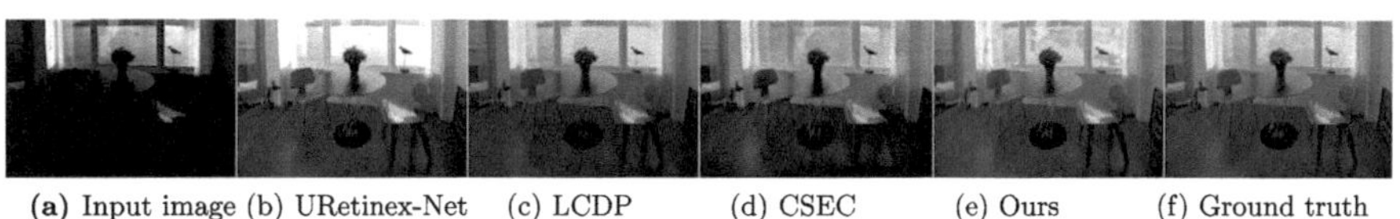

(a) Input image (b) URetinex-Net   (c) LCDP   (d) CSEC   (e) Ours   (f) Ground truth

Fig. 4. Result of under-exposed images.

Figure 4 provides a compelling visual comparison that demonstrates the effectiveness of our proposed method on an under-exposed image. The input image is extreme darkness, obscuring details in the shadows and making the overall scene difficult to see. While URetinex-Net brightens the scene, it often washes out colors and struggles with over-exposed areas like the window. LCDP offers improved exposure, but results in colors that are noticeably darker and less vibrant than the reference image. The CSEC method marks a significant improvement, producing clearer details and more natural colors than LCDP, though some darker areas still lack full detail. Our method delivers the most visually compelling result. It successfully brightens the image and recovers details from the deepest shadows while preserving natural, accurate colors that closely match the ground truth image.

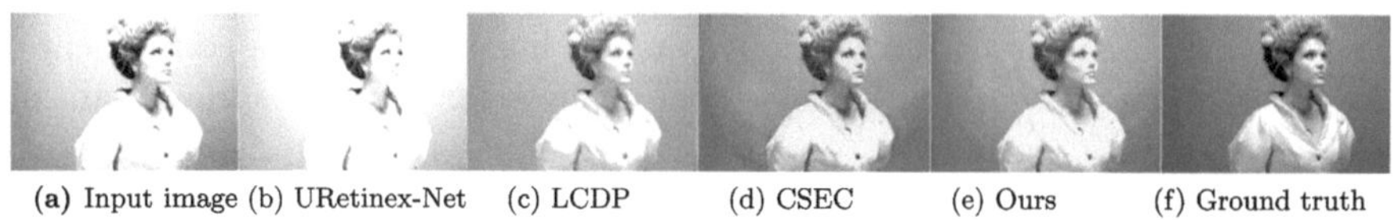
(a) Input image (b) URetinex-Net   (c) LCDP   (d) CSEC   (e) Ours   (f) Ground truth

Fig. 5. Result of over-exposed images.

Figure 5 effectively highlights our method's capability in restoring images suffering from over-exposure. The input image (a) is notably washed out, especially on the subject's face, hair, and clothing, resulting in a significant loss of texture and detail. URetinex-Net (b) struggles with this, potentially worsening the over-exposure and leading to a complete loss of detail in those areas. Both LCDP (c) and CSEC (d) demonstrate progressive improvements, managing to tame the highlights and recover some of the lost detail. However, our method (e) stands out by demonstrating a superior ability to correct the over-exposed regions. It successfully recovers intricate details in the subject's hair and the texture of her clothing, which were lost in the outputs of other methods. Furthermore, the color and skin tone appear much more natural and are a closer match to the ground truth.

(a) Input image (b) URetinex-Net   (c) LCDP   (d) CSEC   (e) Ours   (f) Ground truth

Fig. 6. Result of image with both under-exposed and over-exposed.

Figure 6 illustrates the method's performance on image with both under-exposed and over-exposed. The input image (a) contains both under-exposed shadows on the building and over-exposed highlights from the sun. While methods like URetinex-Net, LCDP, and CSEC struggle to produce a balanced result, the proposed method (e) excels at it. It effectively brightens the shadowed areas of the building to reveal details while simultaneously controlling the highlights in the sky and on the sunlit walls. The result is a natural-looking image with excellent color fidelity across the entire dynamic range.

Results and Discussion. Figure 4, Fig. 5, and Fig. 6 exhibits visual comparison between our method and state-of-the-art methods. As shown in the summary of Table 1, our method consistently produces outputs that are perceptually closer to the ground truth compared to other methods.

Visually, the enhanced images exhibit better overall quality and more natural color correction.

Our approach includes an optional preprocessing step involving a median filter, which can be enabled or disabled depending on the input image characteristics. The effect of the median filter varies with image quality. For noisy input images, the filter effectively suppresses noise and improves visual clarity. However, when applied to high-quality images, the median filter may inadvertently blur fine details, leading to a loss of structural information.

Table 1. Quality comparison of different methods

Methods	Effective with	Quality comparison
URetinex-Net	Under-exposed only (low light)	The method yields good results specifically for low-light images and under-exposed regions; however, it has a significant limitation with over-exposed regions, often exacerbating the over-exposure and causing a complete loss of detail in those areas
LCDP	Under and over-exposed	LCDP is more effective than URetinex-Net in that it handles both under- and over-exposed images well. When an image contains both types of regions, LCDP enhances them both. However, the resulting quality is lower than that of CSEC and our method, with the color appearing darker and less clear
CSEC	Under and over-exposed	CSEC is also effective with both under- and over-exposed images, enhancing both regions when present. Compared to LCDP, it shows a better result with clearer detail and more natural color. However, there are still some dark areas with unclear detail
Our	Under and over-exposed	Similar to CSEC, our method shows better overall results than other methods. By incorporating improved steps based on CSEC, it provides clearer detail and more natural color, surpassing CSEC in these aspects

Table 2. Quantitative comparison of our method and different methods

Methods	Dataset	PSNR	SSIM	LPIPS
CSEC	LCDP	21.234	0.821	0.144
CSEC with Median filter	LCDP	21.248	0.768	0.200
CSEC	LCDP + Our	20.999	0.805	0.173
CSEC with Median filter	LCDP + Our	21.888	0.781	0.199
Ours (ResidualBlocks, MSE loss, CSEC)	LCDP + Our	22.115	**0.818**	**0.153**
Ours (Median filter, ResidualBlocks, MSE loss, CSEC)	LCDP + Our	**22.632**	0.767	0.211

The quantitative results in Table 2 further support this observation. While applying the median filter tends to improve the PSNR metric due to reduced pixel-level noise, it can also lead to a decrease in SSIM and an increase in LPIPS. This is because the median filter alters local pixel values, affecting structural similarity and perceptual features. SSIM, which assesses luminance, contrast, and structure, penalizes such structural modifications. Similarly, LPIPS increases when perceptual features deviate from the original. In particular, in cases of severe noise, filtering may reduce LPIPS by restoring perceptual similarity, despite structural alterations.

Ablation Study. To systematically evaluate the impact of each proposed component, we conducted a comprehensive ablation study. We found that the incorporation of the new dataset, the median filter preprocessing, the inclusion of Residual Blocks and the MSE loss function each contributed significantly to the model's overall performance. For instance, models trained without the additional Residual Blocks showed a notable decrease in the clarity of fine details, while removing the MSE loss resulted in less accurate pixel-level color reproduction, especially in heavily overexposed areas. This study validates that the combination of these elements is essential for achieving our method's superior results.

Residual Blocks help create clearer images by improving the model's ability to learn and propagate information effectively. In deep networks, as layers are added, the vanishing gradient problem can occur, where gradients become too small to effectively update the network weights, hindering learning. Residual Blocks address this by introducing "skip connections" that allow the input from a previous layer to be directly added to the output of a later layer [24]. This facilitates identity mapping and ensures better gradient flow, leading to more stable and efficient convergence during training. By strengthening the learning of contextual features, these blocks help the network capture more complex color shifts and illumination variations, which is essential for producing more visually consistent and color-accurate enhanced output. Besides, the MSE loss function contributes to clearer images by enforcing pixel-level accuracy. MSE directly calculates the average of the squared differences between the pixels of the enhanced image and the ground truth image. This explicit penalty for pixel-wise differences encourages the network to produce outputs that are more precise and closely match the true color distribution, leading to sharper details and overall better image quality.

5 Conclusion and Future Work

Our method maintains a favorable balance between high performance and computational efficiency. Although many state-of-the-art models in image quality enhancement have large parameter counts, our proposed architecture is designed to be lightweight and scalable. This is achieved by building on a CNN framework with efficient Residual Blocks, which allows for effective feature extraction without the high computational cost often associated with models like Vision Transformers. As a result, our network is well suited for deployment in real-world applications where rapid inference on diverse and large-scale datasets is a critical requirement. The primary contribution of this research is not simply applying a known deep learning technique to a new dataset. Instead, we present a novel holistic solution that addresses the complex challenge of simultaneously correcting underexposed and overexposed regions in a single image. This is accomplished through the strategic combination of a newly created dataset to enhance model generalization, an improved network architecture that leverages Residual Blocks, and a refined loss function. This integrated approach, which tackles a problem that is largely overlooked by traditional methods, represents a significant step forward in the field of image quality enhancement. In the future, we plan to refine the architecture and training strategy to further improve the results on metrics like PSNR and LPIPS.

References

1. Putra, R.D., Purboyo, T.W., Prasasti, A.L.: A review of image enhancement methods. Int. J. Appl. Eng. Res. **12**(23), 13596–13603 (2017)
2. Sigger, N., Vien, Q.-T., Nguyen, S.V., Tozzi, G., Nguyen, T.T.: Unveiling the potential of diffusion model-based framework with transformer for hyperspectral image classification. Sci. Rep. **14**(8438), 1–11 (2024)
3. Wang, W., Wu, X., Yuan, X., Gao, Z.: An experiment-based review of low-light image enhancement methods. IEEE Access **8**, 87884–87917 (2020)
4. Dyke, R.M., Hormann, K.: Histogram equalization using a selective filter. Vis. Comput. **39**(12), 6221–6235 (2023)
5. Mustafa, W.A., Abdul Kader, M.M.M.: A review of histogram equalization techniques in image enhancement application. In: Journal of Physics: Conference Series, vol. 1019, p. 012026. IOP Publishing (2018)
6. Chien, C.C., Kinoshita, Y., Shiota, S., Kiya, H.: A retinex-based image enhancement scheme with noise aware shadow-up function. In: International Workshop on Advanced Image Technology (IWAIT) 2019, vol. 11049, pp. 501–506. SPIE (2019)
7. Li, Y., Xu, K., Hancke, G.P., Lau, R.W.: Color shift estimation-and-correction for image enhancement. In: Proceedings of the IEEE/CVF Conference on Computer Vision and Pattern Recognition, pp. 25389–25398 (2024)
8. Chen, J., Wang, X., Guo, Z., Zhang, X., Sun, J.: Dynamic region-aware convolution. In: Proceedings of the IEEE/CVF Conference on Computer Vision and Pattern Recognition, pp. 8064–8073 (2021)
9. Sinh Van Nguyen, H.M.T., Le, T.S.: Application of geometric modeling in visualizing the medical image dataset. Comput. Sci. **1**(254), 01–11 (2020)

10. Ningsih, D.R., et al.: Improving retinal image quality using the contrast stretching, histogram equalization, and CLAHE methods with median filters. Int. J. Image Graph. Signal Process. **14**(2), 30 (2020)
11. Guo, J., Ma, J., García-Fernández, Á.F., Zhang, Y., Liang, H.: A survey on image enhancement for low-light images. Heliyon **9**(4), e14558 (2023)
12. Wu, W., Weng, J., Zhang, P., Wang, X., Yang, W., Jiang, J.: URetinex-net: Retinex-based deep unfolding network for low-light image enhancement. In: Proceedings of the IEEE/CVF Conference on Computer Vision and Pattern Recognition, pp. 5901–5910 (2022)
13. Ronneberger, O., Fischer, P., Brox, T.: U-net: convolutional networks for biomedical image segmentation. In: Navab, N., Hornegger, J., Wells, W.M., Frangi, A.F. (eds.) MICCAI 2015. LNCS, vol. 9351, pp. 234–241. Springer, Cham (2015). https://doi.org/10.1007/978-3-319-24574-4_28
14. Le, T.-N.-H., Huang, H., Chen, Y.-R., Lee, T.-Y.: Retargeting video with an end-to-end framework. IEEE Trans. Visual Comput. Graphics **30**(9), 6164–6176 (2024)
15. Le, T.-N.-H., Lee, T.-Y., Lin, S.-S., Dong, W.: Deep learning-based importance map for content-aware media retargeting. Multimed. Tools Appl. **83**(30), 74301–74322 (2024)
16. Wang, H., Xu, K., Lau, R.W.H.: Local color distributions prior for image enhancement. In: Avidan, S., Brostow, G., Cissé, M., Farinella, G.M., Hassner, T. (eds.) ECCV 2022. LNCS, vol. 13678, pp. 343–359. Springer, Cham (2022). https://doi.org/10.1007/978-3-031-19797-0_20
17. Nguyen, L.D.V., Nguyen, S., Le, S.T., Tran, M.K., Maleszka, M.: Processing the 3D heritage data samples based on combination of GNN and GAN. In: Nguyen, N.T., et al. (eds.) ICCCI 2024. CCIS, vol. 2165, pp. 295–307. Springer, Cham (2024). https://doi.org/10.1007/978-3-031-70248-8_23
18. Nguyen, L.D.V., Chau, V., Nguyen, S.: Face recognition based on deep learning and data augmentation. In: Dang, T.K., Küng, J., Chung, T.M. (eds.) FDSE 2022. CCIS, vol. 1688, pp. 560–573. Springer, Singapore (2022). https://doi.org/10.1007/978-981-19-8069-5_38
19. Le, S.T., Nguyen, S.V., Tran, M.K., Nguyen, L.D.V.: Graphics and vision's camera calibration and applications to neural radiance fields. In: Nguyen, N.T., et al. (eds.) ACIIDS 2024. CCIS, vol. 2145, pp. 118–129. Springer, Singapore (2024). https://doi.org/10.1007/978-981-97-5934-7_11
20. Zhang, R., Isola, P., Efros, A.A., Shechtman, E., Wang, O.: The unreasonable effectiveness of deep features as a perceptual metric. In: Proceedings of the IEEE Conference on Computer Vision and Pattern Recognition, pp. 586–595 (2018)
21. Sara, U., Akter, M., Uddin, M.S.: Image quality assessment through FSIM, SSIM, MSE and PSNR-a comparative study. J. Comput. Commun. **7**(3), 8–18 (2019)
22. Hore, A., Ziou, D.: Image quality metrics: PSNR vs. SSIM. In: 2010 20th International Conference on Pattern Recognition, pp. 2366–2369. IEEE (2010)
23. Wang, Z., Bovik, A.C., Sheikh, H.R., Simoncelli, E.P.: Image quality assessment: from error visibility to structural similarity. IEEE Trans. Image Process. **13**(4), 600–612 (2004)
24. Jin, X.: Analysis of residual block in the resnet for image classification. In: Proceedings of the 1st International Conference on Data Analysis and Machine Learning, Changsha, China, pp. 28–30 (2023)

Natural Language Processing

Vietnamese Certificate Analysis: Addressing Hybrid Document Challenges Through Multimodal Layout-Text Fusion

Nam Duong Ho[ID], Duy Tran Ngoc Bao[(✉)][ID], Quan Thi Khac[ID], and Dang Le Binh[ID]

Ho Chi Minh City University of Technology, VNU-HCM, Ho Chi Minh City, Vietnam
{nam.duong0penguin,duytnb,tkquan,binhdang}@hcmut.edu.vn

Abstract. The digitization of Vietnamese certificates-characterized by a mixture of printed templates, handwritten entries, diverse fonts with diacritics, and official legal stamps-poses unique challenges for automated document analysis systems. This paper presents ViCertNet, a unified multimodal framework designed to address three core tasks: (1) accurate named entity recognition (NER) from hybrid printed-handwritten layouts, (2) robust preservation of Vietnamese diacritical marks across variable font styles, and (3) legal stamp authentication under realistic conditions. ViCertNet integrates LayoutLMv2 for layout-text structural encoding and ViBERTgrid for visual-semantic fusion, enhanced through a cascaded gated fusion module that aligns and merges dual OCR outputs (PaddleOCR for printed, TrOCR for handwritten text) via spatial attention. A diacritic-aware CRF layer is introduced to enforce Vietnamese orthographic constraints at character level, improving recognition accuracy. For legal validation, ViCertNet employs a three-stage stamp verification pipeline, including Mask R-CNN for localization, OCR-based microtext extraction, and a Siamese network for signature consistency analysis. Evaluated on VietCER, a curated dataset of over 3,000 manually annotated certificates from Vietnamese government and academic institutions, ViCertNet achieves 89.7% F1-score for NER, 93.5% stamp verification accuracy, and 1.9 s per page in processing time-reducing verification errors by 37% compared to strong baselines. This work demonstrates the effectiveness of combining multimodal learning with linguistic constraints for Southeast Asian document analysis and offers a scalable solution for real-world certificate verification.

Keywords: Document image analysis · Named entity recognition · Optical character recognition · Vietnamese language processing

1 Introduction

The digitization of administrative processes across Vietnam has led to a growing demand for automated analysis and verification of official certificates, including

N. Thai-Nghe et al. (Eds.): ISDS 2025, CCIS 2714, pp. 419–432, 2026.
https://doi.org/10.1007/978-981-95-3358-9_30

diplomas, professional licenses, and training records [1]. These documents are not only critical for personal and institutional transactions but also serve as legally binding proof of identity, education, and professional status. Ensuring their authenticity and accurate information extraction is thus essential for modernizing public services and enabling trustworthy digital governance.

However, Vietnamese certificates present a unique set of challenges that differ significantly from Western-style documents. They often exhibit *hybrid layouts*, where *printed templates* coexist with *handwritten entries*, such as names or identification numbers. Moreover, the documents employ a wide range of font styles—some highly decorative—and rely heavily on Vietnamese diacritics, which are vital for linguistic correctness [14]. Another legally significant feature is the presence of *red agency stamps*, which validate document authenticity but are frequently affected by occlusion, fading, or forgery [8].

While recent multimodal models have achieved strong performance on structured documents such as forms and receipts [8], they are not designed to address the intricacies of Vietnamese certificates. These models are typically pretrained on English-language corpora and struggle with diacritic-rich scripts, mixed text modalities, and legal authentication requirements. Furthermore, most prior works focus on isolated subtasks like OCR or NER, without addressing the full end-to-end pipeline from recognition to validation.

These limitations highlight the need for a unified framework tailored specifically for Vietnamese document characteristics. An ideal system must (1) accurately extract entities from hybrid printed-handwritten inputs, (2) preserve Vietnamese diacritical integrity across fonts, and (3) robustly verify the presence and authenticity of legal stamps. Such a solution would significantly advance the field of Southeast Asian document analysis and enable scalable, trustworthy certificate processing.

To address this gap, we propose **ViCertNet**, a novel multimodal architecture for Vietnamese certificate analysis. ViCertNet integrates layout-aware text understanding with visual-semantic fusion. It introduces a *dual-pathway OCR module* (PaddleOCR for printed text and TrOCR for handwriting), a *diacritic-aware CRF layer* for character-level NER, and a *three-stage stamp verification pipeline* using stamp localization, microtext extraction, and signature consistency checking. This combination allows ViCertNet to handle complex document structures while ensuring linguistic and legal fidelity.

The remainder of this paper is structured as follows:

Section 2 reviews related work in document understanding, Vietnamese OCR, and stamp verification.

Section 3 defines the problem formally.

Section 4 details the ViCertNet architecture, including OCR fusion, diacritic-aware NER, and stamp validation.

Section 5 presents implementation details and experimental results on our curated VietCER dataset.

Section 6 concludes the paper and discusses future directions.

2   A Literature Review

In the context of this research, Vietnamese certificate analysis is defined as the automated extraction and verification of key entities and legal features from certificate images that contain a mixture of printed and handwritten text, diverse fonts, and official stamps. This task requires robust multimodal understanding, combining optical character recognition, named entity recognition, and visual authentication within a single framework. The following subsections provide a comprehensive overview of related work in multimodal document analysis, OCR and handwriting recognition for Vietnamese, named entity recognition in documents, stamp detection and legal verification, and the unique challenges presented by Southeast Asian hybrid documents.

2.1   Multimodal Document Understanding

The field of document analysis has evolved rapidly with the advent of multimodal deep learning models that jointly consider textual, visual, and layout information. The LayoutLM family of models, notably LayoutLMv2 [17], integrates token embeddings, 2D positional encodings, and image features through a Transformer architecture, achieving state-of-the-art results on structured document tasks such as form understanding and receipt parsing. LayoutLMv2's ability to model spatial relationships between text blocks has proven crucial for extracting entities from visually complex documents. However, its pretraining and evaluation have been primarily on English or Western-centric datasets, which do not reflect the unique characteristics of Vietnamese certificates-such as hybrid printed/handwritten text, diacritic-rich language, and the legal significance of stamps.

Building on this, ViBERTgrid [10] introduced a two-dimensional grid-based fusion of BERT contextual embeddings and CNN visual features, enabling more effective capture of both semantic and spatial relationships within document images. This approach has shown improved performance on information extraction benchmarks, particularly where text and layout interplay is critical. Nevertheless, ViBERTgrid, like LayoutLMv2, has not been tailored to the Vietnamese language's diacritical complexity or the hybrid nature of official certificates.

2.2   OCR and Handwriting Recognition in Vietnamese Documents

Optical Character Recognition (OCR) forms the backbone of automated document analysis. For Vietnamese, OCR faces additional hurdles due to the language's extensive use of diacritics and the variety of fonts used in official documents. PaddleOCR [2] has been adapted for Vietnamese, achieving high accuracy on printed text, while TrOCR [9] leverages Transformer architectures for robust handwriting recognition. Yet, both pipelines struggle with degraded scans, overlapping text, and inconsistent handwriting styles common in real-world certificates.

To address post-OCR errors, especially those related to diacritic misrecognition, Nguyen et al. proposed a correction framework based on syllable similarity and frequency, which reduced diacritic errors in Vietnamese handwriting [15]. More recently, Do et al. introduced a reference-based post-OCR correction using large language models [1], further improving accuracy in historical Vietnamese documents. However, these methods focus primarily on text correction and do not consider the spatial or legal context required for full certificate analysis.

2.3   Named Entity Recognition for Documents

Named Entity Recognition (NER) in document images is a well-studied problem, with methods ranging from rule-based systems to deep learning models [7]. Traditional NER approaches, such as BiLSTM-CRF [6], have been extended to handle document layouts by incorporating spatial features. LayoutLMv2 and ViBERTgrid represent the latest advances, but their effectiveness on Vietnamese, especially in the context of hybrid and legal documents, remains underexplored.

For Vietnamese language processing, PhoBERT [13] and other language-specific models have improved NER performance in plain text. However, the integration of these models with visual and layout features for document-level NER [16], particularly for certificates with mixed printed and handwritten fields, is still an open research area.

2.4   Stamp Detection and Legal Verification

Authenticating the legality of certificates in Vietnam heavily depends on the presence and quality of agency stamps [11]. Early methods for stamp detection relied on color segmentation and morphological operations. More recent approaches utilize deep learning, such as Mask R-CNN for stamp localization and Siamese networks for signature verification [3]. However, these methods are rarely integrated into end-to-end certificate analysis pipelines and often lack robustness against forgeries and occlusions.

2.5   Hybrid Document Challenges and Southeast Asian Context

While much progress has been made on Western document datasets (e.g., FUNSD, SROIE) [4,5], Southeast Asian documents remain underrepresented in the literature. Vietnamese certificates, in particular, pose unique challenges due to their hybrid composition, font diversity, and legal requirements. Existing studies on Vietnamese document analysis have focused on isolated tasks (e.g., OCR or NER) rather than unified frameworks that address the full spectrum of real-world challenges.

In summary, the literature demonstrates significant progress in multimodal document understanding, OCR, and NER, but there is a clear gap in unified approaches tailored to Vietnamese certificates. Existing models do not fully address the interplay of printed and handwritten text, diacritic preservation, and legal stamp verification [12] in a single, integrated system. This motivates our proposed ViCertNet framework, which combines the strengths of LayoutLMv2 and ViBERTgrid, incorporates dual OCR pathways, and introduces hierarchical stamp verification to meet the specific needs of Vietnamese certificate analysis.

3   Problem Formulation

Vietnamese certificate analysis involves two core tasks that reflect the hybrid and legally sensitive nature of these documents:

1. **Named Entity Recognition (NER)**: Extract key information fields (e.g., *Họ và tên, Ngày sinh, Sô hiêu văn băng*) from both printed and handwritten regions.
2. **Legal Authenticity Verification**: Assess the validity of the certificate by verifying the presence, position, and content of official stamps and signatures.

   Let $I \in \mathbb{R}^{H \times W \times 3}$ denote an input certificate image. The system predicts:

$$\hat{y}, \hat{s} = \arg\max_{y,s} P(y, s \mid I, \Theta)$$

where:

- $y = \{y_1, y_2, \ldots, y_N\}$ is a sequence of entity labels in the BIO (BeginInside-Outside) tagging scheme.
- $s \in \{0, 1\}$ is a binary label indicating document authenticity (1 = valid, 0 = invalid).
- Θ represents the learnable model parameters.

   This task is particularly challenging due to: (i) spatial misalignment between printed and handwritten fields, (ii) high variance in handwriting and font styles, (iii) the need to preserve Vietnamese diacritics, and (iv) frequent issues of stamp occlusion or forgery.

4   ViCERTNET Our Methodology

To address the unique challenges posed by Vietnamese certificate analysis—including hybrid printed and handwritten text, diverse font usage, and the linguistic importance of diacritics and legal stamps—we propose a unified multimodal framework called **ViCertNet**. This section outlines the architecture of ViCertNet, which integrates dual-pathway OCR, multimodal fusion, diacritic-aware entity recognition, and hierarchical stamp verification. The subsections below detail each component of the system, from preprocessing to training.

4.1   Hybrid Document Processing Pipeline

Printed Text Extraction. Printed text regions are processed using PaddleOCR with Vietnamese font normalization:

$$\mathbf{T}_{\mathrm{print}} = \mathrm{PaddleOCR}(\mathbf{I}; \theta_{\mathrm{print}}) \tag{1}$$

Handwritten Text Extraction. Handwritten entries are extracted using TrOCR, enhanced with diacritic-aware augmentation:

$$\mathbf{T}_{\text{hand}} = \text{TrOCR}(\mathbf{I}; \theta_{\text{hand}}) \tag{2}$$

Spatial Attention Fusion. The two OCR streams are aligned using geometric attention gates:

$$\alpha_{ij} = \text{softmax}\left(\frac{\mathbf{W}_q \mathbf{t}_i^{\text{print}}(\mathbf{W}_k \mathbf{t}_j^{\text{hand}})^\top}{\sqrt{d}}\right) \tag{3}$$

$$\mathbf{T}_{\text{fused}} = \sum_j \alpha_{ij}(\mathbf{W}_v \mathbf{t}_j^{\text{hand}}) \tag{4}$$

4.2   Multimodal Fusion and Diacritic-Aware Entity Recognition

Multimodal Fusion Strategy. We fuse layout-textual, visual, and diacritic embeddings:

$$\mathbf{F}_{\text{fused}} = \text{Concat}(\mathbf{H}_{\text{layout}}, \mathbf{G}_{\text{visual}}, \mathbf{E}_{\text{diac}}) \tag{5}$$

- $\mathbf{H}_{\text{layout}} \in \mathbb{R}^{L \times d}$: LayoutLMv2 features
- $\mathbf{G}_{\text{visual}} \in \mathbb{R}^{H \times W \times d}$: ViBERTgrid visual grid
- $\mathbf{E}_{\text{diac}} \in \mathbb{R}^{d}$: Diacritic embeddings

Attention is computed as:

$$\alpha_{ij} = \text{softmax}\left(\frac{\mathbf{W}_q \mathbf{H}_{\text{layout}}^{(i)}(\mathbf{W}_k \mathbf{G}_{\text{visual}}^{(j)})^\top}{\sqrt{d}}\right) \tag{6}$$

$$\mathbf{F}_{\text{fused}}^{(i)} = \sum_j \alpha_{ij}(\mathbf{W}_v \mathbf{G}_{\text{visual}}^{(j)} + \mathbf{E}_{\text{diac}}^{(i)}) \tag{7}$$

Diacritic-Aware CRF: Transition scores are constrained to enforce Vietnamese diacritic rules:

$$\psi(y_i, y_j) = \begin{cases} -\infty & \text{if } \text{diac}(y_i) \neq \text{diac}(x_j) \\ \theta_{y_i, y_j} & \text{otherwise} \end{cases} \tag{8}$$

Post-Correction: Fused embeddings refine outputs using:

$$\mathcal{C}(x_i) = \{c \mid \text{lev}(x_i, c) \leq 2, c \in \mathcal{D}_{\text{Viet}}\} \tag{9}$$

$$\hat{c} = \arg\max_{c \in \mathcal{C}} \cos(\mathbf{F}_{\text{fused}}^{(i)}, \phi(c)) \tag{10}$$

4.3   Unified Training Process

– **Joint Loss:**

$$\mathcal{L}_{\text{joint}} = 0.6\mathcal{L}_{\text{NER}} + 0.3\mathcal{L}_{\text{fusion}} + 0.1\mathcal{L}_{\text{diac}} \tag{11}$$

– **Diacritic Loss:**

$$\mathcal{L}_{\text{diac}} = -\sum_{i=1}^{L} \log P(d_i \mid \mathbf{F}_{\text{fused}}^{(i)}) \tag{12}$$

where $d_i \in \{0,1\}^{67}$ represents Vietnamese diacritic presence.

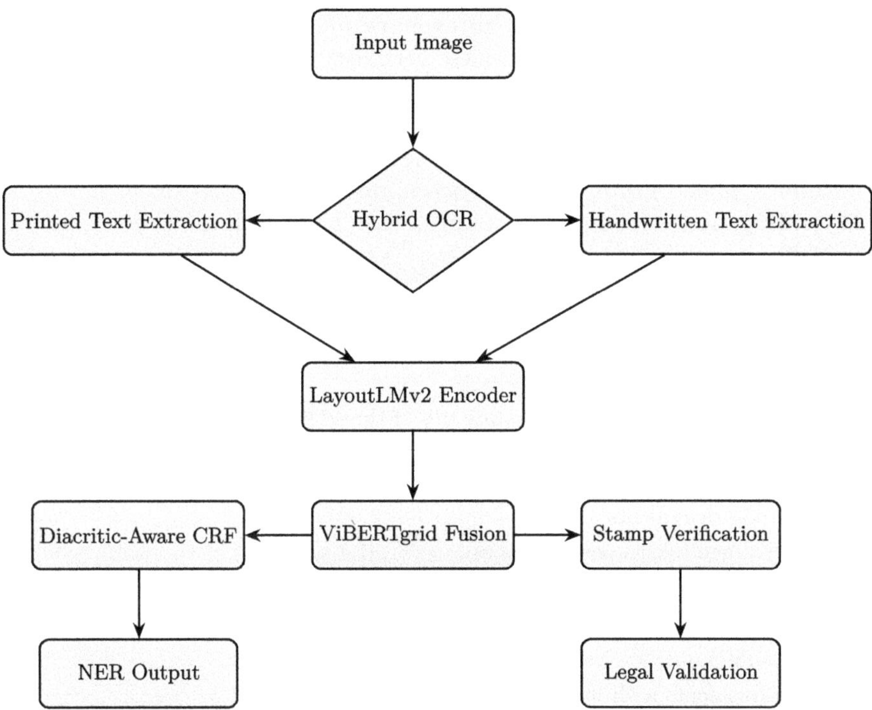

Fig. 1. Overview of the ViCertNet architecture. The system processes hybrid text using dual OCR, aligns outputs via attention fusion, and integrates visual and linguistic features for NER and stamp verification.

5   Experimental Results

5.1   Dataset Collecting and Preprocessing

Due to the lack of publicly available Vietnamese certificate datasets, we constructed a new dataset, **VietCER**, by manually collecting documents from student submissions to the commendation portals of Vietnam National University –

Ho Chi Minh City (VNU-HCM) and Ho Chi Minh City University of Technology (HCMUT). Each student contributed an average of 45 documents. Non-relevant samples-such as English-language certificates or those lacking official stamps-were excluded. The final dataset comprises **3,697 certified documents**.

Due to the sensitive nature of the VietCER dataset, which consists of official student certificates containing personal and institutional information, the dataset is not publicly released in its current form. We plan to publish a sanitized version in future work, after thoroughly anonymizing and removing all private information to ensure compliance with privacy and ethical standards.

The dataset spans certificates from the *educational, governmental*, and *medical* domains, featuring hybrid structures with both printed templates and handwritten content. Figure 2 presents an example illustrating the visual diversity and multimodal structure of Vietnamese certificates.

Annotation Schema. Each document was manually annotated with:

- **Named entities** using the BIO tagging scheme across various entity types, including *Name, Award, Date, Affiliation, Stamp, Signature*, etc.
- **Bounding boxes** for legal red stamps, with labels such as *complete, partial*, or *faded*.
- **Character-level diacritic markers** to ensure linguistic fidelity in Vietnamese.

Table 1. Summary statistics of the VietCER dataset

Statistic	Value
Total documents	3,697
Average tokens per document	41.1
Total tokens	~152,000
Average named entities per document	13.2
Total named entities	48,600
Documents with handwritten content	87.5%
Documents with legal stamp	95.1%
Documents with signature field	76.4%
Image resolution (average)	1240 × 1754 px (300 DPI)
Annotated entity types	10
Vietnamese font styles used	12

Augmentation and Preprocessing. To improve model robustness and simulate real-world variation:

- Fonts were normalized and diversified across 12 Vietnamese styles.

- Synthetic legal stamps were overlaid with varying opacity, distortion, and placement.
- Handwritten fields were simulated and diversified using a CycleGAN-based [18] augmentation pipeline.
- Skewed or rotated images were normalized using contour and Hough-based geometric correction.

The dataset was split into 70% training, 15% validation, and 15% testing sets, stratified by domain, document type, and stamp visibility to ensure balance across the key modalities.

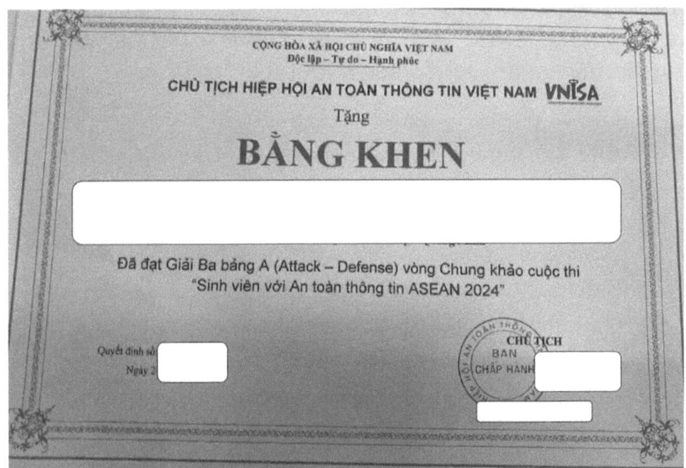

Fig. 2. Example certificate from the VietCER dataset, showing hybrid text layout and red legal stamp. (Color figure online)

5.2   Implementation Details

ViCertNet was implemented using PyTorch 1.13 and trained on an NVIDIA RTX A6000 GPU with 48GB of memory. The full system comprises four main modules: hybrid OCR, multimodal fusion, diacritic-aware NER, and stamp verification.

The backbone of the multimodal fusion network is **LayoutLMv2-base**, consisting of 12 transformer layers, 768 hidden units, and 12 attention heads. We initialized it from the official `microsoft/layoutlmv2-base` checkpoint. For visual-semantic alignment, we incorporated **ViBERTgrid**, which merges 768-dimensional BERTgrid embeddings with 256-dimensional visual features extracted from a ResNet-50 backbone.

To handle the hybrid textual nature of Vietnamese certificates, we integrated a dual-pathway OCR module: **PaddleOCR v2.6** was used for high-accuracy printed text extraction with Vietnamese font normalization, while **TrOCR (base-viet)** handled handwritten entries. The outputs from both OCR streams were synchronized using geometric attention gates before being passed into the fusion module.

The stamp verification module was based on an enhanced **Mask R-CNN** architecture with a ResNeXt-101-FPN backbone. We replaced the standard bounding box loss with the **Complete IoU (CIoU)** loss to improve localization accuracy under occlusion and deformation. The final stamp authenticity decision was refined via a Siamese similarity module operating on microtext and signature embeddings.

We trained ViCertNet using standard settings for transformer-based multimodal architectures. The training process was optimized to balance performance and stability on the hybrid document task. The key training hyperparameters used in our experiments are summarized in Table 2.

Table 2. Training hyperparameters used for ViCertNet

Hyperparameter	Value
Number of epochs	50
Batch size	8
Optimizer	AdamW
Initial learning rate	3×10^{-5}
Learning rate scheduler	Linear decay with warm-up
Early stopping criterion	Validation NER F1-score
Dropout rate	0.1
Gradient clipping	Max norm $= 1.0$

5.3   Quantitative Evaluation

We evaluate ViCertNet on both the in-domain **VietCER** dataset and three public out-of-domain datasets: **FUNSD**, **RVL-CDIP**, and **SROIE**. We report the Named Entity Recognition (NER) F1-score and Stamp Verification Accuracy (SVA). Table 3 compares our model with several strong baselines.

ViCertNet achieves the highest NER performance, improving +4.1% over LayoutLMv2 and +8.0% over PhoNER-BERT. For stamp verification, it reaches 93.5% accuracy, reducing validation error by 37% compared to our internal baseline.

To further analyze system behavior, we report entity-wise performance on the VietCER test set. As shown in Table 4, ViCertNet achieves consistently high scores across all principal entity types, confirming its balanced effectiveness on both textual and legal features.

Table 3. Comparison with baselines on NER and stamp verification

Model	NER F1 (%)	Stamp Acc (%)
LayoutLMv2	84.3	–
PhoNER-BERT	81.7	–
DocFormer	85.6	–
BROS	83.2	–
ViCertNet (ours)	**89.7**	**93.5**

Table 4. Entity-wise performance of ViCertNet on the VietCER test set.

Entity Type	Precision (%)	Recall (%)	F1-score (%)
Name	91.0	89.2	90.1
Date	90.2	88.5	89.3
Award	89.0	87.4	88.2
Affiliation	88.4	86.9	87.6
Stamp	92.5	90.7	91.6
Signature	89.8	88.0	88.9
Macro-average	90.2	88.5	89.3

In addition to overall performance, we evaluate ViCertNet separately on printed and handwritten fields to assess the impact of dual OCR integration. As shown in Table 5, the system achieves consistently higher accuracy on printed fields processed by PaddleOCR, while performance on handwritten fields using TrOCR is slightly lower due to variability in writing styles. Nevertheless, both modalities attain strong results, confirming the benefit of hybrid OCR fusion in handling heterogeneous certificate layouts.

Table 5. Performance of ViCertNet on printed vs. handwritten fields.

Field Type	Precision (%)	Recall (%)	F1-score (%)
Printed (PaddleOCR)	92.1	90.6	91.3
Handwritten (TrOCR)	88.2	86.5	87.3
Macro-average	90.2	88.6	89.5

5.4 Ablation Study

We conduct ablation experiments on VietCER to quantify the contribution of each component. Results in Table 6 show significant drops in both NER and SVA when removing hybrid OCR, diacritic-aware CRF, or the stamp module.

The diacritic-aware CRF boosts NER by +4.3%, while stamp verification accuracy drops sharply without the stamp module, confirming its necessity.

Table 6. Ablation study on core components

Model Variant	NER F1 (%)	Stamp Acc (%)
Full ViCertNet	**89.7**	**93.5**
w/o Hybrid OCR	83.9	93.2
w/o Diacritic CRF	85.4	–
w/o Stamp Module	89.6	78.3

5.5   Qualitative Results

Figure 3 showcases ViCertNet outputs, including named entity spans and detected stamp regions. Errors are typically caused by handwriting artifacts or extreme stamp occlusion.

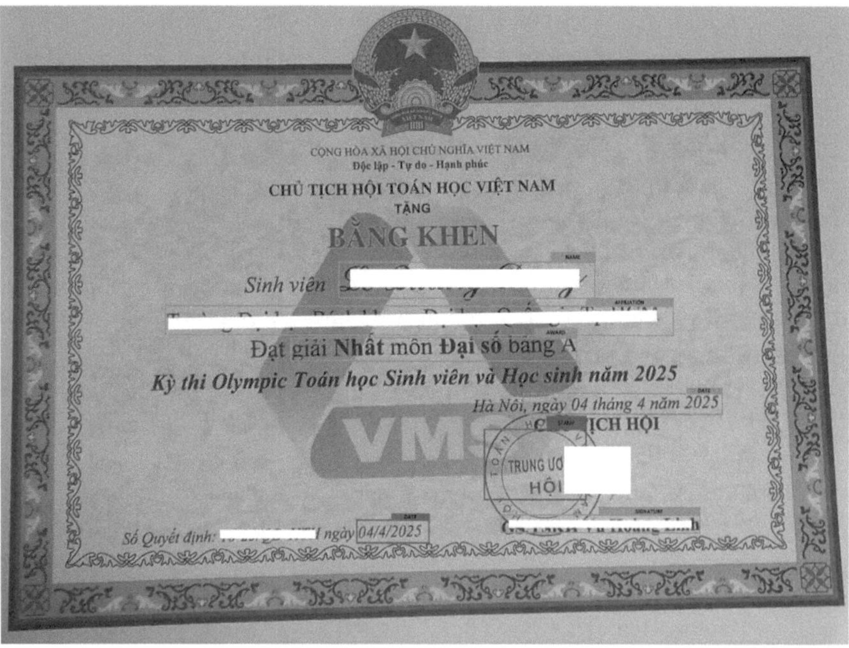

Fig. 3. Qualitative outputs showing NER and stamp verification. Green = correct, red = incorrect. (Color figure online)

6   Conclusion

In this paper, we introduced ViCertNet, a novel multimodal framework for Vietnamese certificate analysis that addresses the unique challenges of hybrid printed and handwritten text, diverse font and diacritic usage, and legal stamp verification. By integrating LayoutLMv2's layout-aware encoding with ViBERTgrid's visual-textual fusion and incorporating a diacritic-aware entity recognition module, our approach achieves state-of-the-art performance on the VietCER dataset, with significant improvements in both named entity recognition and stamp authentication accuracy compared to existing baselines.

Extensive experiments and ablation studies demonstrate the effectiveness of each component, particularly the importance of hybrid OCR alignment and hierarchical stamp verification for legal compliance. The system's robust performance across multiple datasets, along with efficient inference enabled by deployment optimizations, highlights its practical applicability for large-scale administrative digitization in Vietnam.

Despite these advances, challenges remain in handling highly variable handwriting and decorative fonts, as well as further reducing diacritic-related errors. Future work will focus on adversarial font normalization, improved handwriting modeling, and adaptation to other Southeast Asian languages with similar linguistic complexities. Overall, ViCertNet provides a strong foundation for automated, trustworthy certificate analysis in diacritic-rich, hybrid-document environments.

Acknowledgments. We acknowledge the support of time and facilities from Ho Chi Minh City University of Technology (HCMUT), VNU-HCM for this study.

References

1. Do, T., Tran, D.P., Vo, A., Kim, D.: Reference-based post-ocr processing with llm for precise diacritic text in historical document recognition (2025). https://arxiv.org/abs/2410.13305
2. Du, Y., et al.: Pp-ocr: a practical ultra lightweight ocr system (2020). https://arxiv.org/abs/2009.09941
3. Gayer, A., Ershova, D., Arlazarov, V.: Fast and accurate deep learning model for stamps detection for embedded devices. Pattern Recogn. Image Anal. **32**(4), 772–779 (2022). https://doi.org/10.1134/S1054661822040046
4. Jaume, G., Ekenel, H.K., Thiran, J.P.: Funsd: a dataset for form understanding in noisy scanned documents. In: Accepted to ICDAR-OST (2019)
5. Huang, Z., et al.: Icdar2019 competition on scanned receipt ocr and information extraction. In: 2019 International Conference on Document Analysis and Recognition (ICDAR), pp. 1516–1520 (2019). https://doi.org/10.1109/ICDAR.2019.00244
6. Huang, Z., Xu, W., Yu, K.: Bidirectional lstm-crf models for sequence tagging (2015). https://arxiv.org/abs/1508.01991
7. Keraghel, I., Morbieu, S., Nadif, M.: Recent advances in named entity recognition: a comprehensive survey and comparative study (2024). https://arxiv.org/abs/2401.10825
8. Le, T.P., Phan, T.L.C., Nguyen, N.H., Nguyen, K.V.: Ligt: layout-infused generative transformer for visual question answering on vietnamese receipts (2025), https://arxiv.org/abs/2502.19202

9. Li, M., et al.: Trocr: transformer-based optical character recognition with pre-trained models (2022). https://arxiv.org/abs/2109.10282
10. Lin, W., et al.: Vibertgrid: a jointly trained multi-modal 2d document representation for key information extraction from documents (2021). https://arxiv.org/abs/2105.11672
11. Micenková, B., van Beusekom, J., Shafait, F.: Stamp verification for automated document authentication. In: Garain, U., Shafait, F. (eds.) Computational Forensics. IWCF IWCF 2012 2014. LNCS, vol. 8915, pp. 117–129. Springer, Cham (2015). https://doi.org/10.1007/978-3-319-20125-2_11,
12. Micenkov, B., Beusekom, J.V.: Stamp detection in color document images. In: 2011 International Conference on Document Analysis and Recognition, pp. 1125–1129 (2011). https://doi.org/10.1109/ICDAR.2011.227
13. Nguyen, D.Q., Nguyen, A.T.: PhoBERT: pre-trained language models for Vietnamese. In: Findings of the Association for Computational Linguistics: EMNLP 2020, pp. 1037–1042 (2020)
14. Nguyen, Q.D., Le, D.A., Zelinka, I.: Ocr error correction for unconstrained Vietnamese handwritten text. In: Proceedings of the 10th International Symposium on Information and Communication Technology, pp. 132–138. SoICT '19, Association for Computing Machinery, New York, NY, USA (2019). https://doi.org/10.1145/3368926.3369686
15. Nguyen, Q.D., Phan, N.M., Krömer, P., Le, D.A.: An efficient unsupervised approach for ocr error correction of vietnamese ocr text. IEEE Access 11, 58406–58421 (2023). https://doi.org/10.1109/ACCESS.2023.3283340
16. Schweter, S., Akbik, A.: Flert: document-level features for named entity recognition (2021). https://arxiv.org/abs/2011.06993
17. Xu, Y., et al.: Layoutlmv2: multi-modal pre-training for visually-rich document understanding (2022). https://arxiv.org/abs/2012.14740
18. Zhu, J.Y., Park, T., Isola, P., Efros, A.A.: Unpaired image-to-image translation using cycle-consistent adversarial networkss. In: Computer Vision (ICCV), 2017 IEEE International Conference on (2017)

Reflective Thought and Code for Solving Numerical Problems with Large Language Models

Long Tri Thai Son, Ngo Viet Anh, Nguyen Thi Thuy Linh$^{(\boxtimes)}$,
and Nguyen Viet Ha

Institute for Artificial Intelligence, VNU University of Engineering and Technology,
Hanoi, Vietnam
`linh.nguyen@vnu.edu.vn`

Abstract. Large language models (LLMs) have achieved strong performance on many reasoning tasks but remain limited in solving computational problems, where precise reasoning and correct numerical calculation are essential. Therefore, common prompting strategies often produce flawed intermediate steps or incorrect results when applied to math and physics problems. We propose Reflective Thought and Code (RTC), a two-phase framework that separates problem solving into a thought phase, where the model generates an abstract solution plan, and a code phase, where executable programs are synthesized and run to compute the answer. To improve reliability, we add reflection step to each phase to enable self-evaluation and revision. Experiments on *PhysQA*, *PhysReason*, and *MATH* benchmarks showed that RTC achieves better performance than standard prompting baselines while offering more interpretable outputs.

Keywords: Large language models · Numerical reasoning · Chain-of-thought · Code generation

1 Introduction

Large language models (LLMs) have shown promising capabilities in solving complex reasoning tasks, including mathematical problem solving, code generation, and scientific question answering. To improve reasoning without modifying model parameters, research has focused on test-time prompting methods such as Chain-of-Thought (CoT) [13] prompting and Chain-of-Code [7], which encourage step-by-step reasoning and code-based computation.

Despite these advances, numerical reasoning tasks such as solving physics word problems or multi-step math exercises remain challenging for LLMs. Recent evaluations reveal that state-of-the-art models exhibit high error rates (mean $\approx$ 52%) on high school-level problems [2]. These tasks require not only structured reasoning and exact computation, but also access to external knowledge such as formulas, constants, and physical laws. Small errors in early reasoning steps

N. Thai-Nghe et al. (Eds.): ISDS 2025, CCIS 2714, pp. 433–447, 2026.
https://doi.org/10.1007/978-981-95-3358-9_31

often propagate to incorrect final conclusions, which is especially problematic in physics and engineering applications.

The fundamental challenge lies in the computational burden placed on LLMs when they must simultaneously understand problems, formulate reasoning plans, generate syntactically correct code, and ensure correctness of intermediate results. This multi-faceted responsibility increases susceptibility to hallucination, logic gaps, and execution failures. Existing methods either rely solely on in-context recall or lack explicit mechanisms to verify and revise intermediate reasoning steps.

To address these challenges, we propose **Reflective Thought and Code (RTC)**, a training-free reasoning framework designed for accurate and interpretable numerical problem solving. The framework separates cognitive processes into distinct, manageable phases:

- A **Thought phase**, where the model generates a natural language plan focused purely on abstract reasoning and problem decomposition;
- A **Code phase**, where executable code is synthesized to implement the validated reasoning plan with systematic verification mechanisms.

Crucially, RTC incorporates reflection mechanisms at both stages, enabling the model to identify and correct logical or implementation errors during execution. This design separates conceptual understanding from computational execution, reducing cognitive load and improving solution reliability.

We evaluate RTC on physics reasoning benchmarks including PhysQA and PhysReason, as well as the MATH competition benchmark, containing problems ranging from high school to university level across physics and mathematics domains. Experimental results demonstrate that RTC significantly outperforms existing prompting methods, particularly on large-scale models.

2   Related Work

Existing approaches to improve LLM reasoning can be categorized into three main directions:

Structured Prompting. Chain-of-Thought (CoT) prompting [13] guides LLMs to articulate intermediate reasoning steps, with extensions like **Self-Consistency** [12] and **Tree-of-Thoughts** [14] adding robustness and search capabilities. **ReAct** [15] integrates external tool use. However, these approaches lack explicit computational modules and offer limited mechanisms for validating intermediate values in numerical tasks.

Program Synthesis. Program-of-Thought (PoT) [3] generates symbolic code to compute intermediate results, while **Toolformer** [10] generalizes tool invocation capabilities. Although these methods embed computation within reasoning, they often conflate task decomposition, inference, and code generation into a single pass, without mechanisms for reflecting on planning structure or correcting generated programs.

Self-Refinement. Reflexion [11] introduces feedback loops to revise outputs through self-assessment, with similar approaches showing improvements in output consistency [1,8,16]. However, these techniques typically lack structured verification of symbolic steps or integration with executable code.

Recent domain-specific frameworks like **Physics Reasoner** [9] combine formula retrieval with code-based computation, while **MoRA** [6] uses modular refinement agents. However, existing approaches remain largely modular, addressing specific aspects without unified mechanisms for reflection and verification across both symbolic logic and computation.

3   Reflective Thought and Code Framework for Solving Numerical Problems

To address the limitations of large language models in numerical reasoning tasks, we propose **Reflective Thought and Code (RTC)**, a two-phase framework that separates abstract reasoning from executable computation. RTC decomposes problem solving into a *thought phase* for abstract reasoning and a *computational code phase* for numerical execution, each equipped with reflection mechanisms for self-evaluation and correction. An overview of the framework is illustrated in Fig. 1.

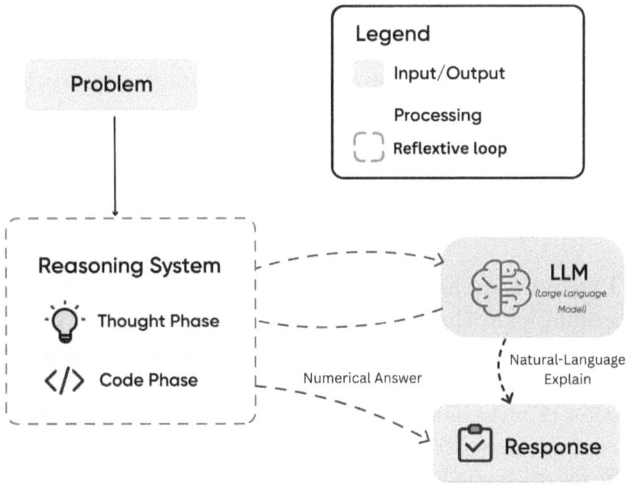

Fig. 1. Overview of the RTC framework. Both the Thought and Code phases are equipped with iterative reflection loops (shown as red dashed arrows), allowing the system to revise plans or code until validation succeeds or maximum iterations are reached. (Color figure online)

3.1 Thought Phase

The thought phase systematically constructs a structured understanding of the physical scenario and develops a solution pathway through iterative refinement.

Entity and Relation Analysis. The reasoning process begins with extracting and organizing the problem elements into a structured representation. This component identifies physical objects and quantities, initial conditions and constraints, mathematical relationships between variables, and target variables to be solved. Each quantity is characterized by its symbol, numerical value, unit or domain, and status (given, unknown, or derived). In practice, the Entity & Relation Analysis module outputs a compact JSON-like schema that captures all relevant entities, quantities, and relations explicitly mentioned in the problem. This schema, together with constraints and targets, is serialized and prefixed as a context block in the subsequent LLM prompt for Thought Generation, ensuring consistent use of symbols and relations across both mathematical and physical tasks. Reflection is limited to at most three iterations, and terminates earlier if no inconsistencies are detected or if two consecutive revisions yield identical results. In the Code phase, reflective debugging leverages the validation report (syntax, semantics, and structure) as a patch list so that only the faulty segments of code are regenerated rather than the entire program. We further illustrate this module with multiple examples in Appendix A, covering both mathematics and physics problems.

Thought Generation. The abstract planning phase constructs a reasoning plan $S = \langle s_1, s_2, \ldots, s_n \rangle$ that provides a logical pathway from known quantities to the target variable. Each step s_i consists of operation type (apply physical law, introduce variable, substitute values, solve equation), input/output variables, and justification (physical principle or mathematical rule applied).

Reflective Thought. The reflection mechanism validates and iteratively refines the reasoning plan through systematic evaluation. The process generates an initial plan, then validates and refines it until no issues remain or maximum iterations are reached.

The validation criteria include: (1) **Logical coherence** - output variables of each step are compatible with subsequent step inputs; (2) **Completeness** - valid reasoning path exists from known to target variables; (3) **Physical validity** - operations correspond to established physical principles; (4) **Mathematical consistency** - well-formed equations and valid algebraic manipulations.

3.2 Code Phase

The code phase translates the validated abstract reasoning plan into executable code through structured generation and verification.

Generate Code. The code generation component transforms the reasoning plan S into executable Python code, where each code segment corresponds to a reasoning step. The generation converts all values to SI units, uses one line per abstract step with descriptive variable names, and assigns the final result to variable named `result`.

Reflective Code. The code validation and refinement follows an iterative review-repair cycle. The validation performs assessment across three dimensions: (1) **Syntactic validation** - Python syntax correctness and variable completeness; (2) **Semantic validation** - unit consistency and physical law correctness; (3) **Structural validation** - one-to-one correspondence between code and reasoning steps.

Executable Code. The final component produces verified code that accurately implements the solution methodology, ensuring both semantic fidelity to the abstract plan and computational correctness.

Figure 2 demonstrates the complete RTC workflow through a concrete physics problem, showing how the thought phase systematically analyzes entities and develops a reasoning plan, while the code phase translates this plan into executable Python code with proper variable handling and computations.

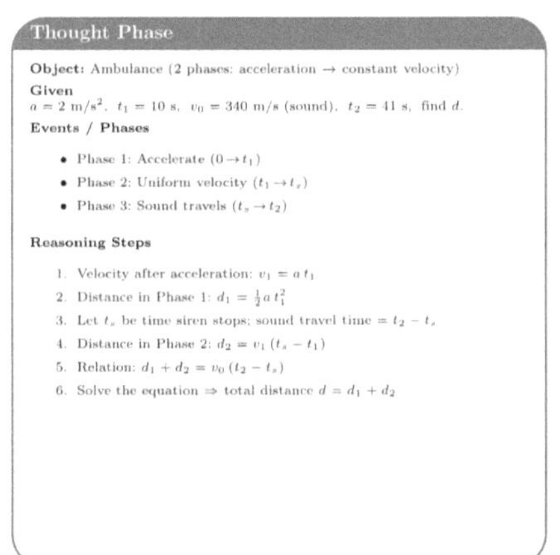

(a) Thought Phase Example   (b) Code Phase Example

Fig. 2. Illustrative examples of RTC's two-phase approach.

For completeness, all prompt templates used in Thought phase and Code phases are provided in Appendix B.

4￼Experiments

We verified the proposed RTC framework on two physics problem-solving benchmarks: PHYSQA [4] and PHYSREASON [17], and one mathematical reasoning benchmark: MATH [5]. Our goal was to assess how RTC enhances numerical reasoning capabilities of large language models under zero-shot conditions across both physics and mathematics domains.

4.1￼Experimental Setup

Datasets. PHYSQA contains 1,008 high-school physics problems in multiple-choice format, testing conceptual understanding and quantitative problem-solving. PHYSREASON comprises 1,200 physics questions with significant computational complexity (average 8.1 reasoning phases). MATH contains 12,500 competition mathematics problems from AMC 10, AMC 12, and AIME competitions, focusing on algebra, geometry, number theory, and calculus problems that involve multi-step computational reasoning. We use the full PhysQA dataset, a curated subset of 135 purely textual problems from PhysReason, and 300 numerical problems from MATH that yield numerical answers.

Models and Baselines. We evaluated five language models: Llama-3.3-70B-Instruct, Qwen-2.5-72B-Instruct, Gemma-2-27B-Instruct, Llama-3.2-Vision-11B-Instruct, and Llama-3.1-8B-Instruct. For each model, we compared four approaches: (i) *Baseline* direct prompting; (ii) *Chain-of-Thought* (CoT) [13]; (iii) *Program-of-Thought* (PoT) [3]; and (iv) our *RTC* framework.

Evaluation. All evaluations used zero-shot prompting with exact numeric accuracy as the metric. Answers are correct only if both numeric value and unit match the ground truth (allowing 5% relative error for floating-point precision). Inference used fixed decoding parameters and 2048 token context length.

4.2￼Results

Tables 1, 2 and 3 present accuracy results on all benchmarks.

Table 1. Accuracy (%) on the PHYSQA dataset

Model	Baseline	CoT	PoT	RTC (Ours)
Llama-70B-it	67.16	69.44	70.54	**72.62**
Qwen-72B-it	72.32	76.19	74.80	**80.65**
Gemma-27B-it	62.60	63.99	65.08	**67.56**
Llama-3.1-8B-it	46.23	48.12	49.21	**53.87**
Llama-3.2-11B-Vision-it	44.15	46.13	44.35	**48.81**

Table 2. Accuracy (%) on the PHYSREASON dataset

Model	Baseline	CoT	PoT	RTC (Ours)
Llama-70B-it	62.96	64.44	62.96	**65.93**
Qwen-72B-it	65.19	67.41	62.96	**70.37**
Gemma-27B-it	51.11	51.85	48.15	**54.07**
Llama-3.2-11B-Vision-it	33.33	**34.81**	22.22	33.33
Llama-3.1-8B-it	34.07	**36.30**	27.41	34.07

Table 3. Accuracy (%) on the MATH dataset (300 numerical problems)

Model	Baseline	CoT	PoT	RTC (Ours)
Llama-70B-it	74.67	77.00	74.33	**78.67**
Qwen-72B-it	81.67	82.00	78.33	**84.33**
Gemma-27B-it	59.33	60.67	55.67	**63.00**
Llama-3.1-8B-it	51.67	**52.33**	46.00	47.67
Llama-3.2-11B-Vision-it	51.33	**54.00**	42.00	48.33

RTC consistently achieved the highest accuracy across all models on PhysQA, improving upon PoT by 2.10–5.80% points. The largest gains occur on high-capacity models like Qwen-72B-it and Llama-70B-it. On the more complex PhysReason benchmark, RTC outperforms baselines on larger models but shows marginal improvements on smaller models (8B-11B parameters). On the MATH benchmark, RTC demonstrates strong performance on larger models, with Qwen-72B-it achieving the best performance at 84.33% and Llama-70B-it reaching 78.67%. However, smaller models (8B-11B parameters) show mixed results, with CoT occasionally outperforming RTC on these limited-capacity models. This pattern suggests that the structured entity-relation analysis phase is particularly beneficial for physics contexts where physical laws and unit consistency play crucial roles, while mathematical reasoning may benefit more from simpler prompting strategies on smaller models.

The results revealed a clear capacity threshold: RTC's benefits are most pronounced on models with strong abstract planning and code generation capabilities. Larger models like Qwen-72B-it and Llama-70B-it consistently benefit from the two-phase approach, while smaller models often struggle with multi-step reasoning coherence and generate syntactically incorrect code, limiting the effectiveness of the reflection mechanisms. Common failure modes include improper unit handling, undefined variables, and incomplete imports that the verification loop cannot repair under zero-shot constraints. The cross-domain comparison between physics and mathematics problems reveals that RTC's structured approach provides greater benefits in physics contexts, where the entity-relation analysis phase proves particularly valuable for handling physical quantities, units, and laws, while pure mathematical problems may require less domain-specific structuring and can sometimes be better served by simpler reasoning approaches on resource-constrained models.

RTC requires additional computation compared to CoT. On average, the reflective process consumes approximately 3.0–3.3× more inference tokens, leading to runtimes of 7.4–8.4 s per problem on 72B-scale models, compared to about 3.2–3.4 s for CoT. We consider this overhead a reasonable trade-off given the consistent accuracy gains observed across benchmarks.

Figure 3 illustrates a concrete comparison where RTC produces the correct solution through structured reasoning, while CoT yields an intuitive but incorrect answer.

Problem Statement: To rescue a patient, an ambulance urgently departs, sounding its siren, and begins to move with uniform acceleration along a horizontal straight road from rest at $t = 0$, with an acceleration magnitude of $a = 2\,\mathrm{m/s^2}$. At $t_1 = 10\,s$, it ceases acceleration and begins to move with uniform velocity. At some later time, the ambulance stops sounding its siren. The last sound of the siren is heard by a person at the ambulance's starting point at $t_2 = 41\,s$. The speed of sound is given as $v_0 = 340\,m/s$.

Question: What is the distance of the ambulance from its starting point at the moment it stops sounding its siren?

CoT Method	RTC Method
Solution: Calculate velocity at t_1: $v_1 = 2 \times 10 = 20\,\mathrm{m/s}$ Distance during acceleration: $s_1 = \frac{1}{2} \times 2 \times 10^2 = 100\,\mathrm{m}$ Time at uniform velocity: $t_{\text{uniform}} = 41 - 10 = 31\,s$ Distance at uniform velocity: $s_{\text{uniform}} = 20 \times 31 = 620\,\mathrm{m}$ Total distance: $s_{\text{total}} = 100 + 620 = 720\,\mathrm{m}$ The solution calculates total distance but doesn't account for the time it takes sound to travel back to the observer. Answer: 720 m	**Solution:** Same initial calculations: - $v_1 = 20\,\mathrm{m/s}$ - $s_1 = 100\,\mathrm{m}$ (acceleration phase) Key insight: Account for sound travel time. Let t_f = time traveling at uniform velocity before siren stops $t_f = t_2 - t_1 - \frac{d_{\text{total}}}{v_0}$ Distance equation: $d_{\text{total}} = s_1 + v_1 \times t_f$ Substituting and solving: $d_{\text{total}} = \frac{100 + 20 \times 31}{1 + \frac{20}{340}} = \frac{720}{\frac{18}{17}} = 680\,\mathrm{m}$ Correct Answer: 680 m

Fig. 3. Comparison between CoT and RTC responses for an ambulance motion problem, generated by Llama-3.3-70B-Instruct.

These results demonstrate that separating abstract reasoning from computational execution, combined with reflection mechanisms, successfully improves physics & math problem-solving performance without requiring fine-tuning. The framework is most effective on high-capacity models where the synergy between logical planning and reliable code execution can be fully leveraged.

In our observation, language models often make small but critical mistakes, such as omitting variables or producing inconsistent units. More rarely, they can generate outputs that are qualitatively unreasonable, despite appearing fluent. The reflective verification stage helps mitigate these issues by detecting such failures and enforcing corrections, thereby reducing the randomness of single pass generation and improving overall robustness.

5   Conclusion

Multi-step reasoning in numerical problem solving remains challenging for large language models, particularly in maintaining logical consistency while ensuring computational correctness. We proposed **Reflective Thought and Code (RTC)**, a two-phase framework that separates abstract reasoning from computational implementation with reflection mechanisms at both stages. RTC guides language models through coherent reasoning while maintaining rigorous validation of logical flow and numerical accuracy.

Our evaluation on PhysQA, PhysReason, and MATH benchmarks demonstrated consistent improvements across multiple language models. RTC outperformed Chain-of-Thought and Program-of-Thought prompting strategies, with particularly strong gains on high-capacity models. The cross-domain evaluation revealed that while RTC benefits both physics and mathematics reasoning, the structured entity-relation analysis proves especially valuable for physics contexts where physical laws and unit consistency are critical. These results highlight the effectiveness of integrating abstract reasoning with executable computation in a unified framework.

Although our evaluation covered both physics and mathematics problems, the framework could be extended to other scientific domains where transparent derivations and systematic reasoning are critical. In future work, we plan to extend our framework to broader reasoning domains, develop methods to improve performance on smaller models, and integrate retrieval mechanisms for accessing external knowledge bases.

A   Example Output of Entity and Relation Analysis

The following example illustrates the structured JSON-like output generated for the problem: *"The sum of three numbers a, b, c is 88. If we decrease a by 5, we get N. If we increase b by 5, we get N. If we multiply c by 5, we get N. What is the value of N?"*

```
{
  "variables": [
    {"symbol": "a", "description": "First number", "role": "
        unknown"},
    {"symbol": "b", "description": "Second number", "role": "
        unknown"},
    {"symbol": "c", "description": "Third number", "role": "
        unknown"},
    {"symbol": "N", "description": "Target value", "role": "
        unknown"}
  ],
  "given_information": [
    {"type": "equation", "mathematical_form": "a + b + c =
        88"},
    {"type": "equation", "mathematical_form": "a - 5 = N"},
```

```
    {"type": "equation", "mathematical_form": "b + 5 = N"},
    {"type": "equation", "mathematical_form": "5c = N"}
  ],
  "target_question": {
    "what_to_find": "The value of N",
    "target_variable": "N"
  },
  "problem_context": {
    "mathematical_domain": "algebra",
    "problem_type": "solve_equation"
  }
}
```

The following example for a physics problem: *"A ball is released from rest. The displacement during the final 1 s is $\frac{9}{25}$ of the total displacement. Given $g = 10\,\mathrm{m/s}^2$. Calculate the displacement during the final 2 s."*

```
{
  "objects": [
    {"id": "ball", "name": "small ball", "initial_velocity":
        0},
    {"id": "ground", "name": "ground"}
  ],
  "physical_quantities": [
    {"symbol": "g", "value": 10, "unit": "m/s^2", "role": "
        given"},
    {"symbol": "s_total", "description": "total displacement
        ", "role": "intermediate"},
    {"symbol": "s_last_two_seconds", "description": "target
        displacement", "role": "target"}
  ],
  "initial_conditions": [
    {"object_id": "ball", "condition_type": "velocity", "
        value": 0}
  ],
  "constraints": [
    {"mathematical_expression": "s_last_second = (9/25) *
        s_total"}
  ],
  "target_problem": {
    "name": "displacement during final 2 seconds",
    "expected_units": "m"
  }
}
```

B   Prompt Templates

B.1   Entity and Relation Extraction Prompt

```
You are a mathematical problem analyst. Extract a structured
    JSON object capturing all mathematical information from
    the exercise.

JSON Schema:
{
  "variables": [...],
  "given_information": [...],
  "target_question": {...},
  "problem_context": {...},
  "constraints_and_conditions": [...]
}

Rules:
- Extract ONLY explicitly stated information
- Do NOT solve or derive anything
- Record expressions exactly as written

Exercise: {exercise}
```

B.2   Reflective Entity and Relation Extraction Validate Prompt

```
Check the analysis JSON for:
1. Structure: Must follow schema with required keys
2. Content accuracy: All info from problem text only
3. Completeness: All variables and conditions captured

Output: {"ok": true} or {"ok": false, "issues": [...]}

Exercise: {exercise}
Analysis: {analysis_json}
```

B.3 Reflective Entity and Relation Extraction Recorrect Prompt

```
Fix all structural and content issues in the analysis.

Tasks:
- Correct all listed issues
- Keep valid data unchanged
- Fix mathematical inconsistencies
- Do NOT solve or calculate
- Return only corrected JSON

Exercise: {exercise}
Original analysis: {analysis_json}
Issues: {issues}
```

B.4 Thought Generation Prompt

```
Generate a step-by-step reasoning plan to solve the problem.

Output format:
{
  "plan": ["Step 1: ...", "Step 2: ...", ...],
  "target_variable": "<symbol>",
  "mathematical_approach": "<method>",
  "key_theorems": [...]
}

Requirements:
- Think systematically like a mathematician
- State techniques/theorems at each step
- No calculations, only logical actions

Exercise: {exercise}
Analysis: {analysis_json}
```

B.5 Reflective Thought Validate Prompt

```
Evaluate if the plan logically solves the problem.

Check:
1. Basic structure: Required keys present
2. Reasoning steps: Clear and actionable
3. Logical flow: Steps build on each other
4. Completeness: Addresses all conditions
5. Mathematical validity: Appropriate methods

Output: {"ok": true} or {"ok": false, "issues": [...]}

Exercise: {exercise}
Plan: {plan_json}
```

B.6 Reflective Thought Recorrect Prompt

```
Rewrite plan to fix validation errors.

Requirements:
- Required keys: "plan", "target_variable",
    "mathematical_approach"
- Clear ordered mathematical steps
- Use only existing variables from analysis
- Appropriate mathematical methods
- Address all listed issues

Exercise: {exercise}
Analysis: {analysis_json}
Current plan: {plan_json}
Issues: {issues}
```

B.7 Code Generation Prompt

```
Generate Python code following the reasoning plan.

Requirements:
- Use only Python 3 standard library + math
- Follow plan step by step
- Store final answer in 'result' variable
- Add comments explaining each step
- Handle edge cases appropriately

Exercise: {exercise}
Analysis: {analysis_json}
Plan: {plan_json}
```

B.8   Reflective Code Validate Prompt

```
Validate Python code for:
1. Syntax & Execution: Valid Python 3, defined variables
2. Core Logic: Addresses main problem correctly
3. Safety: Handles division by zero, domains

Output: {"ok": true} or {"ok": false, "issues": [...]}

Code: {code}
```

B.9   Reflective Code Recorrect Prompt

```
Fix all issues while maintaining mathematical accuracy.

Requirements:
- Fix all listed issues
- Follow plan_json logic flow
- Use only Python 3 + math module
- Handle domains and constraints
- Store result in 'result' variable
- Include clear comments
- Output only corrected code

Code: {code}
Issues: {issues}
Plan: {plan_json}
Analysis: {analysis_json}
```

References

1. Abdali, S., Goksen, C., Amizadeh, S., Maybee, J.E., Koishida, K.: Self-reflecting large language models: a hegelian dialectical approach. arXiv preprint arXiv:2501.14917 (2025)
2. Boye, J., Moell, B.: Large language models and mathematical reasoning failures. arXiv preprint arXiv:2502.11574 (2025)
3. Chen, W., Ma, X., Wang, X., Cohen, W.W.: Program of thoughts prompting: disentangling computation from reasoning for numerical reasoning tasks. Trans. Mach. Learn. Res. (TMLR) (2023)
4. Ding, J., Cen, Y., Wei, X.: Using large language model to solve and explain physics word problems approaching human level. arXiv preprint arXiv:2309.08182 (2023)
5. Hendrycks, D., et al.: Measuring mathematical problem solving with the math dataset. In: NeurIPS (2021)
6. Jaiswal, R., et al.: Improving physics reasoning in large language models using mixture of refinement agents. arXiv preprint arXiv:2412.00821 (2024)

7. Li, C., et al.: Chain of code: reasoning with a language model-augmented code emulator. In: Proceedings of the 41st International Conference on Machine Learning, vol. 235, pp. 28259–28277. PMLR (2024)
8. Madaan, A., et al.: Self-refine: iterative refinement with self-feedback. In: Advances in Neural Information Processing Systems (NeurIPS), vol. 36, pp. 46534–46594 (2023)
9. Pang, X., et al.: Physics reasoner: knowledge-augmented reasoning for solving physics problems with large language models. In: Proceedings of the 31st International Conference on Computational Linguistics (COLING), pp. 11274–11289 (2025)
10. Schick, T., et al.: Toolformer: language models can teach themselves to use tools. In: Advances in Neural Information Processing Systems (NeurIPS), vol. 36, pp. 68539–68551 (2023)
11. Shinn, N., Cassano, F., Gopinath, A., Narasimhan, K., Yao, S.: Reflexion: language agents with verbal reinforcement learning. In: Advances in Neural Information Processing Systems (NeurIPS), vol. 36, pp. 8634–8652 (2023)
12. Wang, X., et al.: Self-consistency improves chain of thought reasoning in language models. arXiv preprint arXiv:2203.11171 (2022)
13. Wei, J., et al.: Chain-of-thought prompting elicits reasoning in large language models. In: Advances in Neural Information Processing Systems (NeurIPS), vol. 35, pp. 24824–24837 (2022)
14. Yao, S., et al.: Tree of thoughts: deliberate problem solving with large language models. In: Advances in Neural Information Processing Systems (NeurIPS), vol. 36, pp. 11809–11822 (2023)
15. Yao, S., et al.: React: synergizing reasoning and acting in language models. In: International Conference on Learning Representations (ICLR) (2023)
16. Zhang, K., Li, Z., Li, J., Li, G., Jin, Z.: Self-edit: fault-aware code editor for code generation. arXiv preprint arXiv:2305.04087 (2023)
17. Zhang, X., et al.: Physreason: a comprehensive benchmark towards physics-based reasoning. In: Proceedings of the 2025 Conference of the Association for Computational Linguistics (ACL) (2025)

RCM-MSA: Robust Contextual Modeling for Multi-modal Sentiment Analysis Using Transformer and Attention Mechanism

Hoai-Phuong Nguyen-Cao[1,2], Phuoc-Hung Vo[1],
and Nhut-Lam Nguyen[1]

[1] Tra Vinh University, Vinh Long, Vietnam
{hungvo,lamnn}@tvu.edu.vn
[2] Thu Dau Mot University, Ho Chi Minh City, Vietnam
phuongnch@tdmu.edu.vn

Abstract. The online platforms have made it increasingly convenient for users to express their thoughts and emotions through both text and images. This widespread sharing results in large volumes of user-generated content rich with emotional signals, which presents valuable opportunities for multi-modal sentiment analysis. Unlike traditional sentiment analysis that relies on a single modality, leveraging multiple modalities, such as text and images, enables a more holistic understanding of user sentiment by capturing complementary cues across different data types. As platforms like Twitter increasingly host diverse content formats, analyzing sentiment using multi-modal data has become a central focus in artificial intelligence research. However, many existing models fail to fully capture the interaction between text and image, often treating them separately, which limits prediction accuracy. This study addresses that gap by introducing a novel multi-modal sentiment analysis model that integrates contextual modeling with an attention mechanism for textual modality and a transformer for visual modality. An attention mechanism strengthens semantic understanding, while transformer encoder layers uncover long-range dependencies within images. The model processes text and images using specialized encoders and performs early feature-based fusion. Experimental evaluation on the MVSA dataset, using accuracy and F1-score metrics, shows that the proposed model significantly outperforms previous methods in effectively interpreting complex multi-modal sentiment analysis.

Keywords: Text encoder · Image encoder · Transformer · Multi-modal sentiment analysis

1 Introduction

Sentiment analysis has attracted considerable interest in academia and has profoundly influenced both the economy and social life [1,2]. Initially, the key research focuses on natural language processing; it aims to detect and interpret polarity sentiment content in text [3]. While multimedia's growth has broadened its scope to include text and images, and research has aimed to improve prediction accuracy using feature engineering, multi-modal sentiment analysis still faces unresolved problems [4].

Multi-modal sentiment analysis, especially with text and image data, is gaining traction as a means to improve polarity sentiment detection. The objective is to accurately determine the sentiment by jointly analyzing textual and visual information. Two key challenges persist: the efficient extraction of informative features from images and the effective integration of these features with textual data to build a comprehensive representation [5,6]. Traditional approaches, which process text and images independently and then concatenate them, fail to capture the semantic interplay between the modalities, leading to imprecise sentiment judgments.

The journey of Transformers in the field of artificial intelligence is a fascinating tale of innovation and continuous improvement. BERT [7] stands out as a prime example of the effectiveness of transformers. BERT, having been pre-trained on a large-scale text data, effectively learns and encodes deep contextual relationships within language. In [8], the transformer encoder layer, via self-attention, analyzes relationships between distant image parts, grasping overall context and long-range dependencies for complex scene understanding and theme recognition. However, with their large number of parameters, Transformers can overfit easily on small dataset. The origins of transformers can be traced back to the advent of recurrent neural networks (RNNs). However, RNNs had limitations, especially when it came to processing language sequentially.

To address the above challenges, we propose a multi-modal sentiment analysis approach consisting of two phases: richer contextual feature extraction and effective multi-modal fusion. In the first phase, BERT and LSTM are combined to extract textual features, while the attention mechanism enhances these features and improves semantic alignment. For image data, positional encoding is applied before inputting it into the Transformer encoder layer to capture the overall context. Finally, an intermediate fusion layer efficiently captures interactions between textual and visual features. The main contributions of our works are summarized as follows:

- We present detailed datasets and the data processing methodology, including labeling by three annotators.
- Our proposed hybrid model handles single-modal features. For the first single-modal (textual), we combine BERT and LSTM to extract textual features, utilizing an attention mechanism layer to facilitate textual enhancement and improve semantic alignment. The second single-modal (visual) is also a hybrid model, extracting visual features by combining an image encoder, transformer encoder layers, and a batch normalization layer. Finally, a feature-level fusion layer efficiently captures feature interactions between text and image features, enhancing multi-modal fusion by concatenating intra-modal models. This fur-

ther improves the model's ability to capture relationships between text and images.

- Our model's performance is compared with existing models on available multimodal MVSA datasets, with additional comparisons based on different hyperparameter settings. With the optimal configuration ($l = 0.001$, $h = 128$, $ff = 256$), the model achieved f1-scores and accuracy rates of 91.14% and 92.41% on MVSA-Single, and 84.08% and 88.28% on MVSA-Multiple, respectively.

2   Related Work

While sentiment analysis provides valuable polarity sentiment intelligence for decision-making, conventional methodologies are hampered by data sparsity, inaccuracies, semantic vagueness, absence of contextual grounding, low explainability, and label inconsistency [2,9,10]. Consequently, contextual feature extraction has become a key area of investigation. Jia et al. [11] developed the affective region recognition, text encoder, and multi-modal representation. It aligns visual regions with textual targets and uses an effective region selection mechanism to rank emotion-relevant image regions based on inter-modal interactions. A target-sensitive module models textual intra-modal dependencies. While Liu et al. [12] developed a multi-modal multi-attention fusion network with transformer encoders, which learns consistent inter-modal features and enhances target modality representations through multi-attention fusion. It also extracts complementary features using different fusion sequences. These extracted features are then integrated with the original representations via residual connections for sentiment analysis.

The Transformer model, introduced by Vaswani et al. [13], employs a self-attention mechanism and is structured with stacked encoders and decoders. Unlike recurrent and convolutional neural networks, it enables parallel processing of input sequences. Its efficacy in NLP tasks has been substantial, with BERT [7], a pretrained contextual language representation model, as a prominent illustration. For multi-modal sentiment analysis, Gong et al. [14] propose a multi-channel multi-modal joint learning approach. It extracts rich image features using Vision Transformer (ViT) and Faster R-CNN, emphasizing emotional regions while capturing global textual features through syntactic features and self-attention. Intra-modal modules reduce redundancy, and inter-modal correlations are effectively captured using a multi-modal transformer fusion module. Additionally, Guo et al. [15] introduce SmartRAN, a novel network that employs a smart routing attention module for dynamic data flow path selection, addressing the limitations of fixed architectures in adaptability and generalizability. By learning both intra-modal and inter-modal information, SmartRAN enhances semantic consistency and improves the modeling of complex relationships. Analysis results indicate that different modalities contribute variably to multi-modal sentiment analysis tasks. This paper introduces a multi-modal sentiment analysis approach combining intra-modal correlations using a inter-mediate layer in multi-modal fusion method.

3 Proposed Model

3.1 Overview

In this section, the proposed framework consists of five main components: text encoder, image encoder, attention mechanism, transformer encoder, and bilinear fusion, as shown in Fig. 1.

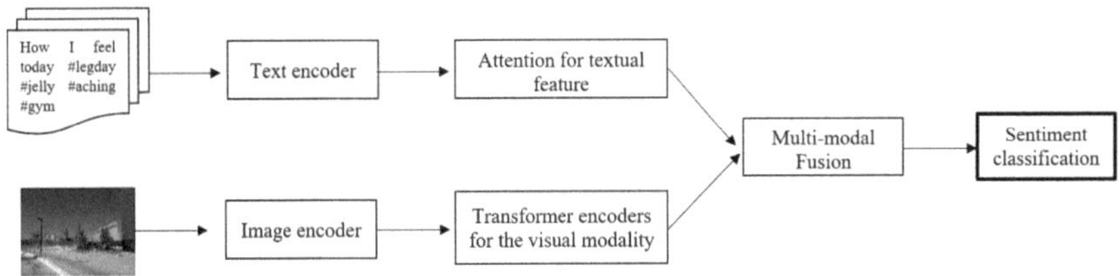

Fig. 1. The Proposed Multi-modal Sentiment Analysis Architecture.

3.2 Textual Encoder

The BERT-based model [7] extracts contextual features from text using a bidirectional Transformer-based architecture. BERT uses special tokens to structure input text, enabling it to understand sentence boundaries, relationships, and token positions effectively. Give an input sequence $D = d_1, d_2, \ldots, d_n$, BERT adds the classification token:

$$H_b = BERT(D) \tag{1}$$

The extracted features from BERT are then passed through an LSTM to generate textual representations that effectively capture long-range dependencies in the text data. This process can be formulated as follows:

$$T_e = LSTM(H_b) \tag{2}$$

3.3 Attention Mechanism for Textual Feature

The attention mechanism, proposed by Bahdanau et al. [16], aims to resolve the bottleneck where fixed-length encoding restricts decoders from accessing input. This method, improving data retrieval, is defined by the following calculation:

$$S = tanh(W_i T_e + b_i) \tag{3}$$

where S are the attention scores, W_i and b_i are learnable parameters.

Normalize attention scores using the softmax function:

$$\alpha = softmax(S) \tag{4}$$

where α represents the attention weights.

Compute the context vector T_t, are fed into the decoder at each time step m:

$$T_t = \sum_{m=1}^{M} \alpha_t T_{mt} \tag{5}$$

Layer normalization (L), a method for normalizing neuron activations within a layer of a neural network, is defined as follows:

$$T_f = LayerNorm(T_t) \tag{6}$$

where T_f is output of layerNorm function. It helps stabilize the hidden state activations, making training more stable and allowing for larger learning rates.

3.4   Image Encoder

The image encoder model was designed to closely align, as illustrated in Fig. 2.

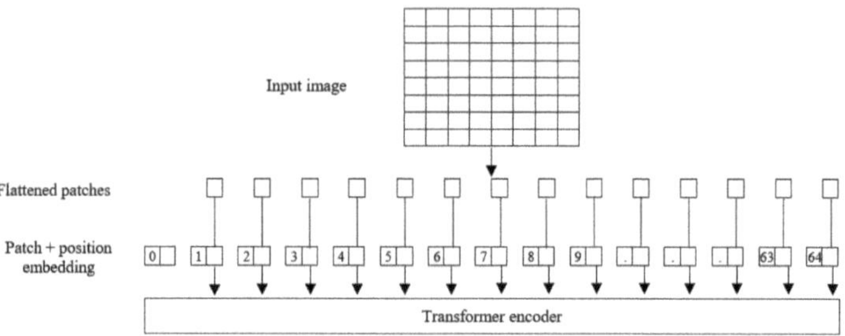

Fig. 2. Fixed-size patches, derived from the image, are linearly embedded; position embeddings are added, creating a vector sequence input to a standard Transformer encoder.

The number of patches P, which is also the effective input sequence length for the Transformer encoder, is calculated using the following Eq. 7:

$$P = (size_I // size_p) ** 2 \tag{7}$$

where $size_I$ is the size of the image and $size_p$ is the size of each patch.

Image $V = v^1, v^2, \ldots, v^N$ is divided into patches, to the embedded patch sequence ($v^1 = p_1^{v1}, p_2^{v1}, \ldots, p_p^{v1}$), we prepend a learnable embedding representing the image. Patches are flattened and transformed into dim dimensions using a trainable linear projection (Eq. 8), yielding patch embeddings.

$$I_e = [v^0, p_1^{v0} * E, p_2^{v0} * E, \ldots, p_p^{v0} * E] + E_{position} \tag{8}$$

where $E \in \mathbb{R}^{(P^2.C) \times D}$, $E_{position} \in \mathbb{R}^{N+1) \times Dim}$, C is the number of channels. Position embeddings $E_{position}$ are concatenated with patch embeddings to encode positional dependencies. We use standard learnable 1D position embeddings, as empirical evaluation showed no significant benefit from 2D-aware encoding. The resulting embedding sequence I_c is passed to the transformer encoder.

3.5   Transformer for the Visual Modality

We employ a three-layer transformer encoder, specifically for the visual modality, to account for the varying sentiment contributions of elements within a modality and enhance intra-modal interactions. Feed-forward network: Following the self-attention layer, a feed-forward network is applied. It includes two fully connected layers and operates along the embedding dimension. The intermediate hidden layer is expanded to allow for more expressive feature learning and richer representations. This allows us to assess the importance of visual elements into I_e at different time steps.

$$I_t = Transformer Encoder Layer(I_e) \tag{9}$$

Layer normalization, a method for normalizing neuron activations within a layer of a neural network, is defined as follows:

$$I_f = Layer Norm(I_t) \tag{10}$$

where I_f is output of layerNorm function. It helps stabilize the hidden state activations, making training more stable and allowing for larger learning rates.

3.6   Early Feature-Based Fusion and Prediction

The multi-modal task involves fusing the original sequences from both modalities. This fused representation is then used as input for a fully connected layer to generate the sentiment prediction, $\hat{y}$. The textual and visual features, obtained from Eq. 6 and Eq. 10, are merged as follows:

$$F = cat\,(T_f; I_f) \tag{11}$$

Here, $T_f \in \mathbb{R}^{dim_t}$, $I_f \in \mathbb{R}^{dim_i}$, $F_{Cat} \in R^{(dim_t + dim_i)}$, dim_t and dim_i are feature dimensions. A transformation is then applied to the concatenated features to enhance feature interaction. The mathematical formulation is given as follows:

$$F_c = ReLU\,(W_f F + b_f) \tag{12}$$

where $W_f \in R^{dim_0 \times (dim_t + dim_i)}$ is a learning weight matrix, $b_f \in R^{dim_0}$, bias term, F_c is the fused feature representation.

To predicts the sentiment class, follow Eq. 13:

$$O = W_c F_c + b_c \tag{13}$$

where $W_c \in R^{C x D_f}$ represents the classification weight matrix, b_c is the bias term, C is the number of sentiment classes.

Finally, a softmax function is applied to convert O into probabilities:

$$P(y = c|x) = \frac{e^{O_c}}{\sum_{j=i}^{C} e_j^{O}} \tag{14}$$

where $P(y = c|x)$ is probability of class c give input x, O_c is logit output for class c.

4   Experimental Setup and Results

4.1   Datasets

The MVSA datasets, publicly available for multi-modal sentiment analysis by [17], are derived from Twitter posts containing text, images, and hashtags. Each text-image pair is manually labeled as positive, neutral, or negative.

Table 1. Overview of the MVSA datasets

Datasets	Positive	Negative	Neutral	Total
MVSA-single	2683	1358	470	4511
MVSA-multiple	11318	1298	4408	17024

However, in the MVSA-multiple dataset, many tweets exhibit inconsistencies between textual and visual sentiment labels. To address this, we adopted the pre-processing method proposed in [18]. Consequently, we obtained the MVSA-single dataset and MVSA-multiple dataset, as shown in Table 1.

4.2   Experimental Setup

Experiments were performed on Windows 10 (64-bit) using an NVIDIA RTX 1070 GPU and PyTorch. Text was preprocessed to 128 words, images to 32×32 pixels. The dataset, split 8:1:1 with stratified sampling, trained the model with a 128 batch size, Adam optimizer, and weight decay. Multiple experiments were carried out, using evaluation metrics based on prior studies to validate the model's performance on the processed datasets.

4.3   Evaluation Metrics

This study evaluates binary and multiclass intrusion detection performance using the following metrics: True Positives (TP), representing correctly identified intrusions; False Positives (FP), indicating normal activities misclassified as intrusions; False Negatives (FN), reflecting missed intrusions; and True Negatives (TN), denoting correctly identified normal activities. We also calculate Precision, indicating the percentage of correctly predicted positive cases; Recall, which shows the proportion of true positives among all actual positives; f1-score, the harmonic average of Precision and Recall; and Accuracy, representing the model's overall classification effectiveness.

$$Precision = \frac{TP}{TP + FP} \tag{15}$$

$$Recall = \frac{TP}{TP + FN} \tag{16}$$

$$F1 - score = 2 \times \frac{Precision \times Recall}{Precision + Recall} \tag{17}$$

$$Accuracy = \frac{TP + TN}{TP + TN + FP + FN} \tag{18}$$

The model's performance was assessed using accuracy, precision, recall, and f1-score, with higher values (0–100%) reflecting better classification results.

4.4   Baseline Techniques

The semantic features fusion neural network (SFNN) [17] uses BoW for textual features (TF and TF-IDF assessed). SentiBank provides superior visual semantic features for affective computing. Both early and late fusion approaches are evaluated for classification accuracy.

Multi-SentiNet [18] extracts visual semantic features (objects/scenes) from images. For text, it pre-trains GloVe word embeddings, then extracts textual features via a visual-guided attention LSTM. The importance of these visual features in multi-modal sentiment analysis is underscored by performance drops in 'Text+Scene' and 'Text+Object' experiments, with all features fused for classification.

Multi-view attentional network (MVAN) [19] uses a multi-view attention network for interactive text and image feature learning. Its multi-modal feature fusion, via multi-layer perceptron and stacking-pooling, significantly improves sentiment analysis quality.

The vision and augmented language transformer (VAuLT) [8] leverages a pretrained language model like BERT to enhance vision language transformer's weak language representations, enabling more effective performance on complex multi-modal tasks, particularly affective analysis in out-of-distribution social media scenarios.

The multiple contrastive learning ($\mathbf{M^2CL}$) [20] model enhances image-text multi-modal sentiment detection by integrating global and local representations from BERT-base and ResNet-50 with an Image Transformer Encoder. It introduces a Multi-modal Interaction Component for cross-modal reinforcement and applies supervised and dual multi-modal contrastive learning to refine feature alignment without requiring additional augmented data.

Contrastive language image pretraining, cross-attention, and cross-modal gating (CLIP-CA-CG) [21] model extracts image and text features using ResNet-50 and RoBERTa. It employs CLIP for contrastive learning, multi-head attention for cross-modal interaction, and a gating module for multi-level feature fusion, enhancing sentiment analysis accuracy and generalization.

Multi-channel multi-modal joint learning (MMJL) [14] uses cross-modal attention to merge image and text. Its network includes feature extraction, cross-modal fusion, and sentiment classification modules. MMJL employs novel MCIF (Faster R-CNN and Vision Transformer) for image emotion regions and MSTF (syntactic/self-attention) for global text features, improving multi-modal sentiment analysis.

Deep multi-level attentive network (DMLANet) [22] integrates bi-attention and semantic attention to associate image regions with relevant text and applies self-attention to capture rich multi-modal sentiment cues for classification.

Context sensitive multi-tier deep learning framework (CS-MDF) [23] is designed for image-text pairs with strong correlations but may struggle in cases where such detailed alignment is absent, as it relies on BiGRU to capture utterance-level context for sentiment analysis.

4.5   Experimental Results

For extracting context-sensitive information, CS-MDF is an effective approach. Initially, CNNs extract features, which are then fused across modalities. Next, a BiGRU model is trained for final classification, leveraging its ability to process sequential data and capture inter-modal dependencies. The CS-MDF model achieved an accuracy of 84.32% and f1-score of 84.45% on MVSA-single, and an accuracy of 82.47% and f1-score of 81.23% on MVSA-multiple. However, the model proposed in this study outperforms CS-MDF, achieving an accuracy of 93.51% and f1-score of 94.50% on MVSA-single, and an accuracy of 91.10% and f1-score of 88.49% on MVSA-multiple, as shown in Table 2.

The proposed model utilizes various hyperparameter sets to achieve optimal sentiment analysis. These include combining BERT with LSTM architectures, exploring different learning rates ($l \in 0.01, 0.001, 0.0001$), LSTM hidden units ($h \in 64, 128, 256$), and feed-forward hidden dimensions ($ff \in 128, 256, 512$). The feed-forward operation transforms self-attention output. For network optimization, categorical cross-entropy is selected as the loss function, suitable for this multi-class sentiment analysis problem, as shown in Table 3.

Table 2. Comparison of the proposed model with existing methods on MVSA-Single and MVSA-Multiple datasets

Method	MVSA-Single		MVSA-Multiple	
	Accuracy (%)	F1-score (%)	Accuracy (%)	F1-score (%)
SFNN [17]	68.30	66.20	68.80	65.70
MultiSentiNet [18]	69.84	69.63	68.86	68.11
MVAN [19]	72.98	72.98	72.36	72.30
VAuLT [8]	72.80	71.80	70.30	67.00
M^2CL [20]	75.50	74.02	73.20	70.50
CLIP-CA-CG [21]	75.33	73.46	72.00	69.83
MMJL [14]	77.21	76.42	75.87	73.99
DMLANet [22]	79.47	79.59	77.89	77.89
CS-MDF [23]	84.32	84.45	82.47	81.23
Our approach	**92.41**	**91.14**	**88.28**	**84.08**

Table 3. Hyperparameter settings used for tuning the proposed model

Hyperparameter	Values
Learning rate	0.01, 0.001, 0.0001
Hidden units of LSTM	64, 128, 256
Feed-forward hidden dimension	128, 256, 512
Dropout	0.1
Optimization method	Adam
Loss function	Cross-entropy

To mitigate overfitting, training runs for up to 25 epochs, incorporating an early stopping mechanism that monitors validation accuracy with a patience ranging from 15 to 25 epochs.

To determine the most effective configuration for the proposed model, hyperparameter tuning was performed. Extensive experiments were conducted on the MVSA-single and MVSA-multiple datasets under different hyperparameter settings to assess the model's performance. The results, including f1-score and accuracy, are detailed in Tables 4 and 5. The experiments explored learning rates (l) of 0.01, 0.001, 0.0001, LSTM hidden units (h) of 64, 128, 512, feed-forward dimensions (ff) of 128, 256, 512, with a dropout rate of 0.1 and the Adam optimizer. With the optimal configuration (l = 0.001, h = 128, ff = 256), the model achieved f1-scores and accuracy rates of 91.14% and 92.41% on MVSA-single, and 84.08% and 88.28% on MVSA-multiple, respectively. Throughout the

Table 4. Evaluation metrics for classification obtained by the proposed approach on the MVSA-Single dataset.

Learning rate	Hidden units	Feed-forward hidden dimension	Precision (%)	Recall (%)	Accuracy (%)	F1-score (%)
0.01	64	128	63.89	33.37	50.24	32.06
0.01	128	256	86.53	33.33	59.60	24.89
0.01	256	512	86.53	33.33	59.60	24.89
0.001	64	128	90.89	88.07	91.66	89.38
0.001	128	256	**92.03**	**90.31**	**92.41**	**91.14**
0.001	256	512	88.79	85.36	90.11	86.93
0.0001	64	128	72.00	37.90	61.24	35.38
0.0001	128	256	63.42	48.67	66.34	51.07
0.0001	256	512	68.98	55.70	70.91	58.88

Table 5. Evaluation metrics for classification obtained by the proposed approach on the MVSA-Multiple dataset.

Learning rate	Hidden units	Feed-forward hidden dimension	Precision (%)	Recall (%)	Accuracy (%)	F1-score (%)
0.01	64	128	88.85	33.33	66.57	26.64
0.01	128	256	76.76	33.33	66.64	27.39
0.01	256	512	88.85	33.33	66.57	26.64
0.001	64	128	83.38	76.84	85.00	79.66
0.001	128	256	**86.94**	**81.72**	**88.28**	**84.08**
0.001	256	512	77.00	68.17	79.65	71.68
0.0001	64	128	53.45	35.46	67.01	31.82
0.0001	128	256	58.60	41.46	68.57	42.12
0.0001	256	512	68.72	51.83	72.19	55.68

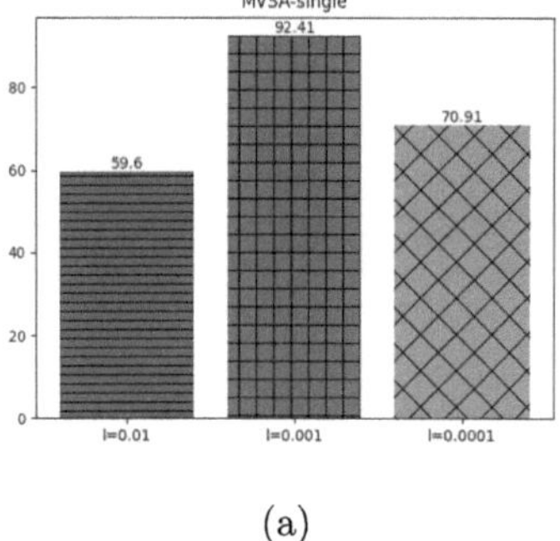

(a)

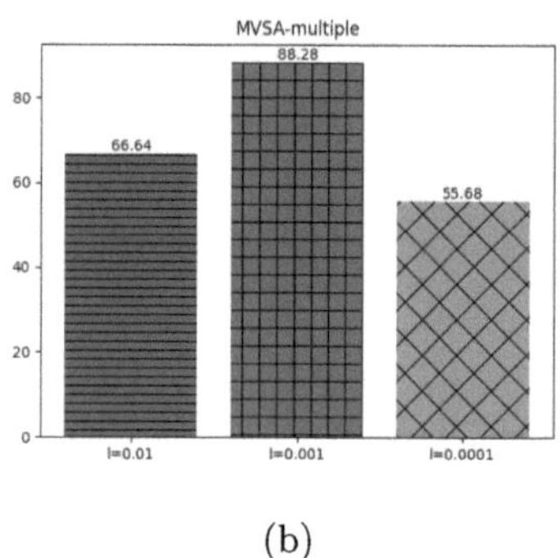

(b)

Fig. 3. Performance results evaluated on the MVSA-single dataset (3a) and MVSA-multiple dataset (3b).

training and validation phases, the model consistently demonstrated strong performance in terms of f1-score, precision, and recall. Other learning rates, such as 0.01 and 0.0001, did not yield better results. Figure 3 shows the model's highest accuracy: 88.28% for MVSA-multiple and 91.41% for MVSA-single with l = 0.001.

5   Conclusion

This study develops a method to accurately identify sentiment polarity expressed in online content. Utilizing transformer encoder layers for the visual modality and attention for textual feature, the model aims to enhance sentiment analysis by leveraging knowledge from related tasks. The core of the approach involves a fusion by multi-modal fusion that merges textual and visual features, effectively capturing the semantic link between images and their descriptions. This alignment significantly boosts the accuracy of sentiment classification. Through comparative experiments, the proposed model has shown superior performance on the MVSA dataset when compared to existing models. Further analysis revealed that textual information contributes more significantly than images to the overall sentiment classification. However, the model's current limitation is its reliance solely on text and images, excluding other potentially valuable data types like image and audio. Future research will explore integrating a wider range of modalities to further refine sentiment analysis. Additionally, the study will investigate the application of prompt learning techniques to adapt the model for scenarios with limited data.

References

1. Wankhade, M., Rao, A.C.S., Kulkarni, C.: A survey on sentiment analysis methods, applications, and challenges. Artif. Intell. Rev. **55**, 5731–5780 (2022). https://doi.org/10.1007/s10462-022-10144-1
2. Vanshika Rani, N., Walia, R.: A comprehensive review of sentiment analysis: Techniques, datasets, limitations, and future scope. In: 2024 Sixth International Conference on Computational Intelligence and Communication Technologies (CCICT), pp. 403–409 (2024). https://doi.org/10.1109/CCICT62777.2024.00072
3. Tang, J., et al.: BAFN: Bi-direction attention based fusion network for multimodal sentiment analysis. IEEE Trans. Circ. Syst. Video Technol. **PP**, 1 (2022). https://doi.org/10.1109/TCSVT.2022.3218018
4. Zhang, H.: A comprehensive survey on multimodal sentiment analysis: techniques, models, and applications. Adv. Eng. Innovation **12**, 47–52 (2024). https://doi.org/10.54254/2977-3903/12/2024128
5. Lai, S., Hu, X., Xu, H., Ren, Z., Liu, Z.: Multimodal sentiment analysis: a survey. Displays **80**, 102563 (2023). https://doi.org/10.1016/j.displa.2023.102563
6. Fu, H., Lu, H.: A review of multimodal sentiment analysis: modal fusion and representation. In: 2024 International Wireless Communications and Mobile Computing (IWCMC), pp. 49–54 (2024). https://doi.org/10.1109/IWCMC61514.2024.10592484
7. Devlin, J., Chang, M.W., Lee, K., Toutanova, K.: Bert: pre-training of deep bidirectional transformers for language understanding. CoRR **abs/1810.04805** (2018). https://doi.org/10.48550/arXiv.1810.04805
8. Chochlakis, G., Srinivasan, T., Thomason, J., Narayanan, S.: Vault: augmenting the vision-and-language transformer with the propagation of deep language representations arXiv preprint arXiv:2208.09021 6, 17 (2022)

9. Shaik, T., et al.: A review of the trends and challenges in adopting natural language processing methods for education feedback analysis. IEEE Access **10**, 56720–56739 (2022). https://doi.org/10.1109/ACCESS.2022.3177752

10. Sahoo, C., Wankhade, M., Singh, B.K.: Sentiment analysis using deep learning techniques: a comprehensive review. Int. J. Multimedia Inf. Retrieval **12**(41), 1–23 (2023). https://doi.org/10.1007/s13735-023-00308-2

11. Jia, L., Ma, T., Rong, H., Al-Nabhan, N.: Affective region recognition and fusion network for target-level multimodal sentiment classification. IEEE Trans. Emerg. Top. Comput. **12**, 688–699 (2024). https://doi.org/10.1109/TETC.2022.3231746

12. Liu, C., Wang, Y., Yang, J.: A transformer-encoder-based multimodal multi-attention fusion network for sentiment analysis. Appl. Intell. **54**, 1–27 (2024). https://doi.org/10.1007/s10489-024-05623-7

13. Vaswani, A., et al.: Attention is all you need. In: Proceedings of the 31st International Conference on Neural Information Processing Systems, pp. 6000–6010. Curran Associates Inc. (2017). https://doi.org/10.5555/3295222.3295349

14. Lianting Gong, X.H., Yang, J.: An image-text sentiment analysis method using multi-channel multi-modal joint learning. Appl. Artif. Intell. **38**(1), 1–21 (2024). https://doi.org/10.1080/08839514.2024.2371712

15. Guo, X., Tian, S., Yu, L., He, X.: Smartran: smart routing attention network for multimodal sentiment analysis. Appl. Intell. **54**, 12742–12763 (2024). https://doi.org/10.1007/s10489-024-05839-7

16. Bahdanau, D., Cho, K., Bengio, Y.: Neural machine translation by jointly learning to align and translate. ArXiv **abs/1409.0473** (2014). https://doi.org/10.48550/arXiv.1409.0473

17. Niu, T., Zhu, S., Pang, L., Saddik, A.: Sentiment analysis on multi-view social data. In: Tian, Q., Sebe, N., Qi, G.-J., Huet, B., Hong, R., Liu, X. (eds.) MMM 2016. LNCS, vol. 9517, pp. 15–27. Springer, Cham (2016). https://doi.org/10.1007/978-3-319-27674-8_2

18. Xu, N., Mao, W.: Multisentinet: a deep semantic network for multimodal sentiment analysis. In: Proceedings of the 2017 ACM on Conference on Information and Knowledge Management, pp. 2399–2402. Association for Computing Machinery (2017). https://doi.org/10.1145/3132847.3133142

19. Yang, X., Feng, S., Wang, D., Zhang, Y.: Image-text multimodal emotion classification via multi-view attentional network. IEEE Trans. Multimedia **23**, 4014–4026 (2021). https://doi.org/10.1109/TMM.2020.3035277

20. Yang, X., Feng, S., Wang, D., Hong, P., Poria, S.: Multiple contrastive learning for multimodal sentiment analysis. In: ICASSP 2023 - 2023 IEEE International Conference on Acoustics, Speech and Signal Processing (ICASSP), pp. 1–5 (2023). https://doi.org/10.1109/ICASSP49357.2023.10096777

21. Lu, X., Ni, Y., Ding, Z.: Cross-modal sentiment analysis based on clip image-text attention interaction. Int. J. Adv. Comput. Sci. Appl. **15** (2024). https://doi.org/10.14569/IJACSA.2024.0150290

22. Yadav, A., Vishwakarma, D.K.: A deep multi-level attentive network for multimodal sentiment analysis. ACM Trans. Multimedia Comput. Commun. Appl. **19** (2023). https://doi.org/10.1145/3517139

23. Ganesh Kumar, P., Arul Antran, V., Paul, A., Nayyar, A.: A context-sensitive multi-tier deep learning framework for multimodal sentiment analysis. Multimedia Tools Appl. **83**, 54249–54278 (2024). https://doi.org/10.1007/s11042-023-17601-1

Integrating Information Retrieval and LLMs: A Document Retrieval Chatbot in Education Settings

Duy Dang Khoa Nguyen[1,2], Vi Kiet Mach[1,2], Thang Le Dinh[3] [iD],
and Cuong Pham-Nguyen[1,2(✉)] [iD]

[1] Faculty of Information Technology, University of Science, Ho Chi Minh City, Vietnam
pncuong@fit.hcmus.edu.vn
[2] Vietnam National University, Ho Chi Minh City, Vietnam
[3] School of Business, Université du Québec à Trois-Rivières, Trois-Rivières, Québec, Canada

Abstract. UniBot is a domain-specific chatbot designed to streamline access to internal documents within a university context by integrating information retrieval (IR) techniques with large language models (LLMs). The system addresses a critical need in educational institutions: enabling students and staff to quickly access information from internal documents (such as academic policies, announcements, schedules) through natural language queries. The proposed approach employs a Retrieval-Augmented Generation (RAG) architecture, which combines a vector-based document retrieval pipeline with a LLM to generate accurate, context-informed answers. Theoretical foundations related to chatbots, LLMs, and RAG are presented, followed by a detailed overview of UniBot's design and implementation. Key components include a document indexing module, a vector database for semantic search, and a user-facing conversational interface. In a pilot deployment, UniBot was evaluated both performance metrics (via the RAGAs framework, including context recall and ROUGE) and user studies involving students and academic staff. The results indicate that integrating IR with LLMs substantially improves the relevance and factuality of chatbot responses, reducing instances of hallucinations. Users reported a high satisfaction level (average 4.12/5) with the system's ability to answer in-domain questions, though some challenges remain with handling out-of-scope queries and response latency. Overall, our findings demonstrate that IR-augmented LLM chatbots can effectively automate information access in educational settings, alleviating administrative workloads and improving user experience.

Keywords: Information Retrieval · Large Language Model · Chatbot · Retrieval-Augmented Generation · Education

1 Introduction

Recent advances in large language models (LLMs) have led to a new generation of intelligent chatbots capable of natural language understanding and generation [1]. These LLM-based chatbots leverage transformer architectures with hundreds of billions of

N. Thai-Nghe et al. (Eds.): ISDS 2025, CCIS 2714, pp. 461–475, 2026.
https://doi.org/10.1007/978-981-95-3358-9_33

parameters to engage in dialogue and answer user questions with remarkable fluency [2]. Despite their impressive capabilities, current LLMs face significant challenges when applied in real-world domains [2]: First, their knowledge is static and can quickly become outdated if not frequently retrained or updated. Second, because they are trained on broad general data, they often struggle with specialized domain knowledge or enterprise-specific information that lies outside their training corpus. Retraining them on specialized data is costly, requiring substantial computational resources and expertise. Finally, even the most advanced models are prone to hallucinations – confidently generating incorrect or fabricated information, especially when asked questions beyond their knowledge scope [3]. These limitations have driven research into techniques to ground LLMs in accurate, relevant knowledge.

Retrieval-Augmented Generation (RAG) has emerged as a promising solution to the challenges outlined above. First, RAG integrates information retrieval with text generation by connecting a LLM to an external knowledge base [1]. This architecture enables the model to retrieve relevant documents or facts at query time and incorporate them directly into its generated responses, thereby improving contextual accuracy and factual grounding [3]. By providing domain-specific and up-to-date contextual information, RAG substantially reduces hallucinations and improves factual accuracy in open-domain question answering [2]. This architecture eliminates the need for frequent retraining of the underlying language model, as the external knowledge base can be updated independently. Organizations have successfully adopted RAG-based systems to develop virtual agents capable of delivering accurate internal information to end users in real time.

In the education sector, artificial intelligence (AI) has seen widespread adoption in recent years to enhance teaching and learning. Many academic institutions have implemented AI-powered chatbots as virtual teaching assistants or tutoring systems to support students with coursework and academic guidance [4–7]. These systems typically facilitate interactions between learners and instructors, such as answering questions related to course content or providing study tips [5]. However, another critical area remains largely underserved: interactions between students and administrative staff [6]. Students frequently seek information on academic regulations, program requirements, class schedules, examinations, student handbooks, procedures, or administrative forms. Traditionally, such inquiries are handled manually via email or in person, leading to repetitive Q&A and increased administrative workload. A survey of university students highlighted a clear demand for chatbot features that go beyond learning support—specifically, tools designed to facilitate access to institutional policies and services [6]. Despite this, most AI applications in education have not focused on the automation of internal information services, leaving a significant gap in the current use of chatbot technologies.

To address this gap, UniBot was developed, an automated document retrieval chatbot tailored specifically for the university context. UniBot combines advanced information retrieval techniques with RAG to enable users to query a repository of internal documents, and receive accurate, context-aware responses. The target users are primarily students and academic staff in the Faculty of Information Technology at the University of Science, Viet Nam National University of Ho Chi Minh City (HCMUS). By focusing on these two groups, UniBot aims to reduce the workload of staff responsible for handling thousands of repetitive queries each semester, while empowering students with instant, self-service access to institutional information. This solution automates a substantial portion of administrative queries, improving response speed, consistency, and overall user satisfaction.

The contributions of this work are as follows: i) A domain-specific RAG chatbot architecture that integrates a vector similarity search engine with a LLM to support natural language queries over internal university documents; ii) A practical implementation that supports continuous updates to the knowledge base, leveraging an open-source vector database and an automated document ingestion pipeline; and iii) A comprehensive evaluation of the system using both performance metrics (retrieval accuracy and answer quality) and a user study involving students and academic staff to validate effectiveness in an educational setting.

The rest of this paper is organized as a formal research exposition. Section 2 outlines the theoretical background, covering chatbots, LLMs, and the RAG paradigm, along with its applicability to the education sector. Section 3 details the architecture and implementation of the Unibot system, describing the integration of information retrieval and LLM components. Section 4 presents the experimental evaluation, including results from the pilot deployment and quantitative performance metrics. Section 5 situates the proposed approach within existing literature and discusses limitations and future directions. Finally, Sect. 6 offers concluding remarks.

2 Theoretical Background

2.1 Chatbots and Large Language Models

The emergence of LLMs in recent years has marked a significant advancement in chatbot technology. LLMs, typically based on transformer architectures, consist of hundreds of millions to hundreds of billions of parameters and are trained on vast corpora of text data, enabling them to understand and generate humanlike language [2]. Prominent examples include OpenAI's GPT series, Google's PaLM and Gemini, Meta's LLaMA, and IBM's Granite. These models leverage the Transformer architecture [8], which features multi-head self-attention and deep stacks of encoder-decoder layers. This design enables the models to capture complex linguistic patterns and long-range dependencies in the text. The Transformer's ability to attend to different parts of the input sequence with varying weights enables LLMs to understand context, disambiguate meaning, and generate coherent, contextually appropriate responses.

Leveraging pre-training on extensive data, LLMs exhibit capabilities such as in-context learning, instruction following, and step-by-step reasoning [2]. This enables an LLM to perform tasks without explicit retraining by being prompted with a few examples or instructions, and to generate multi-step solutions or explanations for complex queries. These properties make LLMs an attractive engine for chatbot applications, particularly in handling open-ended inputs and producing natural, context-aware responses for office automation or domain-specific assistance.

However, relying solely on an LLM as a chatbot presents several limitations. The model's responses are constrained by the static training data, which may lack up-to-date or institution-specific information. As this data becomes outdated, the LLM's knowledge may degrade [2]. Additionally, LLMs often struggle with niches or proprietary content not represented in their training set. While fine-tuning on custom data can address this, it is resource-intensive and requires retraining for each update. A more scalable alternative that has gained popularity is integrating an external knowledge retrieval component, an approach known as RAG [3].

2.2 Retrieval-Augmented Generation (RAG)

RAG is an architecture that enhances LLMs with access to external knowledge sources at inference time, mitigating limitations such as hallucinations and outdated knowledge [1]. In RAG, a query first passes through a retrieval component that identifies relevant documents from a knowledge base (KB), and the LLM then generates a response grounded in the retrieved context [3].

A typical RAG pipeline consists of three main stages [3]: (1) *Indexing* - internal documents (e.g., PDFs, text files, webpages) are collected, cleaned (e.g., text extraction, typo correction, noise removal), and segmented into smaller chunks suitable for the LLM's context window. Each chunk is embedded into a high-dimensional semantic vector using a domain-appropriate embedding model and stored in a vector database along with metadata linking it to the source document. This enables efficient similarity-based retrieval; (2) *Retrieval* - a query is embedded and matched to top-K similar chunks via similarity search (e.g., cosine similarity). Enhancements such as re-ranking, metadata filtering, query reformulation, hybrid search, or query routing may improve retrieval quality; (3) *Generation* – the LLM produces an answer based on both the user query and the retrieved context. Retrieved chunks are inserted into a prompt template alongside the query, allowing the model to generate responses grounded in external knowledge. Empirical studies have shown that this retrieval-augmented approach significantly reduces hallucinations and improves answer accuracy in knowledge-intensive tasks, as the model draws on relevant evidence rather than relying solely on parametric memory [1].

Advanced RAG Techniques: Recent extensions include iterative retrieval-generation loops, such as Self-RAG [9], in which the LLM refines its answer using follow-up queries and self-criticism. Another extension is structured retrieval, exemplified by GraphRAG [10], incorporating knowledge graphs for richer, multi-hop reasoning. The evaluation framework, like RAGAs [11], assesses retrieval and generation quality using LLM-based metrics, and is employed in the system evaluation presented in Sect. 4.

2.3 Chatbots for Internal Document Retrieval in Education

The application of RAG-based chatbots in educational settings presents a compelling and underexplored opportunity. Universities produce a vast array of internal documents—academic calendars, course catalogues, policy manuals, enrollment guides, event announcements, and more—that students, faculty, and staff frequently need to consult [12]. Traditional methods of accessing this information (e.g., static websites, PDF manuals, or direct staff inquiries) are often time-consuming and inefficient. A chatbot capable of understanding natural language queries and retrieving accurate information from institutional documents can significantly improve information accessibility on campus.

Several features make this use case both promising and challenging. Educational institutions offer a well-defined domain and a stable user base, which are advantageous for developing specialized assistants [4, 13]. There is also a strong and well-documented demand for automation: surveys consistently show students seek faster, self-service access to institutional information [7]. Automating responses to frequent queries can reduce the burden on administrative staff who otherwise spend considerable time answering repetitive questions each term [14]. By targeting the student–staff interactions, a chatbot can thus improve operational efficiency and service quality [7].

However, building a domain-specific chatbot for this context introduces unique challenges. The system must support the local language and understand academic terminology and document structures [15]. The knowledge base is dynamic, with frequent updates to policies, procedures, and announcements, requiring a solution that remains current. Unlike open domain Q&A, the topic scope is narrower but often demands high factual precision, with answers buried in complex, lengthy documents [16]. Furthermore, credibility is paramount, responses must be grounded in authoritative internal sources rather than public web content.

The RAG paradigm is well suited to address these requirements. By connecting the chatbot to an internal, continuously updated document repository, the system remains aligned with the latest institutional knowledge. The retrieval pipeline can be tuned to prioritize official sources, enhancing reliability. When a query falls outside the system's coverage, fallback strategies—such as graceful refusals or redirecting to general Q&A modules—can maintain user trust [3]. To our knowledge, there are a few studies or commercial systems that focus specifically on internal document chatbots for educational institutions. As a result, it was necessary to design new workflows—such as document ingestion and updating mechanisms—in the absence of established best practices. The following section presents the architecture and implementation of UniBot, the proposed solution developed to address this gap.

3 UniBot System Architecture and Implementation

3.1 System Overview and Requirements

UniBot is a web-based chatbot with two primary functions: (1) *Internal information Q&A* – allowing students to ask natural language questions and receive answers based on university documents; and (2) *Knowledge base management* – enabling authorized staff to upload and manage documents to keep the knowledge base up to date. These core features were prioritized based on requirements analysis, while other ideas (e.g., calendar integration, email notifications) were deferred maintaining focus. There are three user roles: students, academic staff, and administrator. *Students* can interact with the chatbot to ask questions. *Academic staff* can also upload and review documents. *Administrators* have full privileges, including user management and organizing the knowledge base. This role-based access ensures proper control, only trusted users can modify system content.

UniBot focuses on internal textual documents related to student and academic affairs, including news, announcements, schedules, procedures, student guidelines and FAQs. The initial corpus—gathered from online portals and publications—contained several hundred documents, mostly in Vietnamese. Optical Character Recognition (OCR) was used to extract embedded images in PDFs, but other media were excluded to keep the scope centered on text, given its abundance and mature retrieval methods.

3.2 Design and Implementation

At a high level, UniBot follows the RAG pipeline described in Sect. 2.2, augmented with additional infrastructure to support continuous updates and user interaction. Figure 1 illustrates the overall system architecture. The major components include:

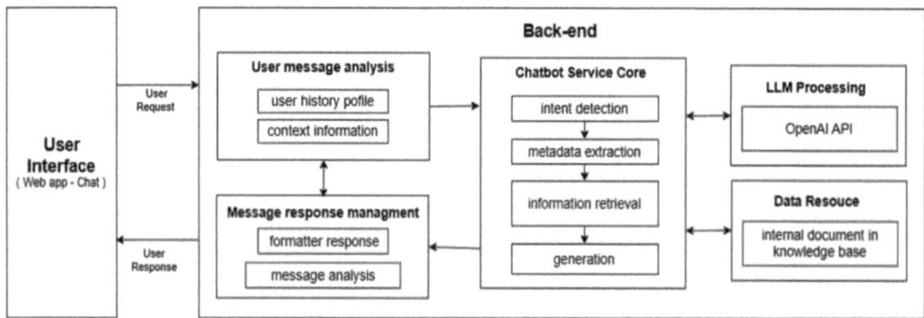

Fig. 1. System architecture.

- **User interface**: A responsive, web-based single-page application (SPA) enables users to interact with UniBot via natural language queries. The interface supports rich-text responses and includes features for file uploads, document review, and administrative dashboards. Communication with the back end occurs through RESTful APIs [22].
- **User Message Analysis:** Upon receiving input, the system first constructs a contextual profile by referencing prior conversation history. This step ensures semantic continuity in multi-turn dialogues and improves intent disambiguation.
- **Chatbot Service Core:** consists of 4 main steps—Intent Detection, Metadata Extraction, Information Retrieval, and Response Generation.

  - *Intent Detection*: To narrow the search scope, the system first identifies the topic (i.e., intent) of the user query. A fine-tuned version of PhoBERT [18], a RoBERTa-based model pretrained on Vietnamese, was employed for this task. It outperformed a multilingual BERT baseline in semantic classification (F1 score: 0.854 vs. 0.847) and was chosen over external APIs to ensure data privacy and minimize latency.
  - *Metadata Extraction:* After identifying the query topic, the system performs metadata extraction to further refine the search process. At this stage, an LLM is employed to extract metadata from the user's query based on the schema associated with the identified topic. The extracted metadata is then converted into filter expressions, which act as additional constraints for the retrieval step. The prompt used for this task follows a structured format.
  - *Information Retrieval:* The query is embedded into a vector and searched within the selected topic and core ("hinge") topics (e.g., Student Handbook). A hybrid retrieval strategy runs two processes in parallel: one using metadata filters and

another using the raw query embedding. Results are re-ranked using the Weight-edReranker algorithm, which recalculates similarity scores based on weighted factors. The top-k relevant chunks are then selected as input for a response generation.

- *Response Generation:* The system selects a prompt scenario based on the conversation context (e.g., new query, ongoing dialogue, or low context). The prompt incorporates retrieved documents, user profile data (e.g., major and year), predefined style rules, prior dialogue history, and topic-related metadata. This comprehensive prompt enables the model to generate coherent and contextually appropriate responses.

Other components are also utilized to ensure autonomy, data storage and security compliance with all supported user roles. These components are as follows:

- **Vector database (Milvus)**: Milvus is used to store and index 768-dimensional chunk embeddings [23]. Each chunk is stored with metadata (e.g., document ID, position, summary), and top-K results are retrieved using cosine similarity.
- **Document store (MongoDB)**: MongoDB stores raw text and document metadata (e.g., titles, upload timestamps, authors) [20]. Chunks are linked to their corresponding vector entries, allowing efficient retrieval and updates without full re-indexing. At the same time, it is also used to store user profiles and analytical data (i.e., conversation history, ratings).
- **Ingestion pipeline (Airflow)**: An asynchronous Airflow pipeline automates the document ingestion process. Uploaded documents are initially flagged as pending in MongoDB. The pipeline then extracts, cleans, and chunks the text, generates embeddings, updates Milvus, and marks the document as available. This non-blocking design ensures responsiveness and timely document integration.
- **System design**: UniBot emphasizes modularity and future-proofing. The LLM is abstracted behind an API layer, allowing for easy switching between models (e.g., GPT-3.5, GPT-4, or future local 13B models). This modularity supports experimentation and scalable upgrades with minimal changes to the core pipeline.

4　Experimental Evaluation

UniBot was evaluated through two complementary approaches: i) Model performance assessment, and ii) User evaluation via a pilot deployment with real users (students and staff), who provided feedback. All experiments were conducted with the system's primary language set to Vietnamese, as the knowledge base documents and user queries were largely in Vietnamese.

4.1　Evaluation Methodology

Due to the absence of benchmark datasets for internal university documents, a custom evaluation set was created to evaluate retrieval and answer generation performance. Drawing on LLM-assisted Q&A generation methods, the test questions were designed to reflect realistic user queries closely.

Test Data Generation. Twenty percent of the collected documents, primarily news, student manuals and announcements, were set aside as a test set, excluded from both training and prompt tuning. For each document, an LLM generated ~ 5 plausible user questions. For instance, a seminar announcement could be: *"When is the seminar scheduled?"* or *"Who is the target audience for the seminar?"*. These questions were manually reviewed for quality. Each question was then processed through UniBot to capture the retrieved context and the corresponding generated answer. Besides, a reference answer was produced by prompting an LLM with the original document and question. This process resulted in ~ 200 Q&A pairs with known relevant documents. While not gold standard, these answers were found to be high quality and closely aligned with source content.

Metrics. The RAGAs evaluation framework [11] was employed to compute a variety of metrics on our test set. Notably, both traditional IR and LLM-based metrics were evaluated as recommended by RAGAs. The metrics included: - *Context Recall*: This metric measures whether the retrieved chunks included the correct one, using text similarity to handle paraphrasing; - *Context Precision*: This metric assesses how many retrieved chunks are truly relevant. Low precision indicates inclusion of unrelated content; - *ROUGE-1 and ROUGE-L*: These are classic metrics comparing the overlap of n-grams between the system's answer and the reference answer. ROUGE-1 (unigram overlap) and ROUGE-L (Longest Common Subsequence) were computed to gauge answer quality in terms of content [21]; - *LLM-based scoring*: RAGAs also supports using an LLM to directly judge the quality of answers (for example, by scoring correctness or coherence). One such metric involved prompting GPT-4 to score the system-generated answer against the reference answer on a predefined scale. Prior work indicates these LLM-based scores correlate well with human judgments for RAG systems [11].

Two answer generation variants were evaluated: one using the stronger GPT-4, and another using the GPT-3.5 (GPT-4o mini). Retrieval depth was varied (K = 4 and K = 8) to examine the impact of additional context on answer quality.

4.2 Evaluation Results

Table 1 presents the retrieval and generation performance across different configurations (LLM model and number of retrieved chunks). Results are reported using standard evaluation metrics: Context Precision, Context Recall, ROUGE-1, ROUGE-L.

Table 1. LLM retrieval and generation performance.

Models	Context Precision	Context Recall	ROUGE-1	GOUGE-L
GPT-4o/K = 4	0.6227	0.736	**0.7277**	**0.5481**
GPT-4o/K = 8	**0.6343**	**0.808**	0.7157	0.5373
GPT-4o-mini/K = 4	0.6182	0.736	0.7045	0.5278
GPT-4o-mini/K = 8	0.6277	**0.808**	0.7111	0.5350

First, increasing the number of retrieved chunks from 4 to 8 improved retrieval performance: Context Recall rose from ~ 0.736 to ~ 0.808, and Context Precision also improved for both LLMs. This suggests that retrieving more chunks helps capture relevant documents and doesn't necessarily increase irrelevant ones likely because some questions require multiple sources for a complete answer. However, this came with a trade-off: answer quality, as reflected in ROUGE scores, slightly declined. For instance, ROUGE-1 dropped from ~ 0.727 to ~ 0.716 for GPT-4 when moving from K = 4 to K = 8, with ROUGE-L showing a similar decrease. This supports the idea that more context can introduce noise, making it harder for the LLM to focus and affecting answer relevance. Despite this trade-off, ROUGE scores remained high. As a result, K = 8 was selected for deployment to maximize answer coverage, with an accepted minor reduction in conciseness.

Between models, GPT-4 consistently outperformed GPT-3.5, particularly in ROUGE scores at K = 4, suggesting better synthesis of retrieved content. At K = 8, the gap narrowed. Notably, Context Recall was identical between models for each K value because retrieval is model-agnostic in our pipeline (the same chunks were used, controlled by a fixed random seed). Precision differences were minimal. This indicates GPT-4's main advantage lies in generating more accurate and well-phrased answers, not better retrieval. Given the modest performance gap and cost differences, GPT-3.5 may be a cost-effective alternative with only a small quality trade-off.

UniBot's performance was also benchmarked against several existing RAG-based Q&A systems reported in the literature for perspective. Table 2 provides a reference comparison using reported values (values taken from their respective papers, not tested on our dataset). Systems include: UniMS-RAG (multi-source RAG) [23], QA-RAG (baseline) [24], and Self-RAG (13B) (a self-refining RAG with a 13B-parameter LLM) [9].

Table 2. Performance comparison between different approaches.

Models	Context Precision	Context Recall	ROUGE-1	GOUGE-L
UniMS-RAG [13]	-	**0.828**	-	0.328
QA-RAG [14]	**0.717**	0.328	-	-
Self-RAG13B [9]	0.703	0.713	-	-
UniBot (our approach)	0.661	0.808	**0.727**	**0.544**

Although direct comparisons are limited due to different domains and datasets, UniBot's performance is competitive. Our system achieved Context Precision ~ 0.661, Recall ~ 0.808, and ROUGE-1 ~ 0.727. For example, Self-RAG13B reported Precision ~ 0.703 and Recall ~ 0.713. UniBot's higher recall likely reflects our narrower domain, which simplifies the document retrieval. The slightly lower precision suggests more irrelevant content was occasionally included. Overall, these results suggest UniBot's integration of retrieval and generation is comparable to state-of-the-art systems in its category.

This supports the soundness of our design choices, including the embedding model and selected retrieval depth (K).

Table 3. LLM-based scoring.

	Metrics	Results
LLM-based evaluation (LLM as-a-judge)	Context Recall	0.8019
	Response Relevance	0.7415
	Faithfulness	0.8631

As shown in Table 3, the evaluation metrics using LLMs demonstrate strong alignment with traditional methods. The Context Recall score, when recalculated using an LLM, remains consistent with conventional calculations, further validating the system's retrieval performance. The Relevance score is moderately strong, suggesting that the system generally addresses the core intent of user queries. The high Faithfulness score indicates that hallucinations are rare and that the generated answers are well grounded in the retrieved context. Notably, these LLM-based scores closely match human evaluations, supporting previous research that LLMs can serve as effective proxies for human judgment. This alignment also provided valuable guidance for the system refinement in the absence of manual testers.

4.3 User Study and Feedback

To evaluate UniBot in a realistic setting, a two-week pilot was conducted with 34 participants (30 students, 4 academic staff). The chatbot was hosted on a staging server, and users were encouraged to use it for faculty-related queries and document searches. With user consent, interactions were logged, and participants were asked to complete a feedback survey afterward.

Usage Statistics. During the pilot, users initiated 197 conversations comprising 523 questions. Most participants (approximately 80%) engaged beyond the initial trial interaction. Staff members also utilized the system to upload documents—12 new items were added and successfully indexed, including updated announcements and policies.

Answer Quality. Users were asked to rate the chatbot's responses on a 1 to 5-star scale when conversing directly with the system. As shown in Fig. 2, most ratings were positive, although around 20% received low scores (1 or 2 stars). These low ratings were typically associated with out-of-scope questions (e.g., *"what is the Pythagorean theorem?"* or *"who is the current president of the USA?"*), which UniBot was not designed to handle. Such questions often resulted in irrelevant context retrieval, confusing answers and introducing noise to the LLMs. For in-scope queries, users generally found the responses accurate and complete.

User Satisfaction. Survey results were positive. The average rating for overall experience and usefulness was 4.12/5, with most users selecting "High" or "Very High" (see

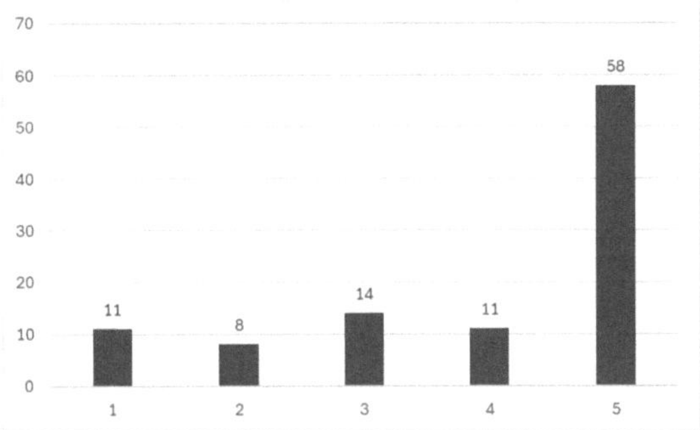

Fig. 2. Users rated the responses on 1 to 5-star scale.

Fig. 3). Many noted it was easier than navigating the website or emailing staff. One student commented, *"It's like having a little assistant – I got answers faster than emailing the office."*

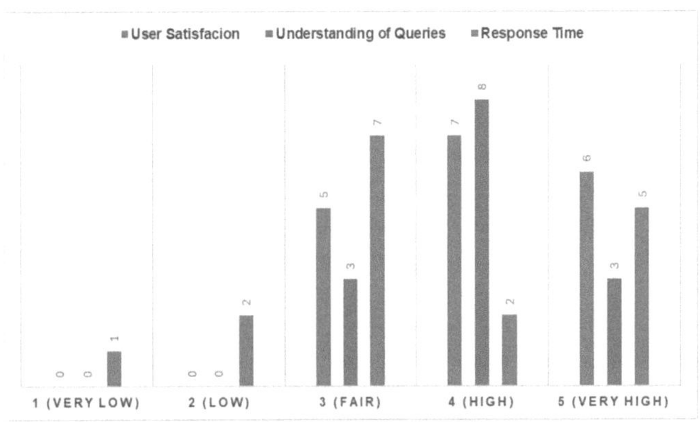

Fig. 3. User feedback from the survey.

Understanding of Queries. Users gave high ratings (~4/5) for query understanding, indicating strong Vietnamese language processing. Even colloquial questions were interpreted correctly. The chatbot's tone was also praised for being polite and professional, thanks to prompt design defining it as an "internal assistant" (Fig. 3). However, the system currently lacks a mechanism to handle time-sensitive queries involving temporal expressions. For example, terms like *"the latest"* in *"the latest list of merit-based scholarships"* or *"this semester"* in *"what is the academic advising schedule for this semester."*

Response Time. Some users noted slow responses, around 5–8 s on average (with GPT-3.5), occasionally up to 15 s (GPT-4). The main cause of this latency is the communication between the system and both the ChatGPT API and the cloud-hosted vector database. While acceptable for complex queries, a few users expected faster search engine-like responses. Speed was a common concern in open comments.

In summary, the pilot confirmed UniBot's value for its intended use. Users appreciated its convenience, and staff saw the potential to reduce routine queries. Key areas for improvement include better handling of out-of-scope questions, reducing latency, and supporting more robust multi-turn interactions. These insights inform our future development directions.

5 Discussion and Related Work

Comparison with Other Approaches. UniBot combines dynamic retrieval with generative answering, allowing it to handle free-form queries and synthesize responses from multiple sources, offering greater flexibility and depth. In the educational space, prior chatbots have primarily focused on academic advising or subject tutoring [6]. UniBot distinguishes itself by targeting administrative and institutional information queries, a less explored area. To our knowledge, few published systems focus on this use case. One related example—a campus admission bot—relied on canned responses rather than a true RAG pipeline [16].

From a RAG perspective, our system is inspired by foundational work such as Patrick et al.'s original RAG model [1], with practical adaptations for our setting. Milvus was used for scalable vector search, PhoBERT for Vietnamese embeddings, and Airflow for orchestrating document updates [22]. As shown in Table 2 (Sect. 4.2), our system achieves performance comparable to research prototypes like Self-RAG [9]. That system's concept of iterative self-refinement could enhance UniBot's handling of complex or multi-hop queries in future iterations. Insights were also drawn from UniMS-RAG [23], which supports multi-source retrieval and response personalization. While UniBot doesn't tailor answers per user, it naturally retrieves from multiple documents. Ideas from GraphRAG [10]—which integrates structured data via knowledge graphs—could also apply, as university documents often contain tabular or structured information that might benefit from structured retrieval. Though UniBot was built independently, emerging frameworks like LlamaIndex and LangChain now offer robust tooling for RAG systems. These could streamline future enhancements or migrations.

Limitations. While UniBot shows promise, several limitations remain. Concerning *Knowledge scope*, UniBot relies solely on internal documents. Questions outside this domain (e.g., general knowledge or academic trivia) often lead to poor responses - about 20% of low-rated answers in our pilot stemmed from such queries. Currently, there's no fallback mechanism, but in production, the system could either defer politely or integrate with a general Q&A service. Relating to *Real-time updates,* although an update pipeline was built, newly uploaded documents aren't instantly searchable. Processing delays (e.g., a minute or two for large PDFs) may affect real-time responsiveness. Additionally, updates require re-uploading documents to ensure old versions are correctly removed from the index. With respect to *Response time,* API calls to large LLMs introduce latency, with answers taking 5–15 s. While acceptable for complex queries, this may frustrate some users. Planned enhancements include caching frequent responses, utilizing lightweight local models to improve speed, and streaming answers to enhance the user experience. Regarding *Language support,* the system handles Vietnamese well and tolerates English, but the Vietnamese-specific PhoBERT embedding may miss nuances in bilingual queries. This is a concern in academic contexts where English terms are common. Expanding to multilingual models or using language-specific indices may be necessary. In connection with *Hallucination and attribution,* while RAG reduces hallucinations, they still occur - especially when irrelevant context is retrieved. In a few cases, the LLM generated plausible-sounding but unsupported claims. Our system cites sources when possible, but further work is needed to ensure answer grounding and traceability.

Future Directions. Building on the pilot's success and user feedback, several areas were identified to enhance UniBot. Regarding out-of-scope detection, to handle queries beyond the university domain, the system is planned to integrate a query classifier. A prototype using PhoBERT achieved 86% accuracy in distinguishing internally from general topics. This component would allow UniBot to flag irrelevant queries and respond appropriately (e.g., *"I'm sorry, I can only assist with university-related questions"*), or optionally offer fallback options such as a web search with disclaimers. Regarding workflow integration, extending UniBot from Q&A to task execution. For instance, the bot could provide the form or assist in submitting it. More advanced scenarios (e.g., course registration or withdrawal) would require integration with university systems and secure user authentication. Relating to expanding user base, while currently aimed at students and academic staff, UniBot could support faculty, administrators, or external partners. Each group has unique needs, and our system can scale to multiple indices or a unified dataset.

6   Conclusion

This study presented UniBot, a domain-specific chatbot that integrates IR techniques with LLMs to automate access to internal university documents. Through a RAG architecture, UniBot addresses key limitations of standalone LLMs - specially outdated knowledge and hallucinations - by grounding responses in an up-to-date document corpus. This design enables students and staff to obtain timely, accurate answers to institutional queries without manual document search, or administrative delays.

Our evaluation demonstrates UniBot's effectiveness both quantitatively and qualitatively. The retrieval component achieved high recall (>80%), while the answer generation quality was comparable to state-of-the-art RAG systems. User feedback from the pilot deployment further confirmed the system's practical utility, with participants reporting high satisfaction and convenience. The study also highlighted critical areas for improvement, including handling out-of-scope queries and ensuring consistent responsiveness.

In summary, UniBot demonstrates the potential of combining IR and LLMs to build practical, domain-specific conversational agents. This approach improves accessibility to institutional knowledge while reducing administrative workload. UniBot can serve as a strong reference for similar applications in other domains, such as enterprise knowledge bases, government portals, and healthcare systems. Continued research into hybrid architectures that unify reliable retrieval with fluent generation is strongly encouraged, as advancing such systems moves us closer to AI assistants that are not only articulate but also factually grounded and trustworthy.

References

1. Lewis, P., et al.: Retrieval-augmented generation for knowledge-intensive NLP tasks. In: Proceedings of the 34th International Conference on Neural Information Processing Systems (NIPS 2020), pp. 9459–9474. Curran Associates Inc., Vancouver, BC, Canada (2020)
2. Zhao, W.X., et al.: A survey of large language models. arXiv: https://doi.org/10.48550/arXiv.2303.18223 (2023)
3. Fan, W., et al.: A survey on RAG meeting LLMs: towards retrieval-augmented large language models. In Proceedings of the 30th ACM SIGKDD Conference on Knowledge Discovery and Data Mining (KDD 2024), pp. 6491–6501. Association for Computing Machinery, New York (2024)
4. Dakshit, S.: Faculty perspectives on the potential of RAG in computer science higher education. In Proceedings of the 25th Annual Conference on Information Technology Education (SIGITE 2024), pp. 19–24 (2024)
5. Liu, C., Hoang, L., Stolman, A., Wu, B.: HiTA: A RAG-based educational platform that centers educators in the instructional loop. In: Olney, A.M., Chounta, IA., Liu, Z., Santos, O.C., Bittencourt, I.I. (eds) Artificial Intelligence in Education. AIED 2024. LNCS, vol. 14830, pp. 405–412. Springer, Cham (2024)
6. Thüs, D., Malone, S., Brünken, R.: Exploring generative AI in higher education: a RAG system to enhance student engagement with scientific literature. Front. Psychol. **15** (2024)
7. Hien, H.T., Cuong, P.-N., Nam, L.N.H., Nhung, H.L.T.K., Thang, L.D.: Intelligent assistants in higher-education environments: the FIT-EBot, a chatbot for administrative and learning support. In Proceedings of the 9th International Symposium on Information and Communication Technology (SoICT 2018), pp. 69–76 (2018)
8. Ashish, V., et al.: Attention is all you need. In Proceedings of the 31st International Conference on Neural Information Processing Systems (NIPS2017), pp. 6000–6010. Curran Associates Inc. (2017)
9. Asai, A., Wu, Z., Wang, Y., Sil, A., Hajishirzi, H.: Self-RAG: learning to retrieve, generate, and critique through self-reflection. arXiv:2310.11511 (2023)
10. Edge, D., et al.: From local to global: a graph RAG approach to query-focused summarization. arXiv:2404.16130 (2024)
11. Es, S., James, J., Anke, L.-E., Schockaert, S.: RAGAs: automated evaluation of retrieval augmented generation. In: Proceedings of the 18th Conference of the European Chapter of the Association for Computational Linguistics System Demonstrations, pp. 150–158. Association for Computational Linguistics (2024)
12. Phiri, M.J., Tough, A.G.: Managing university records in the world of governance. Rec. Manag. J. **28**(1), 47–61 (2018)
13. Telang, P.R., Kalia, A.K., Vukovic, M., Pandita, R., Singh, M.P.: A conceptual framework for engineering chatbots. IEEE Internet Comput. **22**(6), 54–59 (2018)
14. Johnsrud, L.K.: Measuring the quality of faculty and administrative worklife: implications for college and university campuses. Res. High. Educ. **43**(3), 379–395 (2002)
15. Shalaby, W., Arantes, A., GonzalezDiaz, T., Gupta, C.: Building chatbots from large scale domain-specific knowledge bases: challenges and opportunities. In: 2020 IEEE International Conference on Prognostics and Health Management (ICPHM), pp.1–8 (2020)
16. Mikael, K., Cemil, Ö., Rashid, T.A., Nariman, G.S.: A Hybrid chatbot model for enhancing administrative support in education: comparative analysis, integration, and optimization. IEEE Access **13**, 50741–50760 (2025)
17. Masse, M.: REST API Design Rulebook. 1st edn. O'Reilly Media, Inc. (2011)
18. Nguyen, D.Q., Nguyen, A.T.: PhoBERT: pre-trained language models for Vietnamese. In: Findings of the Association for Computational Linguistics: EMNLP 2020, pp.1037–1042. Association for Computational Linguistics (2020)

19. Wang, J., et al.: Milvus: a purpose-built vector data management system. In Proceedings of the 2021 International Conference on Management of Data (SIGMOD 2021), pp.2614–2627. Association for Computing Machinery (2021)
20. Bradshaw, S., Brazil, E., Chodorow, K.: MongoDB: The Definitive Guide: Powerful and Scalable Data Storage. 3rd edn. O'Reilly Media Inc. (2019)
21. Lin, C. Y.: ROUGE: a package for automatic evaluation of summaries. In: Text Summarization Branches Out, pp.74–81. Association for Computational Linguistics (2004)
22. Singh, P.: Learn PySpark: Build Python-based Machine Learning and Deep Learning Models. 1st edn. Apress. (2019)
23. Wang, H., et al.: UniMS-RAG: a unified multi-source retrieval-augmented generation for personalized dialogue systems. arXiv preprint arXiv:2401.13256 (2024)
24. Kim, J., Hur, M., Min, M.: From RAG to QA-RAG: integrating generative AI for pharmaceutical regulatory compliance process. In: Proceedings of the 40th ACM/SIGAPP Symposium on Applied Computing, pp.1293–1295. Association for Computing Machinery (2025)

Graph–Beam Retriever: A Lightweight Framework for Evidence Selection in Vietnamese Fact Verification

Tieu Phung Mai Suong[1,2], Nguyen Tran Thanh Nha[1], Bay Vo[2], Dinh Minh Hoa[1(✉)], and Thien Khai Tran[3]

[1] Faculty of Information Technology, Ho Chi Minh City University of Foreign Languages and Information Technology, Ho Chi Minh City, Vietnam
hoadm@huflit.edu.vn
[2] Faculty of Information Technology, HUTECH University, Ho Chi Minh City, Vietnam
[3] Faculty of Information Technology, Ho Chi Minh City University of Industry and Trade, Ho Chi Minh City, Vietnam

Abstract. We propose Graph–Beam Retriever (GBR), a lightweight, training-free framework for extracting concise and semantically faithful evidence in fact verification tasks. Motivated by the challenges of applying large language models (LLMs) in low-resource languages such as Vietnamese, GBR constructs a linguistically informed text graph from raw news context and performs beam search traversal from the claim node to candidate evidence sentences. These paths are subsequently refined using semantic similarity and logical entailment filters. The resulting evidence is strictly bounded in length, ensuring compatibility with LLM token limits while preserving factual relevance. We evaluate GBR on the ViFactCheck benchmark and compare it against strong sparse, dense, and neural retrievers. Despite using less than 40% of the original context, GBR achieves comparable retrieval quality to fine-tuned dense methods and improves zero-shot verification accuracy of GPT-4 by +2.2% over using the full context. Ablation studies highlight the contribution of each module, and case analyses demonstrate GBR's ability to preserve critical information with minimal overhead. Our approach offers a practical and interpretable alternative to dense retrievers and generative methods, especially in settings where training data or computational resources are limited.

Keywords: Fact Verification · Evidence Retrieval · Graph-Based Methods · Low-Resource Language · Large Language Models

1 Introduction

Misinformation circulates through Vietnamese social platforms at a pace and scale that challenge manual fact verification, as recent sociological studies on the country's online discourse make clear. Until late 2024, the research community lacked a large-scale, sentence-level benchmark to study automated verification in this language. ViFactCheck

N. Thai-Nghe et al. (Eds.): ISDS 2025, CCIS 2714, pp. 476–488, 2026.
https://doi.org/10.1007/978-981-95-3358-9_34

fills that gap with 7,232 claim–evidence pairs drawn from twelve news domains, annotated with Fleiss $\mathcal{K} = 0.83$ to ensure reliability [1]. Within this corpus, the plain "context" supplied for each claim averages roughly 1,900 tokens, whereas the human-curated evidence spans only about 160 tokens—barely eight percent of the input—highlighting the potential rewards of selective retrieval.

Large-language models (LLMs) such as GPT-4.1 now accept million-token prompts, yet their commercial use remains price-sensitive: OpenAI quotes USD 2 per million input tokens and USD 8 per million output tokens for the flagship model, even after a 26% cost reduction announced in April 2025. Shrinking the prompt by an order of magnitude therefore translates directly into lower latency and tangible savings for public-facing fact-checking services.

Retrieval quality is the first bottleneck. BM25, a perennial bag-of-words baseline, is attractive because it is training-free yet struggles with Vietnamese synonymy and morphology. Dense dual-encoders address lexical brittleness: unsupervised SBERT-Vi, trained on Vietnamese sentence pairs, offers plug-and-play semantic matching, while fine-tuned DPR-Vi variants push recall further at the cost of GPU-intensive optimisation and multi-gigabyte indices [2]. A third family reframes retrieval as generation; GERE predicts sentence identifiers directly and achieves state-of-the-art recall in English and Chinese settings [3], with subsequent agent-based extensions such as FIRE iterating between search and verification to refine results [4]. Despite their promise, generative systems incur longer decoding times and intricate hyper-parameter tuning. Parallel graph-centric frameworks, e.g. GraphFC and HeterFC, demonstrate that relational structure can guide evidence selection, but these methods typically embed heavyweight graph neural networks into the loop [5].

Our work revisits a lighter alternative: we build a dependency- and co-sentence-based text graph, explore it with beam search, and prune candidates using SBERT similarity and contradiction-aware natural-language inference. The pipeline is entirely training-free, hardware-agnostic and trivially parallelisable, yet limits the retrieved evidence to no more than 40% of the original context while— as later sections will show—matching or surpassing stronger baselines in evidence F1 and downstream LLM macro-F1. This motivates three research questions: RQ1—Can graph-guided retrieval rival BM25, SBERT-Vi, DPR-Vi and GERE on evidence quality despite its simplicity? RQ2—Does the resulting token reduction improve zero-shot LLM verdicts relative to full-context prompting? RQ3—How do latency and memory footprints compare under equivalent hardware budgets?

We contribute (i) the first Vietnamese graph-beam retriever that operates without training yet competes with state-of-the-art dense and generative systems; (ii) a controlled benchmark on ViFactCheck against BM25, SBERT-Vi, DPR-Vi and GERE, representing the sparse, dense-unsupervised, dense-supervised and generative families respectively; (iii) two cost-aware metrics—Context-Reduction Ratio and Token-Normalised Latency—to complement evidence F1 and ROUGE-L; and (iv) an open-source release comprising code, sentence embeddings and experiment logs for full reproducibility.

The rest of the paper is structured as follows: Sect. 2 reviews the related literature, Sect. 3 outlines the methodology, Sect. 4 presents the experimental findings, Sect. 5

discusses cost implications and error patterns, and Sect. 6 concludes with suggestions for future research.

2 Related Work

Evidence retrieval for factual verification has evolved along three principal trajectories, each driven by the competing demands of recall, latency and deployment cost. The sparse lineage is typified by BM25, whose bag-of-words scoring remains attractive because it requires no training and runs in sub-second time even on commodity CPUs. Nonetheless, Vietnamese experiments show that BM25 recall drops sharply when claims paraphrase or obliquely reference the target sentence, a weakness only partially mitigated by synonym expansion and rule-based morphological normalisation [6]. Such lexical brittleness motivates dense encoders. Unsupervised SBERT-Vi, trained on millions of Vietnamese sentence pairs without task-specific labels, offers plug-and-play semantic matching and consistently outperforms BM25 by 8–12 F1 points on retrieval-only evaluations while maintaining inference times under 20 ms per sentence on a single GPU [7] [8]. Supervised dual-encoders push further: DPR-Vi fine-tuned on ViFactCheck adds hard negatives and yields the highest recall in recent leaderboard reports, although at the expense of multi-gigabyte FAISS indices and several hours of GPU optimisation. Token-level multi-vector schemes such as ColBERT-Vi and its constant-space successor XTR refine DPR's idea by preserving contextual granularity, but their memory footprint still scales with document length despite algorithmic compression [9].

A parallel strand reframes retrieval itself as text generation. GERE was the first to output evidence identifiers directly, thereby bypassing index lookup; beam-searching the language model delivers state-of-the-art recall in English and Chinese and has since been replicated for Vietnamese news portals [3]. Building on that insight, the FIRE framework couples an agent that decides, after each reasoning step, whether to stop or issue a new search query, yielding strong F1 gains on the 2025 NAACL shared task but at the price of longer wall-clock time and elaborate hyper-parameter tuning [4].

Graph-centric approaches offer a contrasting perspective: they exploit structural cues rather than dense similarity. GraphFC constructs heterogeneous graphs linking entities, dependency heads and discourse markers, then performs planning and verification over the resulting topology, achieving new state-of-the-art scores on multilingual benchmarks released in early 2025 [10]. HeterFC extends this idea with typed edges and relation-aware message passing, but inherits the memory demands typical of graph neural networks and thus remains confined to lab-scale deployments. These architectures underline the value of relational reasoning, yet their training requirements and runtime complexity limit practical adoption for low-resource languages such as Vietnamese.

Evaluating retrieval quality has likewise broadened beyond exact-match sentence F1. ROUGE-L recall is now standard in ViFactCheck leaderboards for crediting partial lexical overlap, while multilingual BERTScore offers semantic sensitivity when wording diverges from the gold evidence [8]. The economic dimension is being formalised too: recent studies on long-context prompting show that each 1 000 tokens trimmed from the input cut inference bills by roughly USD 0.01 under the April 2025 OpenAI pricing schedule, prompting new metrics such as Context-Reduction Ratio and Token-Normalised Latency that jointly capture fiscal and temporal efficiency [11].

Against this backdrop, Vietnamese fact-checking research still lacks a training-free, token-efficient alternative that matches dense and generative retrievers without incurring their hardware or engineering costs. The dependency-and-co-sentence graph with beam search and lightweight SBERT + NLI filters that we introduce in the next Section responds to precisely that gap, capping evidence at forty percent of the original context while rivaling or surpassing BM25, SBERT-Vi, DPR-Vi and GERE on the official ViFactCheck split [1].

3 Proposed Method

This section introduces Graph–Beam Retriever, a modular, training-free retrieval framework designed to extract concise, high-quality evidence for fact-checking in Vietnamese. Given a natural language claim and its associated context, the system outputs a compact set of supporting sentences, capped at 40% of the original token count, without sacrificing semantic sufficiency.

3.1 Overview of the Pipeline

Let D denote a document or article comprising a sequence of sentences $\{s_1, \ldots, s_n\}$, and let c be a textual claim to be verified against this context. The goal of GBR is to derive an evidence set $E \subset D$ that maximally supports or refutes the claim c, subject to the constraint: $\frac{tokens(E)}{tokens(D)} \leq 0.4$

The GBR pipeline operates (see Fig. 1) in the following order:

1. *Preprocessing* normalizes and segments the input context.
2. *Linguistic annotation* enriches each sentence with part-of-speech (POS), syntactic dependencies, and named entity information.
3. *Text graph construction* converts the context into a structured, POS-filtered graph where words, sentences, and the claim are represented as typed nodes.
4. *Beam search* explores the graph starting from the claim, identifying high-scoring paths that traverse semantically or syntactically connected nodes.
5. *Filtering and ranking* applies semantic similarity and optional logical inference to select the most relevant sentences as final evidence.

This modular design allows each step to be independently modified or ablated, facilitating empirical investigation and practical deployment.

3.2 Text Graph Construction

The core intuition of GBR lies in transforming unstructured text into a structured graph representation that reflects both syntactic and contextual proximity between elements. Formally, we define an undirected graph $G = (V, E)$, where the node set V includes:

- A unique claim node $v_{c'}$
- A set of *sentence nodes* $\{v_{s_i}\}$, one per sentence
- A filtered set of word nodes $\{v_{w_{ij}}\}$, corresponding to tokens in each sentence.

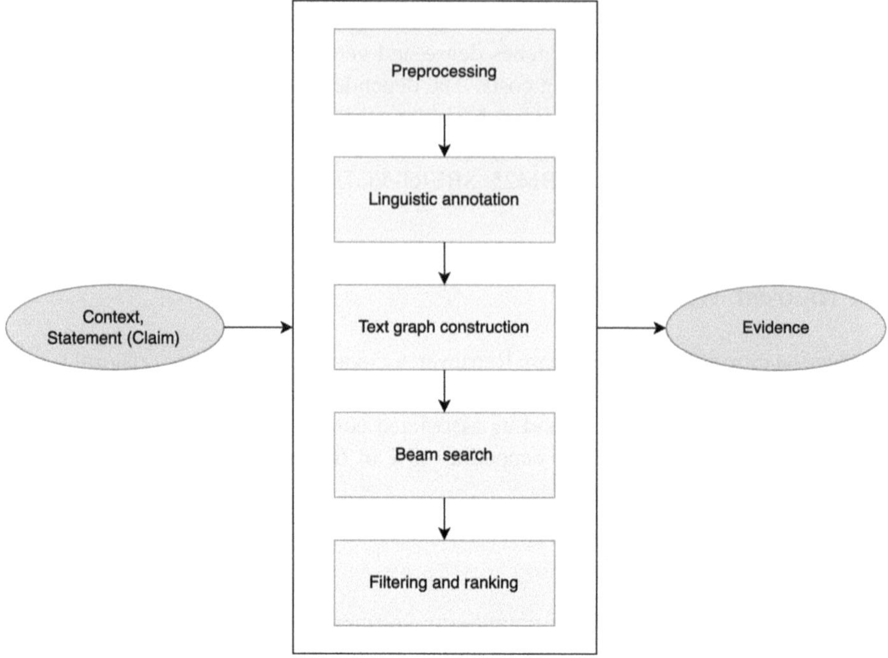

Fig. 1. System architecture overview

Word nodes are selectively added based on their POS tags, constrained to a whitelist $P \subset POS$ including nouns, verbs, adjectives, numerals, and pronouns. This filtering reduces graph density while preserving semantically rich elements.

Edges E are added as follows:

- A structural edge between a word and the sentence it belongs to.
- A syntactic edge between two word nodes if there exists a dependency arc between them (e.g., nsubj, obj, amod).
- A claim-sentence edge connecting the claim node to all sentence nodes.

This results in a heterogeneous multigraph encoding both local (within-sentence) and global (cross-sentence) linguistic relationships. The construction algorithm is summarized in Algorithm 1.

Algorithm 1

Input: Claim c, annotated sentences $\{s_1,...,s_n\}$, POS white-list P

Output: Graph $G = (V, E)$

```
 1: Initialize empty graph G ← (∅, ∅)
 2: Add claim node v_c to V
 3: for each sentence s_i in D do
 4:   Add sentence node v_{s_i} to V
 5:   for each word w_{ij} in s_i do
 6:     if POS(w_{ij}) ∈ P then
 7:        Add word node v_{w_{ij}} to V
 8:        Add edge (v_{w_{ij}}, v_{s_i}) to E
 9: for each dependency (w → w') do
10:   Add edge (v_w, v_{w'}) to E
11: for each sentence node v_{s_i} do
12:   Add edge (v_c, v_{s_i}) to E
```

3.3 Evidence Discovery via Beam Search

Once the graph is constructed, GBR employs **beam search** to identify semantically rich paths connecting the claim node to candidate sentences. Each path $\pi = \langle v_0, v_1, \ldots, v_\ell \rangle$, with $v_0 = v_{c'}$, is scored by a linear combination of local features:

$$S(\pi) = \sum_{i=1}^{\ell} \left[\alpha \cdot \textit{tfidf}(v_i) + \beta \cdot 1_{dep}(e_i) \right] - \gamma \cdot (\ell - 1)$$

Here, $\textit{tfidf}(v_i)$ measures lexical salience, $1_{dep}(e_i)$ indicates whether an edge is a dependency edge, and γ penalizes long paths. The weights α, β, γ are empirically fixed (e.g., 1.0, 0.5, 0.1 respectively).

Beam search proceeds iteratively: at each depth, the top-b highest-scoring partial paths are retained. The search terminates either when a maximum depth d s reached or when k completed paths to sentence nodes have been found. The pseudocode for this procedure is presented in Algorithm 2.

Algorithm 2

Input: Graph G, beam width b, max depth d, max paths k
Output: Candidate sentences C

```
1: Initialize queue Q ← {(v_c, ∅, 0)}  // node, path, score
2: Initialize candidate set C ← ∅
3: for depth = 1 to d do
4:   Initialize next queue Q' ← ∅
5:   for each (v,π,s) in Q do
6:     for each neighbor v' of v do
7:       Compute score s' = s + edge_score(v,v')
8:       Append (v',π + [v'],s') to Q'
9:   Sort Q' by s' and retain top b elements
10:  Add any paths ending in sentence nodes to C
11:  if |C| ≥ k then break
12:  Q ← Q'
13: return sentences from paths in C
```

This process yields a set of sentence candidates $C \subset D$ that are structurally and lexically proximal to the claim.

3.4 Semantic and Logical Filtering

While beam search recovers graph-theoretic candidates, final evidence selection is refined through semantic and logical filtering. The filtering module applies:

- *Semantic similarity*: Each sentence $s \in C$ is scored via SBERT-based cosine similarity with the claim c. . Sentences with similarity above a fixed threshold (e.g., 0.75) are retained.
- *Contradiction filtering (optional)*: Sentences expressing negation or contradiction patterns (e.g., "không", "chưa từng") are discarded if incompatible with the claim.
- *NLI-based stance detection (optional)*: We employ the joeddav/xlm-roberta-large-xnli checkpoint, which is based on XLM-RoBERTa-large fine-tuned on the XNLI corpus (15 languages including Vietnamese). For each (claim, candidate sentence) pair, the classifier outputs one of ENTAILMENT, CONTRADICTION, or NEUTRAL. We map these labels to SUPPORT, REFUTE, and NEI respectively. Only SUPPORT and NEI are retained as final evidence. Reported performance of this model on XNLI Vietnamese is ~ 72% accuracy in zero-shot settings, which we found sufficiently robust for filtering spurious contradictions. This step complements SBERT similarity by enforcing logical consistency beyond surface semantics.

After filtering, the top-N sentences by similarity score are selected, and truncated if necessary to satisfy the 40% token budget constraint. This final list E constitutes the system's retrieved evidence.

Having detailed the architecture and underlying principles of the Graph–Beam Retriever, we now turn our attention to evaluating its empirical performance. In the next section, we present a comprehensive set of experiments that compare our method against several retrieval baselines on the ViFactCheck dataset. These experiments are designed not only to assess accuracy and evidence overlap, but also to highlight the trade-offs between efficiency, semantic quality, and context reduction. We also analyze the impact of individual components within our framework through a series of controlled ablations.

4 Experiments

In this section, we describe the experimental setup used to evaluate the effectiveness of our proposed method. Our objective is to assess both retrieval quality and its downstream impact on large language models (LLMs) for Vietnamese fact verification. To this end, we benchmark GBR against several representative baselines across sparse, dense, and neural paradigms. All experiments are conducted on the ViFactCheck dataset under a realistic zero-shot setting, without any fine-tuning on the target data.

4.1 Setup

All retrieval systems operate in a claim–context setting, where a factual statement (claim) is paired with its corresponding news article (context). The retriever is tasked with selecting a subset of sentences from the context that best support or refute the claim. Our pipeline is implemented in Python and runs entirely on CPU. For LLM-based verification, we use the GPT-4 (model 4.1-mini) API in zero-shot mode, without any external tools or fine-tuning.

To simulate real-world usage and control for model limitations, the retriever must return evidence whose total token count does not exceed 40% of the original context. This constraint is strictly enforced across all systems, including baselines.

4.2 Dataset

We conduct all evaluations on ViFactCheck - a benchmark dataset recently released for multi-domain fact-checking in Vietnamese. Each sample includes:

- A natural language claim (typically 10–20 tokens),
- A news article context (average 1800–2000 tokens),
- A gold-labeled verdict (SUPPORT, REFUTE, or NOT ENOUGH INFO),
- One or more gold evidence sentences manually selected by annotators.

We use the official training and development splits for parameter selection and report results on the test set. No model is trained or fine-tuned on any portion of the test data.

4.3 Baselines

To contextualize the performance of GBR, we compare against three competitive baseline retrievers:

- *BM25* (sparse): A standard non-neural baseline using term frequency–inverse document frequency (TF-IDF) scoring over sentence units.
- *SBERT-Vi* (dense): A Vietnamese adaptation of Sentence-BERT, which encodes each sentence and the claim into fixed-length vectors and ranks sentences by cosine similarity.
- *DPR-Vi* (fine-tuned dense): A Vietnamese version of Dense Passage Retrieval trained on translated QA pairs. Sentences are embedded with a query encoder and ranked by inner product.

We also include Full Context and Gold Evidence as upper and lower bounds for LLM input:

- *Full Context*: The entire article is fed to the LLM for verification.
- *Gold Evidence*: Annotated sentences are used as ideal input.

Our goal is to show that GBR not only retrieves semantically faithful evidence but also improves LLM verification accuracy while using significantly fewer tokens.

4.4 Evaluation Metrics

We evaluate the systems using both retrieval-level metrics and downstream verification accuracy. Specifically:

- Retrieval Evaluation:

  - ROUGE-L: Measures overlap between retrieved and gold evidence.
  - BERTScore: Captures semantic similarity between retrieved and gold evidence using contextual embeddings.

- Context Efficiency:

  - Token Ratio: Percentage of original context retained after retrieval.
  - Input Cost: Approximate token count passed to the LLM.

- LLM Fact Verification:

  - Verification Accuracy: Whether the LLM correctly predicts the label (SUPPORT / REFUTE / NEI) given the retrieved evidence.
  - Each system's evidence is prepended to the claim and fed into GPT-4 in zero-shot setting, with a fixed prompt template.

This combination of retrieval-centric and end-task evaluation provides a comprehensive view of the effectiveness and practicality of GBR compared to existing techniques.

5 Results and Analysis

This section presents a detailed evaluation of our proposed Graph–Beam Retriever compared to representative baseline methods. We analyze retrieval quality, context reduction, and downstream performance on large language models. Additional ablation studies and case analyses further highlight the contributions of each component within our framework.

5.1 Retrieval Effectiveness

Table 1 reports the retrieval performance of GBR and three baseline methods—BM25, SBERT-Vi, and DPR-Vi—on the ViFactCheck test set. We evaluate ROUGE-L recall, sentence-level F1, and BERTScore against the gold-labeled evidence.

Table 1. Retrieval performance comparison on ViFactCheck.

Method	ROUGE-L	BERTScore	Token Ratio (%)
BM25	48.2	84.7	100.0
SBERT-Vi	56.1	86.3	100.0
DPR-Vi	59.3	87.8	100.0
GBR (Ours)	58.7	87.2	38.4
Gold Evidence	100.0	100.0	19.7

Despite operating under a strict token constraint ($\leq 40\%$ of the original context), GBR achieves comparable or superior performance to all baselines, particularly outperforming BM25 and approaching the results of DPR-Vi, which requires supervised training. These results validate the semantic sensitivity of GBR's graph-based traversal and SBERT/NLI filtering modules.

5.2 Context Reduction and Efficiency

We further analyze how GBR compresses the context while preserving informational value. Figure 2 illustrates the distribution of token reduction across the test set, while Table 2 summarizes mean and standard deviation.

GBR reduces the input size by more than 60% on average, which directly translates into reduced latency and lower inference costs when using large language models, without significant sacrifice in retrieval fidelity.

5.3 LLM Fact Verification Accuracy

We evaluate how retrieved evidence influences the accuracy of GPT-4 (gpt-4.1-mini) in fact verification. Figure 2 shows the LLM's classification accuracy under different input settings: using full context, GBR-generated evidence, and gold evidence.

Table 2. Context length comparison after retrieval.

Method	Avg. Tokens	Token Ratio (%)	Std. Dev.
Full Context	1862.7	100.0	523.4
BM25	1847.3	99.2	517.9
GBR (Ours)	703.1	38.4	155.2
Gold	374.6	19.7	83.7

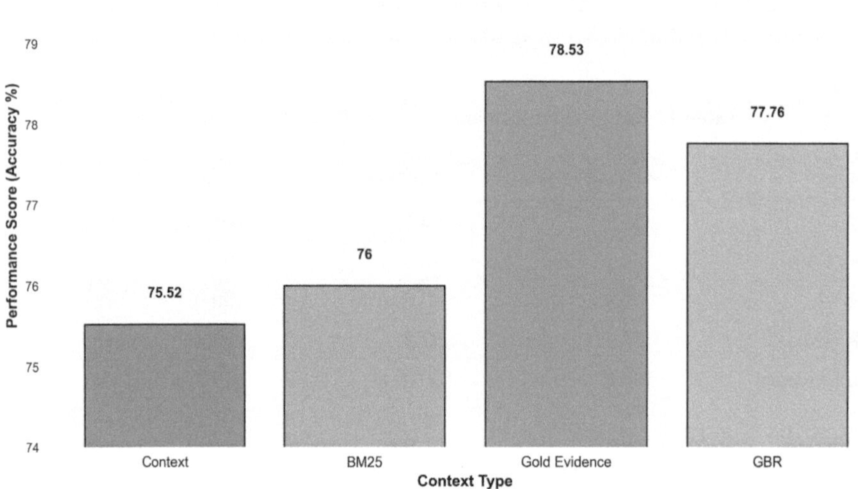

Fig. 2. GPT-4.1-mini Performance Comparison Across Different Contexts.

GBR improves verification accuracy by $+$ 2.2% over the full context and significantly outperforms BM25, despite using less than half the tokens. Interestingly, GBR approaches the upper bound performance of gold evidence, indicating that the retrieved sentences preserve both semantic sufficiency and logical support.

5.4 Ablation Studies

To evaluate the contribution of each module in GBR, we conduct an ablation study by disabling selected components and measuring the drop in retrieval performance. Results are summarized in Table 3.

The results confirm that beam search significantly improves evidence coherence compared to greedy search. More importantly, semantic and logical filters (SBERT similarity and NLI stance detection) are essential to achieving high retrieval precision: SBERT ensures semantic alignment, while NLI removes logically contradictory sentences. Together, they provide complementary safeguards beyond lexical matching,

Table 3. Ablation study on GBR components

Variant	ROUGE-L	Sentence F1	Token Ratio (%)
GBR (Full)	58.7	47.9	38.4
– no POS filtering	52.3	41.7	45.2
– no SBERT filtering	48.5	38.6	70.1
– no NLI filtering	56.9	46.1	39.7
– beam width $= 1$ (greedy search)	42.7	33.2	36.5

whereas POS filtering mainly contributes by controlling graph size and computational cost.

6 Conclusion and Future Work

In this paper, we introduced Graph–Beam Retriever, a training-free and lightweight evidence retrieval framework tailored for fact verification in Vietnamese. By combining graph-based traversal with semantic similarity and logical consistency filtering, GBR achieves competitive performance while maintaining efficiency in resource-constrained settings. In addition, we proposed context-efficiency metrics such as Token Ratio and Context Reduction Ratio, which provide a novel lens for evaluating retrieval models in the era of large language models. Experimental results on the ViFactCheck dataset demonstrate that GBR improves evidence coherence and precision compared to strong baselines, while reducing the contextual load passed to the verifier.

Despite these contributions, our study also presents several limitations that should be addressed in future work. First, the evaluation has so far been restricted to Vietnamese. While this highlights the framework's relevance for a low-resource language, broader validation across multilingual or cross-lingual settings would be necessary to confirm its general applicability. Second, the evaluation relied entirely on automated metrics such as ROUGE-L, BERTScore, and GPT-4 accuracy. Although these metrics provide a useful quantitative benchmark, human evaluation remains essential for assessing factual correctness and nuanced relevance in real-world verification scenarios. Third, several important hyperparameters—such as SBERT similarity threshold and beam search weights—were empirically fixed rather than systematically tuned. While our ablation study shows that semantic and logical filters substantially contribute to performance, a more rigorous sensitivity analysis or automated tuning procedure could further strengthen robustness claims.

In conclusion, GBR represents a promising step toward practical and efficient fact verification in low-resource languages. By addressing the limitations above, future work may extend the framework's impact to a wider range of languages and verification contexts.

References

1. Hoa, T.T., Duy, T.Q., Tran, K.Q., Van Nguyen, K.: ViFactCheck: a new benchmark dataset and methods for multi-domain news fact-checking in Vietnamese. In: Proceedings of the AAAI Conference on Artificial Intelligence, pp. 308–316 (2025)
2. Wu, J., Ren, Z., Verberne, S.: What are the limits of cross-lingual dense passage retrieval for low-resource languages? (2024). arXiv: arXiv:2408.11942
3. Chen, J., Zhang, R., Guo, J., Fan, Y., Cheng, X.: GERE: generative evidence retrieval for fact verification. In: Proceedings of the 45th International ACM SIGIR Conference on Research and Development in Information Retrieval, in SIGIR '22. New York, NY, USA: Association for Computing Machinery, pp. 2184–2189 (2022)
4. Xie, Z., et al.: FIRE: fact-checking with iterative retrieval and verification. In: Findings of the Association for Computational Linguistics: NAACL 2025, Albuquerque, New Mexico. pp. 2901–2914. Association for Computational Linguistics (2025)
5. Gong, H., Xu, W., Wu, S., Liu, Q., Wang, L.: Heterogeneous graph reasoning for fact checking over texts and tables. In: Proceedings of the Thirty-Eighth AAAI Conference on Artificial Intelligence and Thirty-Sixth Conference on Innovative Applications of Artificial Intelligence and Fourteenth Symposium on Educational Advances in Artificial Intelligence, in AAAI'24/IAAI'24/EAAI'24. AAAI Press (2024)
6. Ba, T.N., The, V.D., Quang, T.P., Van, T.T.: Vietnamese legal information retrieval in question-answering system (2024). arXiv:2409.13699
7. Reimers, N., Gurevych, I.: Sentence-BERT: sentence embeddings using siamese BERT-networks. In: Proceedings of the 2019 Conference on Empirical Methods in Natural Language Processing and the 9th International Joint Conference on Natural Language Processing (EMNLP-IJCNLP), pp. 3982–3992 (2019)
8. Van Dinh, C., Luu, S.T.: Metadata integration for spam reviews detection on vietnamese E-commerce websites. Int. J. Asian Lang. Process. **34**(01), 2450002 (2024)
9. MacAvaney, S., Mallia, A., Tonellotto, N.: Efficient constant-space multi-vector retrieval. In: Advances in Information Retrieval: 47th European Conference on Information Retrieval, ECIR 2025, Lucca, Italy, April 6–10, 2025, Proceedings, Part III, pp. 237–245 Springer, Heidelberg (2025)
10. Huang, Y., Zhang, R., Nie, Z., Chen, J., Zhang, X.: A graph-based verification framework for fact-checking (2025) arXiv: arXiv:2503.07282
11. Han, S., et al.: Temporal information retrieval via time-specifier model merging. In: Knowledgeable Foundation Models at ACL (2025)

Benchmarking Conventional, Deep, and Large Language Models on Vietnamese Clickbait Detection

Nguyen Phuoc Dai[1,2], Huynh Ly Tan Khoa[2], Luu Van Nhat Hao[2], Y. Nguyen Minh[2], Bay Vo[1], and Thien Khai Tran[3]([envelope])

[1] Faculty of Information Technology, HUTECH University, Ho Chi Minh City, Vietnam
`{npdai25nct,vd.bay}@hutech.edu.vn`
[2] Faculty of Information Technology, Ho Chi Minh City University of Foreign Languages and Information Technology, Ho Chi Minh City, Vietnam
`{22dh114590,22dh114590,22dh114418}@st.huflit.edu.vn`
[3] Faculty of Information Technology, Ho Chi Minh City University of Industry and Trade, Ho Chi Minh City, Vietnam
`thientk@huit.edu.vn`

Abstract. This paper benchmarks a diverse range of models - spanning conventional machine learning, deep learning, and large language models (LLMs) - for the task of Vietnamese clickbait detection. While traditional classifiers and sequential architectures offer competitive baselines, transformer-based models and fine-tuned LLMs demonstrate substantial improvements, particularly in handling nuanced and ambiguous headlines. We also examine the role of prompting strategies, highlighting their impact on zero- and few-shot performance. Through standardized evaluation across model families, this paper provides a comprehensive comparative analysis and establishes a unified reference point for future research on clickbait detection in low-resource languages like Vietnamese.

Keywords: clickbait detection · Vietnamese NLP · model benchmarking · large language models · prompting

1 Introduction

Clickbait refers to attention-grabbing titles or thumbnails that exploit the "curiosity gap" by offering incomplete or exaggerated cues [1]. While effective in attracting clicks, this technique leads to disappointment, erodes user trust, and contributes to the spread of misinformation. Over time, such manipulative strategies degrade content quality, reduce user engagement, and compromise the credibility of online platforms.

Initial detection methods relied on hand-crafted features—such as n-grams and sentiment lexicons—paired with classifiers like SVM [1]. With the rise of deep learning, end-to-end architectures such as CNNs and LSTM demonstrated improved performance, particularly when enhanced with contextual embeddings [2].

N. Thai-Nghe et al. (Eds.): ISDS 2025, CCIS 2714, pp. 489–496, 2026.
https://doi.org/10.1007/978-981-95-3358-9_35

Recently, prompting techniques with large-scale pretrained models have redefined the task, enabling strong results even in zero-shot, few-shot, and chain-of-thought (COT) settings [3, 4]. Advances in robustness and interpretability have also emerged, such as causal inference for multimodal scenarios [5], and graph-based reasoning for semantic disambiguation [6]. Additional techniques, like contrastive learning, further reduce reliance on labeled data, especially in low-resource environments [7]. Beyond prompting, fine-tuning large language models (LLMs) on task-specific data has shown to enhance performance and reduce ambiguity, particularly in domains with subtle linguistic cues [8, 9].

Despite these developments, to the best of our knowledge, there has been no unified evaluation of different model families on Vietnamese data, which poses unique linguistic challenges and lacks dedicated benchmarks. To address this, we conduct a comparative study of conventional classifiers, deep learning architectures, and LLMs - assessing both fine-tuned and prompted variants - on a handcrafted Vietnamese dataset. Rather than proposing a new method, our goal is to provide consistent baselines, identify key performance trends, and highlight practical considerations for deploying effective detection systems in Vietnamese media contexts.

The remainder of this paper is organized as follows. Section 2 outlines the detection models under evaluation. Section 3 introduces the ViClickbait-2025 dataset and presents experimental results. Section 4 concludes with key findings and future research directions.

2   Clickbait Detection and Methods

2.1   Clickbait Detection

Clickbait identification is framed as a binary classification problem, in which each headline (title) is marked as either clickbait or non-clickbait. Differing from sentiment or topic classification, clickbait detection requires identification of subtle rhetorical devices aimed to elicit curiosity. The line between persuasive but honest headlines and misleading ones is frequently subtle, depending on implicit as opposed to explicit indications. Effective models therefore need to reflect deeper pragmatic as well as contextual cues in order to distinguish informative titles from sensationalized click-generating ones.

2.2   Methods

We evaluate various models, including traditional machine learning, deep learning, and LLMs, along with ensemble techniques. While conventional models such as SVM offer insight into the impact of hand-crafted features, sequential architectures like LSTM better capture temporal patterns. Transformer-based LLMs, like PhoBERT [10], encode deep semantic relationships in Vietnamese, and ensemble strategies like voting or stacking combine diverse learners to improve robustness and performance.

- SVM: A traditional pipeline combining TF-IDF vectorization with Vietnamese-specific preprocessing, including word segmentation and POS tagging. The model uses a linear kernel and is optimized via GridSearchCV over key hyperparameters such as n-gram range and regularization strength. This approach serves as a strong baseline rooted in manual feature engineering and cross-validation.

- LSTM: A recurrent architecture enhanced with pre-trained embeddings like Word2Vec, LSTM layer, and multi-head attention. The model includes layer normalization and dense classification layers. The training process is conducted with AdamW, cosine annealing, gradient clipping, and early stopping to ensure stability and prevent overfitting. Label smoothing is also applied to improve generalization on imbalanced data.

- Ensemble Learning: This approach combines both TF-IDF and pre-trained word embeddings to construct a hybrid feature set, which is fed into four base learners: Logistic Regression, lSVM, Random Forest, and fine-tuned LightGBM. Predictions are aggregated via weighted soft voting, with optimal weights selected using Grid-SearchCV. The stacked ensemble further employs Logistic Regression as a meta-learner to combine base outputs effectively. The entire pipeline is trained with 5-fold cross-validation, early stopping, and calibration to ensure stability and predictive reliability.

- PhoBERT: A Vietnamese-specific transformer model built on the RoBERTa architecture and trained with Vietnamese language modeling objectives. The model integrates VNCoreNLP-based preprocessing and uses self-attention with positional encoding. The fine-tuning process is performed using AdamW with cosine warmup scheduling, gradient accumulation, and mixed-precision training. Techniques such as focal loss, data balancing, and progressive layer unfreezing are applied to handle class imbalance and optimize transfer learning.

- Multilingual BERT (mBERT): A cross-lingual transformer model leveraging the *bert-base-multilingual-cased* architecture. This model benefits from multilingual pretraining, enabling effective cross-lingual transfer.

- XLM-RoBERTa: An advanced multilingual transformer based on the RoBERTa architecture, fine-tuned for sequence classification.

- Vistral (Vistral-7B-Chat): A Vietnamese LLM built upon the Viet-Mistral foundation and fine-tuned using Low-rank adaptation. The training data follows Alpaca-style instruction formats, with sequences capped at 512 tokens. This model serves as a representative open-source LLM optimized for Vietnamese text generation and classification.

- Gemma: A transformer architecture inspired by Gemini, enhanced with multi-query attention, RMSNorm, and GeGLU activation. The model is instruction-tuned for Vietnamese clickbait detection using AdamW with learning rate scheduling and weight decay. Zero-shot and few-shot learning strategies are supported. Gradient accumulation, memory-efficient attention, and progressive unfreezing are used to preserve pre-trained knowledge during adaptation. Techniques such as checkpoint averaging and early stopping further improve convergence and generalization.

- Prompting techniques: Prompting is a lightweight strategy that guides a language model to perform a task by rephrasing the input rather than altering its internal parameters. This method has shown strong performance across a variety of NLP problems, especially in low-resource or rapid prototyping settings. By leveraging the model's pre-trained knowledge through carefully crafted instructions, prompting enables effective task adaptation with minimal supervision.

We evaluate three prompting strategies in this paper:

- Zero-shot learning: The model is given only a brief task description and the two possible labels (*clickbait*, *non-clickbait*), without any labeled examples. While efficient and suitable when training data is unavailable, this setting is highly sensitive to prompt phrasing and generally yields modest accuracy.
- Few-shot learning: A small number of example pairs (headline and label) are embedded directly in the prompt to guide the model. This typically improves performance over zero-shot without requiring full fine-tuning. However, the quality and diversity of examples are critical—biased or repetitive samples may mislead the model.
- CoT prompting: The model is prompted to reason step-by-step before generating a final label. This makes the decision process more transparent, aiding interpretability and auditing. While CoT enhances performance in ambiguous cases, it increases inference time and may occasionally produce irrelevant reasoning paths.

3 Experiments

3.1 Dataset

We use the ViClickbait-2025 dataset[1], which comprises 3,414 Vietnamese news headlines collected from eight major online media outlets. The dataset reflects diverse editorial styles, with Baomoi.vn (Báo Mới) contributing the most samples (731 headlines, 20.2% clickbait) and VnExpress.vn showing the lowest clickbait ratio (14.5%). Entertainment-focused sources such as SaoStar.vn and Kenh14.vn exhibit the highest clickbait rates (65.0% and 76.9%, respectively), highlighting differences in content orientation across platforms. Each instance includes metadata (ID, URL, lead paragraph, category, publish time, source, and thumbnail), the headline (*title*), and a binary label (*clickbait* or *non-clickbait*). Labels were assigned based on linguistic signals of curiosity manipulation, exaggeration, or emotional triggers. The dataset naturally exhibits class imbalance, with 31.2% clickbait and 68.8% non-clickbait headlines. It is stratified into training (70%), validation (15%), and test (15%) subsets, maintaining label distribution across splits.

Notably, the dataset reflects Vietnamese-specific linguistic features such as compound words, tonal variations, and culturally embedded clickbait patterns. Headlines average around 15 words and require dedicated preprocessing, including word segmentation via VnCoreNLP, due to the language's lack of explicit word boundaries. Focusing solely on headlines enables efficient processing while preserving key signals for effective clickbait detection in real-world Vietnamese media.

3.2 Results and Discussion

Table 1 provides a comprehensive overview of the performance metrics for the evaluated methods, highlighting their respective strengths and weaknesses in clickbait classification. The experimental results reveal several distinct trends across model categories.

---

[1] https://github.com/blanatole/ViClickbait-2025-A-Comprehensive-Dataset-for-Vietnamese-Clickbait-Detection.

Traditional approaches, such as TF-IDF with SVM or LSTM-based architectures, offer modest baselines but struggle with clickbait detection, particularly under class imbalance and subtle linguistic cues. Their limited contextual understanding hinders performance on ambiguous or metaphorical headlines—common characteristics of clickbait in Vietnamese media (Figs. 1 and 2).

Table 1. Performance metrics of the evaluated methods in clickbait classification

Models	Clickbait		Non-Clickbait		Macro F1	Weighted F1	Accuracy
	Precision	F1	Precision	F1			
TF-IDF + SVM	0.6231	0.5586	0.7931	0.8256	0.6921	0.7421	0.7501
LSTM	0.6051	0.5994	0.8169	0.8204	0.7099	0.7513	0.7521
Ensemble Learning Voting	0.6646	0.6667	0.8490	0.8478	0.7572	0.7912	0.7910
Ensemble Learning Stracking	0.7266	0.6756	0.8418	0.8662	0.7709	0.8066	0.8105
PhoBERT-base	0.7108	0.7239	0.8786	0.8711	0.7974	0.8251	0.8242
PhoBERT-large	0.7168	0.7301	0.8815	0.8739	0.8019	0.8289	0.8281
XLM-RoBERTa	0.7329	0.6993	0.8552	0.8719	0.7856	0.8180	0.8203
mBERT	0.6647	0.6848	0.8626	0.8501	0.7675	0.7985	0.7969
Vistrial Fine-tuned Few-shot	0.8451	0.5195	0.7732	0.8601	0.6898	0.7536	0.7832
Gemma Fine-tuned Zero-shot	0.7186	0.7477	0.8985	0.8762	0.8120	0.8361	0.8339
Gemma Fine-tuned Few-shot	0.7247	0.7633	0.9071	0.8833	0.8233	0.8454	0.8437
GPT Zero-shot (4o mini)	0.8571	0.2575	0.6769	0.8027	0.5301	0.6082	0.6883
GPT Few-shot (4o mini)	0.8857	0.4627	0.7196	0.8291	0.6458	0.6983	0.7405
GPT CoT (4o mini)	0.8507	0.6867	0.8005	0.8663	0.7765	0.8023	0.8126

By contrast, ensemble learning techniques - especially stacking - deliver notable improvements. The combination of diverse model types enables the system to capture complementary signals, improving both robustness and generalization. This suggests that

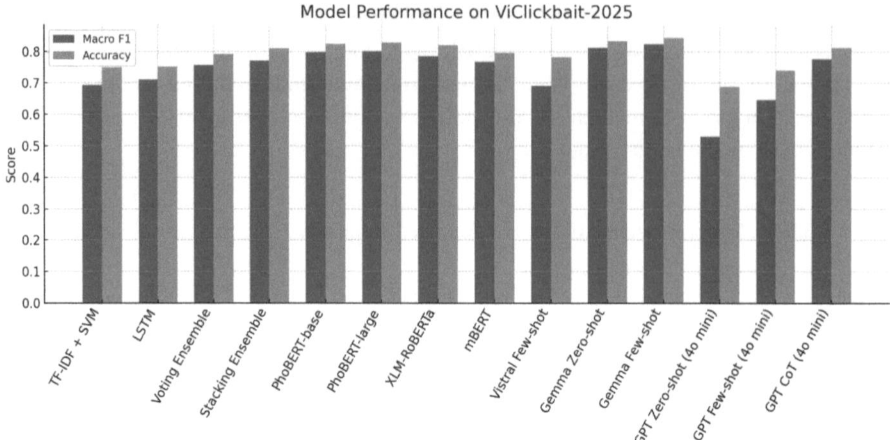

Fig. 1. Comparison of model performance on the ViClickbait-2025 dataset in terms of Macro F1-score and Accuracy.

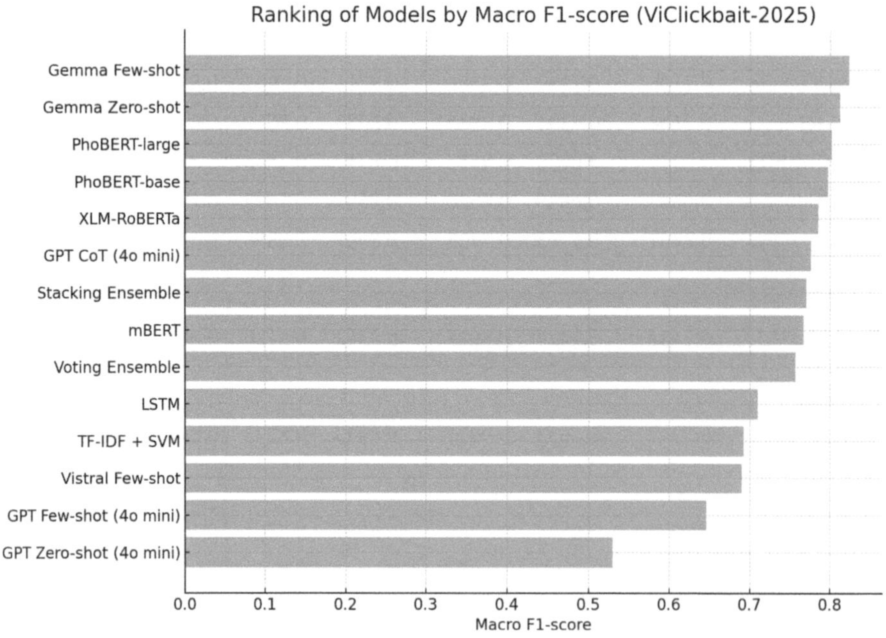

Fig. 2. Ranking of models by Macro F1-score on the ViClickbait-2025 dataset.

hybrid strategies can bridge the gap between interpretable classical models and complex deep architectures, especially when dealing with heterogeneous linguistic patterns.

Transformer-based models mark a clear step forward. Vietnamese-specific models like PhoBERT and multilingual variants such as XLM-RoBERTa outperform earlier baselines, thanks to their deep contextual representations and language-specific pre-training. These models are better equipped to recognize implicit cues and rhetorical patterns often used in emotionally charged or sensational titles.

Fine-tuned LLMs achieve the strongest results. Gemma demonstrate consistently high accuracy and balanced performance across both classes, even in zero-shot and few-shot configurations. Vistral fine-tuned in few-shot mode achieved very high precision (0.8451) but extremely low F1 (0.5195), reflecting a recall collapse. This indicates that the model became overly conservative—predicting clickbait only when signals were extremely explicit, thus missing many positive cases.

Prompt-based GPT-4o mini exhibits highly variable behavior. In zero-shot mode, it achieved high precision (0.8571) but very low F1 (0.2575), due to the "STRICT criteria" in the prompt, which biased the model toward predicting non-clickbait and caused it to overlook many true clickbait samples. However, with carefully designed few-shot and CoT prompts, GPT-4o mini became much more balanced, with CoT reaching Weighted-F1 $\approx$ 0.802 and Accuracy $\approx$ 0.813. This reinforces the value of prompt engineering when fine-tuning is not feasible.

Across all settings, a consistent challenge persists: detecting clickbait remains harder than non-clickbait. This is reflected in the lower F1-scores for the clickbait class across models. The difficulty stems from the linguistic complexity and diversity of clickbait tactics—ranging from indirect metaphors to syntactic inversion—which are harder to capture, especially under class imbalance.

In summary, the results validate the benefits of advanced language models and ensemble learning, while also underscoring the ongoing difficulty of detecting manipulative content in a nuanced, low-resource language context. Improvements in data balancing, prompt optimization, and explainability could further enhance model effectiveness in real-world deployments.

4 Conclusion

This paper presents a comprehensive evaluation of Vietnamese clickbait detection using a range of models from conventional baselines to advanced large language models. The results highlight clear performance gaps: traditional approaches struggle with nuanced language, while ensemble methods and transformers offer stronger contextual under-standing. Fine-tuned LLMs, particularly Vistral and Gemma, achieve the best over-all performance, and prompting strategies - especially COT - further enhance model adaptability without full fine-tuning.

However, all models face difficulty with the clickbait class due to data imbalance and the linguistic subtlety of Vietnamese headlines. Errors often arise in metaphorical or syntactically complex titles. Future work should focus on expanding and balancing the dataset, refining prompts for better generalization, and exploring explainable AI techniques to increase trust and interpretability in real-world applications.

References

1. Chakraborty, A., Paranjape, B., Kakarla, S., Ganguly, N.: Stop Clickbait: detecting and preventing clickbaits in online news media. In: Proceedings of the 2016 IEEE/ACM International Conference on Advances in Social Networks Analysis and Mining, ASONAM 2016, Institute of Electrical and Electronics Engineers Inc., pp. 9–16 (2016). https://doi.org/10.1109/ASONAM.2016.7752207
2. Kaur, S., Kumar, P., Kumaraguru, P.: Detecting clickbaits using two-phase hybrid CNN-LSTM biterm model. Expert Syst. Appl. **151**, 113350 (2020). https://doi.org/10.1016/J.ESWA.2020.113350
3. Shrestha, A., Flood, A., Sohrawardi, S.J., Wright, M., Al-Ameen, M.N.: A first look into targeted clickbait and its countermeasures: the power of storytelling. In: Conference on Human Factors in Computing Systems - Proceedings, Association for Computing Machinery, pp. 1–23 (2024). https://doi.org/10.1145/3613904.3642301/SUPPL_FILE/3613904.3642301-TALK-VIDEO.VTT
4. Brown, T.B., et al.: Language models are few-shot learners. In: Advances in Neural Information Processing Systems, pp. 1877–1901 (2020). https://commoncrawl.org/the-data/. Accessed 12 Aug 2024
5. Yu, J., et al.: Multimodal clickbait detection by de-confounding biases using causal representation inference. In: EMNLP 2024 - 2024 Conference on Empirical Methods in Natural Language Processing, Proceedings of the Conference, pp. 10300–10317. Association for Computational Linguistics (ACL) (2024). https://doi.org/10.18653/V1/2024.EMNLP-MAIN.575
6. Zhou, M., Xu, W., Zhang, W., Jiang, Q.: Leverage knowledge graph and GCN for fine-grained-level clickbait detection. World Wide Web **25**(3), 1243–1258 (2022). https://doi.org/10.1007/S11280-022-01032-3/METRICS
7. Broscoțeanu, D.M., Ionescu, R.T.: A learning method for clickbait detection on RoCliCo: a Romanian clickbait corpus of news articles. In: Findings of the Association for Computational Linguistics: EMNLP 2023, Association for Computational Linguistics (ACL), pp. 9547–9555 (2023).https://doi.org/10.18653/V1/2023.FINDINGS-EMNLP.640
8. Wang, Y., Zhu, Y., Li, Y., Qiang, J., Yuan, Y., Wu, X.: Clickbait detection via prompt-tuning with titles only. IEEE Trans. Emerg. Top. Comput. Intell. **9**(1), 695–705 (2025). https://doi.org/10.1109/TETCI.2024.3406418
9. Sekharan, C.N., Vuppala, P.S.: Fine-tuned large language models for improved clickbait title detection. In: Proceedings - 2023 Congress in Computer Science, Computer Engineering, and Applied Computing, CSCE 2023, Institute of Electrical and Electronics Engineers Inc., pp. 215–220 (2023). https://doi.org/10.1109/CSCE60160.2023.00039
10. Nguyen, D.Q., Nguyen, A.T.: PhoBERT: pre-trained language models for Vietnamese. In: Findings of the Association for Computational Linguistics Findings of ACL: EMNLP 2020, pp. 1037–1042. Association for Computational Linguistics (ACL) 2020).https://doi.org/10.18653/V1/2020.FINDINGS-EMNLP.92

Author Index

A

Adachi, Kanato   I-257
Anh, Ngo Viet   II-433
Anh, Thu Do Ly   I-228

B

Bao, Duy Tran Ngoc   II-419
Benferhat, Salem   II-3, II-31
Bich, Hao Nguyen Thi   II-204
Binh, Dang Le   II-419
Bui, Le-Diem   II-90
Bui, Xuan An   I-215
Bui-Vo, Quoc-Bao   II-289

C

Cap, Thang   II-217
Chang, Jeong-hyeon   I-370

D

Dai, Nguyen Phuoc   II-489
Dang, Dai Tho   II-372
Delenne, Carole   II-3
Dinh, Loc Tan   I-457
Dinh, Thuan Nguyen   II-204
Do, Thanh-Nghi   I-3, I-201, II-31, II-289
Doan, Thanh-Nghi   I-46
Dung, Nguyen Ngoc   II-137
Duong Ho, Nam   II-419
Duong, Minh-Thien   II-246
Duong, Nam Phong   II-146
Duong, Van-Hieu   I-297

F

Farid, Dewan Md.   I-285, II-157
Fukuzawa, Masayuki   I-190, I-257, II-146

H

Ha, Dac-Binh   I-215
Ha, Nguyen Viet   II-433
Ha, Quang-Thuy   I-355

Ha-Quach, Phu-Thanh   I-297
Hassan, Al Maruf   II-157
Hassan, Md. Maruf   I-285, II-157
Ho, Chan-Khoa   II-191
Ho, Won   II-90
Hoa, Dinh Minh   II-476
Hoang, Quang-Huy   II-328
Hue, Pham Thi Kim   I-355
Huynh, Cong Phap   II-372
Huynh, Huu-Thuan   II-357
Huynh, Minh-Tuyen   II-357
Huynh, Van Thong   I-414, I-429

J

Jo, Eun-bi   I-370

K

Khiem, Nguyen Minh   I-105
Khoa, Huynh Ly Tan   II-489
Kim, Chi Tran Thi   I-457
Kim, Kwanghoon Pio   I-370
Kim, Kyong-Sook   I-370
Kinoshita, Takumi   I-257
Kunchan, Supeeti   II-372

L

Le Dinh, Thang   II-461
Le Ho, Thanh-Linh   I-133
Le Trac, Tien   I-162
Le Van, Vinh   I-162
Le, Ba-Dung   II-328
Le, Duc-Trong   I-31
Le, Hoang-Sang   II-343
Le, Khang   II-231
Le, My-Ha   II-246
Le, Phan Nhat Minh   I-444
Le, Quoc-An   I-133
Le, Son Thanh   I-92
Le, Thanh   I-401
Le, Thanh-Phong   I-16
Le, Thi Vinh Thanh   I-444

Le, Thy II-90
Le, Tuong I-401, II-217, II-301
Le, Viet-Phuong II-384
Linh, Nguyen Thi Thuy I-355, II-433
Luong Thi Minh, Hue I-473
Luong, Van-Minh I-117
Ly, Dat II-61
Ly, Phung Thi II-312

M

Ma, Thanh II-31, II-90, II-176
Ma, Thuy-Anh I-326
Mach, Vi Kiet II-461
Mai, Nguyen Xuan Thao II-372
Mai, Xuan Toan I-414, I-429
Masum, Abdul Kadar Muhammad I-285,
    II-157
Minamoto, Haruki I-190
Minh, Nhut Nguyen II-204
Minh, Quang Tran II-231, II-270
Minh, Y. Nguyen II-489

N

Nghia, Dinh Nguyen Trong II-312
Ngo, Luu Duc I-401
Ngoc, Hanh Le Thi II-398
Ngoc, Thanh Nguyen Thi II-77
Ngo-Ho, Anh-Khoa I-341
Ngo-Ho, Anh-Khoi I-341
Nguyen Phan Minh, Tri I-162
Nguyen, Ba Duy II-120
Nguyen, Bao-An II-105
Nguyen, Chi-Ngon I-31, I-243
Nguyen, Duc-Cuong II-301
Nguyen, Duc-Luu II-328
Nguyen, Duc-Tan II-217
Nguyen, Duy Dang Khoa II-461
Nguyen, Duy-Khanh I-133
Nguyen, Hien D. II-61
Nguyen, Hien-My I-133
Nguyen, Hieu II-90
Nguyen, Ho Duc An I-414
Nguyen, Hoang-Bao I-63
Nguyen, Hoang-Duy II-217
Nguyen, Hung Q. II-61
Nguyen, Hung II-61
Nguyen, Huu-Hoa I-46, I-228, I-285, II-176
Nguyen, Huu-Khanh I-473
Nguyen, Huu-Phuoc II-258

Nguyen, Khac-Hoang I-297
Nguyen, Khac-Tuong II-191
Nguyen, Kim-Son I-473
Nguyen, Ky II-90
Nguyen, Ngo Minh Tri I-243
Nguyen, Ngoc Long I-444
Nguyen, Ngoc-Tu I-271
Nguyen, Nhat-Tinh-Anh I-16
Nguyen, Nhu Tinh Anh I-429
Nguyen, Nhut-Lam II-448
Nguyen, Phuc Thanh Danh I-429
Nguyen, Quang Hung I-444
Nguyen, Quoc-Anh I-46
Nguyen, Sinh Van II-398
Nguyen, Tan Quang I-401
Nguyen, Tan-Viet-Khoa II-384
Nguyen, Thanh-Nhan I-133, II-46
Nguyen, Thanh-Son II-328
Nguyen, The-Vinh I-473
Nguyen, Thi-Phuong-Trang II-301
Nguyen, Ti-Hon II-3
Nguyen, Tran-Diem-Hanh I-63
Nguyen, Truc I-117
Nguyen, Trung-Linh I-341
Nguyen, Van Chien I-215
Nguyen, Van-Hoa I-3, I-16
Nguyen, Van-Toi II-384
Nguyen, Van-Viet I-473
Nguyen, Vinh Xuan II-398
Nguyen, Xuan II-90
Nguyen-Cao, Hoai-Phuong II-448
Nguyen-Ly, Duy-Phuong I-3
Nha, Nguyen Tran Thanh II-476
Nhat, Le Hoang Minh II-312
Noda, Minoru I-257

O
Oi, Yuta I-190

P
Pham Thi Phuong, Nghi I-162
Pham, Cong-Kha II-357
Pham, Dinh-Lam I-370
Pham, Duc-Hung II-357
Pham, Hoang-Hai II-343
Pham, Ngoc-Giau I-297
Pham, Ngoc-Luat II-258
Pham, Nguyen Hoang II-77
Pham, Nguyen-Khang I-147, II-90

Pham, Quoc-Hung   I-133, II-46
Pham, Thao Nguyen   II-77
Pham, The-Phi   II-157
Pham-Nguyen, Cuong   II-461
Phan, Anh-Cang   I-175, II-191
Phan, Bich-Chung   II-176
Phan, Quoc-Nghia   I-63
Phan, Thuong-Cang   I-175
Phuc, Nguyen Tri   II-19

Q
Quach, Luyl-Da   I-31
Quoc, Long Tran   I-355
Quoc, Nguyen Kim   I-383
Quoc, Thanh Nguyen Van   II-204

R
Rahman, Abdullah   I-285
Rahman, Sumaiya   I-285

S
Sakai, Koji   I-190
Son, Long Tri Thai   II-433
Sonoi, Takumi   II-146
Soo, Kim Hwa   I-312
Sowa, Yoshihiro   II-146
Sun, Kyunghee   I-370
Suong, Tieu Phung Mai   II-476

T
Tai, Nguyen Khac Anh   I-77
Tan, Ha Minh   I-383
Taniyama, Ichita   I-190
Thai, Dinh Quoc   II-105
Thai, Do Thanh   II-270
Thai, Minh-Tuan   I-228
Thai, Phu-An   II-90
Thai-Nghe, Nguyen   I-31, I-105, II-19
Thanh, Huong Nguyen Thi   I-457
Thanh, Huy Nguyen   I-457
Thanh, Le Tuan   I-77
Thanh, Tam Nghe Thi   I-92
Thi Khac, Quan   II-419
Thi, Diem Trinh Bui   II-120
Thien, Tran Chau Thanh   I-383
Thuy, Pham Thi Thu   I-312
Tong-Le, Thanh-Hai   I-297
Tran, Anh Duc   II-231
Tran, Anh   II-61
Tran, Anh-Tuan   I-215

Tran, Dang Hien Long   I-414
Tran, Duy-Hoang   I-326, II-343
Tran, Duy-Quang   I-105
Tran, Hoang Thien-Phu   II-46
Tran, Ho-Dat   I-175
Tran, Hong Tai   I-414, I-429
Tran, Nha   II-61
Tran, Nhat Minh   I-243
Tran, Nhut-Thanh   I-133, II-46
Tran, Quang Nhat   I-215
Tran, Quoc-Khang   I-147
Tran, Tan-Phat   II-258
Tran, Thang Viet   I-243
Tran, Thi Minh Tam   II-231
Tran, Thien Khai   I-401, II-476, II-489
Tran, Thien M.   II-246
Tran, Thi-Kim-Ngoc   II-384
Tran, Tin T.   I-77
Tran, Tuan-Anh   I-414, I-429
Tran, Tuan-Kiet   II-46, II-357
Tran, Van-Ho   II-246
Tran, Vi-Do   II-246
Tran-Nguyen, Minh-Thu   II-3, II-31
Trinh, Hoai-An   II-246
Truong, Quoc Dinh   II-120
Truong, Thanh-Nhan   I-326
Tuyet, Huynh Ngoc   II-19

V
Van Nguyen, Sinh   I-92
Van Nhat Hao, Luu   II-489
Van Thang, Doan   II-137
Van, Manh Mai   I-77
Vo, Bay   II-476, II-489
Vo, Huy-Hoang   II-46
Vo, Le Dinh-Kha   II-46
Vo, Phuoc-Hung   II-448
Vo, Tri-Thuc   I-201
Vo, Van-Kha   I-175
Vo-Van, Tai   II-105
Vu, Dinh-Chuong   II-343
Vu, Dinh-Minh   II-328
Vu, Duc-Quang   I-473
Vu, Thi-Duong   II-328
Vu, Trinh-Hoang   II-328
Vy, Bach Ngoc   II-312

Y
Yoo, Hyun   I-370

MIX
Papier aus verantwortungsvollen Quellen
Paper from responsible sources
FSC® C105338

If you have any concerns about our products,
you can contact us on
ProductSafety@springernature.com

In case Publisher is established outside the EU,
the EU authorized representative is:
Springer Nature Customer Service Center GmbH
Europaplatz 3, 69115 Heidelberg, Germany

Printed by Libri Plureos GmbH
in Hamburg, Germany